CONCISE ENCYCLOPEDIA OF
MEASUREMENT &
INSTRUMENTATION

ADVANCES IN SYSTEMS, CONTROL AND INFORMATION ENGINEERING

This is a series of Pergamon scientific reference works, each volume providing comprehensive, self-contained and up-to-date coverage of a selected area in the field of systems, control and information engineering. The series is being developed primarily from the highly acclaimed *Systems & Control Encyclopedia* published in 1987. Other titles in the series are listed below.

ATHERTON & BORNE (eds.)
Concise Encyclopedia of Modelling & Simulation

MORRIS & TAMM (eds.)
Concise Encyclopedia of Software Engineering

PAPAGEORGIOU (ed.)
Concise Encyclopedia of Traffic & Transportation Systems

PAYNE (ed.)
Concise Encyclopedia of Biological & Biomedical Measurement Systems

PELEGRIN & HOLLISTER (eds.)
Concise Encyclopedia of Aeronautics & Space Systems

SAGE (ed.)
Concise Encyclopedia of Information Processing in Systems & Organizations

YOUNG (ed.)
Concise Encyclopedia of Environmental Systems

NOTICE TO READERS

Dear Reader
If your library is not already a standing order/continuation order customer to the series **Advances in Systems, Control and Information Engineering**, may we recommend that you place a standing order/continuation order to receive immediately upon publication all new volumes. Should you find that these volumes no longer serve your needs, your order can be cancelled at any time without notice.

CONCISE ENCYCLOPEDIA OF
MEASUREMENT &
INSTRUMENTATION

Editors
L FINKELSTEIN
City University, London, UK

K T V GRATTAN
City University, London, UK

Series Editor-in-Chief
MADAN G SINGH
UMIST, Manchester, UK

PERGAMON PRESS
OXFORD • NEW YORK • SEOUL • TOKYO

UK	Pergamon Press Ltd, Headington Hill Hall, Oxford OX3 0BW, England
USA	Pergamon Press, Inc, 660 White Plains Road, Tarrytown, New York 10591-5153, USA
KOREA	Pergamon Press Korea, KPO Box 315, Seoul 110-603, Korea
JAPAN	Pergamon Press Japan, Tsunashima Building Annex, 3-20-12 Yushima, Bunkyo-ku, Tokyo 113, Japan

First edition 1994

Library of Congress Cataloging in Publication Data
Concise encyclopedia of measurement and instrumentation / editors, L. Finkelstein and K. T. V. Grattan. — 1st ed.
        p.   cm. — (Advances in systems, control and information engineering)
    Includes index.
    1. Engineering instruments—Encyclopedias.   2. Engineering—Measurement—Encyclopedias.   I. Finkelstein, II.   Grattan, K. T. V.   III. Title: Concise encyclopedia of measurement and instrumentation.   IV. Series.
TA165.C62   1993
681'.2—dc20   93–10617

British Library Cataloguing in Publication Data
A catalogue record for this book is available from the British Library.

ISBN 0–08–036212–5

∞™   The paper used in this publication meets the minimum requirements of the American National Standard for Information Sciences—Permanence of Paper for Printed Library Materials, ANSI Z39.48–1984.

Printed and bound in Great Britain by BPCC Wheatons Ltd, Exeter.

CONTENTS

HONORARY EDITORIAL ADVISORY BOARD

FOREWORD

With the publication of the eight-volume *Systems & Control Encyclopedia* in September 1987, Pergamon Press was very keen to ensure that the scholarship embodied in the Encyclopedia was both kept up to date and was disseminated to as wide an audience as possible. For these purposes, an Honorary Editorial Advisory Board was set up under the chairmanship of Professor John F. Coales FRS, and I was invited to continue as Editor-in-Chief. The new work embarked upon comprised a series of Supplementary Volumes to the Main Encyclopedia and a series of Concise Encyclopedias under the title of the Advances in Systems, Control and Information Engineering series. This task involved me personally editing the series of Supplementary Volumes with the aim of updating and expanding the original Encyclopedia and arranging for the editing of a series of subject-based Concise Encyclopedias being developed from the Main Encyclopedia. The Honorary Editorial Advisory Board helped to select subject areas which were perceived to be appropriate for the publication of Concise Encyclopedias and to choose the most distinguished experts in those areas to edit them. The Concise Encyclopedias were intended to contain the best of the articles from the Main Encyclopedia, updated or revised as appropriate to reflect the latest developments in their fields, and many totally new articles covering recent advances in the subject and expanding on the scope of the original Encyclopedia.

The *Concise Encyclopedia of Measurement & Instrumentation* covers a subject of major importance to all those who work in laboratory or research and development environments. It approaches the subject from fundamental principles and brings together what might appear a very diverse field by emphasizing the commonalities in principles and techniques for various applications and the purposive development of measurement and instrumentation systems.

The Encyclopedia comprises almost entirely new material specially commissioned for it and covers its field in comprehensive detail. It is an excellent complement to the *Systems & Control Encyclopedia* and to the other volumes in the Advances in Systems, Control and Information Engineering series and will be of interest to a broad audience working in research and industrial environments.

Madan G Singh
Series Editor-in-Chief

PREFACE

The primary purpose of this Concise Encyclopedia is to take a systematic view of the field of measurement and instrumentation, describing the state of the art and investigating recent advances. The volume concerns itself with a broad overview of the topic of measurement and instrumentation, ranging from the history and essential nature of the subject through to a discussion of specific transducer types, and finally looking at applications in measurement and instrumentation today. There are a number of specialist areas which are investigated within this encyclopedia and the reader should find within it a comprehensive and systematic approach to the various subjects whichare considered.

This work would not have been possible without the contributions of a large number of key authors, who gave up a considerable amount of time and expended a lot of effort in producing the articles contained in the Encyclopedia. Their names are given at the foot of the relevant articles together with their affiliations, and they also appear in the alphabetical list of contributors at the end of the volume, where they are given with their full postal addresses. The editors are grateful to them and in particular to others who have helped the work along. The encouragement of Peter Frank at Pergamon Press has been very helpful in the progress of this work and the editors are particularly grateful to the editorial executive, Mirjam Finkelstein-Wiener, for her administration of their work and the encouragement she provided.

L Finkelstein
K T V Grattan
Editors

GUIDE TO USE OF THE ENCYCLOPEDIA

This Concise Encyclopedia is a comprehensive reference work covering all aspects of measurement and instrumentation. Information is presented in a series of alphabetically arranged articles which deal concisely with individual topics in a self-contained manner. This guide outlines the main features and organization of the Encyclopedia, and is intended to help the reader to locate the maximum amount of information on a given topic.

Accessibility of material is of vital importance in a reference work of this kind and article titles have therefore been selected not only on the basis of article content, but also with the most probable needs of the reader in mind. An alphabetical list of all the articles contained in this Encyclopedia is to be found on p. xiii.

Articles are linked by an extensive cross-referencing system. Cross-references to other articles in the Encyclopedia are of two types: in-text and end-of-text. Those in the body of the text are designed to refer the reader to articles that present in greater detail material on the specific topic under discussion. They generally take one of the following forms:

...as is described in the article *Analog-to-Digital and Digital-to-Analog Conversion.*

...the applications of these techniques (see *Instrumentation in Systems Design and Control*).

The cross-references listed at the end of an article serve to identify broad background reading and to direct the reader to articles that cover different aspects of the same topic.

The nature of an encyclopedia demands a higher degree of uniformity in terminology and notation than many other scientific works. The widespread use of the International System of Units has determined that such units be used in this Encyclopedia. It has been recognized, however, that in some fields Imperial units are more generally used. Where this is the case, Imperial units are given with their SI equivalent quantity and unit following in parentheses. Where possible, the symbols defined in *Quantities, Units, and Symbols* published by the Royal Society of London have been used.

All the articles in the Encyclopedia include a bibliography giving sources of further information. Each bibliography consists of general items for further reading and/or references which cover specific aspects of the text. Where appropriate, authors are cited in the text using a name/date system as follows:

...as was recently reported (Smith 1990).

Jones (1988) describes...

The Introducion describes in more detail the organization of the Encyclopedia and provides a systematic breakdown of the contents by subject area.

The contributors' names and the organizations to which they are affiliated appear at the ends of all the articles. All contributors can be found in the alphabetical List of Contributors, along with their full postal addresses and the titles of the articles of which they are authors or coauthors.

The most important information source for locating a particular topic in the Encyclopedia is the multilevel Subject Index, which has been made as complete and fully self-consistent as possible.

ALPHABETICAL LIST OF ARTICLES

INTRODUCTION

This Encyclopedia is intended to be a concise but comprehensive reference covering the science and technology of measurement and instrumentation. It is intended as a complementary companion to the *Systems & Control Encyclopedia*, edited by M. G. Singh and published by Pergamon Press, and to the other volumes of the Advances in Systems, Control and Information Engineering series.

Measurement, and the instrumentation by which it is implemented, is the basis of natural science, a key enabling technology of automatic control and the management of machines and processes, an essential requirement of trade and an essential tool of quality monitoring and assurance. It is, to an ever increasing extent, the means of medical diagnosis and control of treatment and an essential means of monitoring and preserving the natural environment. It has substantial economic significance.

Modern instrumentation is the integration of many different technologies. They range from the technology of information and knowledge processing systems to the engineering of sensors involving the advanced use of physical and chemical effects and the application of optical, sonic and semiconductor device technology.

The diversity of the technologies which measurement and instrumentation integrate, as well as the diversity of applications, may lead to a view that the area is a disjointed collection of equipment and techniques developed as a by-product of research in basic science or from work on particular applications. While this was the origin of much of measurement and instrumentation, it is not the present reality. Modern instrumentation technology is concerned with the purposive engineering of devices and systems to meet a need or to exploit advances in enabling means. It is based on generic principles. There is a wide range of general purpose techniques and equipment. Technology devised for particular needs is applicable over a wide spectrum of other uses.

The science of measurement and instrumentation is based on the treatment of measurement as an information process and of instruments as information machines. It is founded on the concepts and principles of the sciences of information, knowledge and systems engineering and on a design methodology, which involves the need for a systematically organized knowledge base of established methods and solutions.

The basis of this Concise Encyclopedia is the presentation of the technology of measurement and instrumentation on the foundation of such general principles and concepts. The article *Measurement Science* gives a more detailed account of the principles.

The organizational structure of the volume is presented in the form of an analytical list of contents in Table 1 on p. xvi.

The first of the groups of articles presents the general theoretical principles of measurement and instrumentation. As was mentioned previously, an overview of these principles is given in the article *Measurement Science* which the other articles of the group extend and amplify.

The second group of articles presents instruments and instrument systems in relation to their design and life cycle.

The third group of articles presents instruments and instrument systems in terms of their building blocks and the technologies by which they are integrated. This recognizes that a very great and diverse variety of instrumentation can be built up from a limited range of building blocks and architectures. The articles deal in particular with transducers, the core, distinctive, technology of measurement and instrumentation, and cover signal conditioning and information handling technology.

The traditional presentations of the technology of measurement and instrumentation review familiar measurement systems classified by measurand showing the tested methods of measurement for standard problems. Such a presentation is undertaken in the fourth group of articles.

The fifth group of articles presents some specific fields of application of measurement

Table 1 Analytical list of contents

and instrumentation in general review, both because of their intrinsic significance and because of their value in illustrating general principles and techniques.

In the last section, an article is included that presents the historical development of measurement and instrumentation.

Finally, the treatment of the various topics is integrated by cross-references at the end of articles which refer readers to other articles which are complementary or view the topic from a different perspective.

L Finkelstein
K T V Grattan
Editors

Amplifiers

The way in which voltage amplifiers are specified, their properties and the uses of operational amplifiers are discussed in this article. Three specialized amplifiers are considered:

(a) instrumentation amplifiers, which are high-input impedance differential voltage amplifiers having variable differential gain and high common mode rejection ratio;

(b) isolation amplifiers, which provide galvanic isolation between their front-end circuitry and subsequent stages; and

(c) low-voltage offset and drift amplifiers, employing chopper stabilization and commutating autozero techniques.

In addition to providing single-ended or differential ac or dc voltage gain, amplifiers are used as the signal processing element in instrumentation systems for a wide variety of purposes.

1. Voltage Amplifier Specification

Voltage amplifiers are primarily specified by their input–output configuration, gain, bandwidth, common mode rejection ratio, input and output impedances, input offset voltage, input bias current, and noise. These parameters are defined in Sects. 1.1–1.8, and are usually measured with the amplifier operating in its linear regime.

1.1 Input–Output Configuration
Amplifier inputs and outputs can be either single ended, in which case the signals are applied or measured with respect to ground, or differential, where the signals are applied or measured differentially between two inputs or outputs. Thus, there are four possible input–output configurations for a voltage amplifier and these are shown in Fig. 1. The most common configurations met in instrumentation are the single-ended input, single-ended output (SISO) and differential input, single-ended output (DISO) amplifiers.

1.2 Gain
The gain A of a voltage amplifier is defined as $\Delta e_o/\Delta e_i$, where Δe_o, is the small change at the output caused by a small input voltage change Δe_i, and where the magnitudes of Δe_i and Δe_o are restricted to maintain linear operation. The gain is usually measured by applying a sinusoidal signal of the form $e_i(\omega)\sin(\omega t)$ to

the input of the amplifier, and measuring the output of the amplifier which will be of the form $e_o(\omega)\sin[\omega t + \phi_A(\omega)]$, where $\phi_A(\omega)$ is the phase shift of the amplifier. The gain at an angular frequency ω, $A(\omega)$, is defined as $e_0(\omega)/e_i(\omega)$.

For an amplifier having differential inputs as shown in Fig. 1c,d, it is necessary to define two gains. The differential gain $A_{diff}(\omega)$ is the gain to signals applied between the differential inputs of the amplifier and the common mode gain $A_{cm}(\omega)$ which is the gain of the amplifier measured by applying the same signal (common mode signal) to both inputs of the amplifier. The ability of a practical amplifier to reject common mode signals is measured by its common mode rejection ratio (CMRR) defined as

$$CMRR = 20 \log_{10}\left[\frac{A_{diff}(\omega)}{A_{cm}(\omega)}\right]$$

The CMRR is a function of frequency and is usually specified at a given level of amplifier source impedance. At low frequencies a good differential amplifier will have a common mode rejection ratio in the region of 100–120 dB.

1.3 Bandwidth
The gain of an amplifier does not remain constant over the entire frequency range and thus it is necessary to specify some range over which it remains within certain limits. For a dc amplifier, whose gain is defined

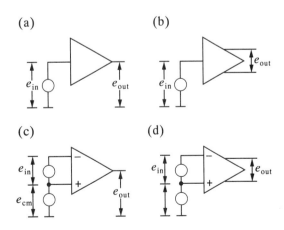

· *Figure 1*
Amplifier configurations: (a) single-ended input, single-ended output; (b) single-ended input, differential output; (c) differential input, single-ended output; (d) differential input, differential output

down to zero frequency, the bandwidth is usually defined as the frequency at which the gain has fallen by 3 dB (i.e., $1/\sqrt{2}$) from its value at low frequency. For an ac or tuned amplifier, defining the bandwidth requires specifying the two frequencies at which the gain has fallen by 3 dB from its mid-band value. The bandwidth is the difference between these two frequencies. These definitions are shown in Fig. 2.

1.4 Input Impedance

The input impedance Z_{in} of an amplifier measures the effect the amplifier will have on the previous stage and it is defined, as shown in Fig. 3a, as

$$ Z_{in} = \left| \frac{e_{in}(\omega)}{i_{in}(\omega)} \right| \ \angle \phi_{Z_{in}}(\omega) $$

In general, $\phi_{Z_{in}}(\omega)$ is nonzero and thus the amplifier will have an input impedance which is complex. For amplifiers having differential inputs, there are two input impedances:

(a) $Z_{in\,diff}$, the differential input impedance measured between the differential inputs of the amplifier; and

(b) $Z_{in\,cm}$, the common mode input impedance measured between each input and ground.

Fig. 3b shows these impedances for a DISO amplifier.

1.5 Output Impedance

The output impedance of an amplifier is a measure of the effect that the following circuit will have on the output of the amplifier as current is drawn from it. For the amplifier shown in Fig. 3a,

$$ Z_{out} = \left| \frac{e'_o(\omega)}{i_L(\omega)} \right| \ \angle \phi_{Z_{out}}(\omega) $$

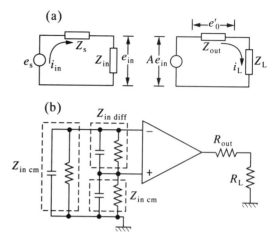

Figure 3
Input and output impedances: (a) equivalent circuit; (b) differential amplifier input and output impedances

Most amplifiers have an output impedance which can be represented as a resistor in series with its output as shown in Fig. 3b.

1.6 Input Voltage Offset

Differential dc amplifiers do not provide zero output voltage for zero input voltage, as a consequence of component mismatch at the front end of the amplifier. Thus, the real amplifier is represented by an ideal amplifier with a voltage source in series with its noninverting input as shown in Fig. 4. The value of the voltage source V_0, which gives the real amplifier zero output is called the input offset voltage. Input voltage offsets of differential amplifiers are typically in the region of 0.1–10 mV, unless special techniques such as chopper stabilization or commutating autozeroing are used, in which case they can be lower.

1.7 Bias Current

Direct current coupled amplifiers have front-end circuitry which needs to be fed with bias current and for which a dc path must be provided. In the case of a

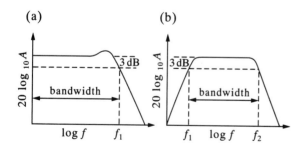

Figure 2
Bandwidth: (a) dc amplifier; (b) ac or tuned amplifier

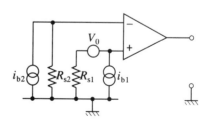

Figure 4
Voltage offset and input bias currents

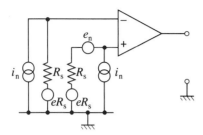

Figure 5
Noise considerations

Table 1
Characteristics of amplifiers using series and shunt feedback

	Noninverting amplifier	Inverting amplifier
Gain	$(Z_1 + Z_2)/Z_1$	$-Z_2/Z_1$
Input impedance	∞	Z_1
Output impedance	0	0

differential amplifier, these bias currents can be represented as two current generators as shown in Fig. 4. These bias currents provide additional input voltage of $(i_{b1}R_{s1} - i_{b2}R_{s2})$. Minimization of this effect is usually achieved by keeping $i_{b1}R_{s1}$ and $i_{b2}R_{s2}$ small and by matching the source resistance at both inputs.

1.8 Noise

The noise generated by a differential amplifier can be represented by a voltage noise generator with rms value $\bar{e}_n$ and two current noise generators with rms value $\bar{i}_n$ at the input of the amplifier as shown in Fig. 5. These noise generators are assumed to be independent. The total rms equivalent noise generator $\bar{e}_{n_{tot}}$ at the input is thus given by

$$\bar{e}_{n_{tot}} = \sqrt{\bar{e}_n^2 + 2\bar{i}_n^2 R_s^2 + 8kTR_s\Delta f}$$

where the first two terms represent the effects of the voltage and current noise generators of the amplifier, and the third term represents the Johnson noise associated with the source resistances R_s. T is the absolute temperature, k is the Boltzmann constant and Δf is the bandwidth.

2. Operational Amplifiers

An operational amplifier is simply a high-gain dc coupled amplifier which is used with feedback to achieve a wide variety of signal processing functions.

It is usually a differential input, single-ended output device. The ideal operational amplifier has infinite gain, bandwidth, CMRR and input impedance and zero output impedance, input offset voltage and bias current, and is noiseless.

Two feedback configurations are widely used with operational amplifiers and these are shown in Fig. 6. In Fig. 6a, series feedback is employed and noninverting gain is provided. In Fig. 6b shunt feedback is employed and the closed-loop amplifier provides inverting gain. Table 1 shows the characteristics of the two amplifiers. By using linear and nonlinear components for Z_1 and Z_2 a large number of circuits can be realized.

2.1 Instrumentation Amplifiers

An often encountered problem in instrumentation is the amplification of low-level dc signals such as the output from a strain gauge bridge, or from a thermocouple pair in the presence of a large dc or line frequency common mode signal. The instrumentation amplifier shown in Fig. 7 provides variable differential gain with high-input impedance and CMRR together with low offset voltage and drift. Amplifiers 1 and 2 provide unity gain to the common mode signal and amplifier 3 acts as a differential amplifier. The differential gain of the amplifier A_{diff} is given by

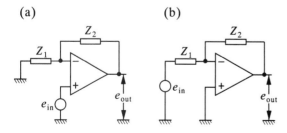

Figure 6
Operational amplifiers: (a) noninverting amplifier; (b) inverting amplifier

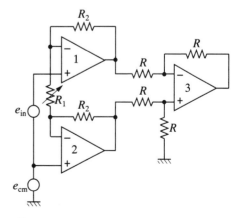

Figure 7
Instrumentation amplifier

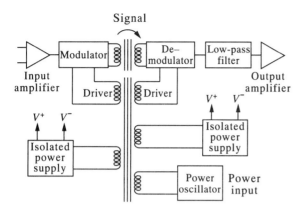

Figure 8
Isolation amplifier

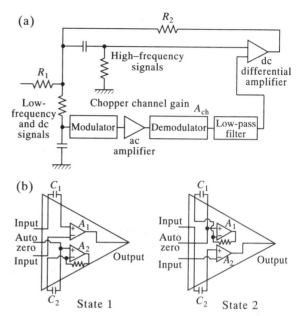

Figure 9
Low-drift amplifiers: (a) chopper stabilized amplifier;
(b) commutating autozero amplifier

$(2R_2 + R_1)/R_1$. The gain can be adjusted by adjusting R_1 without affecting the CMRR.

2.2 Isolation Amplifiers

For amplification where common mode signals are in excess of the supply rails or when, from a point of view of safety, isolation is required, isolation amplifiers provide galvanic isolation between the front end and subsequent stages. Isolation is commonly provided by means of transformer, capacitative and optical coupling. Figure 8 schematically depicts an example of an isolation amplifier, where transformer coupling is used, and this provides the power for the front-end circuitry and also enables the transfer of the signal, in a modulated form, between the front end and subsequent stages.

2.3 Chopper Stabilized and Commutating Autozero Amplifiers

The requirement to provide amplifiers with low dc offset voltage drift has led to two designs of amplifier. Figure 9a shows the operation of the chopper stabilized amplifier. The input signal is split into two components on a frequency basis. High-frequency components are passed directly into the main amplifier, whereas low-frequency components are modulated, amplified and demodulated before being injected into the main amplifier. The low-frequency chopper channel ideally provides gain without offset since the gain is provided by an ac amplifier. The offset in the main amplifier is effectively reduced by A_{ch}, the gain of the chopper channel. Additional offset voltages are introduced by imperfections in the modulation demodulation of the chopper channel.

Another technique for the production of amplifiers with low offset voltages and drift is that of the commutating autozero amplifier shown in Fig. 9b. Two low-bias current amplifiers are built on the same

substrate and the system commutates between two states. In state 1, amplifier 1 is being used as the amplifier and the offset on amplifier 2 is being measured and stored on C_2. The system then commutates to state 2 in which amplifier 2 is now being used, with the offset on amplifier 1 being measured. Such amplifiers can provide offset voltage of the order of a few microvolts, with a temperature coefficient of less than $0.1\,\mu\mathrm{V}\,^\circ\mathrm{C}^{-1}$. Commutation between the two states typically occurs at a frequency of 160 Hz, which generates spikes at the commutation frequency. The amplifiers, which are used with feedback, are suitable for dc and low-frequency signals.

See also: Analog Signal Conditioning and Processing; Analog-to-Digital and Digital-to-Analog Conversion

Bibliography

Burr-Brown 1989 *Integrated Circuits Data Book*, Vol. 33. Burr-Brown, Tucson, AZ, pp. 4-1-4-5

M. L. Sanderson
[Salford University, Salford, UK]

Z. Y. Zhang
[Changsha Institute of Technology, Changsha, China]

Analog Signal Conditioning and Processing

The input quantity for most instrumentation systems is nonelectrical. In order to use electrical techniques for measurement the nonelectrical quantity has to be converted into an electrical signal by use of a transducer.

Transducers may be classified according to the electrical principles involved: passive transducers require external power to furnish a voltage or current signal; active transducers do not require external power, as all the electrical energy at the output signal is derived from the physical input.

1. Passive Transducer Conditioning

1.1 Physical Principles of Passive Transducers

Passive transducers act just like electrical impedances. The measurand may produce a variation in geometrical parameters (volume, surface, length, etc.), or a change of electrical property (resistivity, permittivity, permeability, etc.).

In the inductive transducer the measurement of force is performed by the change in the inductance of a coil. This principle of operation is exploited in the manufacture of many displacement transducers. The linear variable differential transformer (LVDT) is an example of an inductive displacement transducer.

1.2 Passive Transducer Interfacing

The impedance variation of a passive transducer may be converted into:

(a) a voltage V_m representing measuring signal—in this case the signal conditioner may be a voltage divider, a high-impedance current source or a bridge circuit; and

(b) frequency information f_m of the measuring signal, where the signal conditioner is an oscillator, which is convenient for use in telemetry applications.

1.3 Signal Conditioners for Voltage Information

(a) *Voltage divider circuit and current source.* These circuits are undoubtedly the simplest signal conditioners. The measurand, in effect, modulates the excitation, which can be used to provide an increased output level, as shown in Fig. 1.

Consider a simple resistance temperature detector (RTD). For accurate resolution of small temperature changes, with correspondingly small resistance changes, it would be useful to increase the excitation source E or I. When E and I are sufficiently great, the power dissipated in R_m will cause the temperature to rise perceptibly, introducing a measurement error. Further, this type of signal conditioner cannot work with very small resistance changes of strain gauges,

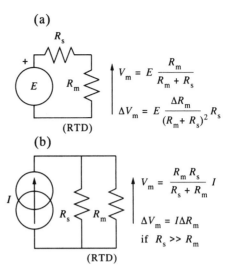

(a)

$$V_m = E \frac{R_m}{R_m + R_s}$$

$$\Delta V_m = E \frac{\Delta R_m}{(R_m + R_s)^2} R_s$$

(RTD)

(b)

$$V_m = \frac{R_m R_s}{R_s + R_m} I$$

$$\Delta V_m = I \Delta R_m$$

if $R_s \gg R_m$

(RTD)

Figure 1
(a) Voltage divider, (b) current source

which can reach 10^{-6} (about $1/10\,000\,\Omega$) if the strain gauge resistance is equal to $120\,\Omega$, as this resistance would have to be measured to an accuracy of seven significant figures. A possible solution to this problem would be to use a bridge circuit.

(b) *Bridges.* In its simplest form, a bridge consists of four elements, an excitation source and a detector, as shown in Fig. 2. It measures the impedance of a passive transducer indirectly, by comparison with a similar element. When the bridge is balanced, the potential difference V_m is $0\,V$; this condition occurs when $R_4 R_2 = R_1 R_2$.

The deviation of one (or more) impedance in the bridge from an initial value must be measured as an indication of a change of the measurand.

Figure 2 illustrates a bridge with all resistances nominally equal, but one of them (R_1) is variable by a

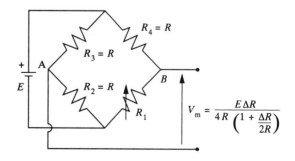

$$V_m = \frac{E \Delta R}{4R \left(1 + \frac{\Delta R}{2R}\right)}$$

Figure 2
Bridge used to read the deviation of a passive transducer (R_1)

factor $R + \Delta R$. When the bridge is unbalanced, the following equation is obtained:

$$V_m = V_A - V_B = \frac{E\Delta R I}{4R\left(1 + \dfrac{\Delta R}{2R}\right)} \qquad (1)$$

As Eqn. (1) indicates, the relationship between V_m and ΔR is not linear, but for small ranges of ΔR it is sufficiently linear for many purposes.

1.4 Signal Conditioner for Frequency Information

This class of signal conditioner converts the impedance variation of a passive transducer into a frequency variation of an oscillator circuit.

There are various oscillator circuits with designs that depend on the frequency they are required to produce. Low-frequency oscillators operating roughly in the 1 Hz to 20 kHz range are often based on relaxation oscillator circuits; for higher frequencies, LC oscillators can be used.

2. Active Transducer Conditioning

2.1 Physical Principles of Active Transducers

The electrical signal at the output of an active or self-generating transducer is only derived from the physical input; examples include thermocouples, photodiodes, piezoelectric transducer, and so on. Since the electrical output is limited by the physical measurand, such transducers tend to have low-energy outputs requiring amplification. In order to design adequate signal conditioners, it is essential to understand the nature of the transducer output signal.

2.2 Active Transducer Interfacing

Active transducers can be modelled as:

(a) a voltage source (thermoelectric effect, Hall effect, electromagnetic effect, photovoltaic effect); or

(b) a current source (photoelectric effect, piezoelectric effect), which requires a special network for signal conditioning.

(a) *Voltage source transducers.* Figure 3 shows the

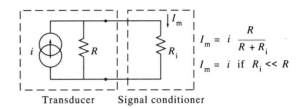

Transducer   Signal conditioner

$$I_m = i \frac{R}{R + R_i}$$
$$I_m = i \text{ if } R_i \ll R$$

Figure 4
Equivalent circuit of a current source transducer

equivalent circuit of a voltage source transducer. It usually requires amplifiers for buffering, isolation and gain. Most of these functions can be performed by operational amplifiers. This is an excellent unity gain follower. If the feedback is attenuated, a follower with gain is obtained, where the gain is accurately set by a resistance ratio.

(b) *Current source transducers.* Figure 4 shows the equivalent circuit of a current source transducer which provides a current output I_m through the signal conditioner input impedance R_i. Usually these transducers require a current-to-voltage conversion which can be performed by a very simple circuit (see Fig. 5). The output voltage is independent of R. It is interesting to note that the output of a piezoelectric transducer may be modelled as a current source in parallel with a small capacitor. The forces to be measured are dynamic (i.e., continually changing over the period of interest). Charge amplifier configurations are required for signal conditioning. In general, manufacturers of piezoelectric devices furnish calibrated charge amplifiers and cables.

3. Linearizing and Processing

A linear transducer is one for which measurand and output signal are proportional. Figure 6 shows an input–output plot of an ideal linear relationship, a nonlinear relationship, and the difference between the two (nonlinearity).

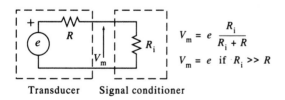

$$V_m = e \frac{R_i}{R_i + R}$$
$$V_m = e \text{ if } R_i \gg R$$

Transducer   Signal conditioner

Figure 3
Equivalent circuit of a voltage source transducer

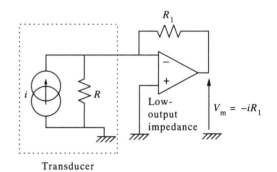

Transducer

$$V_m = -iR_1$$

Figure 5
Current-to-voltage converter

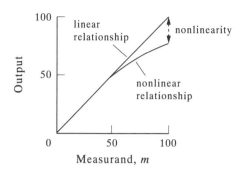

Figure 6
Typical nonlinear relationship

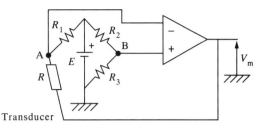

Figure 7
Feedback provided by the use of operational amplifier to linearize the bridge output

To linearize the signal from a transducer, two principal techniques can be employed:

(a) modification of the transducer circuitry; and

(b) processing of the transducer output signal.

3.1 Modification of the Transducer Circuitry

An example of modifying transducer circuitry is the use of networks involving thermistors and resistors to obtain an output that is linear over limited ranges. Thermistor manufacturers use this technique to provide linearized devices having linearities to within 0.2 °C, over ranges such as 0–120 °C.

Another example of this technique is illustrated in Fig. 7, where an operational amplifier provides a feedback signal that balances the bridge, to obtain an output that is proportional to the impedance change of the passive transducer R.

At balance

$$R = R_1 = R_2 = R_3$$

When R changes to $R + \Delta R$

$$V_B = \frac{E}{2}$$

$$V_A = E\frac{R + \Delta R}{2R + \Delta R} + V_m\frac{R}{2R + \Delta R}$$

Because V_A is always equal to V_B

$$V_m = -\frac{E\Delta R}{2R} \tag{2}$$

3.2 Processing of the Transducer Output Signal

This technique can be used to compensate for nonlinearity in the transducer or the associated signal conditioner. It depends on the nature of the nonlinearity.

(a) *Signal processing using a logarithmic or exponential converter.* A logarithmic converter as shown in Fig. 8, provides an output voltage that is proportional to the logarithm of the input voltage. If V_m is an exponential function of the measurand m, this logarithmic device will provide a linear output; the output voltage of which can be written as:

$$s = -K\log\frac{V_m}{V_{REF}} = -K\log\frac{V_0 e^m}{V_{REF}}$$

if $V_{REF} = V_0$

$$s = -K_m$$

where K is a constant.

Usually a logarithmic device can be connected for exponential operation; for a logarithmic function input, an exponential converter will give a linear output.

(b) *Signal processing using an analog multiplier or divider.* An analog multiplier is a device which provides an output voltage that is equal to the product of two input voltages, multiplied by a scale constant.

An example of nonlinearity correction for a large-deviation off-null bridge using an analog multiplier is shown in Fig. 9. The output voltage of single transducer bridge (Eqn. (1)) is not directly proportional to ΔR. In Fig. 9, the output of the summing device can be written:

$$V_s = b\frac{V_m V_s}{K} + aV_m \tag{3}$$

Figure 8
Signal processing using a logarithmic converter

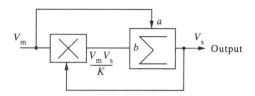

Figure 9
Nonlinearity correction using an analog multiplier

Substituting Eqn. (1) into Eqn. (3)

$$V_s = \frac{aE\Delta R}{4R\left[1 - \dfrac{\Delta R}{2R}\left(1 - b\dfrac{E}{4K}\right)\right]} \qquad (4)$$

If we adjust b (gain of summing device) in order that $b = 4K/E$, we can obtain

$$V_s = a\frac{E\Delta R}{4R}$$

Note that measurement accuracy depends on the power supply stability, therefore it is possible to abolish this deficiency by using a divider. It can be shown that the output V_s is independent of E.

See also: Amplifiers; Analog-to-Digital and Digital-to-Analog Conversion; Signal Processing

T. Lang
[École Supérieure d'Électricite,
Gif-sur-Yvette, France]

Analog-to-Digital and Digital-to-Analog Conversion

1. Digital-to-Analog Converters

The rapid development of new digital-to-analog converters (DACs) has been made possible by the arrival of integrated analog switches. The analog quantity that is the output of a DAC, representing the input digital data, may be a gain (multiplying DAC), a current, and/or a voltage. In a data acquisition and processing system, a DAC may be used as an output interface of a computer. These devices reconstitute, in analog quantity, the original data after processing, storage, or even simple transmission from one location to another in digital form.

1.1 Digital-to-Analog Converters with Weighted Resistances

This basic converter consists of:

(a) an internal reference voltage source E_{ref};

(b) a set of analog switches driven by a digital input code stocked in a binary register; and

(c) an arrangement of binary-weighted precision resistors that develops a binary-weighted output current.

In order to obtain a substantial voltage output at low impedance, an operational amplifier ("op amp") is required. In this example (see Fig. 1), an operational amplifier holds one end of all the resistors at zero volts. The switches are operated by the digital logic, open for "0," closed for "1." Each switch that is closed adds a binary-weighted increment of current E_{ref}/R_j, via the summing bus connected to the amplifier's negative input. The negative output voltage is proportional to the total current, and thus to the value of the binary number.

In the example shown in Fig. 1, for an input code 01100100, the total output current is given by:

$$i_a = E_{ref}\left(\frac{1}{2R} + \frac{1}{4R} + \frac{1}{32R}\right)$$

In practical applications, say for 12-bit digital-to-analog conversion, the range of resistance values would be 4096:1, or 40 MΩ for the least significant bit (LSB). If discrete resistors are used, cost and size are increased.

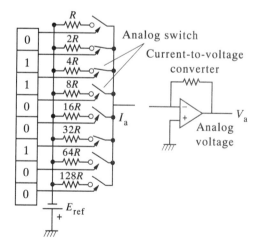

Figure 1
Principle of a digital-to-analog converter with weighted resistors

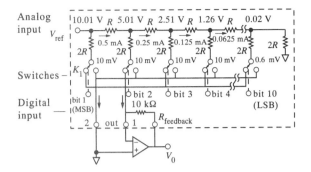

Figure 2
Example of a digital-to-analog converter using R–$2R$ ladder: MSB, most significant bit; LSB, least significant bit; $R = 10\,k\Omega$

1.2 Digital-to-Analog Converters with Resistance Ladders

A way to reduce the resistance range is to use a limited number of repeated values. One convenient, and very popular, form is the use of a R–$2R$ ladder. Figure 2 shows an example of a DAC converter using a R–$2R$ ladder network with an inverting operational amplifier.

The switches are operated by the digital input. The resultant voltage output is proportional to the total current flowing towards the amplifier's negative input.

2. Analog-to-Digital Converters

Analog-to-digital converters (ADCs) convert analog input voltage into its equivalent digital form.

There are a vast number of conceivable circuit designs for ADCs, but there are a much more limited number of designs available on the market in small modular form at low cost and especially designed for incorporation as components of equipment. The most popular of these are parallel types, successive-approximation types, and pulse counting types.

2.1 Parallel Analog-to-Digital Converters (Flash Converters)

The technique of parallel ADCs consists of comparing the analog input voltage E_X with n reference voltages, simultaneously. Figure 3 shows an example of a parallel three-bit converter; it has eight different binary output numbers, seven comparators, and seven reference voltages generated by resistance voltage dividers. For a zero input, all comparators are off. As the input increases, it causes an increasing number of comparators to switch state. The smallest voltage that can switch "on" the first comparator is:

$$\frac{1}{2}\,V_{LSB} = \frac{1}{2}\,\frac{E_{ref}}{7}$$

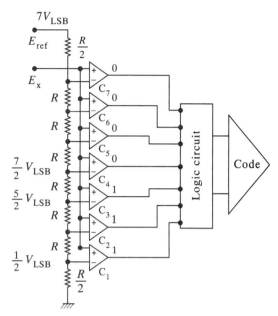

Figure 3
Parallel analog-to-digital converter

Taking an example of an input voltage E_X,

$$\frac{5}{2}\,V_{LSB} < E_X < \frac{7}{2}\,V_{LSB}$$

Comparators one to three are "on," the others are "off." The outputs of these comparators are then applied to the gates, which provide a set of outputs that fulfil the appropriate condition for natural binary output (or another code such as a Gray code).

2.2 Successive-Approximation Analog-to-Digital Converters

Successive-approximation ADCs are quite widely used, especially for interfacing with computers, because they are capable of high resolution. Conversion time, being independent of the magnitude of the input voltage, is fixed by the number of bits in the register and the clock rates. Each conversion is unique and independent of the results of previous conversions, because the internal logic is cleared at the start of a conversion.

Modern IC converters include three-state data outputs and byte controls to facilitate interfacing with microprocessors. A three-state output has, in addition to the normal "I" and "0" states, a "not enabled" condition, in which the output is simply disconnected via an open voltage switch. This permits many device outputs to be connected to the same bus; only the device that is enabled (one at a time) can drive the bus.

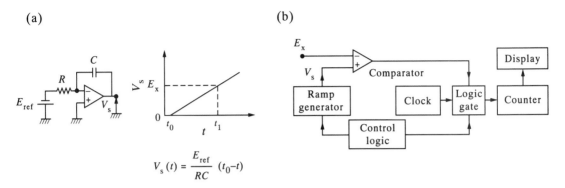

(a)

(b)

$$V_s(t) = \frac{E_{ref}}{RC}(t_0 - t)$$

Figure 4
Single-slope converter: (a) ramp voltage generator; (b) functional diagram

The conversion technique consists of comparing the unknown input against a precisely generated internal voltage at the output of a DAC. The input of the DAC is the digital number at the ADC's output. The conversion process is strikingly similar to a weighing process using a chemist's balance, with a set of n binary weights (e.g., ½ lb, ¼ lb, ⅛ lb, . . ., for unknowns up to 1 lb).

After the conversion command is applied, and the converter has been cleared, the DAC's MSB output (½ full scale) is compared with the input. If the input is greater than the MSB, it remains "on" (i.e., "1" in the output register), and the next bit (¼ full scale) is tried. If the input is less than the MSB, it is turned "off" (i.e., "0" in the output register), and the next bit is tried. The process continues in order of descending bit weight until the last bit has been tried. When the process is completed, the status line changes state to indicate that the contents of the output register now constitute a valid conversion. The contents of the output register form a binary digital code corresponding to the input signal.

2.3 Impulse Counting Converters

Analog-to-digital conversion by the counter method requires the least components, and high accuracy can be attained with relatively simple circuits. However, conversion time is considerably longer than with the other methods. It is usually between 0.1 ms and 100 ms, although this is quite sufficient for many applications. The counter method is therefore the most widely used and the variety of different circuits is extensive. The most important ones of these are single-slope converters, voltage-to-frequency converters, dual-slope converters, and quad-slope converters.

(*a*) *Single-slope converters.* The operating principle of the single-slope ADC is based on the measurement of the time it takes for a linear ramp voltage to rise from 0 V to the level of the input voltage E_X, or to decrease from the level of the input voltage to zero. This time

interval is measured by counting the pulses of a quartz oscillator, and it is proportional to the input voltage, as shown in Fig. 4. At the start of the measurement cycle (time = t_0), a positive ramp voltage is initiated from 0 V. At the same time, the gate is opened, thus allowing clock pulses from an oscillator to get through into a number of decade counting units which totalize the number of pulses passed through the gate.

(*b*) *Voltage-to-frequency converters (VFCs).* Voltage-to-frequency conversion is a simple and low-cost method; analog voltage level is converted into pulse trains, square wave or sawtooth waveforms in a logic-compatible form, at a repetition rate that is accurately proportional to the amplitude of the input signal.

The most popular VFC designs (see Fig. 5) contain an integrator which charges at a rate proportional to

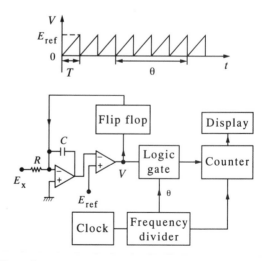

Figure 5
Voltage-to-frequency converter

the value of the input signal E_X. Each time the integrator's charge reaches a certain potential E_{ref}, a comparator triggers a charge dispenser which reduces the integrator's charge. The next pulse is triggered when the net integral has again reached the threshold. Since the time required to reach the switching threshold is inversely proportional to the analog input, the frequency is therefore directly proportional to it:

$$E_{ref} = \frac{1}{RC} E_X T$$

Hence

$$E_X = RCE_{ref}f$$

where the pulse rate f is determined by counting the number of pulses N during a fixed time interval θ.

(c) *Dual-slope converters.* This conversion method, in which not only the reference voltage but also the input voltage is integrated, is illustrated by the block diagram in Fig. 6. The dual-ramp type is especially suitable for use in digital voltmeters.

At the beginning of the conversion, the input signal is applied to an integrator; the integration being carried out for a constant time interval t_1. At the end of the integration interval t_1, the output voltage of the integration is given by:

$$-\frac{1}{RC} \int_0^{t_1} E_X \, dt = -\frac{E_X}{RC} N_1 T$$

where N_1 is the number of clock pulses predetermined (e.g., 1000) and T is the period of the clock generator.

The reference voltage is then applied to the integrator. The polarity of the reference voltage is opposite to that of the input voltage, so that the magnitude of the integrator output voltage decreases, as can be seen in Fig. 6. At the same time, the counter is again counting from zero. The comparator and the counter are now used to determine the time interval required for the integrator voltage to reach zero again.

$$\frac{E_X}{RC} N_1 T = \frac{E_{ref}}{RC} NT$$

Hence, the end result is given by:

$$E_X = E_{ref} \frac{N}{N_1}$$

Dual-slope integration has many advantages. Conversion accuracy is independent of both the capacitor value and the clock frequency, because they affect both the up-slope and the down-ramp in the same

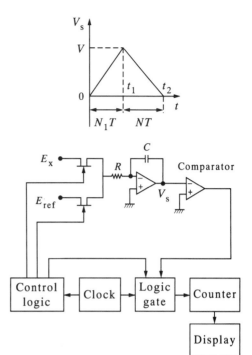

Figure 6
Dual-slope integration analog-to-digital converter

ratio. The only condition is that the clock frequency remains constant during the interval t_2. This short-term stability presents no problem, even for simple clock generators.

(d) *Quad-slope converters.* The basic operation of a quad-slope converter is explained in Fig. 7.

The integrator has four modes of connection:

(a) clamped, when no conversion is in process;

(b) grounded input;

(c) reference input (E_{ref}); and

(d) analog input (E_X).

Voltage $E_{ref}/2$ is continuously applied to the positive input of the integrator.

After the start pulse is applied, E_{ref} is connected to the input, and the integrator output is ramped to comparator zero crossing. This reset phase has a duration equal to the integrator time constant (RC). Different inputs will be applied in sequence to the integrator in order to create four operation phases.

Phase 1. Analog ground is connected to the integrator input. When E_{ref} is applied to the positive input the voltage $-E_{ref}/2$ is integrated for a fixed interval $T/2$. On zero crossing of reset phase, phase 1 is initiated and a counter starts counting clock pulses.

11

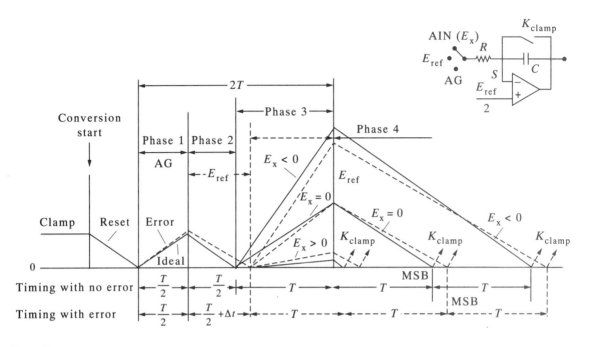

Figure 7
Illustration of a quad-slope principle: AG, analog ground; E_X, analog input (AIN); ——, no error; ———, error

Phase 2. E_{ref} is applied to the integration input. Voltage $(E_{ref} - E_{ref}/2)$ is integrated; the integrator output ramps down until zero crossing is achieved. If there is no offset error, the time for phase 2 will be the same as for phase 1 ($T/2$). Any error will increase or decrease this time by an amount Δt.

Phase 3. E_X is applied to the integration input. The analog input (AIN $- E_{ref}/2$) is integrated. The integrator output ramps upwards with a proportional slope. Phase 4 is initiated when the counter that started at the end of the reset phase reaches a count equivalent to $2T$.

At the beginning of phase 3, a second counter starts at $T \pm \Delta t$.

Phase 4. E_{ref} is applied to the integrator input. Voltage $(E_{ref} - E_{ref}/2)$ is again integrated. The integrator output ramps down at the rate $E_{ref}/2$, plus any error, until zero crossing is once again achieved. The second counter is stopped. Conversion is now complete. The counter output is a 2's complement representation of the analog input.

See also: Digital Instruments; Signal Processing

T. Lang
[École Supérieure d'Electricité,
Gif-sur-Yvette, France]

Analytical Physical Measurements: Principles and Practice

The field of instrumental analysis is one which has developed rapidly in recent years with, for example, the increased requirement for the accurate assessment of the environment, both air and water, the workplace, both from the point of view of those working in it and the quality and potential for pollution of its output, the domestic scene and the meeting of the needs of analysis in the scientific laboratory. As a result, the impact of such measurements can be very wide, both from the costs involved in their making being passed on to the consumer and the pressure of legislative needs, worldwide, which are driving the field. Thus, a comprehensive overview of the subject is necessary to see and evaluate the impact of the methods through a wide range of physical, chemical and biological measurements, and to determine the most suitable and economic way to obtain the desired data on a particular system. The technology is widely used by physicists, chemists, biologists and engineers and often the choice of method for a particular analysis will be determined by the discipline (or prejudice) of the experimenter, rather than by the optimum choice of method through an assessment of the fundamental principles of analytical measurement itself.

With such a wide variety of methods available to the modern analyst, it is important that there is a clear understanding of the principles of the analytical method, and how they are applied in practice in the obtaining of data on the system under investigation. On the basis of this, the analyst can appreciate more fully the value of a comprehensive review of the major methods available, their fundamental physics and chemistry of operation, their advantages and disadvantages for any particular situation, and their limitations. With this knowledge, the analyst can then be referred to manufacturers' catalogues, to assess such factors as price, size and portability, and the need for associated peripheral equipment (such as power supplies and computer support), which may influence or even limit the operation of the analytical technique in the field situation.

1. Principles of the Analytical Method

With the breadth of techniques available, as is seen by a perusal of a modern text on analytical methods (e.g., Willard *et al.* 1988), it is important that the analyst has to hand an understanding of the basic principles underlying the analytical method itself, to enable such an analyst to make the most appropriate choice of

Table 1
Techniques used in instrumental analysis

Spectroscopic techniques
    ultraviolet and visible spectrophotometry
    fluorescence and phosphorescence spectrophotometry
    atomic spectrometry (emission and absorption)
    infrared spectrophotometry
    Raman spectroscopy
    x-ray spectroscopy
    radiochemical techniques
    nuclear magnetic resonance (NMR) spectroscopy
    electron spin resonance (ESR) spectroscopy

Electrochemical techniques
    potentiometry (pH and ion selective electrodes)
    voltammetry
    voltammetric techniques
    amperometric techniques
    conductance techniques

Chromatographic techniques
    gas chromatography
    high performance liquid chromatographic techniques

Individual techniques
    thermal analysis
    mass spectrometry
    kinetic techniques

Hyphenated techniques
    GC–MS (gas chromatography–mass spectrometry)
    ICP–MS (inductively coupled plasma–mass spectrometry)
    GC–IR (gas chromatography–infrared spectroscopy)
    MS–MS (mass spectrometry–mass spectrometry)

Technique A \ Technique B	Gas chromatography	Liquid chromatography	Thin-layer chromatography	Infrared	Mass spectroscopy	Ultraviolet (visible)	Atomic absorption	Optical emission spectroscopy	Fluorescence	Scattering	Raman	Nuclear magnetic resonance (NMR)	Microwaves	Electrophoresis
Gas chromatography	■	□		■	■	■		□	□		□	□	□	
Liquid chromatography	□	□	□	□	■	■	□	□	■	■	□	□		□
Thin-layer chromatography	□	□	■	■	□	□		□	■		□			□
Infrared				□							□	□		
Mass spectroscopy				■										
Ultraviolet (visible)					□				■	■	□			
Atomic absorption					□			□	□					
Optical emission spectroscopy					□		□		□					
Fluorescence					□	■	□	□	■	□	□		□	
Scattering					□			□	□		□			
Raman					□					□	□			
Nuclear magnetic resonance (NMR)					□							■		
Microwaves					□									
Electrophoresis	□	□			■				■	■				■

Figure 1
Schematic showing hyphenated techniques: ■, established; □, feasible in the state of the art

method. This cannot be done merely by reviewing a catalog of methods, as often measurement conditions will vary from day to day, and with that, change the factors influencing the optimum choice of method. To show this clearly, Table 1 gives a list of techniques applied widely in instrumental analysis. They are subdivided into spectroscopic techniques, electrochemical methods and chromatographic approaches, and in addition several individual methods are shown. Further, the list may be enhanced by the employment of hyphenated techniques, where a combination of one or more of the above (or other) methods may be used to obtain the required data. The matrix shown in Fig. 1 illustrates the actual and potential for the state of the art in hyphenated techniques. In each case, the sample is interrogated by the interaction of the analyst through the analytical method(s) chosen, and data extracted may be processed to obtain the desired measurand information. Illustrative examples of these techniques are given later.

2. Evaluation of the Analytical Method

The first step in the choice of the analytical method involves an evaluation of the problem by the analyst responsible for the commissioning of the approach. This involves at least four basic steps to be undertaken in the process. These are as follows.

2.1 How the Method Works

The true evaluation of the method requires an understanding of the fundamental physical and chemical principles involved in the technique, and of the functions of the instrumental components used. The objective of the work is the transformation of information from one form to another, so that it can be assessed more easily or point more directly to the measurand, through the action of the operation of the method itself. Thus, for example, an interrogation of a material by shining white light through it, which in turn will result in the transmission of longer wavelengths in the 600–700 nm band is a simple, but clear illustration that the material was red in color, indicative of one important aspect of measurand information which may be required.

2.2 Advantages and Limitations

Most methods involve a trade off of advantages and disadvantages, for example, simple methods may be destructive, and for precious samples this may be inappropriate, or they may be limited in the data obtainable. Thus, the analyst will need to assess fully these features for the method under consideration, in addition to economic, ergonomic and logistical factors. Further, there will be a need to be concerned with the most appropriate presentation and interpretation of the data obtained, and major factors in quantitative analysis (by far the most common form) are the accuracy, precision and limits of detection of the method. Further, the types of sample handled will be a consideration, for example, if the sample needs to be in gaseous form, as this can present problems for materials which are solids at room temperature, especially where nondestructive methods are required.

2.3 Representative Instrumentation

With the above information, the analyst will need to consider several systems representative of current instruments, to view the practical aspects of the method, in the light of the cost of each analysis, the operator training/retraining and skills base required, the interpretation of the results with the degree of expertise required to achieve this and the overall time taken and feasibility of the approach, especially where repeated and regular analysis is being undertaken.

2.4 Applications of the Method

The analyst may need to consider the implications for his choice for not only specific but perhaps also for further, more complex applications. Thus, it may be necessary for the analyst to consider the wider use and longer term need of the analysis to be undertaken, with a view to the anticipation of any future need and the acquisition of instrumentation of sufficient flexibility to meet both these current needs and to be compatible with future developments in the technology. A particular application area here is the incorporation of computer-based data processing and instrumental intelligence. Thus, instruments which are of themselves state of the art in terms of the physical or chemical methods used may be largely inflexible to the incorporation of new data handling methods and operator interfaces, through features of their design. The anticipation of such needs and the developments of the technology can lead to considerable cost savings for the future, through the purchase of equipment designed with such considerations in mind.

3. Solution to the Analytical Problem

The solution to the analytical problem may be expressed in general terms through an analysis of the steps in the determination method, as illustrated in Fig. 2. The steps are shown in such a way as to reflect as broad an approach as possible—the specific measurand situation may enable one or more of the steps to be eliminated (e.g., a chemical treatment stage) but the method represents a general approach to the analytical method. This starts with the definition of the problem, a nontrivial aspect of the task undertaken by the analyst and the customer. At this stage, the requirements and expectations must be clearly defined for the optimum solution to be considered, and the economic factors developed to the best advantage. The questioning of the customer and analyst is a vital element in the maximizing of the efficiency of the process, which ends with the presentation of results to the customer. It should be stressed that the full analytical method includes the stage of data interpretation—the data which the analyst prepares and which may be fully comprehensible to him must be transformed into results which can be interpreted by the customer. Thus, for example, the data representing the degree of absorption of a material at say 500 nm, could be interpreted by the analyst to present an optical extinction coefficient, and this presented as the result of the analysis to the customer, who may then, through an awareness of a further economic or legislative need, use these results to develop useful knowledge of the system under analysis. The analyst may or may not aid in this process, depending on his degree of expertise in the field.

As can be seen from Fig. 2, the full analytical procedure is considerably more detailed than the simple operation of the instrumentation on the sample. The number of associated "laboratory stages" may be great or small, depending on the nature of the method, but the vital element is the transformation of information which yields the desired knowledge to the customer, who ultimately supports the work being done. If the analyst fails in the noninstrumental aspects, such as problem definition and the delivery of

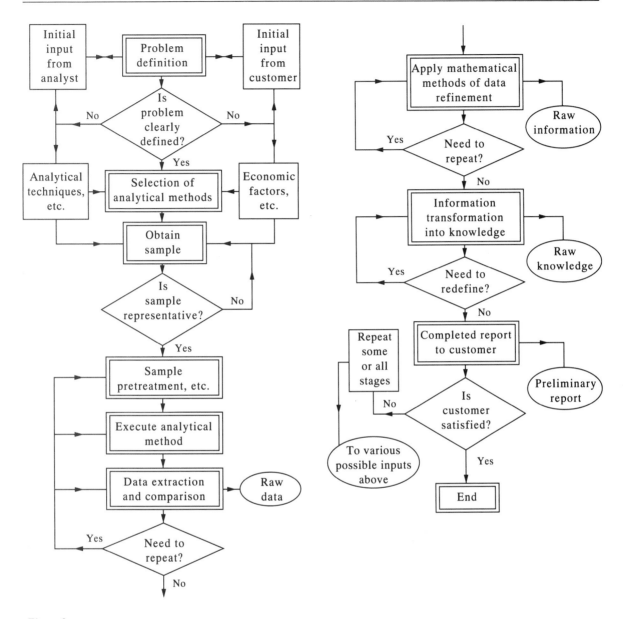

Figure 2
Steps in the determination method for analytical instrumentation

the correct aspect of knowledge to the customer, in spite of the quality of the instrumental procedures, the operation may be a failure, either scientifically, economically or in the absence of the delivery of information on time. This illustrates the way in which the analytical method can be complete—the dialogue which began with the definition of the problem results in useful knowledge, fulfilling the task of the analyst.

4. Information Transformation in the Analytical Method

Through the operation of the instrument, in the interrogation of the sample by physical or chemical means, an important feature of the operation is that information is transformed into a form more useful to the analyst and the customer, in a successful method.

Every analytical instrument may be divided into at least four basic components to represent the operation of this procedure. These are: signal generator, input transducer, electronic signal modifier, and output transducer. The functions of these elements are discussed in the following sections.

4.1 Signal Generator

The signal results from the direct or indirect interaction of the analyte or sample with some form of energy. This may be of the form of either electromagnetic energy, electricity or thermal energy, via heating, for example. The former is the most familiar, with optical methods in the visible, ultraviolet and especially the infrared being particularly useful for spectroscopic analysis. Further, the recent developments in the field of lasers, optical fibers and optoelectronics has enhanced the applicability of this aspect of nondestructive analysis.

Acoustic, especially ultrasonic methods have been especially popular for nondestructive testing, and x-ray and gamma-ray methods have found their uses in structural determination, again in a nondestructive mode. Electrical methods include the use of applied electrical and magnetic fields to induce effects to be measured in the sample. Thermal effects include the use of localized heating, which may or may not be nondestructive in their application. Essentially, the methods operate via a transformation process—the energy which is delivered to the sample is altered in some way, so that a signal may be generated as a result of the interaction that happens. At the next stage of the process, the detection of that signal occurs.

4.2 Input Transducer

The input transducer functions by changing the energy, which is input in one specific form to interrogate the sample, into another form for ease of signal processing. Most frequently, this results in the end product of the transduction process being an electrical signal, although optical or pneumatic signals may be used for some applications. However, an electrical signal is much more compatible with other stages of the information retrieval process, and is used almost exclusively in modern instrumentation, except in special cases where electromagnetic interference at a high level is experienced or signal levels may be particularly small.

A common example of this transduction process is the use of a photodetector, such as a *p–i–n* diode, where an optical signal is transformed into an electrical current. In a similar way, a thermistor can transform temperature into a change in resistance, measured in an electrical circuit, or a thermocouple can transduce the same parameter into a voltage change. In the electrochemical domain, the ion-selective electrode may alter an ion activity level into a voltage or a polarographic cell will be used to transform a concentration of an electroactive species into a current.

4.3 Electrical Signal Transformations

Often the current or voltage produced at the transduction stage is very small or noisy, so a series of electrical transformations must take place in order to produce a signal which can interface to the output transducer and thus be meaningful to the analyst. These include such simple operations as amplification, attenuation, logarithmetic conversion and counting through to more complex procedures such as analog-to-digital (A/D) conversion, filtering and integration, and the application of mathematical techniques, usually in software, such as Fourier transformations. This procedure has been aided by the greater availability of electronic components with such features as wide bandwidth and lower power consumption, and offering complex circuit integration, together with the incorporation of a microprocessor as an integral part of many modern instruments, for control and data processing. A detailed discussion of signal processing and the data analysis methods which may be used is beyond the scope of this article.

4.4 Output Transducers

The range of possible output forms for data has increased considerably over recent years, from the analog display and the paper chart recorder, to various types of video display (e.g., LCD), digital meters, and via the internal or associated computer to disks, both hard and floppy. Such changes have helped to incorporate some of the knowledge that may be stored in an expert system into the instrument, and the display can often be more meaningful than a simple number. Thus, comparisons with previous calculations, data processing and so on, can be carried out and the stored knowledge in the system can, in many cases, enable the operator of the equipment to be competent in knowledge presentation with only limited training, representing a considerable level of cost saving to the customer, especially for routine work.

5. Illustrations of Modern Analytical Instrumentation

5.1 Functional Decomposition of the Instrument

In order to understand the operation of the instrument, and thus to be able to assess fully its capabilities for a particular application, a functional decomposition of the device may be necessary. In such a procedure, the elements of the device are analyzed in terms of their operation and role in the instrument, as well as the physical and chemical principles which underlie their function. In such a way, the limits of operation of a device can be better understood, and its potential for operation both inside and outside the normal range of investigation can be seen. Such a detailed evaluation is beyond the scope of this text, but any comprehensive volume on instrumental analysis (e.g., Willard *et al.* 1988) gives such details for many modern analytical methods. However, in the succeeding discussion of several subdivisions of types of analytical methods (from Table 1), the value of the functional decomposition can be seen in the determination of the "fitness for purpose" of the types of instrumental method discussed.

5.2 Spectroscopic Techniques

Spectroscopic techniques are among the most widely used techniques in modern instrumentation, and different optical sources can give rise to information that is so distinctive that it is particularly valuable for the identification of species. Further, there is a considerable body of literature in the field of spectroscopy, which is the measurement and interpretation of electromagnetic radiation absorbed or emitted when the molecules, atoms or ions of a sample move from one state to another. Every species has a unique characteristic relationship to the interrogating electromagnetic radiation. Thus, in the visible, ultraviolet and infrared parts of the spectrum, the energy from the source used causes changes in the rotational vibrational or electronic energies of the species in a distinctive way. Ultimately chemical instrumentation does not create information, but it refines the information present in the signal from the transducer.

A detailed study of the subject will require a knowledge of the principles of spectroscopy and the fundamental solid-state physics underlying the subject. Many suitable texts are available (e.g. Herzberg 1940). However, as the interaction with matter occurs over the entire electromagnetic (em) spectrum, this represents a change in wavelength from as short as 10^{-9} nm to that in excess of 1000 km. Associated with this wavelength change is a frequency change in the radiation, and with it the energy associated with each quantum or photon of the em energy. Thus, as may be expected from the very breadth of the range, the effects they produce may be quite different, and these em wavelengths are produced by several sources (moving from short to long wavelength) yielding gamma rays, x rays, far, middle and near ultraviolet (uv), the visible spectrum, infrared (ir) radiation (both near and far) and microwaves (and beyond). Thus, in the radio-frequency (rf) range (low photon energy), the energy transitions are associated with the re-orientation of nuclear spin states of the substance (in a magnetic field). In the infrared, the use of this form of radiation is important to change the vibrational and rotational energy of the molecule. Visible radiation is of sufficient photon energy to influence electronic transitions of the loosely held outer electrons of the substance. X rays will raise the inner electrons to excited states, and gamma rays are associated with transitions inside atomic nuclei. Figure 3 depicts the various regions of the em spectrum, associated with these energy changes.

Illustrative of the spectroscopic instrumental field is the spectrophotometer, a commercial device for determination of absorption of samples, usually over a range of the uv, visible and ir parts of the spectrum. Over such a wide wavelength region, it may often be preferable to select a range of components for optimum performance, and hence the value of the functional decomposition of such an instrumental system can be seen. Such a decomposition is shown in Fig. 4, for a wide-range spectroscopic instrument, illustrative of the main features of any such system, regardless of its specific region of operation. The purpose of the elements is as follows.

(a) Source. A range of sources may be used, such as lamps, arcs, lasers, magnetrons, rf oscillators, depending on the interaction to be studied. The main features to be considered include output power, spectral coverage, physical size and configuration, and reliability for long-term measurement use. Further, laser or arc lamp sources may be used in pulsed mode, to study transient events.

(b) Sample compartment. Samples may be solids, liquids or gases, and may be chemically or physically dangerous, with consequent effects on the instrumental construction. Thus, chemically corrosive or hydroscopic samples require a particular environment, as for example do radioactive or biologically hostile materials. The design of a suitable sample compartment to accommodate these features and the state of the material (or a change of state during the measurement) often imposes considerable difficulties for the system designer. Further, the spectroscopist may wish to work with an imposed external condition, such as an electrical or magnetic field (as in the use of hyphenated techniques), create a chemical reaction in the sample compartment or simply undertake a temperature or pressure change, adding further complexity to the device.

(c) Analyzer. The analyzer allows radiation of different wavelengths to fall on the detector. In the visible region, a diffraction grating may be used, whereas for the infrared or ultraviolet, such a device may be unsuitable, and a prism of the appropriate material preferred. Simpler devices such as filters are available, varying in design and operation over a wide spectral range.

(d) Detectors. Again, the range of detectors is wide, and the choice must be reflective of the wavelength coverage of the spectral features to be studied. Some detectors have a coverage over a wide spectral region, but may be lacking in sensitivity. Others, such as the photomultiplier, are highly sensitive over a relatively narrow region but are physically more fragile and require high operational voltages. The optical com-

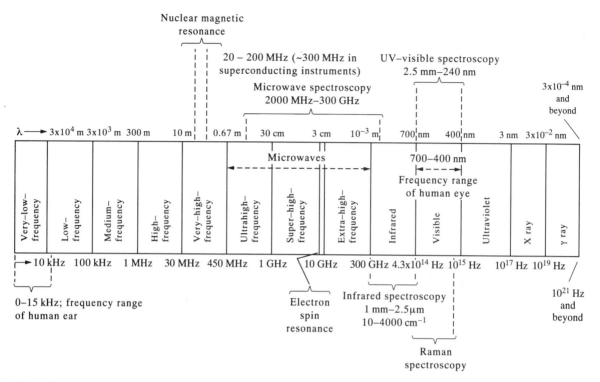

Figure 3
Regions of the electromagnetic spectrum showing the associated energy and wavelength, with indications of associated analytical techniques

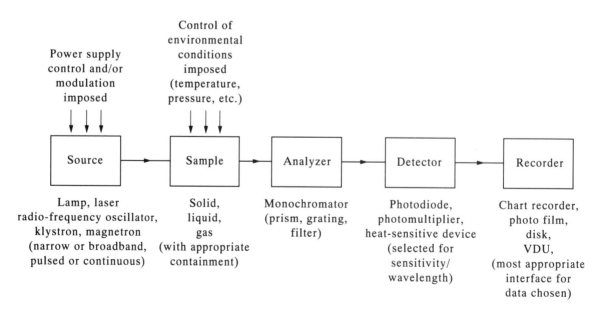

Figure 4
Functional decomposition of a spectroscopic instrument

munications market has led to the development of a range of useful solid-state, low-voltage detectors operating in the near infrared, a region particularly important due to the use of spectral features of ir absorption for species identification. Some detectors, for example scintillators, work by conversion of radiation (e.g., x rays) to visible radiation, which is detected by devices sensitive for that spectral region, or with adequate bandwidth for the investigation of transient events, studied using pulsed sources.

(e) Recorders. The traditional chart recorder and the photographic plate have been supplemented by a wide range of other output transducers, as discussed earlier. The selection of this component depends strongly on the subsequent stages of data analysis or processing to be undertaken.

5.3 Electrochemical Techniques

There are a number of electrochemical methods and their applications are wide. Potentiometric methods embrace two major types of analysis—direct measurement of an electrode potential from which the activity or concentration of an active ion may be determined, and the changes in electromotive force brought about through the addition of a titrant. The field has recently been supplemented through the development of novel ion-selective electrodes, taking their place alongside the established pH electrode. A range of different electrodes is available, including glass electrodes, solid-state electrodes, liquid–liquid membrane electrodes and enzyme- and gas-sensing electrodes.

Ion-selective electrodes are used to measure ion activities, important in determining reaction rates and chemical equilibria. Their advantages of use include the fact that they do not affect the solution being studied, they are portable, comparatively inexpensive and suitable for a direct determination of the parameter under consideration. They are, however, subject to interference, either when some characteristic of the sample prevents the probe from selecting the ions of interest, or when the electrode responds to other ions present in the sample.

Potentiometric titrations offer an approach to measurement in fluorescent or turbid, opaque or colored solutions and, in contrast to the previous direct method, they offer an increase in accuracy and precision, however, at the expense of time and difficulty.

Polarography marked a significant advance in electrochemical methods when it was introduced in the 1920s, with an element of selectivity through the control of the electrode potential, and the subsequent developments in electronics, especially since the 1960s, have contributed to the utility and desirability of the methods for "fingerprint" purposes in analytical applications. The associated technique of voltammetry has great sensitivity to environmentally important materials such as the heavy-metal pollutants, lead and cadmium and the technique is used in forensic work

in, for example, the determination of the presence of drugs in urine samples.

5.4 Chromatographic Methods

The term chromatography describes a series of related separation methods, based on experimental analysis originally undertaken in the early years of the twentieth century. The method has advanced considerably over the years to embrace gases or liquids for the mobile phase of the procedure, with liquids or solids for the stationary phase. Thus, a significant feature of chromatography is that it is a separation method where one of the two mutually immiscible phases is stationary, the other mobile, and they are brought into contact. The sample mixture, introduced into the mobile phase, undergoes a series of interactions as it is being carried through the system by the mobile element of the mixture. The physical or chemical properties of the sample determine the different rates of migration of the individual components under the influence of a mobile phase moving through the column containing the stationary phase. Separation of the components occurs, and this becomes useful to the analyst when one component is sufficiently retarded with respect to the others, for ease of its identification. The value of the technique is in the versatility of the method—due to the wide choice of materials for the stationary and mobile phases, it is possible to accomplish separations of materials which differ only slightly in their physical or chemical properties. The choice of state of matter for the mobile and stationary phases results in different experimental techniques, appropriate for different analytical procedures, as shown schematically in Fig. 5. Thus, liquid column chromatography (LCC), for example, shows several variations as a method. When the separation involves two immiscible liquids, the technique is referred to as liquid–liquid chromatography (LLC). Liquid–solid chromatography (LSC) makes use of the physical

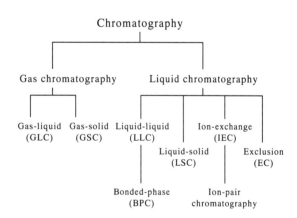

Figure 5
Schematic of various chromatographic methods

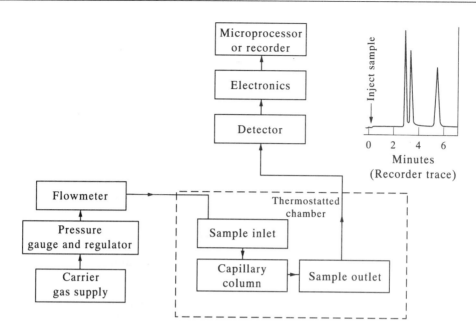

Figure 6
Schematic of a gas chromatograph, showing the decomposition into its essential elements

surface forces associated with the stationary phase, and its retentive ability. In ion-exchange chromatography, ionic components of the sample are separated by use of an appropriate stationary phase, and exclusion chromatography (EC) uses exclusion packings as the stationary phase to bring about a separation of molecules based on molecular geometry and size. If the mobile phase is gaseous, then gas–liquid (GLC) and gas–solid (GSC) chromatographic methods are possible.

Chromatographic methods may be quantitative or simple qualitative. For quantitative measurements, a calibration may be performed using known standards, and care must be taken in the operation of the procedure to ensure that reliable quantitative data are extracted. More detailed texts on the subject give guidance on procedures and methods to ensure that this occurs (Giddings 1965). The choice of phase used in the method, that is, gas or liquid chromatography, will be determined by the ease of presenting the sample in an appropriate phase for analysis. However, only about 15% of known compounds lend themselves to analysis by gas chromatography, because of insufficient volatility or thermal instability. Thus, liquid methods are the most frequently used and the interchange or combination of solvents can provide special selectivity effects that are absent in a gaseous mobile phase. High-molecular-weight species are conveniently analyzed by liquid column methods. Gas methods, either GLC or GSC, both rely on the fact that solutes

travel at their own rate through the column. The output achieved for such a system is a plot of time versus the composition of each gas carrier stream, and the time of emergence of a peak is characteristic of each component.

A functional decomposition of the instrumentation for each method used shows common features, and Fig. 6 shows such a diagram for a gas chromatograph. The main features seen are: carrier gas supply, sample injection system, separation column, detector, recorder and data analysis, and temperature regulation.

The liquid chromatography method shows similar features, with the exception that a solvent pump is used to force the mobile phase through the system, to replace the sample injection system. A precolumn may be used to presaturate the mobile phase, and temperature regulation is not needed if the material is liquid at room temperature, although column temperature can have an influence on the rates of separation.

The applications of chromatography lie in those areas of analysis where information on the composition of gas and liquid mixtures of large organic molecules is required. In this way, data which are difficult to achieve by other methods can be determined through well established analytical procedures.

5.5 Individual Methods

Other methods are widely used, including mass spectroscopy, thermal analysis, and nuclear magnetic resonance (NMR) methods. The mass spectrometer dates back to the early part of the twentieth century. The device produces charged particles consisting of the parent and ionic fragments of the original molecule. The spectrometer produces such fragments which are indicative of the ionic components of the molecule. As a technique, it is very sensitive in identifying species in the presence of others. The characteristic fragmentation patterns of molecules can give information on weight and structure, and very small samples may be used, as long as a gaseous sample can be provided. The mass spectrum becomes a "fingerprint" of the compound analyzed, and no two molecules will be fragmented and ionized in exactly the same manner. Thus, the size and structure of the original compound can be reconstructed by the analyst, even in the presence of other species. The experienced analyst can apply a series of rules for fragmentation patterns, which enable general features of mass spectra of compounds to be predicted. Thus, applications involving the determination of the nature of unknown substances are common; however, although the system uses small quantities of material it is destructive, which may be a limitation in some cases.

Thermal analysis includes a group of techniques in which the physical property of a substance is measured as a function of temperature, when it is subjected to a controlled temperature programme. The device can measure, for example, temperatures of transitions, weight losses, energies of transitions, viscoelastic properties and dimensional changes, among others. Applications include environmental measurements, product reliability and dynamic properties of materials.

Nuclear magnetic resonance (NMR) spectroscopy is a technique using the characteristic absorption of energy by a nucleus in a strong magnetic field. The technique has not, in the past, found ready acceptance by the analytical chemist, as it is less sensitive than other methods, such as chromatography. However, with computer methods, Fourier transform NMR has changed the application of the technique in analytical chemistry and in medical diagnosis the magnetic resonance imaging (MRI) technique has revolutionized aspects of diagnosis and replaced x rays for many applications. Relying on the response of the molecules of the body to NMR, and coupling the output of a series of data to a high-power computer for image analysis and construction, a "slice" of the body can be obtained, revealing the organs present in the volume analyzed. Needless to say, the technique is nondestructive. Electron spin resonance is a branch of absorption spectroscopy where microwave irradiation of the sample is used to induce transitions between magnetic energy levels of electrons with unpaired spins.

Other analytical techniques are often applied to surfaces, for example, ESCA (electron spectroscopy for chemical analysis), Auger electron spectroscopy and ion scattering spectroscopy (ISS). Surface analysis has many applications, including the evaluation of contaminants on a surface, and the assessment of corrosion. Thin-film composition can be related to electrical, transport and emission characteristics of the film.

5.6 Hyphenated Methods

A hyphenated method is a combination of any of the previously discussed methods (and indeed others) in order to gain the maximum amount of information about the species under consideration through the joint use of analytical methods. Frequently coupled methods are chromatography with mass spectroscopy, or with infrared spectroscopy. In this way the benefits of two totally different approaches may be used, and the deficiencies of one compensated by the attractions of another. By coupling the two techniques in a single instrument, the needs of sample preparation for both can be simplified, and the analyst can be sure that the data obtained are indeed on comparative samples. With the increasing use of computer techniques to analyze and process data, such hyphenated methods have considerable scope for future development.

6. Computer-Aided Analytical Methods

Much has been said about the role of the computer in supporting the analytical method. However, this role has been evolving since the 1960s and will continue to do so, with the development of cheaper, more powerful and flexible machines. This evolution can be seen by considering four essentially historical aspects of the use of the computer. They are the use of the microprocessor off-line, on-line, in-line and finally intra-line (including the use of the expert system), after the approach of Willard *et al.* (1988).

The off-line use of the computer is the application familiar to mainframe users of the past decades, where results were collected on tape or disk, and subsequently processed. The time delay in this procedure may be an important factor limiting the analytical method, and such an approach is now only used where very complex calculations are needed, requiring access to a large external computer.

The on-line approach involves the use of an incorporated microprocessor for data acquisition and instrument control functions, for example, and reduces the need for the analyst to interact with the instrument for every action of the procedure. With the expansion of low-price microprocessors, the use of in-line methods has developed, where the processor is an integral element of the device, used, for example, to maintain precise control of the parameters of the instrument, especially where repetitive analyses are performed. Such systems are also more amenable to the upgrade of the system by the replacement of memory elements to perform new tasks.

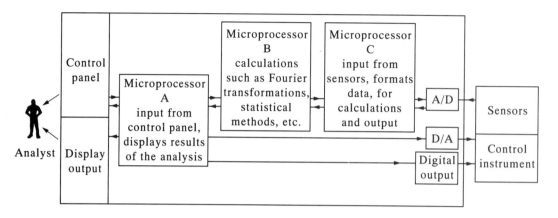

Figure 7
The interaction between the analyst and the analytical instrument incorporating an expert system

The use of the intra-line processor allows for evolution and development of the instrument. Several processors may be included to constitute subsystems that have the capability to change the nature of the measurement system. For example, hardware functions such as basic processing, robot arm control, interface buses and so on, may be incorporated, together with software for data acquisition, instrument control, database management, spreadsheet analysis and graphics output, and including word processing. This enables the final report to the customer to be produced from the instrument. With the incorporation of the knowledge of the experienced analyst and customer, the expert-system-based instrument can offer increasing reliability and flexibility, through, for example, built-in diagnosis, regular internal calibration and the prompting of the inexperienced analyst where necessary. As shown in Fig. 7, this places the analyst as a peripheral aspect of the instrumental system, able to concentrate on the noninstrumental functions such as problem definition and customer support, and leave the routine functions to the intelligence in the device itself.

7. Concluding Remarks

The variety of analytical techniques is large, and a full list is well beyond the scope of the text. The bibliography provides a source of such information. The choice of the most appropriate technique, for any application, is particularly important, in view of the high cost of the equipment. The analysis, and the careful definition of the analytical problem cannot be overstressed, particularly in light of the economic and time implications of an inappropriate decision. However, the principles of the analytical method are universal and the value of the functional decomposition of the analytical equipment can be seen, to enable the most appropriate choice to be made for a particular application.

See also: Chemical Analysis, Instrumental; Optical Measurements; Process Instrumentation Applications; Spectroscopy: Fundamentals and Applications

Bibliography

Giddings J C 1985 *Principles and Theory—Dynamics of Chromatography, Part 1.* Dekker, New York
Herzberg G 1940 *An Introduction to Spectroscopy.* Dover, New York
Willard H H, Merritt L L, Dean J A, Settle F A 1988 *Instrumental Methods of Analysis,* 7th edn. Wadsworth, Belmount, CA

K. T. V. Grattan
[City University, London, UK]

Angular Rate Sensing

This article describes the design and operation of mechanical and optical gyroscopes, which sense angular rates with respect to inertial space. These instruments are used for sensing and for stabilization. To enable precise angle measurements, techniques have been developed for compensating errors and for integrating angular rate inputs. Optical and mechanical gyros used for navigation can detect $0.01°\,h^{-1}$. Less accurate, less costly units are used for vibration sensing.

Mechanical gyros sense changes in angular momentum or, alternatively, Coriolis acceleration. Common mechanical units include floated single-degree-of-freedom and dry-tuned gyroscopes. Micromachined silicon and quartz oscillating sensors and the hemispherical resonator gyro are being developed.

Optical gyroscopes, which utilize the Sagnac effect, include the ring laser, and the emerging interferometer and resonant-cavity fiber-optic gyros.

1. Mechanical Gyroscopes (Angular Momentum and Coriolis Acceleration)

1.1 Floated Gyroscopes

A cross section of a high-performance floated single-degree-of-freedom (SDOF) gyroscope is shown in Fig. 1. The wheel is supported by gas bearings whose support pressure is internally generated by grooves on the bearing surfaces and is spun at constant rate by a synchronous hysteresis motor. The axis of revolution is known as the spin axis.

The wheel spins in a sealed cylinder known as the float or torque summing member (diameter is 0.02–0.04 m). Ideally, the float's only degree of freedom is rotation about the cylinder or output axis. In high-performance units, the float is supported by fluids and magnetic suspensions to minimize friction torque. In response to angular rates about the input axis, the float rotates about the output axis (Fig. 1) with respect to the instrument case. This motion is detected with a soft iron, variable reluctance sensor. Soft iron cores, similar to the position sensors, or permanent magnet devices exert either rebalance or command torque as described below.

In lower performance units, the wheel is supported by ball bearings and the float is supported by pivot and jewel bearings.

1.2 Principle of Operation

The principle of operation of the single-degree-of-freedom gyro is shown schematically in Fig. 2 and

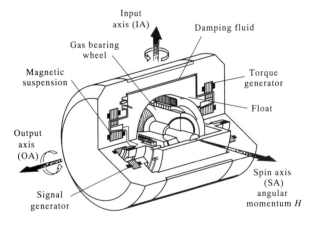

Figure 1
The single-degree-of-freedom floated integrating gyroscope

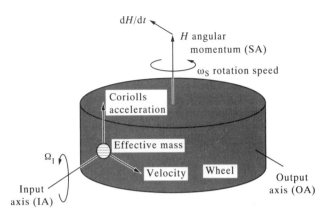

Figure 2
Mechanical gyroscope operating principles

mathematically below. The spinning wheel defines an angular momentum H along the wheel's axis of rotation. A case rate about the input axis causes a change in angular momentum along the output axis as indicated in Fig. 2. Since Newton's law states that the time derivative of angular momentum equals the applied torque, the float rotates with respect to the case about the output axis.

Coriolis accelerations offer an alternative view of the gyro's response to angular rates. When a particle is moving with velocity V in a reference frame such as the float's and the frame is rotated at an angular rate Ω with respect to inertial space, the particle experiences the Coriolis acceleration described by the vector cross product $2\Omega \times V$. An effective mass along the input axis as shown in Fig. 2 accelerates upward while a similar mass along the negative input axis accelerates downward so that the float is twisted about the output axis with respect to the case.

1.3 Equations of Motion

The differential equation describing the motion of the float about its output axis is given by

$$I\ddot{\theta} + C\dot{\theta} + K\theta = H(\Omega_I + \dot{\theta}_I) + M_R - I\dot{\Omega}_O + M_E \quad (1)$$

where θ is the angle of the float with respect to case about OA; θ_I is the float angle about IA (very small for SDOF gyros); Ω_I and Ω_O are the case angular rates with respect to inertial space about IA and OA, respectively; I, C and K are the float inertia, damping, and spring stiffness about OA respectively; M_R in the rebalance or command torque (discussed below); and M_E is the error torque (discussed below).

1.4 Rate Versus Integrating Gyros

In a rate gyro, the stiffness K in Eqn. (1) is significant; however, the gyro will accurately indicate rate when transients have passed. In many applications, accurate integrated rate or angle is desired so that the stiffness term must be small (derived from Eqn. (1)), closed loop control is used, and the gyro is known as integrating. For accuracy, low stiffness is again desired to minimize rate errors caused by pick-off errors. The integrating gyroscope operates through transients and short interruption of the torque rebalance since angular data is stored and recovered in the damping term $C\dot\theta$.

1.5 Platform Stabilization

In platform stabilization, the float output θ controls a torque motor which rotates the platform on which the gyro is mounted about the gyro's input axis. Command torque applied to the float controls the rotation rate of the platform that is, $-H\Omega_I = M_R$. The summation of command torque and angular rate is done mechanically on the gyro float.

Often, the command torque is derived from the output of an accelerometer to realize a pendulum with an 84 min period, a situation known as Schuler tuning, which reduces many errors associated with navigation over the earth's surface. This mechanization offers the long oscillation period while simultaneously the platform is very stiff against interfering torque such as friction, a remarkable situation.

1.6 Error Models and Compensation

In addition to the desired rate terms in Eqn. (1), other torque must be accounted for. A simple model for error torque is

$$M_E = D_F + D_I f_I + D_S f_S + \text{higher order terms} \quad (2)$$

where D_F in the fixed term or bias, D_I and D_S are the mass unbalances about output axis, and f is the specific force (acceleration minus gravity along indicated axis).

If the terms are constants or known functions of temperature, compensation is often done. Low-frequency change in the bias is known as stability or repeatability while higher frequency (typically 0.1–60 Hz) is known as resolution and may be defined in terms of angle. Since errors in rate lead to large errors in angle, updates from external sources such as LORAN allow lower quality intruments. High quality SDOF floated gyros possess unrivalled stability for spans of minutes to several hours.

1.7 Dynamically Tuned Gyroscope

The dynamically (or dry) tuned gyroscope consists of a synchronously driven motor driving a shaft (Fig. 3). Attached to the shaft are a Hook's joint and gimbal that allow the rotor (of typical diameters 0.015–0.04 m) to rotate about the lateral axes while restraining translation. Magnetic position sensors and torque drivers are fixed to the instrument case.

The gimbal's inertia moments cancel the stiffness of the Hook's joint; thus the stiffness in Eqn. (1)

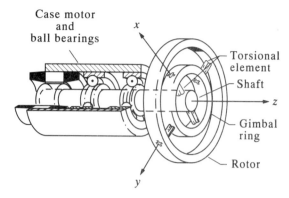

Figure 3
The dynamically tuned gyroscope (after Craig 1972, © IEEE, New York. Reproduced with permission)

becomes small and an integrating gyro, whose performance for periods longer than one day approaches the SDOF, is realized. By symmetry, the dry-tuned gyro senses two axes of angular rate simultaneously while avoiding flotation fluids and magnetic suspensions. Equation (1) applies to both axes but the coupling of $H\dot\theta_I$ and low damping requires that the rebalance torque of one axis be generated by the angle (rotor to case) of the orthogonal axis.

1.8 Micromachined Oscillating Gyro

Micromachining applies the techniques of solid-state electronics, such as photolithography, masking and etching, to carve mechanical shapes. Considerable effort is being devoted to micromachining gyroscopes. Presently, high-speed low-power wheels cannot be micromachined, so the emphasis has been on oscillating or tuning fork gyros.

Figure 4 shows the operation of a silicon tuning fork gyro. Elecrostatic (capacitor) forces in the combs move the masses in opposite directions in a self-excited oscillator loop. Angular rates applied to the substrate cause Coriolis accelerations that rotate the structure at the drive frequency. Capacitor plates below the proof mass sense the differential motion which is proportional to the input angular rate.

Tuning fork gyros are also being made of crystalline quartz and use the piezoelectric effects for both drive and sensing.

While performance is poorer than for the wheel units, the low cost and small size (die dimensions of the order of 0.003 m) of microfabricated gyros should open new applications for rate sensors.

1.9 Other Mechanical Gyroscopes

The hemispherical resonant gyro (HRG) is a vibrating wineglass machined from amorphous quartz driven by electrostatic forces. Electrostatic forces cause the glass to vibrate so that the cup becomes elliptical. As the

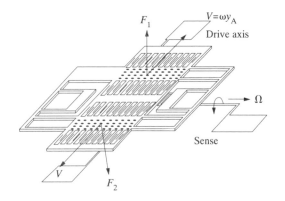

Figure 4
Schematic drawing of the comb drive tuning fork gyro
showing input rate Ω, Coriolis forces F_1 and F_2, and
horizontal drive velocities $V - \omega y_A$

gyro is rotated about the stem, the mode shape moves
in a known relation with respect to the instrument
case. The HRG is being developed for performance
approaching the spinning gyros at costs reputed to be
lower.

In the electrostatic gyro (ESG), a slightly unbalanced beryllium ball is supported and spun electro-
statically in a vacuum. Because of the unbalance, the
ball's position with respect to the case, a function of
case angular rate, can be detected. The ESG is used
for high-accuracy navigation over long time periods on
submarines. Its response to vibration and shock limit
applications in other areas.

2. Optical Rate Sensing

2.1 Sagnac Effect
When a beam of light is travelling around a closed
path, the beam sees an effective length change, a
Doppler shift, in the optical path $\Delta L = \pm 2 A \Omega_1 / nc$
where A is the area enclosed by the path (minimum
projected onto any plane. The input axis is the normal
to the projected plane), n is the index of refraction, c
is 3×10^8 m s^{-1}, and Ω_1 is the input rate. When the
input rate is in the same direction as the light, the
effective path length increases. The difference in
effective path length, which is derived using special
relativity, is sensed by techniques described below.

2.2 Laser Gyroscopes
Ring laser gyros, which are now used for aircraft
navigation, are made of a triangular, sealed cavity
machined from a low thermal expansion coefficient
material such as Cervit. The optical path (Fig. 5),
0.05–0.08 m per side, is defined by two fully and one
partially reflecting very-high-quality dielectric mirrors

which define the beam polarization as normal to the
plane. To effect closure of the beam path, one mirror
is slightly spherical. The cavity is filled with a gas,
often helium neon, which lases when the anode and
cathode are excited with appropriate voltages. The
lasing causes beams, which resonate within the cavity,
to propagate in opposite directions. The difference in
resonant frequencies is

$$\Delta \omega = \frac{8 \pi A}{n L \lambda} \Omega_1 \qquad (3)$$

This sensitivity is improved by increasing the radius
and by decreasing the wavelength of the light in air
(λ). Because of the resonance, the wavelength must
adjust so that the scale factor is insensitive to cavity
length to first order. For high performance, a
piezoelectric element is also used to adjust cavity
length.

The input is read by allowing a small portion of the
circulating light to escape from the ring. The two
beams impinge at a small angle on the detector so that
input rate results in fringes on the detector that
includes two sensitive elements so that the direction of
rotation is obtained. Per Eqn. (3), each fringe is
equivalent to an angle input so that the gyro is
integrating. An alternate view is that a standing wave
is set up in inertial space. The viewer fixed to the gyro
case rides over the peaks and valleys of the standing
wave.

Because of back scatter in the mirrors, the two
waves lock frequencies unless a frequency bias is
introduced. This bias is generally introduced by
mechanically dithering the gyro about its input axis.

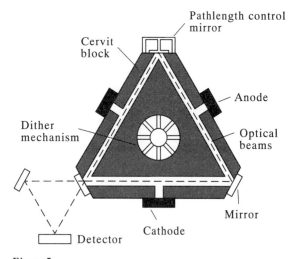

Figure 5
Schematic of ring laser gyroscope. Input axis is
perpendicular to the plane of page

The ring laser gyro with dithered drive has been very reliable. Although the Sagnac-induced length changes are small (typically 10^{-15} m deg^{-1} h^{-1}), the counter rotating waves reject common mode errors. Bias stability, limited by Langmuir flows of the lasing gas and shot noise, and cost are similar to those for dry-tuned gyroscopes; however, scale factor stability is better than all other gyros. Although the gyro integrates, no method exists for applying command torque so that the summation for stabilization must be performed in precision electronics rather than in the gyro as for spinning units.

2.3 Fiber-Optic Gyroscopes

By defining the path by optical fibers, instead of precise mirrors, costs are expected to become lower than for ring laser gyros of similar quality. By conditioning the beams in integrated optics (solid-state devices that can phase shift, guide, and split light), the effect of the fiber's higher back-scatter can be negated. Fiber-optic gyros may be classified as interferometers or resonant fiber-optic gyros.

In the basic interferometer, light is generated by a diode laser. The beam is split and driven into opposite ends of a kilometer-long fiber coil. Propagating in opposite directions, the beams' phases are shifted by the Sagnac effect (Sect. 2.1). After emerging from the coil, the beams are superimposed to produce on a single detector a fringe pattern similar to the ring laser. Wide-linewidth diode lasers reduce backscatter effects.

The resonant-cavity fiber-optic gyro (RFOG) splits light from a diode laser into two counter-propagating beams which are coupled (fiber cores brought into contact) into a short closed loop of fiber that is a high-finesse optical cavity. Rotation shifts the resonant frequencies of the counter travelling waves (Eqn. (3)), and varies the light leaving the ring. An optical coupler and integrated optics chip direct the two light beams onto separate photodetectors. Heterodyning permits the resonant frequency of each direction, proportional to input rate, to be detected. The two directions are maintained at resonance by integrated optics which adjust the frequency of light entering the ring. Because of the resonant structure, the RFOG must be driven by a laser of very narrow linewidth. Presently, development effort focuses on minimizing optical scatter effects.

See also: Position Inertial Systems

Bibliography

Barbour N, Elwell J, Setterlund R 1992 Inertial instruments—where to now? *Proc. AIAA Guidance, Navigation, and Control Conf.* American Institute of Aeronautics and Astronautics, New York
Britting K 1971 *Inertial Navigation Systems Analysis.* Wiley, New York

Craig R J G 1972 Theory of operation of an elastically supported, tuned gyroscope. *IEEE Trans. Aerosp. Electron. Syst.* **8**(3), 280–8
Greiff P, Boxenhorn B, King T, Niles L 1991 Silicon monolithic micromechanical gyroscope. *Proc. Transducers '91.* IEEE Electron Devices Society, New York
Jacobs S F *et al.* 1984 Physics of optical ring gyros. *Proc. Soc. Photo-Opt. Instrum. Eng.* **487**, 1–137
Wrigley W, Hollister W M, Denhard W G 1969 *Gyroscopic Theory, Design, and Instrumentation*, MIT Press, Cambridge, MA

M. S. Weinberg
[Charles Stark Draper Laboratory, Cambridge, Massachusetts, USA]

Artificial Intelligence in Measurement and Instrumentation

Intelligent instrumentation can be defined for the purposes of this article as an instrumentation incorporating advanced information and knowledge processing. A more precise meaning of the definition will emerge from the article. Instrumentation in this context is that concerned with measurement.

1. Intelligence and Artificial Intelligence

Intelligence is an important concept in psychology. It has received much attention, but remains a complex idea with many and diverse connotations and perspectives. It is seen as being manifested in reasoning, judgement, adaptability, learning, abstraction, concept formation, and so on.

The term artificial intelligence has gained a wide currency in information technology (Shapiro 1992).

Machines can be said to exhibit artificial intelligence when performing such complex tasks as natural language understanding, machine vision, problem solving and the like, commonly by imitating the way in which humans carry them out. Key areas of artificial intelligence technology are logic programming, machine learning and knowledge acquisition, knowledge representation, knowledge elicitation, expert systems, computational reasoning and deduction, natural language understanding, planning and problem solving, machine vision and pattern recognition.

2. Intelligent Instrumentation

Intelligent instrumentation has been defined as instrumentation incorporating advanced information and knowledge processing. Some further discussion is needed at this stage.

The term intelligent instrumentation has common, though loose, usage. It is used in a wide sense, wider than the term artificial intelligence. While at the top end of intelligent performance we have instrumentation with a high degree of artificial intelligence, at the other end devices are termed "intelligent," or "smart," when they incorporate no more than limited signal processing.

All instrumentation processes information and knowledge and, to the extent that it performs functions of the human senses and brain, it is intelligent. However, instrumentation which has limited information-processing capability, typically analog and predominantly linear, or which uses only simple functions such as limiting and multiplication, is conveniently distinguished from intelligent instrumentation. For lack of a better term it may be called "conventional."

3. Information and Knowledge Processing Functions of an Instrumentation System

It is most effective to discuss intelligence in instrumentation in terms of the information and knowledge processing functions of an instrumentation system.

The object under observation interacts with a sensing subsystem. The object may simply act on the sensor which converts the action into information in symbolic form to be further processed in the system. In some systems the object under observation is interrogated by the sensing subsystem and the response of the object is sensed.

The information from the sensor is input to an information and knowledge processing subsystem which performs transformations on it. The output from the information and knowledge processing subsystem either is fed to a human–machine interface or is input to other information and knowledge processing systems, which may involve effectuation of the information, that is, the conversion of the generation of an action functionally related to the information carrying symbol. The information and knowledge processing subsystem may be said to perform the functions of cognition: the use and handling of knowledge. One of the cognition functions is perception—assigning meaning to sensed signals.

The functions of cognition are based on knowledge. In machine information and knowledge processing, knowledge consists of symbolic formulae in an appropriate language. It may be descriptive or declarative, representing facts, beliefs and conventions about the real world, or else procedural, representing methods for performing tasks. We may distinguish between surface or empirical knowledge and deep or theoretical knowledge using a model of a particular domain.

Processing knowledge involves a reasoning mechanism or inference engine. The operations performed on knowledge may be either algorithmic or heuristic in nature.

The knowledge and information acquisition and transformation processes are controlled by a set of control functions.

The instrumentation has a human–machine interface through which the information and knowledge output of the instrumentation is passed to a human operator and through which the operator can control the instrumentation.

Computer-aided knowledge and information processing in measurement systems may be on-line, with the processing being in real time. This is intelligent instrumentation proper. However, it is possible to partition the knowledge and information processing functions so that some are performed on-line and some are handled off-line by more powerful computers undertaking a variety of tasks.

While intelligent instrumentation is strictly concerned with the replacement of human operators, it does not necessarily mean total automation. Indeed, it is often most effective to partition tasks between those handled by machine intelligence and those performed by a human operator, who may undertake such tasks as teaching.

4. Hardware and Software of Intelligent Instrumentation

The nature of intelligent instrumentation is (essentially) determined by the technology of the hardware.

Intelligent instrumentation is digital and digital instrumentation increasingly intelligent. Digital technology offers immensely powerful information and knowledge processing capabilities, implementation in general-purpose hardware and software, accuracy, noise immunity, and increasing cost-effectiveness (see *Digital Instruments*).

The use of electrical information processing means that the sensors of modern instrumentation must have electrical outputs. It is essential in the context of intelligent instrumentation to mention the rapid development of modern silicon sensors which combine sensing and information processing in a single block.

Arguably one of the greatest advantages of digital instrumentation, and an important aspect of intelligence, is the capability of the human–machine interface (see *Operator–Instrument Interface*).

Intelligent instrumentation is implemented by an appropriate balance of hardware and software. To an increasing extent, software lies at the heart of intelligent instrumentation systems.

5. Capabilities of Intelligent and Knowledge Based Instrumentation

5.1 Sensing

The information processing functions of intelligent instrumentation enable such systems to correct or compensate for the inadequacies and errors of the input sensors. Typical sensor correction and compensation functions are linearization, deconvolution, feedforward compensation and automatic calibration. Sensors with such functions are commonly called intelligent, or more usually smart, although they are not really intelligent in a psychological or artificial intelligence sense.

Conventional instrumentation commonly uses only a single sensor in a single channel. In intelligent instrumentation it is common to use the fusion of information from multiple sensors.

5.2 Information Output

A human–machine interface with significant information processing power makes a substantial contribution to the performance of intelligent instrumentation.

First, it enables the information to be displayed in a form convenient to the human operator, for example, in the form of text, graphs, charts, mimic diagrams and so on. Second, advanced information processing provides great flexibility in the nature of the information which is output; it need not merely be the raw information that was input, but it may comprise some inferences drawn from it.

5.3 Control, Configuration, Adaption and Learning

In conventional instrumentation the control of the information and knowledge processing, if any, is performed by a human operator through the human–machine interface. The configuration of the system, that is, the transformations performed on the information and knowledge, is generally fixed, though it may to a limited extent be altered by the operator by, say, switching ranges, altering gains, switching between filters and so on. Intelligent instrumentation offers greatly enhanced flexibility, enabling the system to be substantially reconfigured by simple operations from the human–machine interface.

The reconfiguration may be performed automatically, with the instrumentation having an adaptive capability, that is, the capability to modify its function in order to improve performance. Thus, for example, when a signal has been acquired, its characteristics may be examined and used to determine the form of the subsequent information processing and output. The adaption may use an extensive knowledge base.

Intelligent and knowledge based instrumentation may have the capability of learning, which is the capability of acquiring and storing knowledge, leading to a relatively permanent change of performance capability as a result of experienced stimuli. This is related to adaption. The instrumentation may, either automatically or through the action of the operator, incorporate acquired knowledge in its knowledge base or use it to alter that base. Similarly, it may, by either of the above modes, alter its configuration.

5.4 Perception

Perception—the assignment of meaning to sensed signals by the application of knowledge—is a most significant capability of intelligent instrumentation. We may distinguish two forms: filtering and pattern recognition.

Filtering, which is the processing of noise-contaminated signals to detect and extract the information carrying component, may only involve very simple signal processing. However, with more powerful information processing techniques it is possible to enhance performance. Filters which perform more complex operations may be implemented. It is possible to build filters which use deep empirical or theoretical models of the signal or noise, or both. Filters may be adaptive and learn from experience.

Pattern recognition is the use of the features of a signal or signals and the relation between them to identify a source object or its attributes. The information processing capability of modern intelligent instrumentation enables a wide variety of pattern recognition methods to be implemented. Typical applications are object recognition in machine vision, defect detection and identification, diagnosis of machine condition, medical diagnosis and the like. The recognition process may vary from one based on simple rules of classification to processes applying an extensive knowledge base.

It is to be noted that all instrumentation, including the simplest, involves perception in the sense that it assigns meaning to the input and almost always involves some filtering. In the case of perception there is no strict distinction between intelligent and conventional instrumentation—it is a matter of degree.

5.5 Inferential Measurement

The characteristics of the observed objects may not be simply related to those variables which can act directly on a sensor. It is therefore necessary to infer the value of the measurands from the relation or set of relations (model) which they bear to other variables, which can be directly sensed or observed. Such a process will be termed inferential. It is possible to distinguish two forms of inferential measurement: measurement in which the relation between measurand and observables is explicit, and measurement in which the measurand is related to the observables implicitly as a variable or parameter of a model.

Inferential measurement, when measurand and observables are explicitly related, is relatively simple. A typical example is the measurement of mechanical power by the sensing of force and velocity, and their multiplication. Such processes are generally not intelligent in any meaningful sense but, although they can commonly be performed by conventional means, advanced digital techniques offer the advantages of simplicity and flexibility of implementation as well as enhanced accuracy.

Inferring a measurand from observables using an

implicit model is the identification process. In general, it uses an interrogation of the observed object by the sensing system, and observes the response, inferring the measurand from measurements of the interrogation and response variables.

The process of identification has been extensively developed and there is an immense literature on the topic which can only be mentioned here. The principal features of the process are the establishment of an appropriate model, the choice of interrogation signal or signals, the choice of inference or estimation signals, and the problems imposed by noise and measurement uncertainty (see *Identification in Measurement and Instrumentation*).

For many applications identification can be performed by analog techniques, or now more commonly by digital data collection and off-line processing and with substantial operator interaction. With intelligent instrumentation, in many applications the process can be made on-line, and automatic.

Intelligent and knowledge based instrumentation enables the use in inferential measurement of nonnumerical information and models, and of heuristic techniques.

5.6 Concept Formation

An important feature of intelligence is the formation of concepts or models from acquired information.

Concept formation may be defined as the determination of rules by which an object is attributed to a class and by which it may be assigned a symbol.

This is related to the formation of models—in this context the formulation of a set of abstract symbolic expressions, homomorphic with empirical objects.

Abstraction, that is, the removal of irrelevant detail in a description, is a key feature of any concept of intelligence. It is in the context of the formation of concepts or models by instrumentation that the capability to abstract particularly manifests itself.

The formation of concepts or models is the principal objective of many measurement systems. In conventional measurement, concept or model formation is performed by humans, or by off-line processing with human interaction. In intelligent instrumentation, such concept or model formation may be made on-line and be wholly or substantially automatic.

Examples are the learning of patterns or the identification of the structure of models relating observed quantities.

5.7 Inferences from Acquired Information

A total information system does not in general require merely a measure of attributes of the object or phenomena observed, but rather an interpretation of those measures in terms of the objectives of the supersystem which uses the information. Examples are alarms, fault diagnosis and the like. Intelligent and knowledge based instrumentation enables such inferences to be drawn from both numeric and other

symbolic acquired information, using a stored knowledge base if appropriate.

5.8 Measured Information and the Supersystem

As previously mentioned, the information and knowledge output of instrumentation is required, in general, not for its own sake but for the purposes of some supersystem. In the most general sense it is required for control, using the information required to alter the supersystem in some way in order to make it assume a desired form or behavior. The partition between reasoning about acquired information and the application of intelligent control rules is to some extent arbitrary. The two must be compatible.

6. Applications

Intelligence is, to an increasing extent, applied in measuring instrumentation. Smart sensors are becoming commonplace. Medical diagnostic and allied instrumentation, other diagnostics and image processing are outstanding examples of advanced applications.

See also: Digital Instruments; Image Processing and its Industrial Applications; Operator–Instrument Interface

Bibliography

Gregory R L (ed.) 1980 *The Oxford Companion to the Mind*. Oxford University Press, Oxford, pp. 48–50
Ohba R (ed.) 1992 *Intelligent Sensor Technology*. Wiley, Chichester, UK
Shapiro S S (ed.) 1992 *Encyclopedia of Artificial Intelligence*. Wiley, New York

L. Finkelstein
[City University, London, UK]

Automotive Applications of Measurement and Instrumentation

Electronic sensors have been used in road vehicles almost from their inception. The earliest sensors were essentially switches, used to measure the crankshaft position for ignition timing purposes. An early example was Lenoir's gas engine of 1865, which used a rotary switch connecting batteries to a coil to generate the spark (Newcombe and Spurr 1989). Later internal combustion engines used a rigidly coupled magneto to sense the engine cycle position. In the 1920s the familiar distributor, contact breaker and coil arrangement evolved which has persisted to the present day. The cam and contact breaker system combines the two functions of position sensing and current switching.

There are always problems of wear and contact surface deterioration with such a system, and most vehicles now rely on an electronic arrangement in which the functions of position sensing and current control have been separated, with noncontact sensing techniques being adopted.

The next form of sensor to be adopted was used for fuel level measurement. In early vehicles the fuel tank was often placed behind the dashboard, allowing the engine to be gravity-fed. As long as the tank remained in this position mechanical sensing devices such as manometers could be used. However, the fuel tank was soon moved to its current position at the rear of the vehicle, and a pumped fuel supply adopted. This led to the introduction of electrical methods for measuring and displaying the fuel level.

Other automotive potentiometer applications followed later. In the late 1950s Bendix developed and patented their "electro-injector" fuel injection system, in which the throttle position was sensed by a potentiometer. This was subsequently refined by Bosch to form the basis of the well-known "D-Jetronic" system, used extensively by Volkswagen and others in the 1960s.

Potentiometer sensors are now frequently used for sensing throttle and brake pedal position, steering wheel motion and suspension displacement, for automatic gearbox control, and for many other applications. Potentiometers are cheap and reasonably reliable for many applications, but suffer from the major disadvantage common to all devices which rely on a sliding contact, namely wear. While they may be adequate for, say, throttle position transduction, they tend to give rise to problems if used for applications such as shock absorber motion sensing. This is because a car body tends to remain close to one position relative to the wheels throughout a journey, but undergoes large numbers of small excursions around the "mean" position. This phenomenon is known as dither, and unless special precautions are taken it can cause parts of the potentiometer track to become badly worn or even destroyed locally. For this reason many manufacturers are beginning to consider alternative, noncontact forms of displacement sensor, such as inductive, magnetic, capacitive or optical types.

The need to know the speed of the vehicle arose at an early stage, and speedometers became mandatory with the introduction of speed limits in the 1920s. The method chosen was to sense the speed of a rotating magnet driven from the gearbox, by means of the drag effect of eddy currents induced in an aluminum or copper "cup" enclosing the magnet and working against a spring. This arrangement has survived almost unchanged for 70 years, and it is only recently that electronic systems have begun to appear in which a variable reluctance or Hall sensor is used to measure rotation rate by means of a toothed wheel.

Until the 1980s oil pressure was measured mechanically by a pressure pipe connection passed from the oil pump to the back of the dashboard. This arrangement has an unfortunate propensity to leak, which not only endangers the engine but is also injurious to the driver's trousers! Micromachined silicon or thick-film pressure transducers are now becoming widespread. Oil pressure is not usually displayed nowadays since modern bearings are very reliable. Lubricant pressure measurements in modern vehicles are normally used simply to illuminate a dashbaoard warning in the event of catastrophic oil loss.

1. Powertrain Transducers

A complex electromechanical system such as a motor vehicle has to be controlled by the operator in an environment which is constantly changing, and which can generate an almost unlimited amount of input data. To control a vehicle successfully in traffic as many as possible of the mechanical functions of the vehicle need to be automated, leaving the driver free to determine the vehicle's speed and direction. The need for reliable, low-cost instrumentation in a vehicle is consequently very great. To give an example of the wide range of devices required, Table 1 lists typical specifications required for the sensors used to control the engine and transmission (the powertrain). A comprehensive powertrain control system might include all of the devices listed, although a more basic strategy can be implemented successfully with only a few transducers. The critical quantities which have to be measured are ignition timing, airflow into the engine, throttle position and transmission speed.

1.1 Ignition Control

As noted in the introduction, early ignition control systems used mechanical sensing devices to control spark plug firing. Inlet manifold vacuum pressure is also measured in a mechanical system and used to infer engine load. The manifold pressure changes are used to alter mechanically the time at which a switch is closed to create the spark.

Modern timing sensors use electromagnetic, Hall effect or optical approaches to detect the motion of a projection attached to a shaft geared to the crankshaft. A certain amount of error is inevitable in these systems owing to vibration and torsion (wind-up) in the geared drive. It is likely that in the future this will be eliminated by making timing measurements directly on the crankshaft.

Once engine load and speed have been measured, the required ignition timing can be determined from a three-dimensional table relating load and speed to ignition advance. This function is implemented in a rather crude manner by the mechanical techniques described above, but in modern vehicles the optimized data is stored in a microprocessor memory in the form of a look-up table. A typical example is shown in Fig. 1.

Table 1
Automotive powertrain transducers

Sensor/type	Proposed sensing method	Range	Accuracy	Temperature operating range	Response time
Inlet manifold absolute or differential pressure sensor (petrol engines)	piezoresistive silicon strain gauged diaphragm or capacitive diaphragm	0–105 kPa	±1% at 25 °C	−40 °C to +125 °C	1 ms
Inlet and exhaust manifold pressure sensor (diesel engines)	as above	20–200 kPa	±3%	as above	10 ms
Barometric absolute pressure sensor	as above	50–105 kPa	±3%	as above	10 ms
Transmission oil pressure sensor	differential transformer + diaphragm or capacitive diaphragm	0–2000 kPa	±1%	−40 °C to +160 °C	10 ms
Inlet manifold air temperature sensor	metal film or semiconductor film	−40 °C to 150 °C	±2% or ±5%	−40 °C to 150 °C	20 ms
Coolant temperature sensor	thermistor	−40 °C to +200 °C	±2%	as above	10 s
Diesel fuel temperature sensor	thermistor	−40 °C to +200 °C	as above	−40 °C to +200 °C	as above
Diesel exhaust temperature sensor	Cr–Al thermocouple	−40 °C to +750 °C	as above	−40 °C to +750 °C	as above
Ambient air temperature sensor	thermistor	−40 °C to +100 °C	as above	−40 °C to +100 °C	as above
Distributor mounted timing/ trigger/speed sensor/s	Hall effect or optical digitizer or eddy current	zero to maximum engine speed	±1%	−40 °C to +125 °C	N/A
Crankshaft mounted timing/ trigger/speed sensor/s	optical digitizer with fiber-optic linkage or eddy current			−40 °C to +160 °C	N/A
Road speed sensor (speedo cable fitting)	optical digitizer or reed switch or Hall effect	as above	±5%	−40 °C to +125 °C	N/A
Inlet manifold air mass flow (unidirectional)	vane meter or hot wire or vortex shedding	10–200 kg h^{-1} or 20–400 kg h^{-1} (two ranges)	±2%	−40 °C to +125 °C	35 ms for vane only, but target is 1 ms
Inlet manifold air mass flow (bidirectional)	ultrasonic or corona discharge or ion flow	±200 kg h^{-1}	±2%	as above	1 ms
Accelerator pedal position sensor	potentiometer	0–5 k from min. to max. pedal travel	±1%	−40 °C to +125 °C	N/A
Throttle position sensor	potentiometer	0–4 k from closed to open throttle	±3%	−40 °C to +125 °C	N/A
Gear selector position sensor	cam-operated switch or potentiometer	8-position selection or 0–5 k	N/A or ±1%	−40 °C to +125 °C	N/A

Table 1—*continued*
Automotive powertrain transducers

Sensor/type	Proposed sensing method	Range	Accuracy	Temperature operating range	Response time
Gear selector hydraulic-valve position sensor	optical encoder	as above	±2%	−40 °C to +100 °C	N/A
EGR valve position sensor	linear displacement potentiometer	0–10 mm	±2%	−40 °C to +125 °C	N/A
Closed-throttle/ wide-open-throttle sensors	microswitches	N/A	N/A	−40 °C to +125 °C	N/A
Engine knock sensor	piezoelectric accelerometer	5–10 kHz g-range TBE	N/A	−40 °C to +125 °C	depends on resonant frequency
Engine knock + misfire sensor	ionization measurement in cylinder or exhaust manifold	TBE	TBE	−40 °C to +150 °C (externally) probe must meet combustion or exhaust temperatures	TBE
Exhaust gas–oxygen sensor for stoichiometric operation	zirconium dioxide ceramic with platinum surface electrodes or titanium disks in aluminum	less than ½ one A/F ratio used as a switch between lean and rich A/F ratios	not known	300 °C to 850 °C (tip operating temperature)	15 ms
Exhaust gas–oxygen sensor for lean-burn operation	zirconium dioxide oxygen-pumping device with heater	14:1 to 30:1 A/F ratio	TBE	as above	15 ms

Source: M. Westbrook

1.2 Fuel Control

The use of a three-dimensional look-up table as a means of optimizing engine operation has been extended by the introduction of electronically controlled fuel injection systems. Solenoid-actuated fuel injectors are again controlled by a microprocessor, with variations in the injector opening time being used to control the amount of fuel delivered to the engine. With this system both the quantity of fuel injected and the air mass flow rate into the engine are critical, and an airflow sensor is essential. The Bosch air vanemeter was the first into service and is still widely used. It consists of a spring-loaded flap which is placed in the airstream. The flap angle is related to mass flow rate and is transduced by a potentiometer.

An alternative which is becoming popular is the hot-wire anemometer. Automotive versions of these were also first developed by Bosch (Westbrook 1988). They have the advantage of having no moving parts and of giving increased reliability, but require correction for air temperature changes and can be susceptible to contamination of the hot-wire surface.

1.3 Emission Control

Increasingly stringent restrictions are being placed on the amounts of polluting gas which a car exhaust can emit. To reduce these so-called exhaust emissions two approaches are adopted. First, the air–fuel ratio entering the engine is controlled to ensure complete combustion. Air–fuel ratio is inferred from measurements of the amount of oxygen in the exhaust. Second, three-way catalytic converters are placed in the exhaust to remove the critical pollutants of carbon monoxide (CO), unburnt hydrocarbons and nitrogen oxides (NO_x).

Exhaust gas oxygen (EGO) sensors make use of the fact that the migration of oxygen ions across a membrane separating two gases is a function of the partial pressure of oxygen in the two gases. At the stoichiometric air–fuel ratio (when sufficient oxygen is present to burn all the fuel) the partial pressure of oxygen in the exhaust gases equals that of the atmosphere. If suitable electrodes are placed on both sides of the barrier a voltage output appears only when the air–fuel ratio departs from stoichiometry.

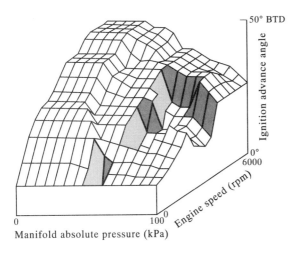

50° BTD

Ignition advance angle

0°

6000

Engine speed (rpm)

0 100

Manifold absolute pressure (kPa)

Figure 1
Typical three-dimensional control surface for ignition
advance

2. Driver Information and Diagnostics

A variety of electronic measurement systems are fitted
to a vehicle for functions other than powertrain
control. Many of these are concerned with warning the
driver of faults, fuel shortage or maintenance requirements.

2.1 Water Temperature

The temperature of the water in the engine cooling
system is generally measured by a simple thermistor.
Often a red warning light is fitted to the dashboard,
rather than a display of the temperature reading itself.

2.2 Fuel Quantity

By the late 1920s remotely operated petrol gauges
were becoming standard, in which a float on a lever
arm in the petrol tank moved the wiper of a wire-
wound potentiometer. This in turn controlled a
remote (dashboard-mounted) voltmeter, which was
provided with a pair of windings known as the
deflection and control windings. The control coil
replaced the hairspring found in an ordinary voltmeter. The two windings were interconnected in such
a way that changes in the battery voltage did not affect
the reading. For example, a decrease in battery
voltage decreased the deflecting force, but also decreased the controlling force. The meter reading which
resulted was a function of the ratio of the two forces,
and was independent of battery voltage.

Modern fuel gauges are not usually voltmeters,
since these have too fast a response and give a reading
which fluctuates owing to the fuel sloshing when the
vehicle is on hills and when cornering. Instead a meter
is used which contains a bimetallic component and a
heating coil. The deflection of the pointer depends on

the current in the heating coil, which is in turn
controlled by the resistance of the float sensor in the
fuel tank. A voltage regulator is usually fitted to
remove the effect of supply voltage variations. The
thermal inertia of the system is made sufficiently large
to smooth out most of the effects of fuel slosh.

Many alternative ways of measuring fuel level have
been attempted. One of the most promising uses the
capacitance change which occurs when a pair of
conducting plates are dipped into the fuel. Other
reported systems used optical techniques. A number
of aircraft fuel gauging systems use ultrasound for tank
level sensing. This development was introduced because it offers the possibility of noninvasively (and
therefore safely) measuring the amount of liquid in a
fuel tank. Sophisticated systems have been developed
in which measurements from a number of transducers
are integrated to give a reading which is independent
of the motion of the fuel ("sloshing"). It seems likely
that, as has happened before, a technology originally
developed for aerospace will eventually be used in
automotive engineering.

2.3 Oil Level

One measurement which it would be very desirable to
improve is that of engine oil level. Current devices
generally operate on the hot-wire principle, and use a
thermistor as both the heat-generating and sensing
device. The thermistor is self-heating, and its resistance changes when the oil level drops sufficiently to
expose it to air. The difficulty is that the system only
works when the engine is stopped (and has been
stopped for long enough to let the oil drain back into
the sump).

3. Suspension Control

Many vehicles are now being fitted with adaptive
suspensions, in which the characteristics of the springs
and/or dampers are adjusted to suit the road conditions and the way in which the car is driven. To
control an adaptive suspension system, measurements
of the dynamic variations in wheel–body distance are
required. Accelerometers are often fitted for this
purpose, although potentiometric displacement transducers have also been used. Ideally four sensors
should be provided, one for each wheel, but cost
constraints usually mean that one centrally positioned
accelerometer has to suffice.

The accelerometer(s) used to control dynamic operation of an adaptive suspension cannot usually also
sense slow changes in vehicle attitude and height. This
"load-levelling" function (so-called because it is
mainly used to compensate for the effect of placing
asymmetric loads in the vehicle) is often provided by
means of adjustable airsprings and ultrasonic or radar
rangefinders, which measure the distance between the
vehicle body and the road at each corner.

4. Future Developments

In the future, research will undoubtedly by directed towards improving the sensors and measurement systems described in this article. The move towards silicon will also continue, with micromachining and thick-film hybrid techniques being used to create transducer architectures on a very small scale.

The development of "smart" sensors (in which much of the signal conditioning is carried out within the transducer housing) will provide standardized digital outputs, which are likely to be transmitted via a communciations bus to the central control system. Smart sensors will probably linearize their own outputs, compensate for environmental changes and include self-calibration and diagnostic functions both for themselves and for the systems to which they are applied.

Unfortunately, silicon devices cannot cope with the highest temperatures found on a vehicle, especially around the engine, so the use of alternative semiconductor materials seems likely. One suitable candidate may be gallium arsenide, which is currently the subject of much research.

Bibliography

Newcombe T P, Spurr R T 1989 *A Technical History of the Motor Car*. Hilger, Bristol, UK
Westbrook M 1988 Automotive transducers: an overview. *Proc. Inst. Electr. Eng. D* **135**(5), 339–47

J. D. Turner
[University of Southampton, Southampton, UK]

B

Biological and Biomedical Measurement Systems

Measurement for the life sciences is complicated by the fact that there is a very considerable variation in results from subject to subject, no matter what parameter is being examined. It is often said that one cannot expect to establish a numerical value for a given biological or biomedical parameter to much better than ±10%, allowing for intersubject variability. Even when measurements are made on the same subject, but under different environmental stimuli, variability of the same order can be expected. Not only does this make the planning of a measurement more complicated, it also tends to mask the very effect that may well be sought, for example, the diagnosis of a disease or the detection of a trend in terms of the effect of treatment.

In a brief article such as this it is impossible to cover in detail the whole area of biological and biomedical measurements. It is felt, therefore, that the major emphasis should be on measurement on the human body, and to this end the approach is to focus on the various human systems and to examine the measurement approaches adopted to look at structure and function. Where appropriate, digressions into more general biological applicability of a given measurement will be made.

1. General Requirements for Measurement in the Life Sciences

As indicated before, most measurements are concerned with elucidating information concerning the structure and/or function of some part of a living system. One approach adopted can be either to remove a small sample of the biological system and subject it to analysis; this is often termed an *in vitro* approach. Alternatively, similar information can be sought, but by subjecting the whole living organism to the measurement system, thus carrying out an *in vivo* measurement.

An example that might help to explain this concerns the analysis of gases in the blood. An *in vitro* measurement could be made by removing a sample of blood from a patient, taking it to a laboratory and then performing a set of analytical determinations for oxygen and carbon dioxide. The *in vivo* approach involves employing a modified Clark cell in conjunction with a heating element. This device is placed upon the skin and the skin is heated up to some 3–4 °C above normal temperature. The response of the blood vessels in the skin is to dilate, thus providing more

blood from which, by diffusion through the tissues of the skin, sufficient gases arrive at the electrode for a determination to be made. Both techniques are widely used and further developments of the *in vivo* method will almost certainly be carried out in the near future.

If the measurement technique adopted is of an *in vivo* nature, then there are a number of additional requirements that need to be satisfied in most existing systems. Measurements generally should be noninvasive and nonhazardous. Noninvasive is generally taken to mean "no need to puncture the skin," whereas nonhazardous implies not only a safety requirement (see Sect. 4), but also that the energy form used to gather information does not itself constitute a hazard, or at least that the hazard is small when compared to the risk of not making the measurement. It is this balance of hazard and risk that must be considered each time a measurement involving, for example, x irradiation is contemplated. The result of any measurement should be information rich and it should be stable and reproducible. It should also be capable of standardization and calibration; it should be simple to administer, pain free and it should not be subject to other hazardous procedures, for example, anesthesia. Finally, there is considerable pressure in most cases for the measurement system evolved to be economic in both capital expenditure and the subsequent cost of ownership.

2. Measurement Techniques

2.1 Cells
Cells are the fundamental building blocks of all living systems, and measurements that can reveal their size and structure are routinely used as an aid to diagnosis and patient management. These measurements may use conventional light microscopes, scanning electron microscopes, transmission electron microscopes, or scanning acoustic microscopes. Measurements such as these are fundamental for research into disease processes and new treatments. Techniques are now available that can reveal the detailed structure of the proteins attached to the surfaces of cells.

2.2 Musculoskeletal System
This system provides the means for support, movement and locomotion for the body, and there are numerous reasons for making measurements when defects in the system occur. Under the broad heading of gait analysis, detailed analysis of the manner in which the foot strikes the ground, how the weight is distributed, how even the gait is, and the angles which the various joints move through, are made routinely.

Commercial systems exist to facilitate many of these measurements, and there is active research to improve techniques in many of the areas mentioned.

Soft tissue, such as tendons and skin, have also been subjected to much detailed analysis, and measurements of mechanical properties have tended to follow techniques developed in materials testing. Additionally, though, in considering skin, both dimension and function are now routinely measured, often with the use of noninvasive techniques such as ultrasound.

2.3 Nervous System

The nervous system is made up of cells known as neurons, which in some cases are extremely elongated and whose terminal fibers may either communicate with another neuron via a synapse or they may act as receptors in association with various sensory systems of the body. Communication through the nervous system is carried out via the nerve impulse, which is a brief electrical pulse of some 1 ms in duration and about 130 mV in amplitude. Measurements of this signal can be made via electrodes introduced through the skin and connected to the particular neuron under investigation. An alternative way of determining whether parts of the nervous system are intact is to stimulate a neuron by injecting an electrical signal and observing the response, which may be to move a limb, for example. Recently, noninvasive techniques have become available for carrying out these measurements. Instead of puncturing the skin and connecting an electrode to the neuron, it is now possible to achieve stimulation of the neuron electromagnetically and, by timing the arrival of the stimulation pulse at some distant point, accurate and reliable measurements of the nervous system properties can be achieved.

2.4 Circulatory System

The circulatory system comprises the heart, which is a two-stage pump, the arteries, arterioles, venules and veins. One part of the heart (the right side) pumps blood through the pulmonary circulation in order to exchange gases via the lungs. The left side pumps oxygenated blood through the systemic circulatory system, which comprises the rest of the body.

In addition, there is a further complex circulatory system, the lymphatic system, using vessels rather similar to the veins. This system is associated with the collection of liquid products (not containing red blood cells) that have left the blood circulatory system, a fluid known as lymph. This system returns the lymph to the venous circulation after removal of various biochemical components produced by tissue metabolism.

(a) *Blood characteristics.* Blood serves numerous functions within the body. It takes part in the respiratory process by conveying oxygen from the lungs to the cells of the body and carbon dioxide from the cells back to the lungs. In addition, blood contains within it nutrients such as glucose, amino acids, fats and vitamins, which it conveys from the digestive tract to the cells of the body. Blood takes part in the excretory function of the body by removing products of metabolism, and it assists in the regulatory functions of the body by maintaining the body's water content and temperature. Blood is a fluid within which there are many solids and cells suspended; by far the most common are the red blood cells, whose function is to facilitate the respiratory role of the blood. White blood cells, or leukocytes, are far less numerous and their role is to assist in fighting bacterial infection. They may also play a role in repairing damaged tissue. Measurements of blood properties are commonly made using small samples taken from the body. Such measurements include the estimation of the number of cells per unit volume and the ability of the blood to coagulate. In addition, a wide range of specialist clinical chemistry techniques, such as the determination of metabolites, enzymes and antibodies, are employed. These are outside the range of this present article.

Measurements of blood gas content are sometimes made using small samples of blood removed from a patient. However, there are also noninvasive techniques for blood gas measurement which have come into common use.

(b) *Pressure, velocity and flow.* Blood pressure is one of the most commonly used measurements and is routinely carried out using a sphygmomanometer. Pressure in the blood changes in phase with the pumping action of the heart; the systolic pressure is the maximum pressure during this period and the diastolic pressure the minimum. Typical values are 120 mmHg and 80 mmHg, a non-SI unit that has been retained. More sophisticated instruments capable of performing this measurement, but removing the subjective nature associated with the inflation and deflation of the sphygmomanometer cuff, have appeared. These usually rely on detecting the flow of blood using ultrasound Doppler techniques, coupled with automated inflation and deflation of the cuff.

Measurement of the velocity of blood can be made using either ultrasound or infrared based techniques. In both cases the Doppler effect is employed and in the case of the use of infrared energy this is usually generated using a small solid-state laser source. The technique is really only feasible for the peripheral circulation down to about 1 mm below the skin surface. If the requirement is to measure blood velocity in much deeper vessels, then Doppler ultrasound techniques are often employed.

Frequently, volume flow is a measurement requirement, for example, the stroke volume of the heart. In this case, a number of techniques using cardiac radiography are available and, in addition, measurements of this type can be made using Doppler ultrasound in association with an imaging technique to obtain dimensional information on the aorta. Information on the way in which the heart valves operate is also frequently required and, again, this may be done using ultrasound imaging or Doppler techniques. More recently, electrocardiogram (ECG) gated magnetic resonance imaging has been applied to evaluation of the heart's performance.

In common with all muscles, the heart generates significant electrical signals which can be picked up on almost any part of the surface of the body. The study of the ECG has been undertaken for many years and this is a conventional technique for diagnosing problems with the heart. Electrodes are usually attached to the arms and left leg of the patient and various forms of ECG are therefore available to the clinician. More recently, automated analysis of the ECG signal has become possible and work over the years has led to a highly sophisticated level of diagnostic information being capable of being deduced from an ECG.

2.5 Respiratory System

Although conventional radiography is frequently used to detect problems associated with the lungs, caused perhaps by abnormal collections of fluid, in the main, functional testing is associated with measuring the ability of the lungs and the associated airways to move air in and out. In addition to measuring the resistance to flow, the total capacity of the lungs is also frequently measured, as is the residual volume. The most commonly used flow transducer in this context is the pneumotachograph, a device in which a fine mesh or set of small tubes is placed across the airway and the pressure difference across this is measured and calibrated to give flow volume.

2.6 Sensory Systems

(a) *Visual.* Measurements to obtain information concerning the refractive power of the eyes can be made by seeking active cooperation from the patient. However, for more detailed studies, frequent use is made of A- and B-scan ultrasound. From such measurements, data such as lens dimensions and the distance from the cornea to the retina can be very accurately obtained, and this can lead to accurate prescription for replacement lenses. In the past, the measurement of intraocular pressure has been made by gently pushing a probe up against the surface of the anesthetized eye. More recently, however, techniques involving a small puff of air which is used to indent the surface of the eye have been introduced, the indentation being measured using optical techniques.

(b) *Auditory.* In the main, the measurement of hearing loss is accomplished by producing tones at given frequencies for each ear and asking the patient whether he or she can hear the tone. This rather subjective form of measurement has been used to then prescribe the type of hearing prosthesis to be fitted.

More recently, equipment has become available which enables the measurement of hearing to become more objective; this is achieved by picking up the electrical activity generated within the ear and using this as a means of judgement on whether hearing is effective or not.

Although measurements of the tactile, olfactory and taste senses have been described and are used in research, olfactory and taste measurements are not employed in a routine way.

2.7 Digestive System

Considerable advances have occurred in the use of fiber-optic endoscopes to examine various parts of the digestive system. Light can be introduced by one part of an optical fiber bundle and then an ordered set of fibers may be used to collect and deliver an image of the interior of the gut to the clinician. This somewhat invasive form of measurement has now been supplemented by the use of instruments to measure the myoelectric activity of the gut. There is good evidence to show that the signals which can be collected from the skin surface are useful for assisting in the diagnosis of digestive tract defects.

2.8 Renal System

The function of the kidneys is to produce urine, together with some important hormonal secretions that play a part in the endocrine system. Kidney function measurements are often carried out by the use of isotope clearance techniques or by radionuclide imaging.

This latter technique involves attachment of a radioactive species such as technetium-99m (^{99m}Tc) to a ligand that will be preferentially taken up by the organ of interest. The degree of takeup or absorption can be imaged by the use of a gamma camera, thus indicating the functional status of the organ.

In the case of acute kidney failure, artificial kidneys are often employed and these take the form of a dialysis system capable of carrying out many of the functions of the kidneys, but outside the body. There are also efforts directed at the design of portable and wearable artificial kidneys which may bear fruit in the relatively near future.

Associated with the renal system is the urinary system and the functional measurement of the performance of the urinary system has become a routine part of the clinician's activities with, today, relatively sophisticated instruments capable of measuring, for example, pressure, volume and flow.

3. Measurement Techniques Having Wide Applicability

3.1 Temperature Measurement

Measurement of skin temperature and core temperature are widely used as ways of assisting in patient diagnosis. The technique is also very old, but it was not until the mercury in glass thermometer was introduced that the technique became both simple and precise. However, mercury in glass thermometers cannot satisfy all the requirements for temperature monitoring and therefore many other devices have been employed, including thermistors, thermocouples, platinum wire, *p–n* junction diodes and quartz crystal resonators. Most of these devices can be fabricated in such a way that they can act as surface temperature monitors for skin or, in certain circumstances, they can be miniaturized so that they can be fitted inside a hypodermic needle in order to invasively monitor the temperature of tissue such as muscle.

The use of contact devices such as those mentioned earlier has limitations in the sense that the peripheral circulation is very sensitive to very small pressures which contact devices may exert. A noncontact method of making temperature measurements is to use the fact that the body acts as an almost perfect blackbody radiator and therefore infrared detection over wavelengths somewhere between 4–30 nm can be employed. Numerous commercial devices exist, based on the use of thermopiles, photoconductive detectors or pyroelectric systems. In many cases, an image of the area under examination is produced and quite remarkable resolution can be achieved ($<0.1\,°C$), particularly with cryogenically cooled detectors. Microwave radiation has also been employed to detect the temperature of the body. In this case, attempts are made to measure temperature subcutaneously, whereas the previous examples quoted are mainly aimed at measuring skin temperature.

Recently, liquid crystal indicators have been used in a contact mode to indicate skin temperature, and there is now a wide range of commercially available systems. Another recent development has been the introduction of irreversible disposable thermometers in which a chemical indicator is used to provide a color change at specific temperatures. Commercial devices usually employ about 50 spots of chemical indicator covering a range from about 35–41 °C, giving rise to a 0.1 °C resolution.

3.2 Imaging

Imaging aimed at microscopic studies, that is, the structure of cells, has been mentioned earlier.

(*a*) *X-ray imaging*. This remains the most common form of imaging used in medicine and comprises a means of producing an image from a beam of photons that have been transmitted through the patient. Different forms of tissue will attenuate the x-ray beam by different amounts and therefore a shadow image of some region of the body may be formed, either on a photographic film or via an image intensifier and television viewing system.

More recently, digital radiography has been introduced and this can be employed to reveal vascular structure in great detail in combination with contrast agents.

(*b*) *X-ray tomography*. Tomography is a method of reducing structural noise by selective blurring of detail originating from anatomical sections. In order to obtain a tomograph, the x-ray source and the film are moved in unison about some fulcrum point. This technique has been in use in radiography for many years and is well developed. More recently, the computerized axial tomography (CAT) technique has been introduced, the first of which, the EMI scanner, was developed by Hounsfield in 1973. In this technique images of slices through the patient in a transaxial sense are obtained and a digital computer is employed to carry out the necessary processing to obtain a useful image. Generally, the employment of CAT has resulted in improved discrimination of soft tissue, an improvement in the accuracy of the images, and initially the technique revolutionized x-ray imaging in the neuroradiological area. Subsequently, whole body CAT scanners were developed and they have been employed in a wide range of diagnostic and patient management situations.

(*c*) *Isotope imaging*. Whereas most forms of imaging are mainly involved in giving rise to structural information, isotope imaging has its main purpose in providing information on function. This can be of a dynamic nature, for instance, liver or kidney function over relatively short periods of time.

The technique is to administer a selected radioactively labelled compound which collects, at least transiently, in the organ of interest. The most common isotope, or more properly, isomer, is ^{99m}Tc, this being used because it emits gamma rays only at an energy of 140 keV, and it has a half-life of only six hours. Detection of the emitted gamma rays is usually accomplished with thallium sodium iodide and the detector is scanned over the area of interest. The original rectilinear scanners have now been almost totally replaced by gamma cameras, which again use collimators and large sodium iodide detectors, but in this case the detectors are coupled to an array of photomultipliers.

(*d*) *Nuclear magnetic resonance imaging*. Nuclear magnetic resonance (NMR) imaging is an imaging application of a technique that has been well-known in spectroscopy for many years. It is particularly successful for imaging living systems, since these contain a large proportion of hydrogen, and the hydrogen nucleus, the proton, is particularly suitable for NMR signal generation. In contrast to the use of x rays, magnetic resonance imaging is particularly sensitive to subtle changes in soft tissue.

The generation of signals and their use to provide images is a fairly complex topic and is well explained by Allen (1991).

As mentioned above, the origin of this imaging

technique came from spectroscopic applications of the NMR phenomenon. A more recent development has been to employ spectroscopy, but of living tissue and as part of an imaging investigation. In other words, the actual chemistry taking place within, for example, a muscle, may be studied alongside the collection of images.

(e) Ultrasound imaging. The use of acoustic energy to obtain information about living tissue is dependent on the mechanical properties of the tissue being interrogated. It therefore provides an alternative and fundamentally different view of the material from that obtained by the use of x irradiation or magnetic resonance imaging. In addition, ultrasound at energy levels employed for medical diagnosis appears to be much safer than, for example, x irradiation, and the systems commonly employed in medicine are certainly more cost effective.

Two major modalities are in common use. The first, using pulse echo ultrasound, is employed for general purpose imaging of the soft tissues of the body. The second, using the Doppler effect, is used for studying movement, in particular, blood flow and the movement of the heart and its valves.

Pulse echo imaging relies on the fact that when acoustic energy meets an interface between two different media, then part of the incident energy is transmitted and part is reflected, the intensity of the reflection being dependent on the difference in acoustic impedance between the two media. Since ultrasound travels relatively slowly (compared with the speed of light, for example) the wavelengths in biological soft tissue are short (300 μm at 5 MHz). This means that ultrasound can be focused and delivered to the body in narrow beams. The attenuation of ultrasound in biological tissues is dependent on frequency and is often quoted as around 1 dB cm^{-1} MHz^{-1}. Something in excess of 70 dB can be tolerated in terms of round trip attenuation of a signal reflected from an interface and, therefore, a penetration of some 230 wavelengths can be achieved. This sets an upper limit to the frequency that can be used, particularly for studies of deep abdominal and cardiac structures.

Considerable advances have occurred in the application of Doppler methods in medical ultrasound. These have arisen primarily due to the employment of pulsed Doppler, in which a toneburst of ultrasonic energy at a very carefully controlled frequency is transmitted into the body and then, by the use of time-delay techniques, studies can be made of moving structures such as red blood cells or heart valves at known distances into the body. Indeed, by using parallel processing techniques, a range of delays can be employed, giving rise to separate information at a set of depths into the body. In this way the flow profile across a blood vessel can be established, which is essential for studies into the early progression of atherosclerosis (blocked arteries). The most recent developments in this area have been to combine Doppler methods of gaining information with the pulse echo techniques for studying stationary structures. In this way, the diameter of a blood vessel can be established at the same time as the flow pattern is measured within it. From this information, the volume flow through the vessel under study can be established.

4. Other Topics

Many important measurement techniques have been developed for other specialities in medicine such as ophthalmology, dermatology, dentistry, and so on. Much use is also made in some areas of the technique of telemetry, thus enabling a patient to be mobile while data on heart rate, for example, is constantly monitored.

In a considerable proportion of the measurement techniques mentioned, the measuring instrument is connected to the patient. This clearly requires that the instrumentation is itself nonhazardous to the patient. Electrical shock hazards can be reduced, if not eliminated, by careful attention to the manner in which the instrument is designed and, in particular, to the way in which leakage currents flow to earth. Attention must also be paid to single fault conditions and to the manner in which earth leakage currents may then flow. In doing this the designer is attempting to ensure that earth leakage currents do not pass through the patient.

Bibliography

Allen P S 1991 Magnetic resonance imaging. In: Payne P A (ed.) *Concise Encyclopedia of Biological & Biomedical Measurement Systems*. Pergamon, Oxford, pp. 251–4
Payne P A (ed.) 1991 *Concise Encyclopedia of Biological & Biomedical Measurement Systems*. Pergamon, Oxford

P. A. Payne
[UMIST, Manchester, UK]

Bridges

Many transducers convert changes in nonelectrical quantities into changes in resistance, capacitance or inductance. Bridge circuits provide a method of comparing the values of unknown components with those of standards, or of detecting changes in the value of a component. There are a considerable number of possible bridge arrangements. Hague and Foord (1971) and Hall (1971) provide detailed analyses and practical details on a large number of these. Two of the most commonly used bridge techniques are the

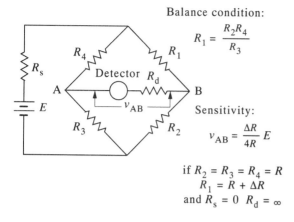

Balance condition:

$$R_1 = \frac{R_2 R_4}{R_3}$$

Sensitivity:

$$v_{AB} = \frac{\Delta R}{4R} E$$

if $R_2 = R_3 = R_4 = R$
$R_1 = R + \Delta R$
and $R_s = 0$ $R_d = \infty$

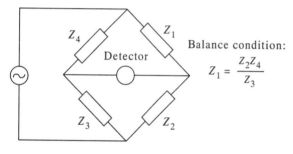

Balance condition:

$$Z_1 = \frac{Z_2 Z_4}{Z_3}$$

Figure 1
Four arm bridge circuits: (a) Wheatstone bridge, (b) ac
four arm bridge

four arm bridge and the inductively coupled or transformer ratio bridge.

The four arm bridge is shown in Fig. 1. This is the
well-known Wheatstone bridge with a dc source and
resistance in the four arms of the bridge. The bridge
can be operated in a balanced mode in which a null
condition is obtained on the detector by either manual
or automatic adjustment of the bridge components. In
this case the balance condition is independent of the
bridge voltage E, its resistance R_s, or the sensitivity of
the detector. The ease of determining the balance
does, of course, depend on all of these. Figure 1 gives
the balance condition and the sensitivity of the bridge.

For the measurement of low resistance it is usual to
arrange for the resistance to be in the form of a four
terminal device and the four arm bridge is modified by
the addition of two further arms. Such an arrangement
is known as a Kelvin double bridge and, as such, is
capable of, to a large extent, eliminating the lead
resistances.

The balance condition for an ac four arm bridge is
given by $Z_1 = Z_2 Z_4 / Z_3$ and for the measurement of
capacitance or inductance this leads to a large number
of possible bridge arrangements depending on the
relative positions and nature of the components of the

bridge. The bridges can be broadly categorized as
either ratio bridges in which Z_2 and Z_3 are kept
constant and Z_4 varied, or product bridges in which Z_2
and Z_4 are kept constant and Z_3 varied. Capacitance is
commonly measured using either Wein or Schering
forms of the four arm bridge and inductance measurements employ the Maxwell, Hay or Owen forms. For
further details of these bridges see Hague and Foord
(1971).

The balance condition of four arm bridges is
affected by the presence of earth impedances associated with the components of the bridge. The effects of
these earth impedances can be minimized by using a
Wagner earthing arrangement. The inductively coupled or transformer ratio bridge is capable of providing a high degree of earth impedance rejection. An
example of an inductively coupled bridge is shown in
Fig. 2. The bridge is balanced when there is zero net
flux in the core of T_2, this being detected by means of
the secondary on T_2. The balance condition shown in
Fig. 2 involves the turns ratios of the secondaries of T_1
and the primaries of T_2, and the value of the standard
component. The ratios can be accurately defined and
are also extremely stable, and thus it is possible, by
changing the ratios, to balance the bridge for a wide
range of values of the unknown component using only
a small number of standards. The impedances to earth
at the drive side of the bridge are rejected because of
the high degree of mutual coupling between the
secondaries of T_1, which leads to their ratio being
largely unaffected by loading. The earth impedances
at the detector side are rejected since the balance
condition ensures that there is no net flux in the core

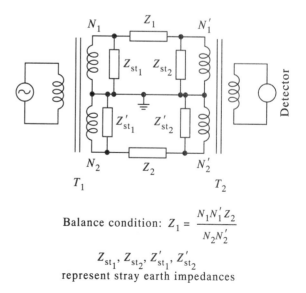

Balance condition: $Z_1 = \dfrac{N_1 N_1' Z_2}{N_2 N_2'}$

$Z_{st_1}, Z_{st_2}, Z_{st_1}', Z_{st_2}'$
represent stray earth impedances

Figure 2
Inductively coupled bridge

of T_2, and thus there is no emf developed across the earth impedances. Inductively coupled bridges are widely used in measurements associated with capacitive and inductive transducers. For detailed analyses of these applications, see Hugill (1983) and Neubert (1975).

See also: Analog Signal Conditioning and Processing

Bibliography

Hague B, Foord T R 1971 *Alternating Current Bridge Methods*. Pitman, London

Hall H H 1971 Impedance measurement. In: Oliver B M, Cage J M (eds.) *Electronic Measurements and Instrumentation*. McGraw-Hill, New York, pp. 264–318

Hugill A L 1983 Displacement transducers based on reactive sensors in transformer ratio bridge circuits. In: Jones B E (ed.) *Instrument Science and Technology*, Vol. 2. Adam Hilger, Bristol

Neubert H K P 1975 *Instrument Transducers*, 2nd edn. Oxford University Press, Oxford

M. L. Sanderson
[Salford University, Salford, UK]

P. M. Clifford
[City University, London, UK]

C

Chemical Analysis, Instrumental

Instrumental chemical analyzers or process analyzers provide quantitative information on the chemical composition of process materials in a timescale that allows such information to be used for closed- or open-loop control and on-line quality assurance. Normally they are restricted, in the context of process operations, to on-line instrumentation—that is capable of being incorporated into an automatic closed-loop or supervisory control system—or on-plant instrumentation, where rapid chemical analysis by plant operatives allows manual corrective actions to be taken. Chemical composition includes the determination of elemental and molecular concentrations, plus a quantitative description of the structure of the process materials, such as the type and concentration of crystalline phase present, or the concentration of water in different forms. Examples of these are: the determination of copper in mineral processing streams; carbon monoxide in flue gas; phase concentrations in cement clinker; and the forms of moisture present in detergent formulations. The distinction between chemical analysis and the determination of physical measurands is not a clear one, for many of the latter are used to infer chemical composition.

Process analyzers have four main roles in process industries and in processes in manufacturing industries:

(a) as on-line analyzers in pilot and other plants to aid process development, modelling and control system design;

(b) as quantitative information providers in automatic or manual control;

(c) as monitors of concentrations of critical components in process plant atmospheres, effluents and stack gases for safety or environmental reasons; and

(d) as part of an on-line or on-plant quality assurance system.

The performance requirements differ greatly for each role, so that different instrumental types or designs may be required when the role of the same analyzer is changed. When used as an aid to process development, the analyzer needs to be flexible for two reasons: (a) it must be readily adaptable to the shifting requirements of the development team; and (b) it must be of use to future developments to maximize return on the capital invested in it. The analyzer must also be readily interfaced to the pilot plant because it may be needed at different points for the process under development, and sufficient time is not normally available for the development of complex sampling systems. However, since such chemical instrumentation is normally used under the supervision of chemical and process engineers for relatively short periods of time, the demands for unattended operation and long periods without maintenance, when used in the other three roles, can be relaxed—indeed, suitably adapted laboratory instruments may adequately fulfill this role. When process analyzers are employed as information providers in a control system, the basic requirements are: (a) that they must provide the required information uniquely, with precision, accuracy and response times dictated by the control system; and (b) that the reliability of the analyzer and its process interface must be such that it does not significantly degrade the reliability of the controlled process. For a control function, repeatability may often be more important than accuracy.

When used as monitors, the chemical analyzers must initially be in a form appropriate to their application, which can range from a personal pocket toxic gas monitor to a remote solar-powered telemetry communicating effluent analyzer on an unmanned treatment plant. Often, high reliability, minimal maintenance and the ability to cope with wide variations in the materials monitored are required. They may have to conform to national or international standards or certification when used in a health and safety role. On-line quality assurance is a relatively new role for such instruments, although on-plant instruments have been used for many years. Here accuracy is a necessary characteristic, as well as acceptability to the supplier or purchaser of the materials under surveillance. One final important characteristic required of process analyzers in all roles is the ability to indicate failure or incipient failure and to "fail safe"—especially since, due to their complexity, many instrumental chemical analyzers carry with them a not wholly deserved reputation for unreliability.

1. Types of Instrumental Chemical Analyzers

Process analysis can be achieved by a surprisingly wide range of instrument types. These will be summarized in broadly related groups.

Electrochemical on-line and on-plant analyzers include pH and pX (selective ion) electrodes, conductometric (both multielectrode and noncontacting), redox, many biosensors and automated chemical analytical systems using polarographic or amperometric techniques. Coulombometric analyzers also fall into this category—one of the commonest being the electrolytic phosphorus pentoxide trace moisture monitor. Generally the analyzing heads of these instruments tend to be compact and externally simple, and they have found ready acceptance in the process and manufacturing industries. Biosensors have, as yet, found limited on-line application due to their relatively short lives and vulnerability to attack. Their likely major role would be in on-plant analysis. Microchip-based electrochemical analyzers have an exciting future, but little current use.

Spectrometric analyzers have also been widely used—ultraviolet, infrared, x-ray, nuclear magnetic and gamma ray techniques are all successfully applied on-line and on-plant. Fiber-optic-based chemical analyzers have a number of on-line applications, and they are likely to increase in importance. Mass spectrometry has had a range of successful applications, especially where rapid response or multicomponent analyses are needed, while microwave spectrometry for gas analysis is one technique that has achieved little success, in spite of its apparent potential. One simple but important application of visible and ultraviolet spectrometry is for colorimetric detection used in automated chemical analyzers. These electromagnetic and magnetic spectrometric techniques often have the major advantage for process analysis of being noncontacting or nonintrusive.

The chromatographic techniques—liquid, gas–liquid, gas–solid and, to a lesser extent, ion—have been very widely used in on-line analysis, particularly in the petroleum refining and petrochemical industries, where the determination of ranges of homologous hydrocarbons and other compounds necessary for control, cannot be achieved by any other method.

While physical property sensors are dealt with elsewhere, any account of process analyzers would be incomplete without a brief summary of their chemical uses. They may be used directly on the material being analyzed—for example, the measurement of dielectric constants at differing frequencies to determine moisture content and forms in foodstuffs—or indirectly, as in the aluminum oxide probe for gaseous moisture monitoring, where moisture selectively adsorbed in the pores of the oxide causes a moisture-dependent and measurable change in its dielectric constant. The physical parameters used include density, viscosity, sound velocity and attenuation, optical rotation, refractive index, dielectric constant, conductivity, microwave power attenuation and naturally occurring radioactivity.

The selection of the correct analyzer for a measurement requirement from this extensive range requires some care. The first step must be the preparation of a detailed and correct performance specification. This would describe the components to be analyzed, concentration ranges, accuracy and precision, response time, ambient conditions, communications, hazardous zone and, finally, as complete a description as possible of the point at which the analyzer is to be used, the process material properties (including hazards such as toxicity and flash point), and the physical state, such as pressure, flow rate and temperature. The latter must include the extremes expected under the operating conditions. With a new process (and many well-established ones) all this information may not be available—however, the successful application of process analyzers depends on the supplier having this range of information available. Correctness is emphasized at the start of this section—not only can an incorrect specification cause problems with the performance and reliability of the installed system, but it can have major financial implications—a factor of two or three too high in specified accuracy could increase the installed cost by an order of magnitude. It may seem that the use of on-plant analyzers short circuits much of this process, but in reality they also require most of this information.

With the specification available, the first step is to find suppliers who are prepared to meet or approach it. The system must be capable of performing the analysis, but one that is stretched to its limits to achieve the required performance is more suspect than one working comfortably within its operating ranges. The complexity of the sampling system is associated with high maintenance costs and doubtful reliability. The user must make a judgement as to the technical and commercial capability of the supplier; for example, does an agent give as good technical backup as a local manufacturer? Maintenance can dominate the cost of ownership of an analyzer, and is an important factor in its selection, which should at some stage include quantification of the cost of ownership against the value of ownership. To assist with the choice of analyzer there are a number of computer-based systems, some using artificial intelligence techniques, which are commercially available or under development.

2. Process–Analyzer Interface or Sampling System

Dominating the performance, reliability and viability of instrumental chemical analyzers is the process–analyzer interface or sampling system. Too much emphasis cannot be placed on the vital role this plays in obtaining analytical information of the appropriate quality for control, environmental monitoring or quality assurance purposes. It is important to appreciate that all instrumental chemical analyzers "see" only a portion of the process stream, whether they are operated on-plant with manually taken samples, at one extreme, or are in-stream probes with the process material flowing by them. The design of the process material–analyzer interface must ensure the following.

(a) The material from which the analyzer gains its information must be adequately representative of the component being measured in the bulk material during the time interval required for measurement or dictated by the short-term variability in the process material composition. It is necessary to stress "adequately representative"—for this is normally concerned with only the parameters being measured. For example, the removal of small concentrations of suspended solids would not affect the adequateness of the sample for determining ethanol in the liquid phase. When quality assurance is one objective of the sampling and analysis sequence, then the requirements for a truly representative sample are generally much more stringent than for process control or monitoring.

(b) The material analyzed must come from the required point both in the process and in time.

(c) It must not introduce unacceptable delays in the availability of the process information, and must respond as quickly as the process demands.

(d) Its reliability must not degrade the overall reliability of the process plant or the analyzer with which it is used. Points to reiterate are that complexity is often related to reliability and maintenance load, and that human factors enter into the reliability of manually operated sampling systems.

(e) It must not significantly alter the concentration of the parameter being measured—this is related to (a), but is an extension of it. Examples of such alteration are the inward leakage of ambient air when determining trace moisture in gases, and the alteration, due to aggressive pumping, of the distribution of moisture forms in a solid or paste, when either the moisture forms are being determined or the analyzer calibration stability depends on constant proportions of moisture forms (as with dielectric measurements).

(f) It must comply with operator and plant constraints, such as accessibility, limited hydraulic head, hazardous zone conditions, and the wear and tear resultant from plant maintenance, or the passage of operators or local transport.

Obtaining the correct design for such a system involves detailed knowledge of the analyzer, process and quality assurance needs, the site of the analytical system and materials handling, chemical engineering, material properties (such as corrosion and abrasion) and maintenance expertise. Freedom to design the most appropriate system varies greatly. With an in-stream probe, such as a pH or dissolved oxygen electrode, siting (to avoid deposition, excessive abrasion or dead volumes), with the possible provision of automatic cleaning facilities, are the only factors that can be varied. However, for the determination, by on-line mass spectrometry, of gas concentrations in a very hot, heavily dust-laden gas stream (such as those encountered in the iron and steel industry), the sampling procedure involves extraction of a sample stream, conveyance to the analyzer, dust removal, cooling, condensed acid water separation and, finally, pressure reduction to meet the needs of the analyzer under very tight time constraints and process conditions, which are rugged to say the least. The design input to such a system is considerable and involves all the skills enumerated above.

When quality assurance is the objective of the on-line or on-plant analysis, the achievement of a truly representative sample with heterogeneous materials must also involve statistical treatment of the validity of the sampling procedure: whether an adequate sample size is taken, a true "cross cut" of the process stream is obtained, and also, if sufficient cross cuts (or their effective equivalents in continuous sampling) have been taken. These considerations have received considerable attention both in international standards and in the works of specialist authors such as Gy (1976) and Cornish *et al.* (1981).

In environmental and safety applications the process–analyzer interface may be identical to those discussed earlier—for example, cross-stack infrared flue gas analyzers and liquid effluent stream pH monitors—or the "process" may be the working atmosphere, the atmosphere surrounding the plant, or a body of water into which the liquid effluent discharges. Generally, the requirements and problems are similar, but there are special cases, such as personal monitors, where lightness and compactness are paramount needs, and in the sampling and analysis of the atmosphere surrounding a plant or site. The latter must take into account current meteorological conditions, such as wind direction and precipitation, and other area variables, such as traffic load, and must not selectively accept or reject coarser particles. The actual positioning is also vitally important for avoiding the sheltering effects of buildings, for example.

There is a second process—chemical analytical instrument interface—the so-called information interface. Essentially this is concerned with the use of the total information generated by the analyzer system, not simply the time-dependent variations in concentration of the components being determined. Users and designers of these systems must be aware of what other information is generated and make use of it as appropriate. For example, turbulence in flowing fluids will introduce "noise" in certain frequency bands from probe type analyzers such as pH sensors. The presence and amplitude of such fluctuations can indicate flow or sensor abnormalities. An infrared spectrum derived from an on-line interferometer can indicate the presence of unexpected suspended solids by the regular increase in scattering at shorter wavelengths. Similarly, a drop in signal energy can indicate window or mirror fouling in similar instruments. The use of such "free" information can be vital in predicting analyzer failure or process abnormalities, and should be fully taken into account in their design and application.

3. Calibration and Stability Checking

Since nearly all instrumental chemical analyzers are, to varying degrees, inferential, means for calibrating them have to be provided at the production and installation stages. Also, the user must be able to check the stability of these calibrations during the operating cycle of the instrument. The initial production calibration by the instrument manufacturer will generally be part of an application study to determine the viability of the instrument for the specified analyses. Hence, limits of detection, sensitivity, precision, and the effects of known or suspected interferences will be of primary concern. The installation calibration will effectively be to relate the output of the analyzer to concentrations in the process material, determined by independent means and to build confidence in its future users. Calibration stability checks are also mainly concerned with confidence maintenance, although some calibration trimming may take place and statistical examination of the data obtained can be used for failure prediction and maintenance demand. They are normally expected to be rapid, easy to implement and preferably fully automated.

Limits of detection, sensitivity and precision can often be determined from synthetic or semisynthetic standards. These would provide the manufacturer with go/no go information at the start of the application study. Ideally the initial calibrations should be completed with samples taken from the actual process, and subsequently analyzed by a method totally different in principle to that employed in the instrumental analyzer. However, this is not always possible—the instrumentation may be required for a new process and the piloting facilities may not be able to provide adequate samples for calibration. A second problem in providing a suite of samples for subsequent analysis and use as standards is that a correctly operating plant may not be able to provide on demand an adequate range of concentrations to fully describe the relationships between the raw instrument outputs and the concentrations of the wanted components or potential or actual interfering components in the process materials. Under these conditions the instrument supplier is forced to adopt synthetic or semisynthetic standards. With process gases or single-phase liquids this presents relatively few problems—the compositions can either be predicted from pilot plant studies or actual process material analyzed for all major components and good synthetic standards made up. Problems can arise even here if the compounds being determined are unstable or highly toxic, although the latter can be dealt with, at a price, by providing adequate protection in the calibration facility.

However, with multiphase materials such as pastes, slurries, emulsions and granular solids—especially those in the food industry—the production of good synthetic standards which mimic both the composition and structure of the process material may present major problems. While a few of the techniques used for the instrumental chemical analysis of solids, such as neutron moderation for proton and hence water content, are largely independent of the forms of material present, with the majority the calibration data obtained from heterogeneous materials will depend on the degree of heterogeneity, which includes particle size, degree of dispersion, moisture forms, voidage and crystallinity. Even the very history of the sample can affect the degree of heterogeneity, for example moisture form shifting with age, and slow processes of crystallization. Therefore, great care is necessary to try to ensure that sampled material used to make up standards is not subjected to processes that will significantly affect its heterogeneity, relative to the material that will be routinely measured.

In practice, after the instrumental performance is demonstrated to be adequate using synthetic or doped standards (doping involves the addition or removal of known amounts of the component being measured to or from base material obtained from the plant), the initial calibration is "trimmed," by adjusting it to fit a range of analyzed samples taken from the pilot or process plant. The trimming may also involve the adjustment of coefficients which quantitatively describe the effects of varying the amounts of interfering components on the basic calibration data. A simple example of this process is the effect that variation of iron and sulfur concentrations have on the determination, by x-ray fluorescence, of copper in mineral processing slurries. Compensation is achieved by measuring the iron and sulfur x-ray intensities and multiplying each by an experimentally determined coefficient that provides a value to subtract from the observed copper x-ray intensity, which can then be almost uniquely related to the copper concentration in the slurry.

After this stage the analyzer is ready to be installed in the process, when a final trimming of the calibration takes place with it working on line as samples are taken for laboratory analysis. This final trimming may also have problems: not only may concentration ranges be too small, but sampling and laboratory analysis may introduce bias or random errors. The former may have only limited consequences for process control but could be disastrous for quality assurance, while the latter can be demonstrated and, to a certain extent, compensated for statistically by taking a large number of samples.

Once a satisfactory calibration has been achieved, periodic checks must be made to guard against drift. Their frequency may be specified by the supplier or by local experience. It is not usually sufficient to simply check instrument stability; although important, the actual analyzer performance should be checked on standard samples. This is relatively easy with gases and single-phase liquids, since synthetic standards can be used (cross-stack analyzers can present problems), but with heterogeneous materials the same precautions should be taken as with the on-process calibration. Some calibration checks use stored standard disks of material; these are convenient but the degree of heterogeneity has to remain constant. An example of calibration drift in the face of a stable response from a standard material is in the determination of water in solids by dielectric measurement, where the calibration may depend on the forms of water present. These can show long-term shifts as a result of changes in the process conditions or the raw materials used.

4. Analyzer Design

The technical design of the analyzer must meet the performance needs of users: specificity, stability, reliability, and value of ownership exceeding cost of ownership. This is generally achieved by a combination of know-how, modelling of critical parts and computer-aided design. The outcome is a robust instrument, operating at design limits only at the lowest concentrations or highest precision. There are design problems associated specifically with operation in a process plant environment. These are caused by widely variable ambient temperature and humidity, corrosion (both from process materials and plant atmosphere), electromagnetic interference, unstable mains supply, dust, vibration, impact, flammable atmospheres, on-plant maintenance, calibration checking and, finally, neglect. Variable ambient conditions, corrosion and dust require that the analyzer operates within a sealed or flushed environment. Here design must minimize the temperature sensitivity of all the components, and the enclosure run at a stabilized temperature above the highest expected ambient value, or temperature measurement and automatic compensation must be built in. The latter has the advantage of warning of temperature excesses.

One of the most common requirements for these analyzers is the ability to operate in an area where flammable gases may be, or are, present (a "hazardous zone"). Zone 2 compatibility is the most common requirement, Zone 1 limitations are often met, while operation in Zone 0, where flammable gas is always present, is less likely with analyzers. Many analyzers, such as pH or dissolved oxygen analyzers can be split into two parts: (a) the measuring head proper, containing the sensor, preamplifiers and line drivers; and (b) a remote readout unit. The measuring head can be made intrinsically safe, using isolating transformers or Zener barriers. With others, such as infrared analyzers, it is preferable to design in an explosion proof form. The analyzer is sealed in an approved design metal case, with armored and sealed connections. Correct design may allow an intrinsically

safe, rather than an explosion-proof analyzer. For example, x-ray tubes, scintillation and proportional counters require high voltages and are difficult to design to be intrinsically safe. By substituting a radioisotope x-ray source, a semiconductor detector and low-power electronics, intrinsically safe design is straightforward. Typical applications are the determination of sulfur and lead in petroleum. The use of "analyzer houses" in hazardous zones can reduce the need for intrinsically safe or explosion-proof analyzers. Such houses have electrical power interlocked to air purging and provide a safe zone where samples can be brought or piped, and one or more analyzers installed. They also provide a weather-protected area for maintenance.

In the food, fermentation and pharmaceutical industries paramount needs are to ensure that no microorganisms can enter the process via the process–analyzer interface, and that analyzers can survive the mandatory sterilization cycles. Equipment such as pH probes must withstand the temperatures and pressures involved, without significant shifts in performance. For other probes it may involve removal either through sterile "locks" or by separation of the analyzer. This favors nonintrusive or noncontacting analyzers, which operate with a rugged and inert barrier between the process materials and the measuring components at all times. Inert means both chemical inertness and that the barrier does not participate in the measurement operations. Such techniques include electromagnetic and magnetic field spectrometry, electrodeless conductivity, activation analysis and dielectric spectrometry. X-ray techniques can also be included when the measured x rays can penetrate a window fulfilling the above barrier requirements. However, when determining lighter elements such as iron or sulfur, these windows have to be very thin (5–50 μm), but they are inert and impermeable. Where containment is imperative, double windows with an interleaved leakage detector are used. Sampling systems can be designed so that no microorganisms enter the process, by the use of automatic steam back purging, microorganism-proof membranes and interlocking valves, but with the hard-to-handle materials found in the food and healthcare industries, the difficulties of scaling down materials handling to analyzer size are often insuperable. Noncontacting techniques are also valuable with highly toxic or radioactive materials, eliminating potential leakage and sample disposal problems.

Analyzers must also be designed to present negligible hazard to their users and maintainers. High voltages (up to 100 kV with x-ray tubes) can be used and toxic materials may be present; beryllium windows with x-rays and thallium-containing infrared windows are examples. Such dangers can be minimized by the use of electrical and mechanical interlocks and permanent warnings. Analyzers using ionizing radiation present a potential hazard and must be designed to comply with national or international standards. These are designed to ensure that, under normal operation and maintenance or defined accident conditions, staff will not be exposed to radiation levels higher than those permitted for ordinary workers. Certain operations, such as source replacement, may be designed so that monitored staff must carry them out. As these operations are infrequent it is generally preferable to use the supplier or a specialist agency for these.

In industries such as cement, iron and steel, or mineral processing, the plant environment demands very rugged instrumentation. Such designs can be avoided by using an analyzer house or control room which is often air conditioned, and to which slurry sample streams are pumped or metal and cement samples brought pneumatically. However, in many plants, remotely sited analyzers are necessary, either for operation, materials handling or financial reasons. These must be designed to work in extremes of humidity, temperature, dust, vibration, impact, interference and a mains supply full of spikes and "brown outs." Evidence from these industries clearly shows that reliable performance can be achieved, but at a significant cost both in development and application terms. A linear amplifier coefficient of $0.01\% \, ^{\circ}C^{-1}$ is inadequate when faced with a $-30-+50\,^{\circ}C$ ambient range.

Maintenance can be a major part of the cost of ownership of analyzers, and design has a significant role in reducing such costs. Designing for high reliability and a minimum of routine maintenance, such as lamp or filter changing, is the first step; the second is ease of maintenance. Electronic boards and other components must be readily removable and, if necessary, capable of basic testing on-plant. Ease of removal of the analyzer head to a safe zone is also important; this favors analyzers that operate from one side of a pipe or duct, and also noncontacting instruments that can be removed without affecting the process.

The final important factor to be considered in analyzer design is the human–instrument interface. This may be masked by a control system with its own displays, but in many cases, and especially with on-plant analyzers used by process operators, it can be a critical factor for acceptance and reliable use. It is a specialist area involving psychologists, ergonomists, engineers and, last but not least, the operators who will be using the analyzers.

5. Computing and Communications

Many instrumental chemical analyzers require a range of input information in order to deduce chemical concentrations. Process temperature and pressure may be required, for example in on-line gas analysis; density, to correct for variations in solid content in slurry analysis; elapsed time, in activation analysis;

and ambient temperature with infrared detectors. Other analyzers, such as gas chromatographs or mass spectrometers, produce a wealth of information which must be stored for use in calculating concentrations. To meet process reliability, internal parameters, such as source intensity, detector noise and board voltages, must be monitored for maintenance, failure prediction and diagnostic purposes. These demand the means to input variables, store them and carry out mathematical operations ranging from subtraction to fast Fourier transforms in real time, and hence, significant computing power. Maintaining 1% precision over a range of 1–100% requires measurement to better than one part in 10^4, possibly in the face of wide temperature variations. This gives an indication of the quality demanded of the analog-to-digital converters, which are the key "front ends" of the computing system. In some analyzers, notably infrared, x-ray and gamma ray, differential measurements are used to extract a small relevant signal from two large ones, requiring high-precision computation. One further use of the analyzer computer may be to monitor and control the process–analyzer interface. This can involve sequencing, flow control, temperature cycling and calibration check procedures. Indeed, some sampling systems approach a miniprocess plant in this respect.

The computational needs of most instrumental chemical analyzers can be met by 16-bit single-board computers. There is an increasing minority where 32-bit and even parallel processing power is necessary to achieve plant compatible response times. Power consumption causes problems with heat dissipation in sealed systems and intrinsic safety, so that low-power systems, such as those based on complementary metal–oxide–semiconductors (CMOSs), are strongly preferred. The requirements for electromagnetic shielding (both to and from the computer) and power supplies are often exacting in a plant environment, and are vital in designing for reliability.

The analyzer's computer has another important role, that of communication. For operator observed instruments, whether on-plant or on-line, a clear display must give operating instructions, display results and give warnings of impending or actual failure. As already pointed out, this is both an ergonomic and instrumental area. The display must also be readable in a wide range of ambient light; direct sunlight obliterates most displays. For on-line communications the analyzer computer has two functions: (a) to receive information from and send instructions to any remote measuring head; and (b) to send analytical and maintenance information to the central control computer or communication network. The trend is to use digital communication throughout, although the majority of sensors on process plants still have analog outputs. Instrument buses such as the Field bus provide the means for stringing together sensors and actuators without the need for individual connections, and analyzers can be used as just another sensor on the instrument bus or in critical situations as a node on the plant communication bus. The advantages of digital communications in measurement are wide dynamic range, rejection of electromagnetic interference and simplicity in cabling. Fiber-optic communication between the analyzer head and computing is becoming more popular for similar reasons, although armored fiber-optic cables can be rather bulky when compared with their metallic counterparts.

See also: Analytical Physical Measurements: Principles and Practice; Electrochemical Measurements; Process Instrumentation Applications

Bibliography

Clevett K J 1986 *Process Analyzer Technology*. Wiley, New York
Cornish D C, Jepson G, Smurthwaite M J 1981 *Sampling Systems for Process Analyzers*. Butterworth, London
Gy P M 1976 The sampling of particulate materials—a general theory. *Int. J. Miner. Process* 3, 289–312
Merks J W 1985 *Sampling and Weighing of Bulk Solids*. Trans Tech, Clausthal–Zellerfield, Germany

K. G. Carr-Brion
[Cranfield Institute of Technology,
Cranfield, UK]

Construction and Manufacture of Instrument Systems

The prime purpose of a measuring instrument or instrument system is to extract specific information about the physical state of a chosen subject, that information being converted into equivalent signals that accurately represent the variable or variables being measured.

Instruments are therefore machines that extract and process information about real-world processes. This type of machine is designed and manufactured very much like most machines. The nature of these information machines does, however, involve—from application to application—a great variety of technologies and practices. They are different from other machines in that they process the nontangible substance "information" that represents knowledge, doing this without adding any more error than can be tolerated.

Putting aside simple manual instruments such as rulers or vernier calipers, the transducer-based forms comprise a well-defined path of information flow that can be portrayed by generalizations on which deep mathematical modelling of behavior can be simulated.

How to apply scientifically based principles to the constructional aspects that are needed to make the actual artifact is not so obvious, and much of this work is carried out from a base of long experience. The following examples portray the diversity involved.

At the very small end of the size scale of instruments is found the miniature microelectronic sensor with its inbuilt circuitry. These are formed in a submillimeter-size die of very pure silicon, mounted in a case that protects it and allows connections and mounting.

More obvious to the eye are more traditional measuring instruments typified by those used in the process industry. An example is a pressure sensor made from a mixture of less exotic materials—common plastics and metals, in the main (see *Pressure Measurement*).

In the science laboratory will be found many forms of analytical measuring instrument, such as the atomic absorption spectrophotometer (AAS) or the electron microscope. These quite complex systems incorporate metals, plastics, optical materials, ceramics, rubbers and other materials in a construction exercise that often requires ultraprecision machining.

At the very large end of the size range is the Australian (*Jindalee*) "over-the-horizon" radar system (see Fig. 1), that has an array of sensing antennae spread over several kilometers. Possibly the largest measuring instrument yet constructed is the very-large baseline (VLB) interferometer arrangement in which the Earth is used as the baseplate of a system of radio telescopes.

As well as requiring use of virtually all disciplines in their manufacture, some instruments are extremely sophisticated. Generally, the greatest complexity is found in defense and aerospace systems. From such activities flows, to other fields of endeavor, the ability to organize and manage the design, manufacture and application support of complex systems.

To describe the alternatives in terms of construction detail would need volumes of text, of which much would be superseded before publication. It is more useful here to consider construction and manufacture at a level of abstraction that covers the key factors which are always involved.

1. Purpose of an Instrument System

The prime consideration for a manufactured artifact having a practical purpose is that it satisfies the need for which it was created. Its degree of excellence depends on its fitness for the purpose. Its constructional detail and manufacturing process play a large part in deciding the final performance. Design defects can seldom be properly overcome at the time of production, for changes are, by then, around a thousand times more costly than if they had been carried out in the earlier design phase.

Purpose may be defined as a set of primary func-

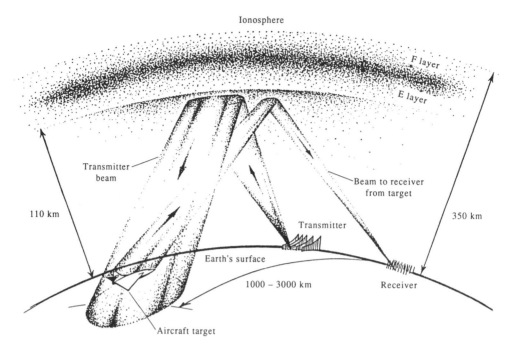

Ionosphere

F layer

E layer

Transmitter
beam

Beam to receiver
from target

110 km

350 km

Transmitter

Earth's surface

1000 – 3000 km

Receiver

Aircraft target

Figure 1
A very large measuring instrument for monitoring movements of objects thousands of kilometres away—*Jindalee*
"over-the-horizon" radar system. (Courtesy of Defence Science and Technology Organisation, Australia)

tions that the product is primarily designed to carry out. These are called the use functions.

Another set of parameters that must also be satisfied pertain to the attractiveness of using it and to the desire to possess it. These are known as the esteem functions.

Fitness for purpose is evaluated in terms of the adequacy of the design to keep performing over both short and long time periods, its fidelity in extracting the required information and how well it performs to the overall design requirement statement in many other respects, such as documentation support and cost. Becoming increasingly important is its ability to satisfy user-domain issues such as warranty, maintenance, service and public liability. Its quality is assessed in terms of its conformance to the stated specifications.

Conformance needs to be assessed both quantitatively and qualitatively. Both kinds of measure are necessary because real difficulties exist in defining the quality of information and of the many subjective parameters relating to esteem and, often, performance.

A measuring instrument is designed and applied to extract certain information that represents an improved state of knowledge. It is formed from a sequence of stages which progressively process the information taken from a subject, until it is in the desired equivalent form. When packaged, these serial stages may not be in any obvious order in the final organic placement.

The information is passed from the subject into the measuring system by use of a convenient communication carrier. This carrier is usually an energy link but it can also involve a transfer of mass. In both cases the information is contained by representation as modulations of the steady-state flow of the carrier used. Stray effects that produce unwanted modulations give rise to error of measurement (see *Transducers: An Introduction*).

2. Problems of Practical Implementation

Once a measurement need has been established, the next step is to select a transducer principle that will connect the measuring system to the subject of interest. Several stages of subsequent signal transduction and modification are usually needed before the signal is in a usable form.

Each stage is built by implementing a "law of nature," that is, a physical principle, into an artifact form. A first example is the scientific law that describes how an electrical resistance varies with temperature change, which is used as the basis of the resistance temperature detector (RTD). A second example is an expression for strength of materials modelling the deflection of an elastic material, this being implemented as the basis of a weighing load cell (see *Mass, Force and Weight Measurement*).

Published descriptions of transducer technology too often describe only this more obvious and easily described first level of appreciation. To obtain adequate conformance, practical application of the chosen conversion effect requires far more understanding and application of effective engineering.

The difficulty is that when the effect is implemented the practical necessity to use available materials at affordable costs means the artifact is also responsive to unwanted sensing effects. These cause errors in the use of the first-level principle.

Errors arise from two main sources. Internally changes will occur to material properties with time, and there will be generators of unwanted forms of energy, such as self-heating of key components and stray electric fields that cause error-inducing feedback energies. Externally there will be influence parameters, such as temperature and relative humidity.

In the RTD example, ideally the sensing resistance and circuitry should respond only to changing temperature of the probe. In reality, relative humidity will alter unwanted shunting electrical resistances of the RTD, vibration will gradually alter its resistance properties and temperature will influence the gain and other properties of the following electronic processing circuitry. Furthermore, temperature gradients across parts of the wiring might also generate significant thermoelectric currents which reduce the accuracy of the signal interpretation. These can be regarded as second-level effects; their complexity rises rapidly as instrument performance targets rise—more of them appear and interaction between them becomes more salient.

Practical implementation only succeeds when viable solutions to second-level problems are satisfactorily implemented. It is in this area that design, construction and manufacturing issues become vitally important.

3. Construction Strategy

It is helpful to consider an instrument design as a set of competing variables that must each be integrated to match a stated performance and manufacturing need.

Each design decision is a degree of freedom of choice until it is frozen by intent or imposed circumstance. There are, therefore, numerous degrees of freedom in which to move, or not to move, as the need may be.

On one hand, the design has to allow certain degrees of freedom, such as allowing a point parameter to move along one line. Examples are obtaining the correct gain of an amplifier as signal frequency changes, and the one-axis linear motion of the sensing mass in an accelerometer in order to reduce its cross-axis sensitivity.

On the other hand, the choice of materials and minor components to make a part will have many degrees of design freedom. An instrument weighing-spring design, for instance, has a choice of variables—namely material, alloy content, annealing, hardening, shape, and temperature of operation. These factors must be juggled to achieve a suitable design specification.

The selection of principles, materials and mode of implementation is made difficult because the characteristics of materials or components are too often far from ideal. For example, the design and manufacture of high-precision mechanical springs, when needed in exacting applications such as in a spring–mass form of gravity meter, require carefully selected material, shape, wire size and manufacturing processes. The product must also be operated at the temperature at which the thermoelastic effect is minimal, this temperature resulting from the melt and annealing history. As we do not yet have a spring metal that has its temperature coefficient of thermal expansion, thermoelastic coefficient, mechanical hysteresis, machining factors and price all at the preferred values, the design task becomes a complicated process of gradual optimization within interacting goals.

The responding mass in that gravity meter must be restricted to one-axis motion, the spring and amplifiers must have conversion sensitivity to a constant 0.1% per month for the whole system, and it must respond to force changes as small as one part in 1×10^8. This has to be achieved in the hostile environment of a deep, very hot and wet, oilwell hole after the instrument has been taken to the site in a helicopter or a four-wheel drive vehicle.

To reach a satisfactory performance, there are five dominant methods for overcoming the problems caused by given limitations. These are applied, often in combination, in accordance with allowable cost, time and proprietary rights factors.

The first approach is to seek the best-suited first-level principle on which to base the instrument, such as the one having the greatest sensitivity to the measurand so that unwanted signals are comparatively smaller, or one that is less sensitive but also less affected by influence variables. As there are hundreds of principles to choose from, and hundreds of combinations for setting up a suitable serial information processing chain, this first stage of selection is not a trivial task, if done well.

A second path is to use better materials or components in key places; examples are a more exotic material with a lower coefficient of the particular parameter causing stability problems, or an electronic amplifier with lower drift over time.

The third possibility is to identify the cause (e.g., temperature) of the unwanted effect. This can then be measured as it changes and compensation carried out by local feedback or by correction of the signal at a later stage. It is often possible to provide this compensation without need for great sophistication—for example, a bimetal strip responds to temperature changes and can be used to make small compensating motions. Recent advances in intelligent materials suggest that some materials can sense the working environment and change properties in a beneficial way.

Another approach is to set up the instrument so that it is sufficiently isolated from the influencing effects that are causing difficulties. In this example one might control the environment of the instrument. In borehole geophysical parameter measuring instruments (down-hole loggers) it is normal practice to place the sensors and circuitry in watertight, temperature-controlled pods.

These options clearly impinge greatly on subsequent construction and manufacturing decisions. Despite application of these concepts it is not always possible to reach the performance needed for a sufficiently long calibration interval. Where no means can be found to overcome the problems of stability, the system is built to allow it to be reset automatically at regular intervals. An example is the common use of automatic daily calibration of chemical pollution sensors by passing standardized pollutant gases through them after they have been purged clean with an appropriate gas.

In some cases parts do not perform properly for long enough, and in some cases automatic or manual replacement at the necessary interval provides a solution. Use of disposable sensors, such as high-temperature dip thermocouples that dissolve almost immediately when they are placed in the ladle of molten metal, is an illustration of this methodology.

4. Materials and their Limitations

Instruments are formed by combining a wide variety of proprietary vendor-supplied components; these are held together to form an instrument system with custom-built structures. Simple sensors (e.g., for temperature, pressure and displacement) tend to be produced in moderate volume and are offered as off the shelf products. Electronic circuitry assemblies for instruments are sometimes made as application specific integrated circuits (ASIC), but are more usually found in instruments as components joined up to form the system needed.

Figure 2 shows the inside workings and construction of a commercial pressure sensor made in medium volume for the process industry.

The performance offered in these assemblies is largely decided by the materials from which they are made.

Materials are used to achieve one of two purposes. Active materials are those that exhibit a reproducible

degree of freedom that changes markedly as the variable of interest changes. The platinum metal film or wire of the RTD provides useful electrical resistance variation with temperature changes, and thus makes a good sensing element.

Passive materials are those that ideally do not respond to any internal or external influence parameter, and therefore make good structural elements.

The difficulty is that materials all have some degree of activity. Indeed, in one application a material might be chosen as the active sensor while in another circumstance it might be the best choice for a passive role. A good example is the use of very pure silicon to form a large range of microelectronic sensors. Minute changes to the doping levels of impurities, variations in the thickness of layers and selection of the geometry allow the same passive structural material to be made into sensors of many chosen variables.

A material with the desired characteristics is often not available because one suited to instrumental applications has not yet been developed. Instrument constructors must make use of materials, techniques and manufacturing methods that are developed for products having larger commercial interest. Carbon fiber, a good example and useful for making very stiff and light structures with almost zero coefficient of thermal expansion, was first developed for use in turbine engine fan blades.

Within the needs of a given instrument the range of properties of materials will be found to be quite limited; use of improvement strategies is essential in high-performance instruments.

Further to these difficulties is the fact that material performance data are often hard to obtain or will not be determined with sufficient accuracy. Important properties are often decided by conditions before, during and after processing, such as the cooling rate after forming, annealing processes, work-hardening in machining and the time allowed for internal stresses to relax sufficiently.

Early instruments were made of several natural materials—bone, glasses, porcelain, ivory, silk, wood and precious metals. Lack of product uniformity, availability and reliable behavior, plus their high cost and machining difficulties, have seen most of these fall out of favor compared with modern, highly processed, manufactured stock materials.

The main groups of materials used today are metals, plastics and ceramics and glasses. Common forms of these well-described materials find straightforward application as structural and sensing elements. However, special care is essential in use of instruments in order to adequately control the various degrees of unwanted freedom. Unfortunately, few published accounts cover design of elements for instrument applications in terms of fidelity of freedoms; instead, they provide information in terms of yield-strength driven design.

As an example, consider the thin sheet diaphragm

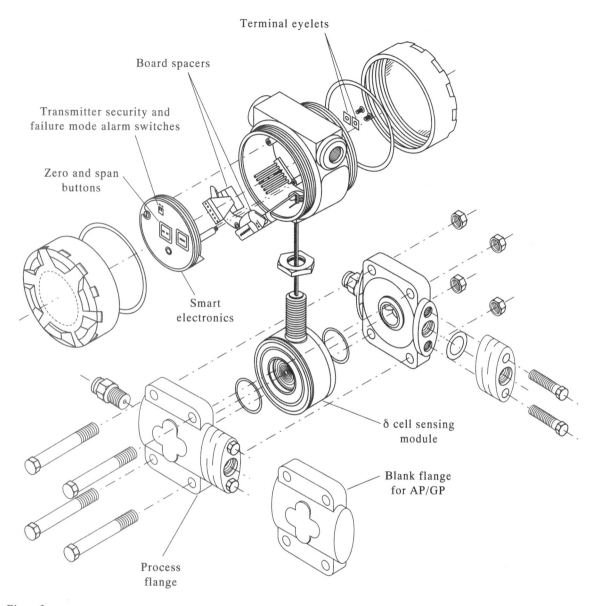

Terminal eyelets

Board spacers

Transmitter security and
failure mode alarm switches

Zero and span
buttons

Smart
electronics

δ cell sensing
module

Blank flange
for AP/GP

Process
flange

Figure 2
Exploded view of process industry pressure sensor labelled to illustrate key points of construction and manufacture
(Courtesy of Rosemount Inc., USA)

element in a pressure sensor. The active degree-of-freedom role is for the diaphragm to deflect in proportion to the pressure exerted on the element. In application of this simple principle to measurement, an important parameter is how well it returns to the original state when a pressure has been applied and removed (its hysteresis). This application requires experimentally determined data that are not available for common sheet metals; for these constructional

materials would be more used in cases where the yield strength would be the property called for in bending operations. If the pressure gauge were allowed to be stressed to the yield point it would then become useless, for its reproducibility of transduction would have been lost.

Plastics have brought about major changes in instrument construction. They offer a vast range of properties that are made to order by adjusting the mix and processing. They are readily made into complex shapes by affordable methods, they are corrosion-free and can be decorated by numerous methods. Their main weaknesses are high thermal expansion, relatively low stiffness, noticeable creep and inability to work at high temperature. However, by the use of composites and careful design, they are used for structures ranging from crude frameworks to miniature precision gearwheels. Plastics find limited use as active elements—mainly as piezoelectric sensing film.

Metals are the main structural and sensing element materials used when precision and temperature factors are demanding; the properties needed for such cases are obtained from special metal alloys. Common metals such as steel, aluminum and brasses are used for boxes, stands and extruded items but they tend to be used less today than in the past because plastics have taken over their roles when a more complex shape of structural element is needed. Their less-known (and less-available) proprietary alloys—such as Ni Span C, invar, phosphor bronze, and suchlike—find increasing use.

Glasses (silicas and optical) have long been dominant instrument materials, but they are now partially displaced by more modern substitutes—plastic lenses, ceramic parts and fused silica being examples. In this area the structural use of ceramics is growing because of their high stability and high working temperature. They also suit certain comparatively new production processes, such as the manufacture of hybrid micro-electronic circuits.

The material that has most transformed the way instruments are made is silicon, in its role as the substrate of integrated electronic circuitry. It was only as recently as 1955 that an eminent physicist suggested that transistors might find limited use for making "walkie-talkies" (portable radio transceivers) smaller but would have little impact otherwise!

It can be expected that silicon and gallium arsenide materials will continue to be used increasingly as the sensing and circuitry parts of instrument systems. Reasons for the speed of adoption in this field being slower than expected are the need for quite large-volume production to cover setup and plant costs, unsatisfactory product uniformity, the need for higher temperatures of operation and the lack of commercially viable research proposals. However, these factors are steadily being overcome.

5. Constructional Regimes

The multidisciplinary nature of instruments requires all materials and manufacturing techniques to be used at some time or other. Development of engineering discipline groupings over the past century has delineated certain key regimes of relevance.

The mechanical regime, which is as old as human life, has had a steady but slow progression. It is typified by assemblies of frameworks, linkages, bearings, fasteners, gears, belts, springs and cams. Many of the techniques currently used can be traced to definitive scientific descriptions of a century or more ago. Advances in recent decades have been more concerned with how they are designed using computer-aided simulation that result in systems being more responsive, less costly and more reliable. This regime is best used for structures and precision transducers.

The next oldest is the optical or radiation regime, that incorporates optical and other radiation concepts and practices. It is typified by assemblies of lenses, prisms, gratings, filters, sources and sensors that are held together by precision mechanical structures. This has its origins some six centuries ago, but defense optics and remote sensing applications have led to significant advances in design, materials and performance since the 1950s. This regime relies heavily on the use of computer power (without which tens of person years of manual effort would be needed to design a crude low-range zoom lens system) to design optical elements for cameras, telescopes, and so on. There are now many special-purpose exotic materials available for infrared and ultraviolet sensing. This area has become known as optical instrument engineering.

Clearly the most used technique is the electronic regime. This is typified by assemblies of wires, switches, resistors, inductors, capacitors and a vast range of integrated circuits combined variously on printed wiring boards, ceramic substrate hybrids, and as silicon chip integrated circuits (ICs) of great complexity. Considerable software to monitor and control the system is almost always present. Advance here is at such a pace that it is typical for an instrument produced by this regime to be out-of-date in around two years—and there are no clear signs of the situation stabilizing. Over just 40 years, design time was reduced from years to months. Cost and size factors are down a 100-fold. Data processing power is up at least 1000-fold.

Instrument creation is dominated by the electronic regime, for this is the one that largely decides the factors that drive market acceptance and system performance. It is highly applicable for memory, data-processing and communications functions, that is information processing roles.

Other regimes exist that find less prolific, but often vital, use. Pneumatic and hydraulic systems, for example, are still used in process control, and in automotive and aerospace use.

It has long been obvious that chemical systems have great promise as the key regime in future instrument manufacture. Potentially, at least, they can be grown from raw materials, and they are reproducible, self-repairing and cheap. Despite much research, these properties have not yet been harnessed to make effective, commercially viable instruments. It is reasonable to assume—for it has already happened in Nature—that one day we shall see the currently used

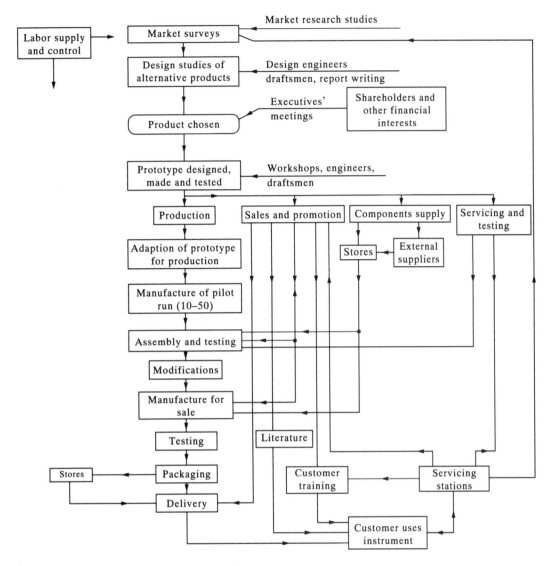

Figure 3
Steps in the creation of an instrument product

electrophysical electronic information technology methods give way to electrochemical forms.

6. Manufacture

Initial stages of development of an instrument product are typified by relatively small scales of operation compared with other (apparently similar) items such as electronic consumer goods. Instruments are mostly made as custom products and systems that suit a particular need. Only a handful of global firms exist that make instruments and systems in large volumes.

The product run for a typical, mature, instrument is in the range of 50–200 items.

The steps of instrument creation are well documented (see Fig. 3). This knowledge enables cost estimation to be carried out with some precision. Commercial risk arises, however, from factors relating to the assumptions that need to be made before manufacture commences. Factors that are particularly hard to control include possible failure to conform with the specified performance, public liability, regulatory issues, government legislation changes and unexpected competitive product releases.

The first stage is a design phase. Here the contributions of intuitive, industrial and engineering forms of design are all integrated to form a realizable product. A tight operational requirement is prepared to start the process of artifact building. That leads to an engineering or technical specification. Selected parts of the full system are next designed and made, in order to see whether the problem aspects, which are less well understood, can be overcome. Some classic books, dating back to around 1900, describe the laboratory practice involved with instruments, much of which is still relevant for the research stage of construction.

A full system is then assembled to reach the proof of concept or, perhaps, the working prototype (also sometimes called the "Alpha test") product stage. Typically some two to five units are made. Here manufacturing methods are expensive because handcraft methods must be used instead of the low-cost automated processes that can later replace them. Tests on these units give the maker the chance to evaluate performance first.

Next follows field testing (sometimes called the "Beta test" phase), in which users see whether the product is likely to measure up to the needs of the market, and thus to gain the confidence of the product's financial backer. In a preproduction run for a volume product, some 50–100 items would be made. However, many instruments may only need to be made as a sole item, or in a limited number.

Because of the many factors involved in instrument production, it is not practicable to generalize the production aspect to any great length. Vagaries include matching the manufacturing method to the expected volume when sales are not yet known.

The complexity of instruments, coupled with their low volume of manufacture, means that often they cannot be adequately tested. Furthermore, the purchaser is usually unwilling to pay the necessary premium for testing, taking the lowest price option. It is a well-established fact that many proprietary instruments do not reach their declared specifications. Independent instrument evaluation helps to detect such deficiencies, but the test process takes several months to carry out and can easily cost as much as the product. However, if the cost of a failed measurement is assessed it will usually be found that evaluation is justified.

Bibliography

Ashby M F 1992 *Material Selection for Mechanical Design*. Pergamon, Oxford
Duckworth W E, Harris D D, Sydenham P H 1992 Active and passive roles of materials in measurement systems. In: Sydenham P H, Thorn R (eds.) *Handbook of Measurement Science*, Vol. 3. Wiley, Chichester, UK

Ichinose N 1987 *Introduction to Fine Ceramics*. Wiley, Chichester, UK
Jones A J 1990 Materials and devices, Pt 1. Sensor technology. Department of Industry, Technology and Commerce, Canberra
Kroschwitz J I 1990 *Concise Encyclopedia of Polymer Science and Engineering*. Wiley, Chichester, UK
Moore J H, Davis C C, Coplan M A 1983 *Building Scientific Apparatus—A Practical Guide to Design and Construction*. Addison–Wesley, Reading, MA
Rubin I I 1988 *Handbook of Plastic Materials and Technology*. Wiley, New York
Stanley W F 1901 *Surveying and Levelling Instruments*. E and F Spon, London
Strong J 1942 *Modern Physical Laboratory Practice*. Blackie, London
Sydenham P H 1979 *Measuring Instruments: Tools of Knowledge and Control*. Peregrinus, Stevenage, UK
Sydenham P H 1986 *Mechanical Design of Instruments*. Instrument Society of America, Research Triangle Park, NC
Sydenham P H, Hancock N H, Thorn R 1989, 1992 *Introduction to Measurement Science and Engineering*. Wiley, Chichester, UK
Sydenham P H, Thorn R 1982, 1983, 1992 *Handbook of Measurement Science*, Vols. 1, 2, 3. Wiley, Chichester, UK

Further information can be found in issues of the *Hewlett-Packard Journal* (descriptions of modern instruments and details of commercial construction and manufacturing methodologies used)

P. H. Sydenham
[University of South Australia,
Ingle Farm, Australia]

Correlation in Measurement and Instrumentation

Cross correlation, often referred to as correlation, is a method of determining a relationship between two sets of data. This relationship may not be sufficiently close to be seen when in a table or graph, but cross correlation—a statistical technique for recognizing patterns—can detect the true relationship and reject unrelated and confusing data.

Cross correlation applications abound. Our brain may do a cross correlation process when we recognize a face in a crowd; this is emulated in computer vision systems. Behavioral and medical data are related to social and dietary causes by cross correlation methods. The dynamic models, which govern the behavior of civil engineering structures in an earthquake, and process plants in response to changes in material flow or quality, can be determined by cross correlation.

1. Properties of Correlation Functions

A general introduction to the principal features of random data correlation will be given in this section. Readers who wish to study the subject in depth may wish to refer to the more extensive texts on the use of random data for measurement of system dynamics (Bendat and Piersol 1980), or its application in the narrower area of flow measurement (Beck and Plaskowski 1987).

1.1 Autocorrelation

The autocorrelation function $R(\tau)$ of random data $x(t)$ is effectively the cross correlation of $x(t)$ with itself. It describes the general dependence of the values of the data at one time t on the values at another time $(t + \tau)$ (see Fig. 1). It is obtained by averaging the instantaneous product of the two values $x(t)$ and $x(t + \tau)$ over a time T approaching infinity:

$$R_{xx}(\tau) = \lim_{T\to\infty} \frac{1}{T} \int_0^T x(t)x(t + \tau)\,dt \qquad (1)$$

In practice, the averaging time (observation time) T must be finite and

$$R_{xx}(\tau, T) = \frac{1}{T} \int_0^T x(t)x(t + \tau)\,dt \qquad (2)$$

The autocorrelation $R_{xx}(\tau)$ is a real-valued even function of τ with a maximum at $\tau = 0$; thus

$$R_{xx}(\tau) = R_{xx}(-\tau) \text{ and } R_{xx}(0) \geqslant |R_{xx}(\tau)| \quad \text{for all } \tau \quad (3)$$

If the data to be correlated do not include a periodic component, then the mean value μ_x of $x(t)$ and the autocorrelation function are related as shown in Fig. 2 by $\mu_x = (R_{xx}(\infty))^{1/2}$. If the data $x(t)$ being correlated have a periodic component, then the autocorrelation function $R_{xx}(\tau)$ will also have a periodic component

with the same period, but phase angle information will be lost (see Fig. 3).

By putting $\tau = 0$ in Eqn. (1), it can be shown that the mean square value of $x(t)$ is given by $\bar{x}^2 = R_{xx}(0)$, that is, the mean square value is equal to the autocorrelation function at zero time displacement.

If $x(t)$ has no periodic components and no dc component, then $R_{xx}(\tau) \to 0$ for $\tau \to \infty$. The value of $R_{xx}(0)$ is used as a normalizing factor to give the correlation coefficient $\rho_{xx}(\tau)$:

$$\rho_{xx}(\tau) = \frac{R_{xx}(\tau)}{R_{xx}(0)} \qquad (4)$$

1.2 Cross Correlation

The cross correlation function of two sets of stationary ergodic random data, such as the data from cross correlation flow sensors (see Fig. 4a), describes the general dependence of the values of one set of data on the other (see Fig. 4b). Hence, the cross correlation

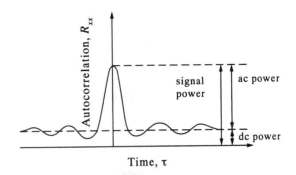

Figure 2
Autocorrelation function of an entirely nondeterministic process, except for a dc component

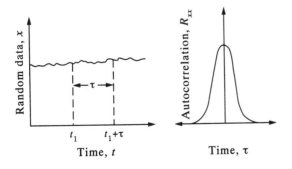

Figure 1
The autocorrelation function

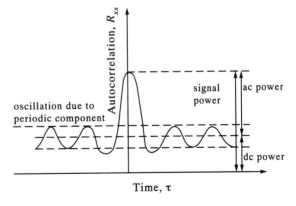

Figure 3
Autocorrelation function of a nondeterministic process with an added periodic component

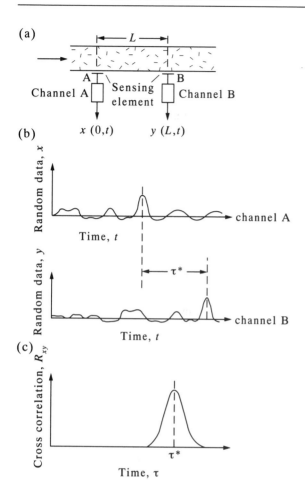

(a)

Channel A | Sensing element | Channel B

(b) $x\,(0,t)$ $y\,(L,t)$

channel A

channel B

(c)

Figure 4
(a) Principle of cross correlation flow measurement; (b) upstream and downstream signals; (c) cross correlation function

function (see Fig. 4c) of the value of $x(t)$ at time t and $y(t)$ at time $t + \tau$ may be obtained from the average of the product of the two values over the observation time T:

$$R_{xy}(\tau) = \lim_{T \to \infty} \frac{1}{T} \int_0^T x(t)\,y(t + \tau)\,\mathrm{d}t \qquad (5)$$

The function $R_{xy}(\tau)$ is always a real-valued function, which may be either positive or negative. The cross correlation function $R_{xy}(\tau)$ does not necessarily have a maximum at $\tau = 0$, nor is $R_{xy}(\tau)$ an even function, as was true for the autocorrelation function. However, $R_{xy}(\tau)$ does display symmetry about the ordinate when x and y are interchanged:

$$R_{xy}(\tau) = R_{yx}(-\tau) \qquad (6)$$

If the mean values $x(t)$ and $y(t)$ are zero, then the correlation coefficient $\rho_{xy}(\tau)$ is given by

$$\rho = \frac{R_{xy}(\tau)}{(R_{xx}(0)\,R_{yy}(0))^{1/2}} \qquad (7)$$

The maximum value of the correlation coefficient is unity, which occurs only when $x(t)$ and $y(t)$ are identical, apart from a relative time shift and a scale/gain change. The correlation coefficient has a significant effect on error, which will be discussed in Sect. 1.3.

1.3 Errors in Correlation Estimates

(a) *Magnitude errors.* For bandwidth limited white noise with mean value zero (i.e., ac coupled data as used for cross correlation flowmeters), bandwidth B and integration time T, the variance of the autocorrelation estimate for all τ is given by (Bendat and Piersol 1980)

$$\mathrm{var}(\hat{R}_{xx}(\tau)) = \frac{1}{2BT}(R_{xx}^2(0) + R_{xx}^2(\tau)) \qquad (8)$$

Similarly, Bendat and Piersol have shown that, for the cross correlation of signals x and y, the variance of the cross correlation estimate is

$$\mathrm{var}(\hat{R}_{xy}(\tau)) = \frac{1}{2BT}(R_{xx}(0)\,R_{yy}(0) + R_{xy}^2(\tau)) \qquad (9)$$

For $R_{xy} \neq 0$ the normalized mean square error is given by

$$\varepsilon^2 = \frac{1}{2BT}\left(1 + \frac{1}{\rho_{xy}^2}\right) \qquad (10)$$

where ρ is the correlation coefficient (Eqn. (7)). Equation (10) shows clearly that the normalized mean square error is reduced in proportion to the bandwidth–time product and that it is increased if the correlation coefficient is reduced.

(b) *Time delay errors.* Referring to Fig. 4, errors in estimating the time delay are relevant to estimating the flow velocity (Sect. 2). This error can be derived from the magnitude error (Eqn. (9)) by using a modelling function related to the shape of the cross correlation function (Beck and Plaskowski 1987), to give the error in measuring the time delay τ^* of the cross correlation peak:

$$\mathrm{var}(\tau^*) = \frac{0.038}{TB^3}\left(\frac{1}{\rho_{xy}^2(\tau^*)} - 1\right) \qquad (11)$$

2. Case Study—Application to Correlation Flow Measurement

Flow measurement is a particularly important aspect of plant instrumentation; industrial quality flowmeters using well-established techniques have been commercially available for many years for the measurement of clean single-phase fluids. However, in the important and growing field of multicomponent flowmetering there is an embarrassing lack of proven equipment.

Cross correlation methods are ideally suited to the measurement of multiphase flows and also to the problem of obtaining reliable measurements at inaccessible locations, for example, the application of correlation techniques to flowmetering (Bentley and Dawson 1966). The basic principle of cross correlation measurement is simply to measure the time taken by a disturbance to pass between two points spaced along the direction of flow (see Fig. 4). In a correlation flowmeter the measurable properties, such as the variation of temperature, pressure, capacitance, electrical conductivity, or other physical parameters, are detected by a sensing element and converted into electrical signals. Referring to Fig. 4, $x(0, t)$ denotes the transducer output signal derived from channel A—the amplitude of the signal is directly related to the instantaneous value of the measured parameter of the fluid at point A. Similarly, $y(L, t)$ is the signal related to the value of the measured parameter at point B.

The cross correlation function $R_{xy}(L, \tau)$ of $x(0, t)$ and $y(L, t)$ is given by

$$R_{xy}(L, \tau) = \lim_{T \to \infty} \frac{1}{T} \int_0^T x(0, t - \tau) y(L, t) \, dt \quad (12)$$

The maximum value of $R_{xy}(L, \tau)$ will occur at $\tau = \beta$, where the correlation delay time is equal to the transit time β of the measured parameter of flow between the points A and B. Thus, the velocity of flow is given by

$$u = L/\tau^* \quad (13)$$

where u is the flow velocity, L is the distance between A and B and τ^* is the value of τ corresponding to the peak value of $R_{xy}(L, \tau)$.

If the velocity profile across the pipe diameter is uniform, then the volumetric flow rate can be obtained by multiplying u, which is determined from Eqn. (13), by the cross-sectional area of the pipe.

In a fully developed pipe flow (i.e., an isotropic flow where nonstationary effects due to nearby bends and disturbances in the flow do not exist), the measured signals at transducers A and B (Fig. 4a) will constitute a stationary and ergodic process. The difference in the measured signal between transducers A and B, which results from the continuous random process of turbulence generation, can be represented by the addition

of a noise signal at the downstream location. This break up of turbulence (i.e., the reduction of $R_{xy}(\tau)$ in Fig. 4c) reduces the correlation coefficient (Eqn. (7)) and, hence, increases the error (Eqn. (11)). The sensor spacing (L in Fig. 4a) deserves consideration. A small spacing reduces the time delay variance (Eqn. (11)), a large spacing increases it, but the actual time delay is proportional to L, so there must be an optimal L to give the most accurate flow velocity measurement. It can be shown (Beck and Plaskowski 1987) that the variance in flow velocity u is modelled by the equation:

$$\text{var}(u) = \frac{K}{LB^3} \left(\frac{1}{\rho_{xy}^2(\tau^*)} - 1 \right) \quad (14)$$

where K is a constant of proportionality.

In practice, the sensor spacing is not critical (from Eqn. (14) and experimental results), and values of L between one and five times the sensing zone size (along the pipe axis) are satisfactory.

3. The Future

An example of future potential is in flow imaging, for the measurement of two-phase flows (like those from oil wells and in many industrial processes), where the phase distribution may vary in an unknown way between fine bubbles or particles and elongated slugs. No currently available flowmeters will measure these flows unless the phases are mechanically separated,

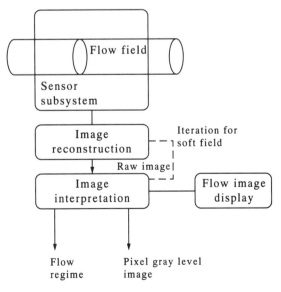

Figure 5
Principal elements in flow imaging

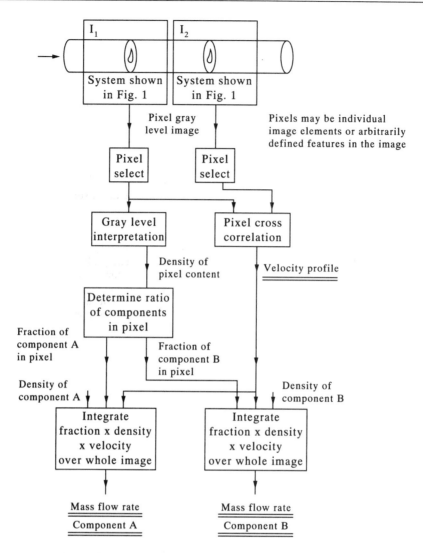

Figure 6
Measurement of velocity profile and component mass flow

which is often impossible and nearly always prohibitively expensive.

A flow imaging system (see Fig. 5) has sensors similar to those for cross correlation flowmeters and uses an array processor to carry out a tomographic image reconstruction algorithm (Huang *et al.* 1989). This is analogous to the methods used in medical computer-aided tomography (Murphy and Rolfe 1980). Figure 6 suggests the type of system that is expected to develop in the late 1990s for the measurement of two-phase flow, having virtually unlimited phase distributions. It relies on obtaining a tomographic image of the flow at two axially spaced locations.

See also: Signal Processing

Bibliography

Beck M S, Plaskowski A 1987 *Cross Correlation Flowmeters—Their Design and Application*. Hilger, Bristol, UK

Bendat J S, Piersol A G 1980 *Engineering Applications of Correlation and Spectrum Analysis*. Wiley, New York

Bentley P G, Dawson D G 1966 Fluid flow measurement by transit time analysis of temperature fluctuation. *Trans. Soc. Instrum. Technol.* **18**, 183–93

Huang S M, Plaskowski A B, Xie C G, Beck M S 1989 Tomographic imaging of two-component flow using capacitance sensors. *J. Phys.E* **22**, 173–7

Murphy D, Rolfe P 1980 Aspects of instrument design for impedance imaging. *Clin. Phys. Physical Meas.* **9** (A), 5–14

Xie C G, Plaskowski, A, Beck M S 1989 An eight-electrode capacitance system for two-component flow identification, part 1: tomographic flow imaging. *Proc. IEE Pt. A* **136**(4), 173–83

Xie C G, Plaskowski A, Beck M S 1989 An eight-electrode capacitance system for two-component flow identification, part 2: flow regime identification. *Proc. IEE Pt. A* **136**(4), 184–90

M. S. Beck
[UMIST, Manchester, UK]

A. Plaskowski
[Micromath International, Warsaw, Poland]

Current Measurement

As outlined in the article *Voltage Measurement*, many techniques and commercial instruments are available for the rapid and accurate measurement of voltage. It is not surprising that these should have been adapted for measuring current by using a suitable resistor as a current-to-voltage transducer. The voltage drop generated across the current carrying resistor by the current in it is almost proportional to the current, but is perturbed by spurious emfs due to thermal effects. Moreover, the resistance of the resistor will change due to self-heating, its ambient temperature and its drift. The uncertainties involved in measuring current depend mainly on the resistor and are considerably larger than the uncertainties in the voltage on which they depend.

1. Four-Terminal Resistances

When connections are made in an electrical circuit there is a small contact resistance in series with the component and connecting leads. The size of this resistance, which depends on the contact materials, the area of contact, the cleanness of the surfaces and the pressure exerted is likely to be at least a few tens of microhms. If the usual two-terminal connection is made, the effective value of the resistor is increased by this indeterminate amount. A four-terminal component is designed to overcome the problem by separating the current and potential circuits (see Fig. 1). The four-terminal system

(a) decreases the contact resistances by having only one connection at each terminal; and

(b) eliminates or reduces the effect of contact resistance as follows: (i) resistances R_{cc} appear in the current circuit and merely increase the total voltage necessary to circulate a given current; and (ii) as the voltage measuring circuit presents a very-high input resistance (ideally infinite), the current through R_{cv} is very small and, hence, the voltage drop across R_{cv} is negligible. (For exam-

ple: if $R_s = 1 \, k\Omega$ and $I = 1 \, mA$, a voltage of $1 \, V$ is generated. If the voltage measuring circuit has an input impedance of $1 \, M\Omega$ the current drawn is $1 \, \mu A$ and, assuming that $2R_{cv} = 100 \, \mu\Omega$, the voltage drop across $2R_{cv}$ is $10^{-10} \, V$.) Any current flowing through the voltage measuring circuit subtracts from the current that ought to be flowing through the main resistor R_s, and the voltage drop is therefore lower than it should be.

2. Power Dissipation

The validity of the assumption that the voltage drop across the resistor is a measure of the current flowing depends on the assumed resistance value. This is influenced considerably by internal Joule heating and, to a lesser extent, by variations in the ambient temperature. The problem is in determining the value under working conditions, as not only is the temperature changing due to self-heating, but the effective temperature is not accurately measurable, so that corrections using a temperature coefficient are not fully valid.

The power dissipated can be quite considerable. Assuming that a voltage drop of $1 \, V$ is required, so that an accurate range on the "voltmeter" can be used, then various combinations of current, resistance and power are:

1 mA	10 mA	100 mA	1 A	10 A	100 A	1000 A
1 kΩ	100 Ω	10 Ω	1 Ω	100 mΩ	10 mΩ	1 mΩ
1 mW	10 mW	100 mW	1 W	10 W	100 W	1 kW

Cooling may be difficult. Resistors dissipating little power and of modest accuracy are often convection cooled using the ambient air. Higher power and more accurate resistors may be oil cooled. Sometimes the oil itself is cooled using water in a cooling coil in the oil container. Air blast cooling is sometimes used. With air cooling the temperature measured depends on the

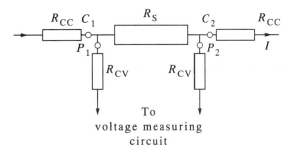

Figure 1
Four-terminal measuring system for current (C) and potential (P)

position of the thermometer and the presence of air currents, and with oil cooling the temperature measured depends on the amount of stirring. In all cases the temperature of the coolant is not that of the resistor, and the difference is influenced by such factors as the time for which the current has been stable and the pattern of any changes in the current.

It may be difficult to reproduce the exact conditions under which the resistor was calibrated, and there may also be some doubt about the validity of the calibration value.

For low values of resistance, construction becomes more troublesome. The effective length becomes shorter, cross-sectional area becomes large, and the resistance becomes increasingly difficult to define. The pattern of the current flow will vary with current level, making the resistance behave nonlinearly. The true current-to-voltage ratio can be determined only by measurement, and there are few laboratories in which currents of more than a few tens of amperes can be measured with sufficient accuracy and stability.

3. Current Transformers and Comparators

Many of the problems outlined above may be considerably reduced if the current can be decreased by an accurately known ratio before measurement. On ac this is achieved traditionally by using current transformers, but recently ac and dc comparators—inductively coupled devices working under zero-flux conditions—have become available.

Using current transformers overcomes the problem of large power losses associated with resistors, and can provide isolation between the current supply and the measuring circuits. Transformers are designed to have standard secondary currents of either 5 A or 1 A for the rated value of primary current.

4. Alternating Current Transformers

Although the construction is different, the basic current–voltage–flux relationships of a power transformer apply equally to a current transformer. It must be remembered that as the primary current is not controlled by the secondary-circuit impedance, the transformer secondary acts as a current source. It is therefore disastrous to open-circuit the secondary if primary current is flowing. The transformer core will be driven hard into saturation, and large peaks of emf produced. Not only is this dangerous but the core will probably need demagnetizing, and will certainly need recalibrating.

Since the power required by the secondary circuit is transferred through the magnetic field, there is a primary exciting current. The existence of this current creates a phase shift through the transformer, and also prevents the ratio between the primary and secondary currents from being constant and independent of

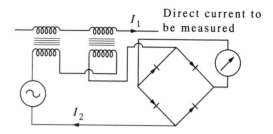

Figure 2
Direct current transformer system

signal level. Even with careful and costly construction, there may be ratio errors greater than $\pm 0.1\%$. The specification of current transformers is outlined in BS 3938 (1982).

5. Direct Current Transformers

These devices are not strictly transformers, but saturable reactors in which an ac signal is controlled by the dc to be measured (see Fig. 2). The total number of ampere turns (ATs) in each core will be the sum of the direct ATs produced by the current I_1 and the alternating ATs produced by the current I_2. The result is a circuit for which the ac impedance, and therefore current, is controlled by the magnetic bias produced by the current to be measured, I_1. The current I_2 is rectified and displayed. The ratio between I_1 and I_2 is only approximately linear, and the accuracy achieved is no greater than $\pm 1\%$.

6. Current Comparators

The errors in current transformers are created by the current required to produce the core flux. If there were no need to transfer power via the magnetic field, there would be no need for a core flux or, consequently, for a magnetizing current. This argument has led to inductively coupled systems being fed with power from both primary and secondary sides operating under zero-flux conditions.

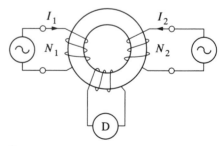

Figure 3
Alternating current comparator system

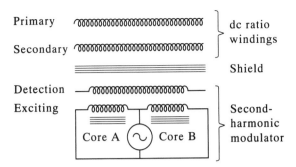

Figure 4
Direct current comparator system

7. Alternating Current Comparators

For no core flux to exist, $N_1 I_1 = N_2 I_2$. This condition can be easily detected by using a third (detection) winding (see Fig. 3). Using high-permeability cores and careful winding and shielding techniques, ratio accuracies better than ± 1 in 10^6 can be achieved.

8. Direct Current Comparators

The statement that for no core flux to exist $N_1 I_1 = N_2 I_2$ is equally true for both ac and dc systems. The difficulty in the latter system is in determining the condition for zero core flux as the presence of a flux cannot be detected in the form of an emf induced in a search coil. Instead a second-harmonic modulation system excited at up to 1 kHz is used.

The basic system comprises two circular, high-permeability cores, each carrying an ac exciting winding being excited so that the ac ATs are acting in

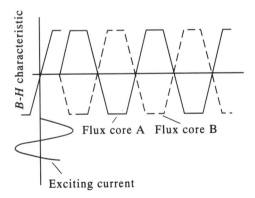

Figure 5
Fluxes induced in transformer cores (*B–H* characteristic) for $N_1 I_1 = N_2 I_2$

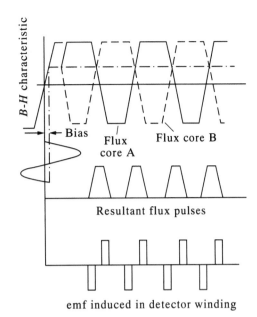

Figure 6
Fluxes induced in transformer cores for $N_1 I_1 \neq N_2 I_2$; emf induced in detector winding

opposite directions round the cores. A third (detection) winding is placed around both cores (see Fig. 4). The cores are of very-high-permeability material and so are susceptible to stray direct fields and to changes in the earth's magnetic field; they are therefore enclosed by a heavy magnetic shield.

The operation of the system is as follows: if the direct ATs produced by the ratio windings are zero, then the cores are magnetically neutral, the alternating fluxes produced are equal and opposite at all times, and the resultant emf induced in the detector winding is zero (see Fig. 5). When $N_1 I_1 \neq N_2 I_2$ the cores are magnetically biased and the alternating fluxes produced are unequal, producing a resultant flux and an induced emf in the detector winding (see Fig. 6). Accuracies up to ± 1 in 10^8 are achieved in commercially available systems. Applications range from the control and measurement of currents up to 10^5 A, to state-of-the-art dc potentiometers and resistance bridges.

See also: Voltage Measurement

P. M. Clifford
[City University, London, UK]

R. Walker
[University of Portsmouth,
Portsmouth, UK]

D

Density and Consistency Measurement

Process variables such as consistency, density and viscosity are often used as intermediate measuring parameters when control of the final product production is being carried out. For example, the control of these variables is, in many cases, necessary to compensate for disturbances caused by the nonhomogeneities of the process materials. The important control variable may be consistency (e.g., in the wood processing industry) or density (in the mineral processing industry). Viscosity control is commonly used in the chemical and food industries as an intermediate control variable of the process. Thus, these aspects are particularly important in the process industry, and it is important to understand their basis.

1. Fundamentals of Consistency and Density Measurement

One key variable in many processes, especially in wood processing, is consistency. However, the problem of measuring and controlling the consistency of pulp has not been fully solved. There are several reasons for this: pulp is not a Newtonian fluid, and the behavior of the pulp thus varies with concentration and different process circumstances. Moreover, the quality of the raw materials affects the measurement and the input to the control system. In the pulp and paper industry, consistency is controlled with particular care at the following points during production (Lundberg 1980):

(a) before grinding and separation;

(b) before bleaching;

(c) before the paper machine, to eliminate variations in the weight of the paper; and

(d) at the headbox of the paper machine.

The pulp consistency is defined as follows:

$$C_s = F/W \times 100\% \qquad (1)$$

where C_s is the consistency (expressed as a percentage), W is the total weight of the pulp and F is the weight of the fibrous material. Although the consistency could be measured directly, by measuring the weight of the dry fibrous material (Lavigne 1972) for control purposes, only indirect measurements are used.

The indirect methods are usually based on the following measurement principles (Walbaum 1980):

(a) pure surface friction (flow resistance in a pipe);

(b) internal friction (e.g., torque of an agitator loosening the stock);

(c) shearing force (e.g., a force generated in a rod shearing the stock); and

(d) combinations of the above methods.

For weak slurries, the following methods have been developed;

(a) absorption of ultrasonic energy;

(b) absorption of light;

(c) polarization phenomena of cellulose fibers; and

(d) matting of fibers on wire.

The choice of method for consistency control depends on the measurement range and the physical dimensions of the process. Discussion of the underlying physical principles occurs elsewhere (e.g., analytical methods, spectroscopy).

Consistency measurement and control are disturbed by variations in several variables, such as flow, temperature, "freeness," pH, filler content and chemicals added, state of the slurry, air content, quality of the fibers, and other stochastic disturbances (pulp slurry is a nonhomogeneous material).

In most cases viscosity variations in the pulp also disturb the measurement of consistency, and thus, its control. As shown, the consistency measurement problem cannot be solved uniquely, which means that many control methods and strategies have been developed. The problems of consistency control are also quite similar in the food industry.

Unlike consistency, density can be measured reliably, accurately, and with reasonably cheap instruments. Density gauges can be located inside the process vessel or in the pipe immediately after the process vessel. No sampling is required and there is no measurement delay. Thus, most density control occurs via simple feedback loops with a proportional–integral–differential (PID) algorithm. Due to these factors, control of density is often used instead of control of variables, which are more difficult to measure. Such variables are consistency, solids content, or the composition of a liquid or gas. The density value is an approximate value for the required variable and this value is often adequate for control. In some cases the value must be calibrated according to laboratory assays to achieve a particular sampling rate.

Density is defined as $p = W/V$, where W is the total

weight of the material and V is total volume of the material. This equation shows how easy it is to measure density by weighing a volume of the material. In addition to being used as a manual sample measurement, this method is also used in a continuous measurement. The fluid to be measured flows through a pipe with flexible ends, and the pipe is weighed. Two other methods commonly used are measurement of the absorption of radiation, or the oscillation frequency of a wire or a tube, which affect the density to be measured.

In mineral processing, density measurement is used to estimate the solids content of a slurry composed of liquid and solid particles. The liquid is mostly water. The solids content (by weight) is defined as follows:

$$S = W_s \times 100(\%) \qquad (2)$$

where S is the solids content of the slurry and W_s is the weight of solids in the slurry.

The equation governing the relationship between solids content and density is:

$$S = \frac{P_s(P - P_l)}{P(P_s - P_l)} \times 100(\%) \qquad (3)$$

where P_l is the density of the liquid and P_s is the specific gravity of the solids. It can be seen that if P_s and P_l are constant, only the density measurement is required to calculate the percentage solids value. In the mineral processing or metal refining industries, density is controlled before certain subprocesses to stabilize the material fed to them. The following subprocesses require a stabilized density or solids content: filters, separators (screens, mechanical classifiers, hydrocyclones), grinding mills and flotation cells.

The density is also of importance to the pumping of material in a controlled system. If the density is too high, a slurry can cause blockages in the pipes or break equipment, whereas, if it is too low, it can cause an increase in the associated pumping and separation costs.

2. Applications of Consistency Control

Pulp consistency control in the paper industry is usually a multistage process in which the control is carried out by diluting the pulp slurry with dilution water (white water). The stock and white water are mixed by a pump, for example, after the latency chest, and the mixing point is a zero capacity process. Moreover, for practical reasons, the consistency measurement point is situated at a distance of several meters after the pump. The pure time delay and control of the zero capacity process make dilution control problematical. For these reasons, and because

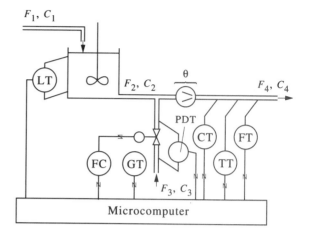

Figure 1
Instrumentation of the consistency control system

of measuring difficulties, no reasonable result can be achieved using feedback control alone. The accuracy of consistency control can, however, be improved by compensating for the effects of most of the variables which cause problems in its measurement. A computer-controlled system, in which the effects of temperature and flow variations on the consistency measurement are compensated for, is illustrated in Fig. 1 (Kortela 1980).

The use of a mathematical model to improve the accuracy of consistency measurement has been described by Duhon (1976). The stability of feedback control of the dilution process can also be increased by including a dead-time control mode and a special method to affect the flow of dilution water in the control strategy of the system (Kortela 1980). Four different in-line consistency control systems have been suggested by Duffy (1975). In these the consistency is measured by determining the flow resistance in the pipe.

It has been shown that an effective consistency control can be obtained by solving the consistency control problem in two stages, both before and after the latency chest (Gavelin and Hesseborn 1976). To attain a constant flow of stock to the screens, good flow control must be carried out.

As mentioned, the pulp flow is nonhomogeneous and of a stochastic nature, and thus, the consistency of the pulp slurry cannot ideally be measured. However, a stochastic model can be formulated for the dilution system, shown in Fig. 1

$$x(t) = \mathbf{A}x(t) + \mathbf{B}u(t) + w(t) \qquad (4)$$

$$y(t) = \mathbf{C}x(t - \theta) + v(t) \qquad (5)$$

where $\mathbf{A}$, $\mathbf{B}$ and $\mathbf{C}$ are matrices, x is the state vector (e.g., the consistencies C, at different points), u is the control vector, w and v are the process and the measurement noise, respectively, and θ is the time delay between the mixing pump and the consistency measurement point. This type of model has been used successfully in consistency estimation and control of a subprocess of a kraft paper machine (see Fjeld 1978).

Another important compensation problem for consistency variations arises in the headbox of the paper machine. Measuring a low consistency, of less than 1%, is not easy and thus in many cases indirect methods to measure and control consistency have been suggested. Different feedback and feedforward control strategies can be used, such as consistency control, using the basis weight valve or the dilution water pump. The effects of consistency variations can also be compensated for by a pressurized headbox.

Some modern control concepts have also been analyzed for this subprocess. Least-squares quadratic nominal control theory has been applied to the control of a pressurized headbox (Fjeld 1978), as well as adaptive multivariable control in a pilot process (Nader *et al.* 1979). A comparison or PID-type and state-space control has been made for a laboratory sealed headbox circuit (Karttunen and Rajala 1977). It has been shown that for this kind of multivariable system state-space control is more successful than PID control.

3. Applications of Density and Viscosity Measurement and Control

In the wood processing industry, density measurement and control is widely used in the pulp mill. Chemical regeneration subprocesses for alkaline "cooking" typically include five different places where density control loops stabilize alkali concentration and dry solids content, optimize energy and chemical costs, and prevent breakage of process equipment (Lavigne 1979). To give an example, the green liquor density is regulated to a uniform alkali concentration by manipulating the amount of weak wash liquor added. The green liquor alkali concentration can only be measured in a laboratory, and thus, density is used as an intermediate measurement for the concentration control loop (Felix 1970). In computer-controlled systems, density is calibrated according to green liquor assays. The dissolving feedback control in Fig. 2 is made in two phases: rough control in the dissolving tank and finer control after the tank or after the green liquor clarifier.

Sedimentation, as applied to mineral processing industries, has been defined (Keane 1979) as the separation of a suspension into clear fluid and a rather dense slurry containing a higher concentration of solid. The continuous sedimentation process is controlled by regulating either the density of the fluid or the slurry, depending on the purpose of the sedi-

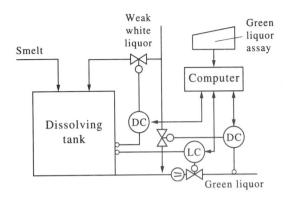

Figure 2
Green liquor concentration control via two density loops

mentation process. A typical example is the thickener preceding the filter. Many filters, like many other subprocesses following the thickener, require a uniform content of solids to operate efficiently. Thickener underflow density control can be implemented with a pinch valve or a variable speed pump. Often the subprocesses following a thickener require, in addition to uniform density, a uniform flow rate. Then density control can be undertaken by water dilution. These density controls in sedimentation are reasonably easy to implement with single feedback loops.

The size separation of particles in industrial processes is carried out by screening or classification. Slurry density has an effect on the performance of screens but it is of much greater importance in classification. Commonly used classifiers are rake classifiers, spiral classifiers and hydrocyclones. In the case of hydrocyclones, density control is even more important. Most mathematical models of the hydrocyclone are very complicated and, thus, not usable for control. Lynch (1977) has developed a regression model which combines the cyclone with the grinding process. This is the normal situation when a cyclone control is implemented.

A one-stage, closed, wet grinding circuit contains a grinding mill, a pump sump, a pump and a hydrocyclone. To stabilize the cyclone overflow particle size, the cyclone feedwater flow rate and solids flow rate have to be controlled. This can be done by density and flow rate control loops. These loops manipulate the dilution water addition and the pump speed. This control strategy only stabilizes fast process disturbances.

The cyclone feed particle size distribution has an effect on the cyclone overflow particle size distribution. When ore feed is increased, the size of particles in the discharge or the mill increases. A coarser particle distribution in the cyclone feed causes more solids in the cyclone underflow, which returns to the mill. This causes the circulating load of solids to increase. The control strategy used is based on the fact that cyclone feed density indicates a circulating load. The flow rate is constant because of the constant speed of the pump and the constant pump sump level. The density, which now indicates the circulating load, is controlled by manipulating the ore feed. This strategy does not stabilize fast process disturbances.

The implementation of control loops in the cyclone overflow is hindered by many different process time constants. The cyclone is fast, the dilution is slower, and the overall grinding circuit with process returns is slowest. A multivariable control has been developed for the few computer-controlled grinding circuits (Jamsa *et al.* 1983). The strong interactions between main control variables are allowed for in these systems.

In processes of the food industry, such as the processing of soups, yeasts, margarine, processed cheese, sugar, and so on, an important variable is the viscosity of the material. Thus, in the food industry, viscosity is often controlled instead of consistency or density. A large number of measuring instruments is available, for measurement of this important parameter, the rotation viscosimeter having proved suitable for industrial plants.

See also: Viscosity Measurement

Bibliography

Duffy G G 1975 The development of two consistency instruments. *Appita* **28, 309**, 15
Duhon R E 1976 Consistency measurement using a mathematical model. *Proc. 76th ISA Ann. Conf.* AVINBP 31(4)
Felix M 1970 Computer control in a kraft mill. *Pap. Eucepa Conf.* **13**, 491
Fjeld M 1978 Application of modern control concepts on a kraft paper machine. *Automatica* **14**, 107; 17
Gavelin G, Hesseborn B 1976 Screening systems improved with consistency and flow control. *Pulp Pap.* **50**, 146–50
Jamsa S-L, Melama H, Penttinen J 1983 Design and experimental evaluation of a multivariable grinding circuit control system. *Proc. 4th IFAC Symp. Automation in Mining, Mineral and Metal Processing.* Pergamon, Oxford, pp. 153–61
Karttunen P, Rajala M 1977 State space control—what it is and what it is able to do (in Finnish). *Sakho Elect. Finl.* **50**(1), 9–14
Keane J M 1979 *Sedimentation: Theory, Equipment and Methods*, Pt. 2, World Mining. Freeman, San Francisco, CA
Kortela U 1980 Pulp consistency measurement and control system. *Proc 4th IFAC Conf. Instrumentation and Automation in the Paper, Rubber, Plastics and Polymerization Industries.* Pergamon, Oxford, pp. 17
Lavigne J R 1972 *An Introduction to Paper Industry Control.* Freeman, San Francisco, CA, pp. 172–93
Lavigne J R 1979 *Instrumentation Applications for the Pulp and Paper Industry.* Freeman, San Francisco, CA pp. 121–5; 131–6
Lundberg K 1980 Consistency and flow measurement in stock lines. *Tappi* **63**(12), 85–90
Lynch A J 1977 *Mineral Crushing and Grinding Circuits.* Elsevier, Amsterdam, pp. 87–114; 204–15
Nader A, Leveau B, Gautier J P, Foulard C, Ramaz A 1979 New developments in multivariable control or paper machine head boxes. *Proc. Int. Symp. Papermachine Headboxes.* pp. 127–49
Walbaum H H 1980 On-line measurements of processes and products in the pulp and paper industry. *Proc. 4th IFAC Conf. Instrumentation and Automation in the Paper, Rubber, Plastics and Polymerization Industries.* Pergamon, Oxford

E. Timonen
[Technical Research Center Finland,
Oulu, Finland]

U. Kortela
[Helsinki University of Technology,
Helsinki, Finland]

K. T. V. Grattan
[City University, London, UK]

Design Principles for Instrument Systems

Instruments form part of a wide class of information machines, and may most usefully be considered as systems. The technology of instruments and instrument systems is based on the principles which govern their design. This article provides a brief survey of general design methodology as applied to instruments and instrumentation, using an approach based on the principles of systems engineering and information technology.

1. Instruments and Instrumentation

Instruments are devices or machines for the acquiring, processing and use (effectuation) of information. They comprise an extensive class, which may be termed information machines, and which embraces devices for measurement, control, communication and computation.

The general principles of instruments and instrumentation are described in the article *Measurement Science*, to which reference should be made. Instruments operate by transforming a physical input into a physical output, maintaining a prescribed functional correspondence between their information-carrying features. This essential function, the realization of a prescribed transformation of physical variables, determines the nature of the construction, operation and design of instruments. The other basic aspect of instruments and instrumentation is that, more than any other class of equipment, they are often combined as systems of distinct functional or constructional components. They are, thus, best described, analyzed and designed by using the methods of systems engineering.

There are very many different kinds of instrument and many applications of instrumentation. Modern instruments are generally complex, with many functional components. They commonly embody, within a single unit, different technologies: electronic, computational, optical, mechanical, physicochemical, and so on. A particular function may generally be realized in many different ways; constructively different, but functionally equivalent. It is therefore impossible to provide a comprehensive descriptive catalog of instruments and instrumentation; the technology is best described by a design methodology for instruments and instrumentation.

2. Design

Design is the transformation of knowledge of a primitive need into knowledge necessary to implement or realize an artifact or system intended to meet that need. For a general review of design methodology the reader is referred to Finkelstein and Finkelstein (1983).

It is generally agreed that design activity may be represented as a process consisting of a sequence of stages starting at the perception of need, and terminating at the communication of the final firm knowledge necessary to implement or realize the artifact or system. Each stage is an elementary design process which starts with an initial concept and refines that concept. The stage is a sequence of steps, subprocesses or operations.

Consider the elementary design process. The initial step of the process is that of definition: collecting and organizing the basic knowledge in the form of the model (see Fig. 1), in an appropriate scheme of representation, of an initial design concept to be refined. The model and related knowledge is, in general, provided by the previous stage of design. Associated with this model is a set of evaluation criteria, or value model, which express the degree to which the concept meets the requirement.

The next step in the process is the generation of a more refined candidate design concept. The generation of candidate design concepts is central to the design process.

The candidate design concept is represented as a configuration, that is, in terms of attributes of its construction. It is next analyzed to determine its performance attributes and evaluated using the value model previously established. A number of such design concepts are normally generated and stored.

The stage terminates in a decision step where the candidate design concept judged to be the most satisfactory is accepted for implementation, realization

Information from preceding stage of process

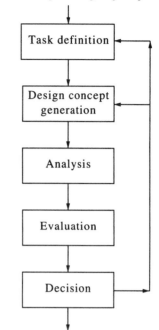

Information to next stage of process

Figure 1
Design process model

or as the initial concept on which the next stage of design will be based.

If none of the candidate design concepts are acceptable it may be necessary to return to an earlier stage in the process, for example, to alter the value criteria, to generate more candidate design concepts or to analyze the existing ones in more detail.

Viewed overall, design proceeds from a global view of the artifact or system to progressively more localized considerations, and from abstract and fluid descriptions to concrete and firm ones.

In instrument design, the design stages typically start with a definition of the overall system which arises from the original requirement for the instrument or instrumentation system. This establishes the main features of the system as a set of interconnected subsystems and also fixes the nature and specifications of the subsystems. In turn, the subsystem design stage uses this information to establish the main features of the subsystem and the nature and the specifications of its simpler components. This sequence of stages proceeds to the design of elementary components. If at any stage it appears that it is not possible to meet the specification generated by the preceding stage, it is necessary to return to an appropriate point of that latter design stage.

3. A Systems Approach to Instrument Design

As mentioned earlier, the design of instruments and instrumentation stems from considering them as systems. The model of the design process described in Sect. 2 is founded on the systems approach.

The following principles or aspects underlie the design of instruments and instrumentation.

First, the instrument or instrumentation is decomposed into simpler constructional or functional components. This enables simplification of complex systems; it also demonstrates that many diverse systems can be built up from a much smaller variety of elementary building block components and structural schemes.

The second principle is that of abstraction, that is, consideration of only an essential restricted set of relevant aspects of the system. This reduces complexity and gives access to analytical tools. For instrument and instrumentation design in particular, the system and its components are considered in terms of function (that is, what they do), rather than construction (what they are). This promotes the consideration of alternative realizations of a particular function.

The third aspect of the systems approach is the use of mathematical models, in particular, those of signal theory, dynamic systems and control theory.

The fourth principle is the application of formal evaluation and decision procedures.

The final, but perhaps the most important, principle is that the system under design must be considered as part of a larger supersystem of which it is a component. Interaction with the environment must also be considered; for example, in an instrument system, not only the instrument itself, but also the system from which it acquires information and the subsequent output processing system must be taken into account.

4. Design Concepts

4.1 General

The core of the design process is the generation of design concepts; its basis is decomposition and abstraction.

When an instrumentation system under design is decomposed into a set of functional components, the main components which handle knowledge and information can be realized by the selection of available functional blocks and the use of essentially algorithmic methods of design. The design methodology of those parts of the system is not specific to instrumentation.

However, in the case of sensors and the interaction between the sensors and the object under observation it is commonly necessary to find an appropriate working principle and an embodiment of the principle.

There are a number of basic methods of determining such working principles and finding their embodiments. They will be discussed in this section. They are useful heuristics rather than algorithmic procedures.

4.2 Established Design Concepts

The obvious starting point for a design is provided by established design concepts.

A formal method of the utilization of established design concepts is based on the use of an appropriately organized knowledge base, constituting an appropriately classified systematic design concept catalog, together with a system of purposeful searching of the knowledge base.

Informally, one may use the literature describing proven design solutions, commercial catalogs, competitive products and the like. In either case systematically organized knowledge is an important prerequisite to successful design activity.

4.3 Physical Effects and Design Concept

Design concepts for the realization of an instrument component which maintains a functional relation between physical variables may be deduced from the relevant physical effects.

One approach is the systematic listing of the physical laws which involve a physical variable of interest, or which relate two physical variables which are to act as the input and output of a device. For example, if we require to generate concepts for sensing electric current we may list the physical laws involving electric current, each of which may suggest a design concept. One law is that of force between current-carrying conductors, which suggests that we can sense current by sensing the force between two coils, as in a dynamometer instrument. The law of force on a current-carrying conductor in a magnetic field suggests the principle of the moving coil ammeter. Ohm's law, relating voltage and current, suggests sensing current by sensing the voltage across a standard resistor. Joule's law of heat generated by a current in a resistor suggests the principle of the hot-wire ammeter. This list is illustrative but not exhaustive.

The second approach is the examination of a single law involving the variables in question which then suggests a design concept. Consider the generation of design concepts for instruments to measure the viscosity η. The relevant law is $F = -\eta A(\mathrm{d}v/\mathrm{d}x)$, where F is the force acting on an area A normal to the velocity gradient $\mathrm{d}v/\mathrm{d}x$ of a fluid. This suggests two viscometer principles: generating a velocity gradient in a fluid and sensing the resultant shearing force, or generating a shearing force and sensing the resultant velocity gradient. From this we can proceed sequentially to elaborate variants of the principles. Taking two solid bodies with fluid between them, we may have the two bodies fixed with fluid moving or one body fixed with the other body moving, or all may move. The mode of motion may be translation or rotation, and the motion may be steady, harmonic or transient. We may then derive the idea of two coaxial cylinders with fluid between them. We may drive one cylinder (inner or outer) at constant angular velocity and measure the torque on either moving cylinder (rotor) or stationary cylinder (stator). Alternatively, we may drive the rotor at constant torque and measure the resultant velocity of stator or rotor. This systematic process then suggests oscillatory motion of the rotor. A complete range of variant design concepts is thus generated.

4.4 Analogies

A powerful method of generating design concepts is the consideration of analogies. When considered in terms of an abstract description, a system may be seen to be essentially similar to another and aspects of the second may suggest design solutions for the first.

Physical analogies, such as those between electrical, mechanical, fluid-flow and thermal systems, are often useful. Thus, in seeking methods of measuring heat flux it may be useful to examine methods of measuring electrical current density.

At its most elementary, design solutions for an artifact may suggest solutions for another in the same domain, but handling different values of physical variables.

One important source of analogies for instrumentation and information processing are living organisms, especially the animal system of receptors, nerves, brain and effectors.

4.5 Divergent or Lateral Generation of Concepts

Design concepts can be generated by creative thinking in which ideas diverge and move laterally. Some of the basic approaches to the promotion of creativity can be listed. Design concepts are considered in abstract form, neglecting irrelevant detail until the latest possible stage. Every attempt is made in the first instance not to consider established practice, authority or apparent obstacles, in order not to allow promising concepts to be abandoned prematurely. Finally, for the same reason, idea generation and evaluation are separated.

One can consider this process as a cooperative functioning through a blackboard, on which partial solutions are placed, of a number of separate knowledge bases, encompassing different knowledge domains.

4.6 Transforming Concepts

Some existing design concept usually forms the first step in the process of generating a design. This may then act as a starting point for generating new concepts by transformation. Starting with a particular instrument design one may proceed to derive new concepts by the following steps.

(a) *Functional description and decomposition.* The instrument is considered as a system of components and each component described in terms of the functions it performs.

(b) *Systematic examination.* The functional description is the basis of a systematic examination of the design. Each component is examined to determine the following.

(i) What secondary functions, ancillary or incidental to the main one, are performed? For example, analyzing the functions of the sheath of a sheathed thermocouple reveals that it smoothes out spatial and temporal temperature fluctuations as well as protecting the thermocouple wire.

(ii) Why is the function being provided? It is necessary to determine what contribution each component makes to the total system function.

(iii) What principle underlies the operation of the component? The operation is described in abstract terms revealing the essential physical principle; instead of saying that the component is a coil moving between the poles of a magnet, one describes it as a conductor moving in a magnetic field, leading to consideration of other forms of conductor and other forms of magnetic field generation.

(iv) What are the fundamental limitations? The basic limitations of the performance of a component imposed either by the incidentals of a particular realization or by the essential principle of operation are examined.

(c) *Systematic variation.* Having decomposed a design concept into a system of components, and examined the components systematically, new designs are generated by systematic variation.

(i) The functional structure may be varied. The system to be designed can be examined to see whether certain functions may be transferred to other components of the larger system of which it forms a part. Thus, in the design of a sensor to work in a computer based system, filtering may be carried out either in the sensor or in the computer. We may consider how a number of separate functions in a structure may be performed by a single component, or alternatively, how a multiplicity of functions, performed by a single component, may each be realized by separate components. The sequence of functions may also be varied. For example, in the sequence sensing, transmission, amplification and display, amplification could be performed before transmission.

(ii) The realization of functions may be varied. Each function must thus be considered to see whether there are alternative forms of realization, either different forms of component may be used which operate on the same principle, or different principles of operation could be considered for the performance of the same function.

(d) *Systematic negation.* A basic form of variation is that of systematic negation, examining, in the case of each function, the possibility of the following.

    (i) Component removal—that is, the extent to which a particular function is essential and the extent to which the elimination of the function affects performance.

    (ii) Component reversal—for example, when an instrument operates by the motion of a coil relative to a fixed magnet one could consider a fixed coil and a moving magnet.

(e) *Systematic analysis of characteristics.* Each attribute of a design is listed to consider the possibility of:

    (i) enhancing desirable characteristics; or

    (ii) removing limitations.

5. Concept Analysis

As stated earlier, a design concept takes the form of a model of a configuration. The concept must be analyzed to determine the performance attributes of the object under design. The form of models for instrument components and systems is discussed in the articles *Instruments: Models and Characteristics* and *Instruments: Performance Characteristics*. Computer methods are used for the analysis of these models.

6. Evaluation

The value model of the object under design relates the attributes of the design concept to the requirements. The evaluation step in the design process calculates a utility of the concept with respect to requirements.

7. Decision

The decision step which terminates the design process is based on the use of decision theory which governs the choice of alternatives based on the consideration of their utility and possible alternative states of the world.

See also: Instrument Systems: General Requirements

Bibliography

Finkelstein L, Finkelstein A C W 1983 Review of design methodology. *Proc IEE* **130**, 213–21
Finkelstein L, Abdullah F, Hill W J 1992 State and prospects of the computer aided design of instruments. In: Steussloff H, Polke M (eds.) Integration of design implementation and application in measurement automation and control. *Proc. INTERKAMA Congr.* Oldenbourg, Munich, Germany

L. Finkelstein
[City University, London, UK]

A. C. W. Finkelstein
[Imperial College of Science, Technology and Medicine, London, UK]

Digital Instruments

Digital instruments range from a simple digital timer to a complex full data logger in a chemical plant. They operate over a wide range of speeds from as slow as a digital voltmeter, reading 10 times per second, to as fast as a logic analyzer, capturing data 2×10^9 times per second. Despite this enormous variety, the basic form and function of all digital instruments is the same. Traditional analog and continuous-function components are replaced by digital and processing components. Once a processor is in the path from variable to display, the problem of sampled data occurs. The instrument is no longer continuous in operation. It samples its inputs prior to processing and displaying results. Now that integrated circuits are dense enough to hold the necessary transistors, nearly all functions of any instrument can be integrated and all modern digital instruments are based on microprocessor technology.

The digital instrument connnects to the variable(s) being measured. This is either direct or is provided by transducers, which then link through signal conditioning circuits, conversion and sampling to provide data which may be further digitally processed prior to display, storage or use as control information. All steps are concerned with obtaining the values needed from the wide, analog world into the narrow, digital, binary form essential for digital processing.

1. Transducers

Some digital instruments, such as voltmeters and oscilloscopes, can connect directly to electrical signals but most real-world variables are not electrical and are also analog. An input transducer, sometimes called a sensor, responds to a physical stimulus, producing an electrical signal with a known relationship to it, which can be acquired and subsequently processed. The range of electrical signal which is output from transducers varies and despite many attempts there is only limited standardization. Integrated circuit complexity has increased to the point where all the analog circuit stages can be included with transducers and digital

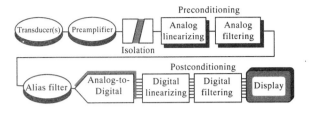

Figure 1
Components of a digital instrument

processing elements in a single chip. Such intelligent transducers provide a route to the standardization of instruments.

A transducer is the one part of a system which has to be trusted, as it is often not possible to self-test, and only by replication can there be any check on correct operation. The variety of transducers is staggering but they can be categorized by the variable being measured, the type and measurement technique of the transducer and its accuracy. Input variables can be grouped by the units in which they are measured. The Système International (SI) shows the extent of this list and two sorts of variable: those for which an absolute measure exists, the basic and derived units of the system, and dimensionless quantities such as chemical content which are measured as ratios.

Active type transducers require no external power source to operate. For example, a current is generated by a flowmeter formed from a paddle wheel incorporating a fixed magnet which moves past a fixed coil. Passive transducers require an external current source, such as a potentiometer to measure rotational position, and may therefore need additional wiring. The measurement techniques define the major categories of transducer: direct, indirect or system; analog or digital; incremental or absolute.

Direct transducers convert the physical stimulus to an electrical signal. Indirect transducers require an additional stage where the input variable is converted to an intermediate form which is then converted to the electrical signal related to the original variable. Pressure cannot be directly converted into an electrical signal, but is converted to an intermediate displacement. The extra stage can give additional error. The system or servo transducer technique requires a perturbation of the system to detect a null. Consider measuring the liquid level in a tank with only a single light source and sensor. A tube allows a sample of the liquid to be admitted, filling it up to the same level as the tank, a valve releases the sample, and the time until the sensor indicates an empty tube is translated to the depth.

Analog transducers give continuous values which require a conversion stage to give discrete, digital values. Transducers which can provide digital encoded values at their output, for example, optical disk encoders, are obviously preferable.

The final division is into absolute or incremental encoding. A transducer which can be read in isolation and yet gives the variables' value is an absolute encoder. On some roads, stones are placed every mile from the center of London. Each stone is marked (in Roman numerals) with its distance from the center, and arriving at a stone with no other information tells you where you are. When motorways were built a post was still put every mile but they were not labelled. Starting from the beginning of the road and counting the markers the same information and accuracy is achieved. This is an incremental encoding scheme. Only a single input bit is needed, or two bits if direction is required, rising to three if a starting point is included. This compares with the full number of bits for the required range and resolution of an absolute encoder.

2. Digital Data Capture

If the input signals to the instrument are already electrical and digital, such as for a logic analyzer, then it is only necessary to consider the threshold value or values which distinguish zero from one, and form the noise margin in between, and the time base for data capture. A choice of free-running instrument clock (asynchronous) or a clock provided with the data (synchronous) must be made. If data is captured asynchronously the time base must be set to give the required resolution, or transitions will be missed. Such missed events and cases where transitions are not made cleanly from one state to the other, can be marked on the display as glitches. Threshold and time base adjustments are then made to resolve these events.

3. Transducer Accuracy

The ability of a transducer to measure the desired value of the variable is specified by its threshold, resolution and full range. The threshold is the minimum measurable value, that is, the lowest point at which a reading can be obtained. The resolution is the minimum measurable value difference, which can be thought of as the quantization step of the transducer even if it is analog, and is the first determining factor of the accuracy. The full range is the maximum measurable value, that is, the limit above which readings cannot be obtained.

Linearity and monotonicity define any nonuniform relationship between the input variable and the output electrical signal. A transducer is monotonic if, for all increases in the input variable, the output either remains the same or increases, but does not decrease. The same applies in the reverse direction. A nonmonotonic relationship with a "dip" in the curve would fail if its algorithm was looking for the direction of changes on which to take some action. Linearity is measured as the maximum deviation of the actual output signal from a straight line joining the end points of the range, and many transducers, for example thermocouples, are nonlinear in operation. Further errors which relate to the longer term ability of a transducer to give consistent results are hysteresis, stability and repeatability. If readings are different when taken with an increasing or decreasing variable then the difference is hysteresis or backlash. It is simple to correct in a digital instrument. The repeatability of a transducer is an error between readings of the same value of input, taken some small time apart. The stability is the repeatability of readings over a long period of time. Drift in output for a given input can be caused by many physical processes; variations in temperature and power supply for active transducers are common culprits. Only calibration, once in operation, can overcome these restrictions in a design.

The final set of errors are all linked to time. Transducers all have some mass or energy requirement and hence inertia, which gives rise to an initial delay, followed by a period of slewing, defined by a maximum slew rate, and can cause an overshoot beyond the correct output value when slewing fast. The alternative of an asymptotic approach to the value is usually slower to settle. Once the value is within the resolution of the transducer the combined response time has passed and the transducer gives the correct output.

4. Signal Conditioning

Prior to the availability of microprocessors, most conditioning operations in instruments were performed by analog electronic circuits before any conversion. Inputs from transducers, including the effects of sampling, need some processing before being used or displayed. Neither a steel mill nor the human brain use digital signals between 0 V and 3 V as exist in a microprocessor. This signal conditioning splits into those parts which must be done before the signal is converted to digital form or are carried out with analog electronic circuits (preconditioning), and those which are done by digital programs and storage (postconditioning). The intervening phase performs analog-to-digital (A/D) conversion if necessary. Preconditioning circuits naturally operate in parallel with the processor, not using any of its time, but are less flexible than postconditioning by program.

5. Isolation

It is fairly common that a processor and its transducers are unable to share a common earth or zero point. An obvious example would be monitoring fluctuations on overhead power supply cables. A transducer would be many (or many hundreds of) kilovolts above the ground to which the microprocessor, and its console presenting results to a human operator, will be connected. An isolator is essential. Metal-cored transformers are low-frequency devices, limited to the audio range. Ferrite cores give much higher frequency performance, typically 3 kHz–1 MHz, with isolation of ±2.5 kV achieved by the separation of the primary and secondary coils. However, no transformer can pass direct current, and therefore low-frequency signals cannot be isolated using one. The discovery of light-emitting diodes (LEDs) and complementary phototransistors made optical isolation the dominant technique. Gallium arsenide infrared-emitting diodes (at about 850 nm) with silicon phototransistors as the initial detector allow dc and low-frequency isolation and pass signals with submicrosecond rise times. Isolation is ±2.5 kV when fabricated in a six-pin dual in-line integrated circuit. Higher isolation is achieved by separating the source and detector by air or a fiber-optic light pipe to increase the dielectric strength. For digital signals the speed may be increased above 20 MHz by using a gallium arsenide phosphide emitter, low currents in the detecting transistor and a Schmidt-triggered direct logic gate in the same integrated circuit.

Optical isolators convey many other benefits, notably protecting the subsequent instrument circuits from damaging voltage transients, surges and malfunctions which may occur in harsh industrial environments, permitting widely varying input voltages via resistor and LED connection, and preventing ground loops occurring due to multiple-point earthing.

6. Preamplification (Preranging)

If an input voltage must be cut down to achieve the correct range then a simple two-resistor potential divider gives the desired attenuation. This is cheap, simple and robust. If gain, conversion between current and voltage or addition or removal of bias (offset) is required, operational amplifiers are used. These are integrated circuit, multistage, dc coupled, silicon transistor amplifiers designed to have high gain ($>10^6$), high input impedance (hence, input current $<10^{-9}$ A), low output impedance, wide frequency response, including dc, low drift and low thermal coefficient. A commercial operational amplifier (op amp) usually has two inputs which have matched gain, with one inverting and the other not, and are summed by the operation of a differential pair input. This is followed by high-gain amplifiers and a power output driver. The most common arrangement for the complete unit is with an input resistor R_1, and another resistor R in a feedback path around the amplifier. It is easy to show that output and input voltages are related by:

$$V_0 = -V_i\left(\frac{R}{R_1}\right)$$

Accurate amplification is arranged simply by the ratio of the two resistors. For instrumentation there are a number of useful twists to the basic arrangements. Current-to-voltage conversion is easy and a unit gain buffer is made by making $R = R_1$. It is often necessary to add or remove a fixed offset voltage to the input. A second input resistor (R_B) is connected to the common junction. A fixed-bias voltage is connected to the other end of the resistor, then:

$$V_0 = -V_i \left(\frac{R}{R_1} \right) - V_s \left(\frac{R}{R_B} \right)$$

V_s can be positive or negative, and the amount of bias is set by R_B. A sum of two inputs would be produced if V_s was variable. Substituting a capacitor for the feedback resistor R produces an integrator with transfer function:

$$V_0 = \frac{1}{RC} \int_0^t V_i \, dt$$

7. Analog Linearizing

Straightening the nonlinear responses of otherwise satisfactory transducers requires an analog circuit with the inverse characteristic of the transducer. These are not easy to obtain continuously, except in a few cases such as logarithmic functions, but reasonable approximations can be fabricated using steps of straight line segments of the form $Y = mX + c$. There are many circuits which can provide straight-line fitted curves, such as using Zener diodes which do not conduct when reverse biased until a chosen voltage is reached when they avalanche and conduct, connecting a resistor in parallel with any others already connected to set the gain m for the segment starting at the chosen c.

8. Analog Filtering

There are many cases when filtering is desirable. Only a small section of frequencies may be of interest for particular measurements such as vibration in a structure. The removal of unwanted frequencies, particularly those of superimposed noise, for example, 50(60) Hz and 100(120) Hz mains hum may increase the accuracy of measurement. A known frequency distortion may be present in the received signal, and the application of a suitable filter can correct this. Analog filter design of both passive and active types is well established and, though the parallelism argument still holds as for all preconditioning, the flexibility and efficiency of digital filtering techniques is continuously making them more prevalent in instrumentation applications.

9. Sampled Inputs

Input data is not continuously taken because the processor also performs calculations, takes data from other channels and outputs display or control signals. To determine the effect of this sampling, consider what is the actual information to be input, and how often must the input be sampled. Also, what is the effect of the finite time needed to take a sample, and what noise is superimposed on the information signal? Considering the errors which could be introduced by these points, additional circuitry is necessary to constrain the input signal to retain accuracy.

A sine wave $a = A \sin(2\pi ft + \phi)$ has three variables: amplitude, frequency and phase. When it is sampled only the amplitude is taken; the frequency and phase are determined inherently by the times at which the amplitude samples are taken. Any real signal is more complex but, if it is a continuous wave, it can be analyzed as a sum of various individual sine waves. There is a relatively long period (sample time T) in between actually taking the samples for a relatively short period (aperture time τ). These are the two most important variables in any sampled system because the sample time T provides a limit on the frequency of any signal which is to be sampled, and the aperture time τ provides a limit to the accuracy of measurement.

The criterion, discovered by Nyquist, providing the limiting sample time is:

"if a signal has its highest frequency f_{max} Hz, then it is necessary to take more than $2f_{max}$ samples per second to enable the signal to be reconstructed."

This means all data from the wave has been acquired. A special case shows the limit. If a wave is sampled exactly twice per cycle ($2f_{max}$), then the samples could all occur at the zero crossing points and therefore no data would be recorded. If the signal is sampled below this rate then another wave at a lower frequency than the original can be fitted to the samples. This is called an alias and appears as a corruption to the original, as noise which cannot subsequently be removed if it overlaps any frequencies in the original. The alias signal frequency is the difference between the actual input signal frequency and some integer multiple, or harmonic of the sampling frequency.

As long as $1/T > 2f_{max}$ then the terms are all separate in the frequency domain and the sampling rate is adequate. If, however $1/T \leq 2f_{max}$, then frequency components from $1/2T$ to f_{max} are folded back on top of those from $1/T - f_{max}$ to $1/2T$. This is called frequency folding and is responsible for the signal errors, and produces the alias.

The obvious solution is to sample at a frequency greater than the Nyquist criterion, but there is a problem: noise. If, for example, sampling at 20% above the Nyquist point was carried out, and there was some superimposed noise at $3f/2$ on the sampled signal then, due to the effect explained, the noise will be folded about the sampling frequency and therefore distort the original signal at $(2f + 20\%) - 3f/2 = 7f/10$. The higher the frequency of the noise the greater the problem. An extremely simplistic view of sampling is what happens to the wheels of a stagecoach in an old cowboy movie. They are filmed (sampled) at 24 frames per second. As the stage pulls out everything appears normal, but as it speeds up the wheels appear to reverse and rotate backwards. The wheels are moving faster than half the sampling rate of 24 frames per second, an alias is seen, and all information on the position of the spokes, which are the highest frequency, is corrupted.

10. Alias Filter

The only satisfactory solution is a two-stage procedure. First, use an adequate sampling frequency, and second, use an alias filter to remove higher-frequency components from the sampled signal. $2.5f_{max}$ is the minimum rate commonly used in digital instruments for continuous waves. There is a loophole for using lower sampling rates in special cases. If, after folding, the folded signals do not overlap the original signal, because of gaps in the spectrum of the original, then the folded parts can be removed and no error results. Repetitions of the same signal may be sampled over a number of cycles.

11. Sampling Aperture

The frequency and phase of an input signal are inherently measured by knowing when the samples are taken. The duration of taking a sample acts either as an averager over the aperture time, or as an uncertainty in when the sample was taken, depending on the electronics of the instrument. If it is simply averaging then a loss of amplitude accuracy results in a more limited resolution. Aperture uncertainty affects the frequency and amplitude accuracies, and therefore has a complex effect on the signal.

Simple rules can be derived for the worst case error to an individual sample and for average error, noting that the worst case can never be met on successive samples. The worst case occurs only to the highest frequency signal and is reduced by the effects of other frequencies in the input signal. If a single sine wave of the highest frequency in the signal is taken and its worst rate of change is found, then this can be used to find the error.

The worst case error is given by:

$$\frac{\Delta V}{V} = 2\pi \tau f_{max}$$

and the worst case aperture time τ is given by:

$$\tau \cong \frac{1.6^*(\text{error})}{f_{max}}$$

If each individual sample must be guaranteed to have a stated accuracy then the worst case should be used. Under any other circumstances, such as averaging algorithms, use of digital filter algorithms and so on, then the normal calculation should be used. This average value is derived by considering the effect of the uncertainty over the whole frequency range of the input signal.

The average error is given by:

$$\frac{\delta V}{V} = 1 - \left| \frac{\sin(\pi \tau f)}{(\pi \tau f)} \right|$$

and the average case aperture is given by:

$$\tau \cong \frac{0.78^*\sqrt{\text{error}}}{f_{max}}$$

Thus, the aperture time τ is the time required to sample a signal until its amplitude is stored and is equivalent to an error in the measured amplitude of the signal. Aperture uncertainty is the time over which it is unknown when the sample was taken and is equivalent to a frequency-dependent error in the sample. The only cure for sampling high-frequency signals is to use some additional electronics in the digital instrument to shorten these two times.

For a simplistic view of τ limitations consider taking a picture of a racing car with a fixed camera using an exposure time of 1/25th s. If the picture is taken at the correct time the car will be blurred (averaging due to long aperture). If the picture cannot be taken at a fixed time then the car may only partially appear or may be missed altogether (aperture uncertainty time). If the shutter speed is increased to 1/1000th s and opened at the right time then a perfect picture will result as the aperture time is now correct for the maximum frequency.

12. Sampling and Holding

A short aperture time, during which the signal value must be determined, is followed by a long sample time between samples. With a mechanism to capture and store an analog value quickly, then all the sample time is available for conversion. A sample/hold circuit is used when the aperture time must be shorter than a converter can achieve, if slow conversion is required to lower the cost, or if the signal is not continuously available, perhaps due to multiplexing inputs.

Sample/holds rely on the ability of a capacitor C to hold a charge over a period of time. This allows a voltage to be captured as $V = Q/C$. The capacitor is then disconnected to hold the voltage. A fast field-effect transistor (FET) switch is needed for sampling high-frequency signals. It has very high OFF resistance and reasonably low ON resistance. A logic signal to its

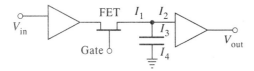

Figure 2
Sample and hold circuit

gate switches the transistor on or off. To ensure minimal disturbance to the charge in the capacitor when it is holding and checked, an operational amplifier with unit gain and a very high input impedance buffers the output, and to minimize the disturbance to the input it is also similarly buffered.

For the charge in the capacitor to hold the voltage sampled accurately, I_4 (the inherent leakage of the capacitor due to noninfinite resistance between the plates) must equal the sum of I_1 (the leakage current through the OFF field effect transistor) and I_2 (the leakage current of the buffer amplifier). In practice it does not, and the capacitor voltage droops. Of course I_3 could be greater than I_4 when the voltage on the capacitor will increase; this causes a negative droop.

While the switch is closed the input voltage is connected (via the ON resistance of the FET) to the capacitor, and its voltage follows the input. When the switch is opened the charged capacitor is isolated and the signal held. The time this takes is the aperture time of the sample/hold and the uncertainty in its occurrence is the aperture uncertainty time. Later the gate signal closes the switch to resample the input. The voltage on the capacitor has to alter at some maximum slew rate to catch up to the input signal. It may overshoot in a similar fashion to the transducer response discussed previously. The total time for equalizing the voltages to within the desired accuracy is called the acquisition time.

Therefore a sample/hold circuit must have high accuracy, probably with unit gain, high input impedance to minimize the effect on the source, a short aperture time, and a long hold time with low droop rate. In addition, a short acquisition time needs a high slew rate.

13. Multiplexing

Sample/holds naturally fit between any preconditioning and the analog-to-digital converter. Expensive analog-to-digital converters give a strong argument for sharing one converter across many inputs. The same argument applies in low-cost, low-frequency systems where a microprocessor carries out all the control operations to perform conversion.

An analog multiplexer selects which signal is connected to the sample/hold, or directly to the converter. It consists of the same address inputs and decoding circuits as are needed for a digital selector or multiplexer. Instead of logic gates enabling the chosen input on the select signal, an analog multiplexer uses a FET with the select signal connected to its gate input. The analog input and output are connected to the source and drain of the FET. This is similar to the use made in the sample/hold circuit, and often the multiplexer and sample/hold are combined by having a single hold part and multiple switches to connect the various inputs for sampling.

14. Analog-to-Digital Conversion (A/D)

Conversion from the bounded infinite number of analog values of the unknown input to the 2^n digital values can be done one level at a time, one bit at a time, or the whole word at a time. The fastest way is a parallel or flash converter. This has one comparator and one voltage reference for each possible digital number—for only six-bit accuracy 63 comparators and 63 reference voltages are required. This limits the word length possible to that which can be packed onto a chip. A resistor chain develops all the intermediate references from a master voltage reference. Comparator outputs are connected to an encoder to give the binary value. Conversion times of less than 10 ns are available with 10-bit accuracy. For greater accuracy, serial techniques use a digital-to-analog converter (D/A) with a controller to generate test values for comparison with the unknown input.

General-purpose, medium-speed converters use successive approximation. Conversion starts by comparing the unknown input with an output from the D/A equal to half the maximum range of possible inputs. If the comparator indicates that the unknown input is greater than the D/A output then the next test is against three-quarters of the maximum. If the test indicates it is not greater then the next test is against one-quarter of the maximum. For each bit in the word the test is on half of the range of the previous test. Accuracy is limited by the accuracy of the D/A (usually its R–$2R$ resistor chain) and only n steps, of one bit at a time, are taken for an n-bit word. It is applicable to both multiplexed and nonmultiplexed inputs directly.

The individual step of one bit change for testing is very fast, it is just the number of steps which makes the converter slow. Arranging to require only one step to find the unknown input gives conversion only marginally slower than a flash converter, but with greater accuracy.

Instead of starting with half maximum voltage, we start from the previous value, changing only the least significant bit(s) to give tracking (or servo) converters. The D/A output must be able to step up or down with the direction determined by the comparator. If the input signal slews faster than the tracking converter can step, then the signal is lost, and then, instead of stepping the least significant bit, the control changes to step in increments of twice the size (the next least significant bit). If this fails then the next bit up is stepped until the signal is reacquired when steps revert down to the smallest increment. The most significant bits of the value in the register are always correct. Its accuracy (in bits) reduces when the input slews too fast.

If a multiplexed set of inputs are to be connected to a single converter then tracking will not work, unless the last value for each channel can be stored and reloaded, hence another approach is required.

15. Intermediate Variable Analog-to-Digital Converters

The problem with the previous converters is that they rely on the accuracy of many components for their own resolution and accuracy. Flash converters need 2^n accurate reference voltages and comparators, and the feedback types require a digital-to-analog converter based on a resistor ladder. If lower-speed operation is acceptable then cheaper, very-high-accuracy approaches are available. The unknown voltage is converted into an unknown time or frequency which is then measured. The most common conversion in slow instruments is the dual-slope technique.

A switch connects either the unknown input (V_{in}) or a fixed negative reference voltage ($-V_{ref}$) to an integrator, the output of which goes to a single comparator. The comparison is with zero volts. Thus, two ramps can be created, one determined by integrating the unknown input, the other by integrating the fixed reference.

Conversion starts with the switch set to connect $-V_{ref}$ to the integrator, thus causing it to ramp down to the starting point. The switch then connects V_{in} to the integrator, which ramps up at a rate determined by the unknown voltage. When the comparator indicates the output of the integrator increasing through zero, a counter is started. After a fixed count T_1 the switch is set back to the negative reference, the counter is cleared and counting restarts. When the comparator again indicates the integrator output passing zero, following the downward ramp towards V_{ref}, the counter is stopped and a count T_2 is noted. The counts of the duration of the up and down ramps are now in proportion to the respective voltages V_{in} and V_{ref}. Simple geometry gives:

$$V_{in} = \frac{T_1}{T_2} V_{ref}$$

Component value drift will be very small during a conversion, therefore the ratio of the two times is extremely accurate. The voltage reference determines the resolution of the converter. The time period of the integration process T_1 is usually chosen to give specific

noise rejection, commonly against mains pick-up. This technique is cheap, if slow, and forms the heart of almost every digital multimeter (DVM or DMM) yielding 20-bit accuracy.

As with successive approximation, dual-slope reconverts the whole signal value each time. Other methods such as delta–sigma conversion can assemble an n-bit output without having to wait for a counting period of 2^n. They employ a single-bit D/A, giving a pulse train to balance the input charge for comparison (instead of the single reference), an additional integrator stage or stages, and a more complex counting algorithm to provide a rolling, weighted average of the comparator inputs.

16. Converter Accuracy

The primary error in an analog-to-digital conversion is the quantization error. An A/D must take an unknown input voltage and approximate to it the nearest of a finite set of digital values. The quantization error is the smallest increment $\pm\delta V = V/2r^n$ of the input voltage V, to which the digital output can be approximated where r is the radix and n is the number of digits. For example, if $r = 2$ (binary) and $n = 8$ then the quantization error is $\pm 1/512$, that is, $\pm 0.2\%$. The signal-to-noise ratio in linear converters varies with the signal size as $(6.02n + 10.8)$ dB.

This is the argument for using a nonlinear conversion if a uniform error is required across the whole range. For this purpose the D/A in a successive approximation converter would be logarithmically weighted. The dynamic range of a converter is the ratio of full range to resolution, that is, $6.02n$. The remaining errors are monotonicity, linearity and differential linearity, offset and gain errors, and temperature coefficients.

Feedback converters based on D/As exhibit the worst differential linearity errors at the major transitions, particularly at one-half of full range when all bits are required to change state (011111 to 100000 or vice versa). The dual-slope technique exhibits very small differential linearity error, but shows ordinary linearity errors due to limitations of the integrator circuit.

17. Postconditioning

Microprocessors are cheap and are continuously becoming faster and functionally more powerful. However, by the early 1990s postconditioning still could not cope with the highest frequencies, due to the complexity and hence speed limitations of some of the algorithms needed. The great advantage of any postconditioning is that it is completely flexible and adaptive if desired, as all parameters are set by stored constants, and functions by choice of stored (in read-only memory) algorithm.

18. Digital Ranging

The accuracy of the digital data cannot be improved, but it can be biased by an additive constant and ranged or scaled by multiplying by a suitable factor. The total range of numbers which can be stored in either fixed or floating point formats must be checked. Digital ranging is simple and very fast but external ranging may also be necessary to match input values to converter input circuits.

19. Digital Linearizing

If ranging only requires a single addition and/or multiplication, digital linearizing is nearly as simple, requiring a case statement to provide all the necessary straight line segments, one segment being:

$$\text{if } V_1 < V_{in} < V_2 \text{ then } V := m_1 V_{in} + C_1$$

A look-up table provides fastest mapping. There is an accuracy problem when linearizing. As input is passed through an A/D converter, its accuracy is limited to the nearest digital representation. Points which should be different, and would have been had the input been linear, will have the same digital representation. The cure is to convert to a longer word, using extra bits, subsequently linearizing and then normalizing to the original accuracy.

20. Digital Filtering

Any calculation can be performed on the sequence of stored samples and therefore the use of algorithms to remove various frequencies is quite practical. Algorithms can be produced which give functions that cannot be made with analog electronics.

The input sampled data is piecewise constant, that is, having taken a sample, the input may be considered to have been at that value since the previous sample. For a simple resistor R and capacitor C low-pass filter with input U and output V and sample time T:

$$U = iR + V$$

also

$$i = C\frac{dv}{dt}$$

therefore

$$\frac{dV(t)}{dt} = -\frac{V(t)}{CR} + \frac{U(t)}{CR}$$

which can be expressed as a final digital equation for the new output value in terms of the previous output and current sample:

$$V_{j+1} = U_{j+1}(1 - e^{-T/RC}) + V_j(e^{-T/RC})$$

When each sample is taken it is multiplied by a constant and added to the scaled previous output to produce the new, filtered value, this is a digital program. Any order of filter could be produced by an entirely similar process to give filters of any analog characteristic, although it is better to cascade first- or second-order filter algorithms (pipeline) rather than compute a single higher-order form directly because of finite word length effects.

21. Word Length

A processor has a fixed word length, which may have limiting effects on sampling and the subsequent processing of data. The main effects of limited word length are the accuracy of each sample, the problems of small differences, and the accuracy of constants.

Processing algorithms can be written in different ways. Those which involve differences between numbers which are close together or some division will result in a loss of significant digits. Constants required in any algorithm cannot necessarily be stored exactly, for example, a "pole" in the response of a closed-loop control system may be shifted (by rounding error) so that the net result is a just unstable rather than a just stable system. Errors in constants may be cumulative, and the accuracy of data can suffer considerably with only minor errors in a few constants.

22. Processing, Display and Overall Control

Any processing is then possible on the conditioned, converted and sampled values. Complex data fusion from numbers of sensors, pattern recognition and fuzzy processing are all found in digital instruments. Overall control of the instrument is handled by the processor, with operator interaction via keyboard and mouse pointer steadily replacing analog controls. The final part of any instrument is a display. This can be as simple as a four-digit, seven-segment liquid crystal display or as complex as multiple video screens providing a full mimic display of the operation of a chemical plant and all its variables, temperatures, pressures and flow rates.

See also: Analog-to-Digital and Digital-to-Analog Conversion; Artificial Intelligence in Measurement and Instrumentation; Signal Processing

Bibliography

Cripps M 1989 *Computer Interfacing: Connection to the Real World.* Edward Arnold, Sevenoaks, UK

M. D. Cripps
[City University, London, UK]

Displacement and Dimension Measurement

Displacement and dimensions are the most frequently measured quantities for the spatial orientation of man and machinery, for production control in machine building, instrument manufacturing, civil engineering and microelectronics, for quality control in conventional and computer-aided manufacturing (CAM), for statistical process control (SPC) and flexible manufacturing systems (FMS), for intelligent quality measurements (INQUAMESS) in computer-integrated manufacturing (CIM), and for a knowledge of relevant economical factors in trade. Such measurement is also applied in medicine and biology to obtain information about natural processes.

The main field of length, displacement and dimensional metrology requires the measurement of more than 10^6 different shapes of spatially defined objects in manufacturing industry. In microelectronics the distances between the upper and lower specification limits (tolerances) are smaller than 1 μm. In machine building the tolerances are greater than 5 μm.

As a result of the increasing application of computerized numerically controlled machine tools with real-time in-process measurements, electrical methods for length displacement and dimensional measurements enjoy an ever increasing interest. Under the influence of universal personal computer instrumentation with standardized data processing hardware and software, a great number of displacement and dimensional sensors with different working principles are used.

1. Definitions and Standards

Displacement and dimension are spatial measurement quantities. The dimension of a measurement object is a length measurement. Position is the spatial location of a point or object with respect to a reference point, expressed as a length measurement. Distance is the spatial difference between two points or objects, also expressed as a length measurement. Displacement is the change in position or distance of a point or object with respect to a reference point. Linear displacement is a displacement with fixed direction (angle) and flexible length. Angular displacement is a displacement with fixed length and flexible direction (angle). The angle is a derived quantity. Proximity is the

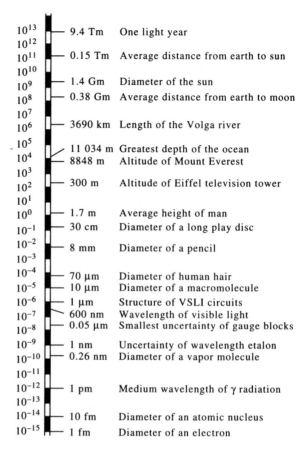

10^{13}	9.4 Tm	One light year
10^{12}		
10^{11}	0.15 Tm	Average distance from earth to sun
10^{10}		
10^{9}	1.4 Gm	Diameter of the sun
10^{8}	0.38 Gm	Average distance from earth to moon
10^{7}		
10^{6}	3690 km	Length of the Volga river
10^{5}	11 034 m	Greatest depth of the ocean
10^{4}	8848 m	Altitude of Mount Everest
10^{3}		
10^{2}	300 m	Altitude of Eiffel television tower
10^{1}		
10^{0}	1.7 m	Average height of man
10^{-1}	30 cm	Diameter of a long play disc
10^{-2}	8 mm	Diameter of a pencil
10^{-3}		
10^{-4}	70 μm	Diameter of human hair
10^{-5}	10 μm	Diameter of a macromolecule
10^{-6}	1 μm	Structure of VSLI circuits
10^{-7}	600 nm	Wavelength of visible light
10^{-8}	0.05 μm	Smallest uncertainty of gauge blocks
10^{-9}	1 nm	Uncertainty of wavelength etalon
10^{-10}	0.26 nm	Diameter of a vapor molecule
10^{-11}		
10^{-12}	1 pm	Medium wavelength of γ radiation
10^{-13}		
10^{-14}	10 fm	Diameter of an atomic nucleus
10^{-15}	1 fm	Diameter of an electron

Figure 1
Length scale illustrating the different orders of magnitude: VSLI, very-large-scale integration

spatial closeness of two points or objects, expressed as a length measurement. Area and volume are combined length measurements.

Length units are additive. The dimensions of the length scale comprise more than 30 orders of magnitude, as shown in Fig. 1.

2. Mechanical Measurement Techniques

2.1 Solid-State Devices

Solids are materials in a condensed state of aggregation; they are amorphous (glass) or crystalline (metal), and have a fixed shape, preserved autonomously by atomic forces under fixed environmental conditions (temperature, pressure, humidity, chemical influences). The shape of a solid or marks on it may thus be used as measures for length measurements.

For mechanical length measurement the sensor must be mechanically connected to the object to be measured. The contact is frictional and sometimes conjugate. The coupling must be accomplished without slippage or backlash.

If an object and a mechanical scale are rigidly coupled, the section on the scale, limited by the boundary lines of the object, is a dimensional measurement of the object:

$$\Delta s_a = \hat{s}_a - \underset{\sim}{s}_a$$

where Δs_a is the length of the object, $\hat{s}_a$ is the upper mark on the scale, and $\underset{\sim}{s}_a$ is the lower mark on the scale.

For small distances it is necessary to amplify the indication with levers, threads or gear transmissions, according to:

$$\Delta s_a = \frac{\Delta s_e}{i}$$

where Δs_a is the deflection of the transmission output (indication), Δs_e is the deflection of the transmission input (measured length), and i is the transmission factor (reciprocal sensitivity).

2.2 Circular Division Devices

A circle is a closed line of length (or diameter) $l = 2\pi r$, where r is the radius of the circle and, thus, 1 rad is $1/2\pi$ of the diameter of the circle. Techniques for circular divisional metrology may then be applied.

2.3 Gravitational Sensors

Solids, liquids and gases are gravitationally pulled to the center of the earth, in proportion to their masses. Free movable pendulums are oriented towards the center of the earth (plumb), and may be used for measurement of deviations (displacements) from the vertical. The natural gravitational standard is the center of the earth. For measurement of deviations (displacements) from the horizontal, the natural standard is the gravitational surface of the earth (level).

3. Optical Measurement

In geometrical optics the measurement of length depends on Abbé's optical law of imagery. The equation for lenses is as follows:

$$\frac{1}{v} - \frac{1}{u} = \frac{1}{f}$$

where v is the distance of the image, u is the distance of the object, and f is the focal length of the system.

The image scale given by:

$$V = \frac{v}{u}$$

where V is the factor of proportionality used when measuring a length with a microscope. The reference quantity is a glass scale with very precise graduations. The scale constants are 8 μm for transmitted light and 40 μm for incident light.

In physical optics, length measurement is accomplished by interference measurements.

Optical length differences shift the light waves as a result of geometrical length differences. These reactions are interference phenomena which can be evaluated in the form of maxima and minima. The displacement is proportional to the wavelength and phase shift, according to:

$$\Delta\varphi_a = \frac{2\pi n \Delta s_e}{\lambda}$$

where n is the index of refraction. Δs_e is the deflection, λ is the wavelength, and $\Delta\varphi_a$ is the phase shift.

4. Acoustic Measurement Sensors

Thickness sensing using beamed sound energy is accomplished by echo ranging techniques. These use either a sound source in combination with the sound receiver or a single-element receiver operating alternately in transmit or receive mode. In both cases the thickness is determined by measuring the time required for a transmitted pulse to be reflected by the interface and to return to the receiver, and by knowing the velocity of sound in the material through which the sound travels.

5. Electrical Measurements

5.1 Resistive Sensors

Resistive sensors work on the principle that the electrical resistance of wires, liquids and thin films depend on their length l, their cross section q and specific resistance ρ, according to:

$$R = \frac{\rho l}{q}$$

Potentiometric displacement sensors with potentiometric transduction elements are relatively simple but not very accurate devices. The output voltage u_a depends on the input length s_e, according to:

$$u_a = \frac{u_b(R_a \| Rs_e/l)}{[R(1 - s_e/l) + (R_a \| Rs_e/l)]}$$

$$u_a = \frac{u_b}{(1/s_e) + R(1 - s_e/l)R_a}$$

81

where R is the total resistance of the potentiometer, R_a is the terminating resistor, u_a is the output voltage, u_b is the operating voltage, l is the total length of the potentiometer, and s_e is the input (tap) length (m).

Strain gauge displacement sensors are bonded to the object or to a beam and connected to a Wheatstone bridge circuit. They may be manufactured as wires, thick films or thin films. The variation of the resistance depends on the variation of the geometrical shape and the internal mechanical stress.

Strain gauges are sensitive to temperature changes. For temperature compensation, two strain gauges are used in the measurement circuit; only one is deformed by length variation and stress.

5.2 Capacitive Devices

Capacitive sensors work on the principle that the electrical capacitance C of a capacitor depends on the area A, the distance s and the dielectric constant ε of the material between the electrodes, according to the following equation:

$$C = \frac{\varepsilon A}{s}$$

Capacitive displacement sensors are built from cylinder or plate capacitors with variable electrodes, variable distances between electrodes and variable dielectric constants.

The capacity of a plate capacitor is given by:

$$C = \frac{\varepsilon_0 \varepsilon_{rel} A}{s}$$

where A is the active area, s is the distance between plates, ε_0 is the absolute dielectric constant, and ε_{rel} is the relative dielectric constant.

When A is constant the capacitance is given by:

$$C = \frac{k}{s}$$

where k is the k factor of the strain gauge ($k \sim 2$ for metallic materials and, $k \sim 50$ for semiconductors). For variation Δs of the distance between the plates the variation of capacitance is given by:

$$\Delta C = \frac{k[s - (s + \Delta s)]}{s(s + \Delta s)} = -\frac{k \Delta s}{s(s + \Delta s)}$$

and for $\Delta s \ll s$ we have:

$$\Delta C_a = -\frac{k \Delta s_e}{s^2}$$

which indicates that a nearly proportional relationship exists between length and capacitance.

In general, capacitive level and thickness sensors exploit the different dielectric constants of the two materials between the capacitor plates to carry out length measurements. The capacitor is positioned between the two plates so that the borderline between the two materials is perpendicular or parallel to the plates.

5.3 Reluctive Sensors

Reluctive sensors work on the principle that the magnetic resistance depends on the length l, permeability μ and cross section q. The magnetic resistance is given by

$$R_m = \frac{l}{\mu q}$$

and the inductance L is given by

$$L = \frac{w^2}{R_M}$$

with w number of turns of the coil are measures of length, displacement or dimension.

Reluctive (inductive) displacement sensors with a crossbar work with an air gap in the magnetic circuit. The permeability of iron μ_{Fe} is much greater than the permeability of air μ_L. Thus, it is possible to measure small variations in length by varying the air gap between the yoke and crossbar.

Reluctive (inductive) displacement devices with plungers vary the magnetic resistance in the magnetic circuit. Variation of the magnetic circuit leads to a change in the inductance. Compared with crossbar sensors, the plunger sensors have a greater measurement uncertainty, but they have a greater working range.

Linear variable differential transformers (LVDTs) consist of a primary winding and two secondary windings. The windings are arranged concentrically and next to each other. A ferromagnetic core (armature) is attached to the sensing shaft. The core slides freely in the bobbin. In operation, ac excitation u_b is applied across the primary winding and the moving core varies the coupling between it and the two secondary windings. As the core moves away from the null position the coupling to one of the secondary windings increases. Coupling factors and output voltages are proportional.

Synchro type displacement sensors are built from a single-phase rotor in a three-phase stator. The stator windings are mechanically spaced at 120° intervals. When a single ac voltage is used to excite the rotor winding the output voltages from the stator windings vary with the angular displacement of the rotor.

If the stator is excited from a three-phase supply, a constant amplitude variable-phase output can be obtained from the rotor.

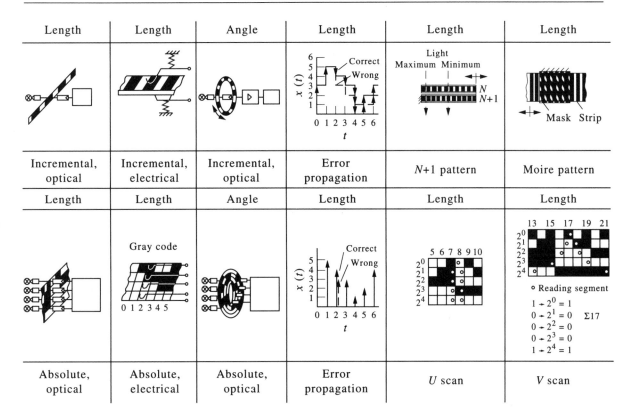

Length	Length	Angle	Length	Length	Length
Incremental, optical	Incremental, electrical	Incremental, optical	Error propagation	N+1 pattern	Moire pattern
Length	Length	Angle	Length	Length	Length
Absolute, optical	Absolute, electrical	Absolute, optical	Error propagation	U scan	V scan

Figure 2
Several techniques for analog-to-digital conversion

Resolver type displacement sensors have another number and smaller spacing between their windings compared with synchro type displacement sensors. Often they have two-phase stators with two stator windings spaced 90° apart, and two rotor windings, also spaced 90° apart. Using the resolver for measurement purposes, one rotor winding must be shorted.

Microsyn type displacement sensors have a four pole stator with two coils each. The rotor has no windings if it is made from ferromagnetic material. Microsyns have a performance similar to differential transformer sensors.

Inductosyn type displacement sensors have a meander shaped printed scale with electrical conductors on an insulator. The scale is supplied with a high-frequency voltage. The sensor is made from two conducting loops with one 90° shifted coil, by comparison with the scale.

The voltages induced in the head of the movable sensor periodically change with the displacement. The phaseshift between the two voltages characterizes the direction of the motion. The system works phase-cyclically.

6. Digital Measurement Devices

The rapid progress of microelectronics, as well as digital and microcomputer technology, demands digital sensors for easy measurement data processing. There are some principles only to generate natural digital signals. The most popular approach is to convert the analog output of an analog sensor into a digital signal using an analog-to-digital converter (ADC). Several techniques are shown schematically in Fig. 2.

Incremental digital displacement sensors have a disk or a straight edge which is divided into a number of equal sectors. The nature of the sectors and the reading device is dictated by the transduction principle utilized. Most of them operate in a contactless mode with optical sensors. They can only measure the displacement with respect to the starting point of the given displacement.

The resolution of optoelectronic, coded and incremental displacement sensors can be improved. One way is by the multiplication of logic operation on the basis of output axis crossing. Another method involves the application of interference patterns produced by a stationary reticle or a mask placed over the moving pattern. The $N + 1$ pattern is mainly used with coded disk elements. The Moire pattern is mainly used in linear displacement sensors.

Absolute digital displacement sensors have a disk or a straight edge which is divided into four or more tracks. The sectors are arranged so that a binary code is produced when using optical sensors. These sensors measure displacement with respect to an internal fixed reference point.

The reliability of scanning can be improved by using special techniques such as the V scan:

$$x_\nu = \pm l_s \nu / 4 \pm 2n l_s \nu, \quad n = 0, 1, 2, \ldots$$

where x_ν is the displacement from the reading line, l_s is the length of a segment of track, ν is the ordinal number of the track, 0 is the signal advancing, and 1 is the signal lagging.

7. Applications

Length measurement instruments are increasingly used for production control and quality assurance.

The instrumental solution must supply the required output information for the application to the user. The instruments must be compatible with other equipment on the market, such as, signal processors, microcomputers, standard interfaces (RS 232C and IEEE 488), plotters and color displays. Measurement instruments must be able to be incorporated into different automated systems.

One of the characteristic trends in instrumentation is the availability of "turnkey" process solutions. Thus, electrical output sensors are generally preferred. Measurement data processing is accomplished using universal personal computer instruments and laptop personal computers.

Recently, single chip microcomputers have been used to make sensors intelligent. In the light of this background it is necessary to know the working principles of sensors, in order to select the most suitable, and to organize the measuring procedures with the help of microcomputers.

See also: Force and Dimensions: Tactile Sensors

Bibliography

Beckwith T G, Buck N L 1969 *Mechanical Measurements.* Addison-Wesley, Reading, MA

Doebelin E O 1975 *Measurement System: Application and Design*, 2nd edn. McGraw-Hill, New York

Farago F T 1968 *Handbook of Dimensional Measurement.* Industrial Press, New York

Hofmann D 1986 *Handbuch Meßtechnik und Qualitätssicherung* (Handbook of Measurement Engineering and Quality Assurance), 3rd edn. Verlag Technik, Berlin

Holman J P 1966 *Experimental Methods of Engineers*, 2nd edn. McGraw-Hill, New York

Hughes T A 1988 *Measurement and Control Basics.* Instrument Society of America, Research Triangle Park, NC

Moore W R 1971 *Foundation of Mechanical Accuracy.* MIT Press, Cambridge, MA

Norton H N 1969 *Handbook of Transducers for Electronic Measuring Systems.* Prentice-Hall, Englewood Cliffs, NJ

Norton H N 1982 *Sensor and Analyzer Handbook.* Prentice-Hall, Englewood Cliffs, NJ

Norton H N 1982 *Electronics Engineers Handbook, Transducers.* McGraw-Hill, New York

Sydenham P H 1983 *Handbook of Measurement Science*, Vol. 2. Wiley, New York

Warnecke H J, Dutschke W 1984 *Fertigungsmeßtechnik* (Plant Measurement Engineering). Springer, Berlin

D. Hofmann
[Friedrich Schiller University, Jena, Germany]

E

Electrochemical Measurements

Electrochemical measurements are important techniques for quantitative analysis of composition or chemical measurands, particularly inorganic species dissolved in aqueous media. While their principle of operation derives from the fields of interfacial science, thermodynamics and mass transfer kinetics, the instruments themselves are often simple devices from the point of view of their operation, construction and associated electronics. Classical electrochemical measurement techniques are routinely employed in research, in medical monitoring, and in industrial monitoring and process control.

The three electrical parameters, potential, current, and resistance (or its reciprocal, conductance), provide a convenient basis for the classification of electrochemical techniques. Common to each is a pair of electrodes in contact with a solution of the measurand (usually in the form of ions). The type of measurement to be performed depends quite simplistically on the instrumentation applied and the nature of the interface between the solution and the electrodes.

This article will discuss the fundamental operating principles of these classical electrochemical techniques and give examples of their application in laboratory and industrial measurement systems.

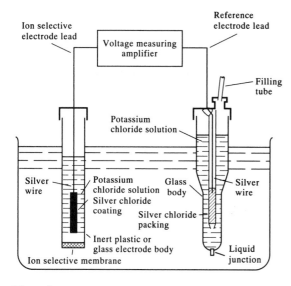

Figure 1
Typical cell used in analytical potentiometry, illustrating ion-selective and reference electrodes immersed in measurand solution

1. Potentiometry

Analytical potentiometry is based on measurement of the electrical potential across an ion-selective membrane. The experimental setup is shown in Fig. 1. Two probes are required and are inserted into a solution of the measurand. Each probe consists of two elements which contribute to the cell potential. The ion-selective electrode probe consists of an internal reference electrode and an ion-sensitive membrane. The reference probe consists of an internal reference electrode connected to the measurand by a liquid junction. The voltage across the cell, which is the algebraic sum of the potentials developed by the reference electrodes, and across the membrane and liquid junction, is read using an accurate voltage measuring amplifier.

An ion-sensitive membrane develops a potential that is quantitatively related to the ratio of concentrations (actually activities) of the measured ions at either interface. Using the cell of Fig. 1, where the reference electrodes are identical (hence their potentials cancel) and ignoring the liquid junction potential, this relationship is stated by the Nernst equation (Bard and Faulkner 1980)

$$E_{cell} = E^0 + \left(\frac{RT}{nF}\right)\ln\left(\frac{c_1}{c_2}\right) \qquad (1)$$

where E^0 is a constant (the standard potential), $R = 8.3141 \, J \, K^{-1} \, mol^{-1}$, $F = 96.487 \, A \, s \, mol^{-1}$, $T = (°C) + 273.15 \, K$, n (in equivalents per mole) is the charge of the measurand ion, and c_1 and c_2 are the concentrations of the measured ions at the outer and inner membrane interfaces, respectively. The concentration of the measured ion in the internal solution is usually fixed so that the measured potential is a function of the concentration of the measurand at the outer surface only. Substituting values for the constants in Eqn. (1) and converting to common logarithms illustrates that an ideal electrode responds with a slope of $0.19841 \, T$ mV per decade in concentration of measurand (monovalent ion). A few values of this slope are shown in Table 1. These slopes would be divided by two for divalent measurands. It is noted that for the sign convention adopted here, the measured potential increases with the concentration of positively charged ions and decreases with the concentration of negatively charged ions.

Ideality of potentiometric response is directly determined by its adherence to Eqn. (1).

Careful cell design can reduce errors and calibration procedures using media of controlled ionic strength are necessary for exact determination of concentration. Unwanted species may interact with the measur-

Table 1
Values of the Nernst slope at a few temperatures

Temperature (°C)	Nernst coefficient (mV)
0	54.197
25	59.157
50	64.118
100	74.038

and, changing it to a form not measurable by the membrane, or may themselves be measured by the membrane, an effect termed selectivity (often published with data on electrode performance).

The pH electrode structure is shown in Fig. 2a. The membrane is a specially formulated silica-based glass fused to the end of an inert glass tube. The inner chamber encloses a filling solution that is buffered to hold the hydrogen ion concentration on the inner surface of the membrane constant, and also contains the electrolyte used to control the potential of the inner reference electrode. To measure pH, the probe is immersed in the test solution along with a reference probe to form a pH measuring cell. The electrical resistance of the glass pH measuring membrane is extremely large (usually greater than $200 \, M\Omega$), and therefore the voltage across the terminals of the cell must be measured with a high-input impedance amplifier. From Eqn. (1), with c_2 constant, $c_1 = [H^+]$ is the measurand solution and, using the identity $pH = -\log[H^+]$, we obtain

$$E_{cell} = E^{0\prime} - 2.3026(RT/F)\,pH \qquad (2)$$

where $E^{0\prime}$ is the standard potential for the pH measuring cell. Eqn. (2) leads directly to an operational definition of pH which is applied to pH measuring instrumentation. Using the cell of Fig. 1 and a standard buffer solution having the value pH_s, the cell potential E_s is determined. The test solution is then placed in the cell and pH_x is determined by measurement of the new potential E_x. Writing Eqn. (2) in terms of pH_s, E_s, pH_x and E_x, and subtracting, gives

$$pH_x = pH_s - (F/2.3026\,RT)(E_x - E_s) \qquad (3)$$

A second adjustment on the pH meter may be used to compensate for changes in the slope term due to temperature, but not for changes in $E^{0\prime}$, or changes of values of pH with temperature.

Important examples of a class of potentiometric sensors based on crystalline membranes, shown in Fig. 2b, are the fluoride electrode that utilizes single-crystal LaF_3 and various polycrystalline membranes composed of silver sulfide mixed with silver halides or other metal sulfides that give membranes having selective response to anions such as Cl^-, Br^-, I^-,

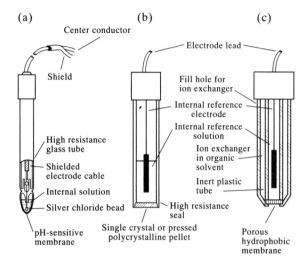

(a)   (b)   (c)

Figure 2
The three major classes of ion-selective electrodes: (a) pH glass electrode; (b) solid-state electrode; (c) liquid ion-exchange electrode

SCN^-, CN^- and S^{2-}, and cations such as Ag^+, Cu^{2+}, Cd^{2+} and Pb^{2+}. A third important type of probe (see Fig. 2c), is based on liquid ion exchange. An ion exchanger is dissolved in an organic solvent that is immiscible with the test solution and is immobilized as the ion-sensitive membrane using a thin porous material such as cellulose acetate. The selectivity of these membranes is determined by the chemistry of the exchange reaction between the exchanger and the measurand.

2. Coulometry and Amperometry

If voltage is applied across two metallic electrodes immersed in an electrolyte, current will flow and several electrochemical processes may occur depending on the magnitude of the applied voltage, the nature of the electrodes, and the chemical makeup of the solution. Since current flow in solution requires the movement of ions, cations will move toward the cathode (the negative electrode) and anions will move toward the anode (the positive electrode). As electroactive ions reach the electrodes, they exchange electrons at the electrode surface; cations take electrons from the cathode (reduction) and anions give electrons to the anode (oxidation). These electron exchange reactions thus give rise to the flow of electrons (current) in the external circuit.

At either electrode, the generalized redox reaction can be written as:

$$Ox + ne = Red \qquad (4)$$

for which the Nernst equation at 25 °C is

$$E = E^0 - (0.05916/n) \log [\text{Red}]/[\text{Ox}] \qquad (5)$$

As with potentiometry this expression shows the quantitative relationship between the electrode potential and the chemical concentrations expressed in the logarithmic term. In potentiometry the value of the logarithmic term establishes the potential which is measured. In amperometry and coulometry, the applied potential determines the value at equilibrium of the ratio [Red]/[Ox]. If the applied voltage is equal to the standard potential E^0, the ratio is unity. If the applied voltage is made only 0.12 V (two times the slope factor in Eqn. (5)) positive of the standard potential, that is, $(E - E^0) = +0.12$ V, the equilibrium shown in Eqn. (4) is driven to the left with the flow of anodic current, until the ratio becomes approximately 0.01 (for a process with $n = 1$). Likewise, if $(E - E^0) = -0.12$ V, the equilibrium is driven to the right with the flow of cathodic current until the ratio becomes approximately 100 (again for $n = 1$).

As with potentiometry, the desired electrochemical information is obtained at only one of the electrodes in the cell. In situations where current flow is allowed, it is convenient to use a three-electrode cell and a potentiostat as shown in Fig. 3. Here the potential of the analytical electrode is controlled relative to a reference electrode and the current is measured in a second loop formed with an auxiliary electrode. This arrangement allows very accurate control of the potential of the analytical electrode, independent of the iR drop that accompanies current flow and the redox process that occurs at the auxiliary electrode. Modern instrumentation based on operational amplifiers is available for this application.

Controlled potential coulometric techniques involve exhaustive electrolysis. Here the analytical electrode,

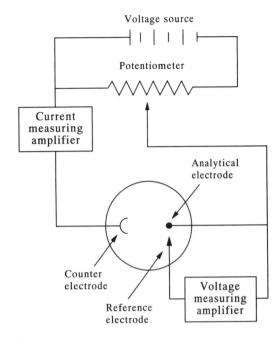

Figure 3
Basic components of the three-electrode potentiostat circuit

called the working electrode, is made large and the solution is rapidly stirred. In cases where 100% of the current goes into the redox process involving the measurand, measurement of the total quantity of electricity (coulombs) and application of Faraday's law provides the analysis.

Many materials, both organic and inorganic, can be analyzed by these coulometric techniques. Several examples are shown in Table 2.

Table 2
Selected examples of analytical methods employing controlled potential coulometry

Measurand	Working electrode	Supporting electrolyte
Alkali and alkaline earth metal ions (Li, Na, Ca, . . .,)	Hg	tetraethylammonium perchlorate in acetonitrile
Heavy metal ions (Ti, V, Cr, Mo, Mn, . . .,)	Hg	various acids in aqueous solution
Fe, Co, Ni and hexacyanoferrates	Pt, Au	acids and neutral salts with added complexing agents
Cu, Ag, Au	Hg, Pt	various acids in aqueous solution
Halides and pseudohalides (Cl^-, Br^-, I^-, CN^-, SCN^-, . . .,)	Ag	neutral and basic buffered aqueous solutions
Compounds containing sulfur (S^{2-}, SO_3^{2-}, $S_2O_3^{2-}$)	Ag	perchlorate salts in buffered basic aqueous solutions
Compounds containing nitrogen (N_2H_4, $N(C_2H_5)_4^+$, NO_3^-)	Pt	acids and buffered acids in aqueous solutions

The other current technique, amperometry, employs a small analytical electrode, often a microelectrode, termed an indicating electrode. The idea here is to measure the absolute value of the electrode current without affecting the bulk concentration of measurand. Two important ways in which this technique is used will be discussed, direct amperometry and amperometric titration.

Direct amperometry uses the current observed at an electrode bathed in the measurand to indicate concentration. Since the current depends on mass transfer of measurand to the electrode, it is necessary to define cell geometry and hydrodynamics precisely. The electrode may be held stationary with the solution forced by it by stirring or placing it in a flowing stream, or the electrode may be put in motion, for example, rotated or vibrated, with respect to the otherwise quiescent sample. The experimental difficulty of defining and maintaining reproducible hydrodynamics is a limitation of this technique, but otherwise it has the advantage that the steady-state current is easily measured and reacts quickly to changes in concentration of the measurand.

Direct amperometry is best suited to continuous monitoring of flowing streams. In cases where the flow is not constant it is necessary to flow the sample through a cell where it can be stirred or, alternatively, to use a rotated electrode. The use of amperometry for titration endpoint detection has widespread use in laboratory situations and in automated instruments such as titrators. A titration involves quantitative and stoichiometric reaction of measurand with a reagent and this may be used to extend the application of amperometry.

The precision and accuracy attainable with amperometric titration is primarily determined by the volumetric errors associated with the titrimetry, rather than the precision of detecting the endpoint. Examples of popular applications include silver ion titrant and a rotated platinum electrode for halides, sulfide, thiocyanate, cyanide and sulfhydryl compounds. Many iodometric and bromometric procedures, again with platinum electrodes, are used for a diverse number of analyses; for example, oxidation–reduction reaction with measurands such as As(III) and Sb(III), oxidizable organic compounds, and addition reactions with olefinic fats and oils to determine unsaturation (bromine or iodine numbers). A final important group of applications that will be noted are complexation and precipitation titrations of various metal ions with organic reagents using mercury indicating electrodes; 8-hydroxyquinoline for bismuth, cadmium, copper, iron, magnesium and zinc, dimethylgloxime for nickel, α-nitroso-β-naphthol for copper and cobalt, ethylenediaminetetraacetic acid (EDTA) for many metal ions, and other complexans having specificity for particular measurands. This subject is treated by Stock (1965).

3. Conductivity

The previously discussed techniques make use of processes occurring at electrodes. Conductivity takes advantage of the bulk ionic properties of the sample: experimental cells and instrumentation are designed to measure the resistance (or its reciprocal, the conductance, measured in siemens (S)) between two inert electrodes in contact with the measurand. Since the mobility or movement of ions under the imposed electric field supports the flow of current in the cell, the observed conductance depends on the number and, hence, the concentration of ions present. However, since all ions do not contribute equally, the measured cell response is only analytically useful in situations where the identity of the measurand is known and is present as the single ionized component contributing to the conductivity.

Because of different mobilities, ions demonstrate characteristic conductances. Values of conductance for several ions, reported for one equivalent weight of the conducting ion, are termed equivalent conductance. Mobilities are affected by temperature and the total concentration of ions in the solution. The mobility of ions, hence the conductance, increases with tempera-

Table 3
Equivalent ionic conductance of several ions at infinite dilution at 25 °C (S cm^{-2} mol^{-1})

Cations	λ^{0a}	Temperature coefficient[b]	Anions	λ^{0a}	Temperature coefficient[b]
H$^+$	349.8	0.0139	OH$^-$	198.6	0.018
K$^+$	73.5	0.0193	Cl$^-$	76.4	0.0202
Na$^+$	50.11	0.0220	NO$_3^-$	71.42	0.020
Ca^{2+}	59.50	0.0230	SO$_4^{2-}$	80.0	0.022
Mg^{2+}	53.06	0.022	CO$_3^{2-}$	69.3	0.02
Cu^{2+}	53.6	0.02	HCO$_3^-$	44.5	
(n-Bu)$_4$N$^+$	19.5	0.02	Picrate$^-$	30.4	0.025

Source: Light and Ewing 1989
[a] Data are on an equivalent basis, that is, for ions of charge z, values are for $1/z$ mole   [b] temperature coefficient = $(1/\lambda^0)(d\lambda^0/dT)$

ture (unlike electronic conductivity), and increases with dilution. Table 3 shows values of equivalent conductances of several ions at 25 °C corrected to infinite dilution, together with temperature coefficients.

The measured conductance of the solution also depends on the geometry of the electrodes; it is directly related to the electrode area and inversely related to the distance between them. Based on the above,

$$L = \left(\frac{a}{d}\right) \sum_i z_i \cdot c_i \cdot \lambda_i \qquad (6)$$

where L is the conductance in ohm^{-1} or siemens, a is the area of electrodes in cm^2, d is the distance between electrodes in cm, c_i is the concentration of the participating ions in equivalents cm^{-3}, λ_i is the equivalent conductance of the participating ions in $S\,cm^2$ $equivalent^{-1}$, and z_i is the charge on the participating ion.

It is not convenient to measure the ratio d/a for each cell but, since it is constant for any given cell, it may be assigned a value θ (in cm^{-1}), termed the cell constant. This constant is determined experimentally using solutions of accurately known concentrations of potassium chloride, for which values of specific conductance (the conductivity of a cube of the solution, 1 cm on each edge) denoted by k, in $S\,cm^{-1}$, have been precisely determined (ASTM 1983). A few values are shown in Table 4.

The cell constant is readily determined using the expression $\theta = k/L$, where k is known from tabulated values and L is measured with the cell being calibrated. Three types of cells are used in composition measurement; two-electrode, four-electrode, and electrodeless. The cells and associated instrumentation are shown in Figs. 4a–c.

Two-electrode cells are best suited for measurement in clean solutions, since coatings and films which form on electrodes may produce resistance and result in error. Since the intent is to only measure the bulk conductivity of the electrolyte, the interfacial imped-

Table 4
Specific conductances of potassium chloride solutions at 25 °C

Approximate normality	Weight of KCl in solution $(g\,l^{-1})$	k $(\mu S\,cm^{-1})$
1.0	72.2460	111342
0.1	7.4365	12856
0.01	0.7440	1408.8

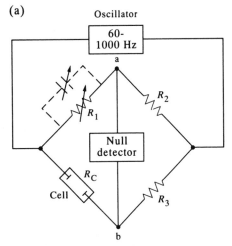

(a)

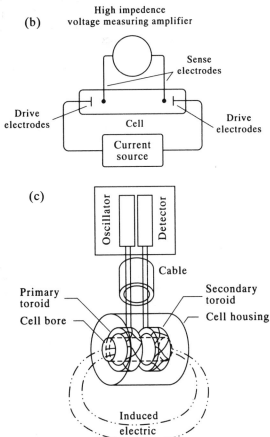

Figure 4
Basic components of conductance circuits: (a) two-electrode cell forming one arm of a standard Wheatstone bridge; (b) typical components of the four-electrode cell and circuit; (c) typical components of the electrodeless cell and circuit

ance of the electrodes with the solution must be minimized. The cell and instrument are designed accordingly.

Four-electrode conductivity is attractive when coating and fouling of electrodes is of concern. Current is imposed across the drive electrodes and the potential drop through the electrolyte is detected between two points in the cell using the sense electrodes (see Fig. 4b). The sense electrodes are monitored with a high-impedance voltage-measuring amplifier so that the current drawn and the associated polarization is minimal.

One way to eliminate electrode polarization effects is to eliminate the electrodes. Techniques to do this are called electrodeless conductivity. The probe, shown in Fig. 4c, consists of two encapsulated toroids. When immersed in an electrolyte, the solution forms a conductive loop shared by both toroids. One toroid radiates an electric field in this loop and the other detects a small induced electric current. Practically speaking, the two toroids form a transformer whose coils are interconnected by the resistance of the electrolyte.

In summary, the electrodeless technique is widely applied in process monitoring and control applications. Two-electrode techniques are best suited to solutions having high resistivity where fouling is neither likely nor has a measurable effect. Four-electrode and electrodeless techniques find their greatest utility in conductive solutions where fouling is likely and would produce error. Typical applications include measurement of salts, acids and alkalis in chemical processes in the mining, metallurgy, pulp and paper, and in aluminum industries where samples often contain solids, oils and corrosive and abrasive materials. Light and Ewing (1989) provide a recent review of this topic.

See also: Analytical Physical Measurements: Principles and Practice; Chemical Analysis, Instrumental

Bibliography

American Society for Testing and Materials 1983 Standard test methods for electrical conductivity and resistivity of water. *Annual Book of ASTM Standards*, Vol. 11.01. ASTM, Philadelphia, PA, pp. 149–56
Bard A J, Faulkner L R 1980 *Electrochemical Methods, Fundamentals and Applications*. Wiley, New York, Chap. 2, pp. 44–85
Light T S, Ewing G W 1989 *Handbook of Analytical Instrumentation*. Dekker, New York
Stock J T 1965 Amperometric titrations. In: Elving P J, Kolthoff I M (eds.) *Chemical Analysis*, Vol. 20. Interscience, New York

K. S. Fletcher
[Foxboro Company, East Bridgewater, Massachusetts, USA]

Environmental Measurement and Instrumentation

The realization that many of the processes that sustain the wealth and well-being of the world's industrialized nations are themselves responsible for serious environmental damage has led to the introduction of legislation which governs the levels of pollutants discharged into, and present in, the environment. As a consequence, numerous measurements are made which aim to control and monitor environmentally polluting gases, liquid chemical species, and physical and biological quantities. These measurements utilize a diversity of portable, field, on-line and laboratory instruments which play a vital role in protecting the health of the aquatic, terrestrial and atmospheric environment.

1. Background

The 1980s are likely to be remembered as the decade in which environmental issues emerged from relative obscurity to become part of the mainstream public and political debate.

It is now widely recognized that activities such as transport, power generation, manufacturing, agriculture and food processing are environmentally damaging and awareness of their consequences, such as acid rain, contaminated water courses, depletion of the ozone layer and global warming have stimulated a profound change in attitudes within government, industry and the public alike. This has led to a critical examination and reappraisal of many aspects of late twentieth-century life and, most importantly in this context, the introduction of an ever-growing body of national and international environmental legislation.

While there is often a lack of agreement regarding the long-term impact and importance of many environmental issues, it is universally recognized that pollutants must be closely monitored and controlled to ensure compliance with legislation and to protect the environment, a function that is undertaken by a diversity of field and laboratory instruments. As an ever-growing number of quantities are implicated as environmentally threatening, novel measuring techniques and instruments will be developed, particularly for field use, which will utilize several advanced sensor technologies. Sensors are a vital enabling technology that will, to a large degree, govern the availability and performance of future generations of environmental monitoring instruments.

2. Environmental Pollutants

While the greatest attention has been paid to gases and liquid chemical species, several microbiological quantities and physical phenomena also pose an environmental threat. Examples include bacteria and viruses in water, and noise and ionizing radiations in the external environment.

In addition to pollutants that pose a direct threat to public health or the environment, several other quantities are measured which act as indicators of environmental health or which characterize the environmental impact of a discharge. Most involve water quality and examples include dissolved oxygen (DO), biochemical oxygen demand (BOD), chemical oxygen demand (COD) and total organic carbon (TOC).

Examples of the many quantities that are measured by environmental instrumentation, and their sources, are illustrated in Table 1.

3. Overview of Environmental Instrumentation and Measurements

Environmental monitoring instrumentation varies greatly according to application and exploits a diversity of sensing and analytical methods and technolo-

gies. These instruments are used to detect and quantify hazardous substances at the point of discharge and in the external environment and measurements include simple spot tests, more accurate field determinations, continuous on-line measurements and high-accuracy laboratory analysis.

While some measurements are undertaken simply to determine the presence or otherwise of a particular quantity, the majority aim to ensure compliance with legislation. More specifically, environmental measurements are made: to quantify hazardous discharges at source; to monitor and control pollution prevention or abatement processes; during environmental audits and environmental research programmes; to monitor hazardous waste sites; to monitor environmentally threatening accidents; to assist in clean-up operations; and to monitor and record the state of the aquatic atmospheric and terrestrial environment on a local, national and global basis.

The instrumentation employed to fulfill these tasks includes fixed-site monitors, on-line devices that con-

Table 1
Environmentally hazardous quantities and their sources

Gases	
sulfur dioxide	fossil-fuelled power generation and vehicle exhausts
oxides of nitrogen	as above
carbon monoxide	as above
carbon dioxide	as above and landfill sites
hydrogen chloride	incinerator emissions, chemical industry
methane	landfill sites, agriculture, hydrocarbon processing and storage, gas pipelines
volatile organics	fuel storage, chemicals industry, vehicles
radon	natural radioactive decay
dioxins	incinerator emissions, chemicals industry
CFCs	propellants, refrigerants, electronics industry
Chemical species in the liquid phase	
toxic metals	many industrial processes, mining, metal refining
organic species	chemical industries, metal cleaning
halogenated organics	metal cleaning, chemical industry, by-products of water chlorination
ammonia	agriculture, industry, human waste
chlorine	drinking water chlorination
slurry/sewage	farming, human waste treatment and disposal
nitrates	agriculture, chemical processes
phosphates	agriculture, sewage treatment, industry
pesticides	residues from agriculture, agrochemicals industry
herbicides	as above
oils and fuels	leakage from production and storage facilities, marine discharges, industrial cooling water discharges
Physical quantities	
ionizing radiations	nuclear power generation, military activities, industry, natural sources
non-ionizing radiations	microwave communications, hv power transmission lines
noise	roads, airports, quarries, construction sites
Biological quantities	
BOD	industrial and water treatment effluents
blue-green algae	eutrophication of water courses
coliform bacteria	sewage treatment effluents, human waste
salmonella	human waste
hepatitis, polio, etc.	as above

tinuously monitor gas or liquid discharges at source, portable instruments that mostly fulfill a screening role and laboratory-based instruments such as chromatographs and spectrometers which are used mostly to analyze field samples.

4. Recent Technological Trends

As with all forms of electronic instrumentation, that used in the environmental context has benefitted greatly from recent developments in electronics, in particular, microprocessor technology. The use of digital signal processing is now cost-effective in all but the most simple portable monitors and has conferred many benefits to environmental instrumentation, including the ability to store data and download it to computer and other data acquisition systems; active compensation of sources of error such as thermal drift or cross sensitivity; automated or greatly simplified calibration routines; and the ability to undertake complex computations.

Recent advances in sensor technology have also exerted a significant impact on this field and have been most evident in the gas monitoring context where the use of optical sensing methods such as ir absorption have yielded instruments that are far more reliable, rugged and accurate than their predecessors. Some of the sensor types that will influence future generations of environmental instrumentation are considered in Sect. 9.

5. Monitoring Liquid Pollutants

Arguably the greatest threat to the environment is posed by the presence of hazardous chemical species in water courses such as rivers, lakes, reservoirs, aquifers and the oceans as these interconnect the entire aquatic biosphere. These emanate from authorized industrial and agricultural discharges, illegal discharges, accidental spillages, runoff from the land and leaching from landfill sites into underlying groundwater. Despite the growing number of on-line and portable instruments now available, the majority of aqueous determinations still rely on off-line laboratory analysis, because of the often very low concentrations of the species involved. Some of the more important quantities monitored in the aqueous phase, together with the dominant detection methods presently adopted, are listed in Table 2.

5.1 Monitoring at the Point of Discharge

Monitoring liquid effluent at the point of discharge involves both off-line determinations and continuous on-line measurements. The former involve subjecting samples to laboratory analysis by instruments based on well-established and accurate techniques such as mass spectroscopy (MS), atomic absorption spectroscopy (AA) and chromatography (HPLC) which are able to determine most organic and inorganic species. These instruments play a vital role where effluents contain complex or unknown mixtures of species but in the case of known and "simple" quantities such as dissolved oxygen, turbidity or pH, which are primarily

Table 2
Example of methods and techniques used to monitor chemical quantities in water

Quantity	Method	Technique
Biochemical oxygen demand (BOD)	microbiological	five-day incubation method (BOD-5)
On-line BOD	microbiological	respirometry
Bacteria and viruses	microbiological	cultivation methods
Dissolved oxygen	electrochemical	membrane electrode
pH	electrochemical	glass electrode
Ammonia	electrochemical	membrane electrode plus ammonium chloride
	optical	photometry
Nitrates	optical	colorimetry
	electrochemical	membrane electrode
Suspended solids	gravimetric	filtration methods
	optical	from turbidity
Turbidity	optical	light transmission or scatter
Oil in water	optical	light scatter
Phosphates	optical	colorimetry
Algae	optical	fluorescence
Chlorine (free and combined)	optical/chemical	photometry
Total organic carbon (TOC)	optical/chemical	photochemical oxidation followed by CO_2 detection
Pesticides and herbicides	analytical	GC, HPLC, GCMS
Organics (halogenated, etc.)	analytical	GCMS, HPLC
Trace metals (Pb, Hg, Cd, etc.)	analytical	atomic absorption spectroscopy
Metals in higher concentrations	optical	colorimetry
	electrochemical	membrane electrodes

indicators of an effluent's capacity to pollute, portable instruments are widely used.

In recognition of the fact that off-line methods involve an often unacceptably long time lag between sampling and the provision of data and are also frequently exceedingly costly, there is a strong trend towards the use of continuous, on-line monitors. These are vital to prevent pollution incidents going undetected and to ensure compliance with permitted discharge levels, and many such monitors are now available. These are based around electrochemical techniques, optical techniques, hybrid wet reagent–optical and electrochemical methods and on-line implementations of analytical methods such as MS. Many suffer limitations such as poor reliability, limited accuracy and selectivity, and/or high ownership costs and many novel sensing methods are therefore under development.

Portable and fixed instruments will be developed in the near future that will offer the ability to quantify a range of waterborne pollutants such as heavy metals, halogenated organics, herbicides and pesticides, that are presently determined in the laboratory. Although unlikely to offer the sensitivity attained by laboratory techniques, these will play a vital screening role and provide early warning of polluting incidents.

5.2 Monitoring in the External Environment

Monitoring waterborne pollutants in the external aquatic environment is a vital function and one that also involves a combination of portable instruments, fixed monitors and laboratory methods. The latter still dominate, principally because the pollutants involved are often unknown, comprise a mixture of species or occur in very low concentrations (low ppm or ppb) and similar techniques to those for the off-line determination of effluents are used, namely, HPLC, AA (atomic absorption spectroscopy) and GCMS (gas chromatography–mass spectroscopy).

Field measurements are undertaken by single- or multiple-parameter portable instruments which respond to quantities such as pH, temperature, conductivity, dissolved oxygen and other "simple" physical or chemical quantities. Colorimetric spot tests using comparators and photometers are also widely employed to determine a range of metals, chlorine, ammonia and other chemical species.

6. Monitoring Polluting Gases

The broad methodology adopted in monitoring polluting gases is similar to that considered above, namely, the use of fixed and on-line monitors, portable instruments and laboratory methods. However, the many requirements for continuous measurements dictate a far greater use of fixed and on-line monitors, both to quantify gases at their point of discharge and to determine ambient concentrations. Many of the measuring techniques employed are standard methods, specified by national and international legislation.

6.1 Ambient Air Quality Monitoring

Monitoring ambient air quality is an important and rapidly growing practice and numerous local, regional and national schemes are in place or being commissioned, particularly in industrialized regions and urban areas prone to pollution from vehicles.

The most frequently monitored gases are those which pose a health hazard or indicate a deteriorating air environment and include ozone (O_3), carbon monoxide (CO), oxides of nitrogen (NO_x) and sulfur dioxide (SO_2). Most ambient air quality monitoring is undertaken by high sensitivity fixed-site gas analyzers which respond to levels up to 1000 or 10 000 ppb and employ detection methods such as chemiluminescence (NO_x), uv fluorescence (SO_2), uv absorption (O_3) and ir absorption (CO). Some are used in a standalone mode but many are interconnected and the signals fed into centralized data acquisition and analysis systems.

6.2 Monitoring Gases around Industrial Sites

Much gas monitoring around industrial complexes involves field sampling, that is, absorbing air samples onto activated charcoal, and subjecting the desorbed gases to sensitive laboratory methods such as GC or wet chemical analysis.

Portable GCs and other analytical instruments are also used but, with the growing recognition that these methods are time consuming and of often limited utility, more sophisticated, real-time techniques are starting to be deployed. These include long-path ir absorption, FTIR (Fourier transform infrared) and LIDAR (light detection and ranging), a costly but highly sophisticated laser-based technique capable of quantifying most polluting gases at parts-per-million or parts-per-billion levels and mapping them in one, two or three dimensions. These techniques are implemented as costly, transportable instruments which make extensive use of computer-based signal conditioning but are expected to decrease in both size and cost as a consequence of developments in optics, optoelectronics and signal processing.

6.3 Monitoring Gases at the Point of Discharge

Measurements of polluting gas concentrations at their point of discharge (e.g., from power plant and incinerator stacks), are made widely as emission levels are governed by strict legislation in most industrialized nations. The instruments used may be either extractive, where the gas sample is drawn from the stack, conditioned and fed to the analyzer, or in-stack, including those that undertake cross-stack measurements.

Many earlier methods such as those utilizing extractive wet chemistry have been superseded recently by single- or dual-beam nondispersive infrared (NDIR) and, to a lesser extent, NDUV techniques. These are sufficiently sensitive and accurate to monitor gases such as SO_2, NO_x, Cl_2, HCl, CO_2 and CO at their points of discharge, where concentrations typically reach several hundred or thousand parts per million. Chemiluminescent analyzers are also used in this context to determine oxides of nitrogen, and the growing concern surrounding volatile organic compounds (VOCs) and other hydrocarbons has led to the use of stack monitors employing flame-ionization detectors. Sampling and preconditioning requirements are vital issues that frequently determine the type of instrument deployed in a particular application.

6.4 Monitoring Landfill Sites

Landfill gas is a mixture of methane and carbon dioxide, together with many trace species, and its monitoring is of rapidly growing importance. The dominant methods involve portable instruments which employ catalytic or ir sensors and which frequently measure the gas concentration as a percentage of the lower explosive limit (%LEL).

More sophisticated ir-based analyzers, including long-path types that can determine average gas concentrations on the surface or around the perimeters of sites, and narrow-gap ir sensors that operate down boreholes, are starting to be deployed. Some use is also made of laboratory gas analyzers to determine trace gas concentrations.

7. Monitoring Microbiological Quantities

Microbiological monitoring, which involves the quantification and identification of bacteria and viruses in water, makes scant use of instrumentation, as the majority of determinations are made by long-established laboratory culture methods which are enshrined in legislation. In the longer term, these methods are likely to be replaced, at least in part, by gene probes.

BOD (biochemical oxygen demand) is a quantity of major importance to the water and sewage treatment industries and quantifies the oxygen required for the biochemical decomposition of a waterborne organic pollutant. The standard test, designated BOD-5, takes five days to complete and efforts are underway to develop a more rapid alternative, based on electronic instrumentation. Technologies under consideration include on- and off-line bioreactors where the oxygen consumption of a colony of microorganisms is monitored, and multiparameter measurements (e.g., pH, uv absorption, turbidity, temperature, etc.), whereby various algorithms equate the signals generated to BOD-5.

8. Monitoring Physical Quantities

The most important physical quantities posing an environmental threat are noise and ionizing radiations (alpha, beta and gamma). Since the Chernobyl accident in 1986, growing importance has been attributed to monitoring ambient ionizing radiation levels but the significance, if any, of nonionizing radiations, such as microwaves, remains a controversial issue.

8.1 Noise Measurements

The majority of environmental noise measurements are made with portable sound meters, most of which use high-accuracy air condenser microphones as the input devices. These instruments are designed to determine compliance with the various national and international occupational or environmental noise standards. Spectrum analyzers are used to examine the frequency components of noise signals.

8.2 Radiological Measurements

Many different radiological quantities are monitored in the environmental context and this is a complex subject that can only be treated superficially here. However, practices are conceptually similar to those adopted in gas monitoring, namely, the use of fixed and portable field instruments, in-stack detectors and laboratory instruments and analyzers.

Fixed, ambient monitoring systems generally aim to detect gamma radiation and use Geiger tubes coupled into data acquisition networks. Gamma spectroscopy is adopted where specific gamma-emitting isotopes are sought and instruments are based around scintillating sodium iodide (NaI) detectors. In the case of alpha or beta radiations, air is sampled onto a filter which is subsequently examined with a solid-state, silicon detector.

Likewise, stack monitors that measure overall gamma levels use Geiger tubes, and specific gamma-emitting isotopes such as krypton and iodine employ gamma spectroscopy. Alpha and beta emissions are measured by silicon detectors, as in the case of ambient monitoring methods.

Gamma spectroscopy is used widely as an analytical laboratory technique and instruments are mostly based on liquid-nitrogen-cooled germanium detectors which, although costly, are far more sensitive than their NaI-based counterparts. Radioactive soil, food samples or marine sediments are examined for alpha and beta emitters in the laboratory by silicon detectors (alpha and beta spectroscopy). Portable instruments mostly employ Geiger tubes for the detection of gamma radiation and silicon detectors for measuring alpha and beta radiations.

9. Future Outlook

While many substances and quantities that are deemed to pose an environmental threat are being phased out or their use dramatically reduced (CFCs, dioxins, certain pesticides, halogenated organics, etc.), it is inconceivable that all environmentally hazardous substances will be eliminated. Thus, needs will exist for

Table 3
Some novel sensor technologies and their perceived applications

Technology	Measurands and applications
Fiber-optic—optrodes	monitoring waterborne pollutants such as ammonia, metals, pH and hydrocarbons in fixed and portable instruments
Fiber-optic—distributed	monitoring the leakage of hydrocarbons, methane, oils and fuels, etc., from storage sites and pipelines
Integrated optic	in portable and laboratory instruments to determine metals, herbicides, pesticides, etc., in water samples
Open-path optical	in fixed instruments for monitoring hydrocarbons, halogenated organics, metals, TOC, etc., in water. Also to monitor gases in portable and fixed instruments (see also below)
LIDAR/DIAL	in transportable instruments for the remote measurements and mapping of most polluting gases, for environmental audits, locating fugitive emissions and tracking hazardous emissions
Lasers	for the remote detection and analysis of oil slicks, and uses in LIDAR and other optical sensors
Silicon ISFETs	in portable and possibly fixed instruments to determine pH and metal ions in water
Silicon RADFETs	to measure ionizing radiation in air, in personal and fixed monitors
Biosensors	detection of herbicides, pesticides, metals, BOD and possibly many other waterborne pollutants, in portable and fixed instruments
Gene probes	uses in the laboratory to identify individual species of microorganism
Live-cell (bio)sensors	monitoring water toxicity, herbicides and pesticides and possibly BOD, in portable and fixed instruments
Thick-film electrodes	in portable and possibly fixed instruments to determine pH, metal ions, dissolved oxygen and other quantities in water
Multisensor arrays	speculative uses to determine metals in water and polluting gases in the atmosphere and around industrial plant

instrumentation that can reliably, accurately and cost-effectively monitor a wide range of chemical, biological and physical quantities for the foreseeable future. As a consequence and because of the limitations of much presently available instrumentation, there is a growing research effort into new and improved measuring methods and technologies, much of which involves sensor developments. Some of the potentially relevant classes of sensor are listed in Table 3, together with the quantities to which they respond and their perceived applications.

See also: Analytical Physical Measurements: Principles and Practice; Chemical Analysis, Instrumental

Bibliography

Abbou R (ed.) 1987 Hazardous waste detection, control, treatment. *Proc. World Conference on Hazardous Waste*. Elsevier, Amsterdam
Bogue R W 1990 The role of advanced sensors in environmental monitoring and pollution control practices. Laboratory of the Government Chemist, London
Briggs R, Grattan K T V 1990 The measurement and sensing of chemical species in the water and water-using industries. *Department of Trade and Industry*, London
Down R D 1991 *Environmental Control Systems*. Instrument Society of America, Research Triangle Park, NC
Elvidge A F 1990 *Process Control Instrumentation Hand-*
book. Institution of Water and Environmental Management, London
Instrument Society of America 1991 Environmental protection, control and monitoring. *Proc. ISA Int. European Regional Conf.* ISA, Research Triangle Park, NC
Johnson S P and Corcelle G 1989 *The Environmental Policy of the European Communities*. Graham and Trotman, London
Knoll G F 1989 *Radiation Detection and Measurement*. Wiley, New York

R. W. Bogue
[Robert Bogue and Partners,
Bere Alston, UK]

Errors and Uncertainty

An error in a measurement is the difference between the result of a measurement and the true value of the measured quantity. All measurements contain an error because procedures and instruments are not perfect. The error in a measurement is not, in practice, precisely determinable, so the result of a measurement is accompanied by a degree of uncertainty. This article considers the nature, sources and methods of description of errors and uncertainty in measurements and measuring instruments

1. Definition of Error

The error ε is the difference between the result of a measurement q_r and the true value of the measured quantity q_m:

$$\varepsilon = q_r - q_m$$

The true value of a quantity—the value that would be obtained from a perfect measurement, as defined by the appropriate measurement scale—is an idealized concept which cannot be empirically determined; even the definitions of the SI scales involve uncertainty. In general, therefore, we use the term "true value" in a conventional sense to mean an approximation to the true value which, for practical purposes, is close enough for any difference to be neglected. For example, in the calibration of an industrial instrument of moderate accuracy the value of a quantity as measured by an instrument of high accuracy is accepted as the conventional true value.

2. Measured-System and Measuring-System Errors

Consider the measurement system shown in Fig. 1. The complete measurement system consists of the object or system under measurement and the measuring-instrument system, which includes both the instrument acquiring the measurement information and the system processing it. The total measurement error consists of two components:

(a) the error originating in the measured system and the interface with the measuring-instrument system; and

(b) the error originating in the measuring-instrument system, also known as instrumental error.

2.1 Errors Arising in the Measured System

Consider the errors originating in the measured system and the interface. Let us take as an example, the measurement of the exhaust gas temperature of a small gas engine at some nominal speed and load.

The first type of error arises from the imperfection of the model of the measured system which underlies the measurement process. For example, if the temperature of the exhaust gases is not uniform, the result of a measurement made assuming a uniform distribution will be in error.

The second type of error arises from the change in the configuration of the measured system produced by introducing the sensor. For example, the introduction of a sensor in the exhaust duct raises the back pressure of the engine, requiring an increase in fuel flow and hence raising the exhaust gas temperature.

The third type of error is caused by the sensor absorbing power from the measured system. For example, heat exchange between the exhaust gases and the sensor alters the temperature of the exhaust gases.

These are the types of error occurring in passive measurement, in which the measured variable is sensed; the errors arising in a measured object in the process of active measurement are considered in Sect. 7.

2.2 Definition and Classification of Instrumental Errors

An instrument functions by maintaining a functional relation between the information carrying characteristics of the signal or signals at its input and the information carrying characteristics of its output. In a measuring instrument a functional relation is maintained between the magnitude of the measured quantity at the input and the numerical value displayed at the output. There is a nominal or intended instrument response law which relates the numerical value displayed at the output to the value of the measured quantity at the input. The nominal response law is

$$q_i = q_m$$

which states simply that the numerical value q_i displayed by the instrument when the input is the measured quantity q_m is equal to the true value of q_m (see Fig. 2).

The nominal response law is determined by calibration or by mathematical modelling and is deemed to

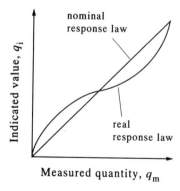

Figure 2
Response laws

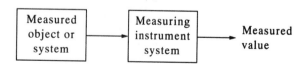

Figure 1
Measurement information system

hold for specified conditions of the environment and the system of which the instrument is part, termed reference conditions.

In practice the real response law of an instrument is such that the application at its input of a true value q_m results in an indicated value q_i, which is not equal to q_m, so that:

$$\varepsilon_i = q_i - q_m$$

where ε_i is the instrumental error. The instrumental error consists of:

(a) intrinsic errors, apparent when the instrument is used under reference conditions; and

(b) influence errors, arising from the action on the instrument of influence variables, that is, physical variables in the instrument's environment, which are not being measured, but which affect the instrument's response; these errors occur when the influence variables depart from their reference values.

(*a*) *Intrinsic instrumental errors.* There are two types of real instrument response under reference conditions:

(i) determinate response, in which repeated applications of a particular input always result in the same response; and

(ii) indeterminate response, in which repeated applications of a particular measured quantity input result in different responses.

If the real response is determinate but differs from the intended one, it results in a determinate error and is termed false. We may distinguish the following sources of error in a false response law.

(i) Errors in the determination of the intended response law arise either from

    (a) calibration, when erroneous or uncertain standards or procedures are used to establish the intended response law; or

    (b) calculation, when the intended response law is calculated using a mathematical model which is inadequate for the purpose.

(ii) Errors result from accepting a nominal response law different from a determined or determinable real one. Thus, for example, a mass produced instrument may be fitted with a linear scale which is correct at two points, even though the overall instrument response is nonlinear.

(iii) Errors result from a change with time of the real response law of the instrument as a result of degradation of material properties, wear, or damage due to maloperation. These are known as secular errors.

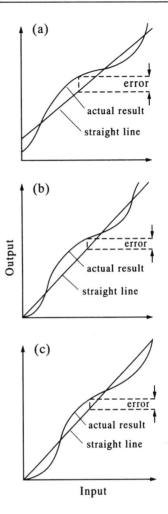

Figure 3
Deviations from (a) independent linearity, (b) zero-based linearity and (c) terminal-based linearity

The following are some of the most common forms of false response law error.

(i) Zero error—the deviation from zero of the indication of a measuring instrument for zero value of the measured quantity.

(ii) Gain error—in a linear real response law, the deviation of the slope from the unit value assumed in the nominal law.

(iii) Nonlinearity error—the deviation of the real response from nominal linear response. There are various ways of expressing nonlinearity (see Fig. 3):

    (a) departure from independent linearity—deviation from a best-fit straight line by least squares for the available data;

(b) departure from zero-based linearity—deviation from a best-fit straight line for the available data but assuming no zero error;

(c) departure from terminal-based linearity—deviation from a straight line joining the responses of the instrument at the terminal points of the span.

(iv) Hysteresis error—the error resulting from hysteresis, the property of a measuring instrument whereby it gives different readings according to whether the value of the quantity being measured has been reached by a continuously increasing or continuously decreasing change in that quantity.

(v) Quantization error—the error which may result from the measurement of a value of a quantity by a process in which response can change only in discrete quantum steps, as in digital measurement.

(vi) Dynamic error—the error in the operation of an instrument used in the dynamic mode, in which the instantaneous value of the instrument reading is required to be a function of the instantaneous value of the measured quantity, where that quantity is varying. A practical instrument always has a memory, or delay, as a result of which an instantaneous reading is a function not only of the instantaneous measured quantity input, but also of the past history of the measured quantity. Typically, a stepwise change in the measured quantity causes the instrument reading either to rise slowly to a new steady-state value corresponding to the new value of the measured quantity, or to oscillate about the steady value before settling down to it. When the measured quantity varies harmonically, the reading of a practical instrument tends to lag behind it in phase, and the gain of the instrument varies with frequency. The difference between the instantaneous indication and the instantaneous measured value is the dynamic error.

Indeterminate response laws originate in such effects as solid friction and varying resistance at contacts, which result in a nonrepeatable relation between measured quantity and instrument reading, Such laws give rise to random errors, considered in Sect. 3.

(b) *Instrumental influence errors.* Instrumental influence errors may be caused by influences arising in the measuring-system environment or in the measuring-instrument system, that is:

(i) the instrument output interface—the influence being the effect on the instrument reading of any power abstracted from the instrument at its output;

(ii) the instrument input interface—the influence being the effect on the measured system of the power abstracted from the measured system by the instrument (the resultant error is strictly a measured-system error rather than an instrumental one); and

(iii) the energizing power supply in active instruments—the power supply being an influence on instrument reading, its departure from reference conditions causing error.

The environment influence variables which can act on an instrument may be classified, as follows, according to their energy form.

(i) Mechanical. Vibrations affect the relative position of elements of an instrument, for example, by introducing variable stresses which affect material properties. Since the force of gravity acts on the components of a mechanical instrument, its response may be affected if its orientation changes with respect to the vertical.

(ii) Electrical. Electrical signals are induced in circuits by various forms of stray coupling.

(iii) Fluid mechanical. High ambient pressures may change the geometry and material properties of an instrument, and hence the instrument's response law. This may be of significance in, say, underwater instrumentation.

(iv) Thermal. As temperature affects all material dimensions and properties, it is a significant influence variable for all instruments.

(v) Chemical. A change in the chemical composition of any components of an instrument as a result of environmental chemical influence may significantly affect the components' behavior. Electrochemical potentials may also be significant.

(vi) Ionizing radiation. Such radiation may alter the material properties of instrument components or induce electrical signals.

Let us denote the measured quantity by q_m, and the indicated value by q_i. Let us further denote an influence (disturbance) variable by $q_{d1}, q_{d2}, \ldots, q_{dn}$. The response law of the instrument may then be written as

$$q_i = f(q_m, q_{d1}, q_{d2}, \ldots)$$

Let the reference conditions be denoted by $\bar{q}_{d1}, \bar{q}_{d2}, \ldots$. Then, in general, the influence error ε_d is given by

$$\varepsilon_d = f(q_m, q_{d1}, q_{d2}, \ldots,) - f(q_m, \bar{q}_{d1}, \bar{q}_{d2}, \ldots,)$$

Then, writing

$$\Delta q_{dn} = q_{dn} - \bar{q}_{dn}$$

we have

$$\varepsilon_d = \left(\frac{\partial q_i}{\partial q_{d1}}\right) \Delta q_{d1} + \left(\frac{\partial q_i}{\partial q_{d2}}\right) \Delta q_{d2} + \ldots$$

where, if the values $\Delta q_{\mathrm{d}n}$ are small, the terms with higher order derivatives may be neglected. In many practical cases these derivatives tend to be negligibly small. The terms $\partial q_i / \partial q_{\mathrm{d}n}$ are called the influence error coefficients. (It is more usual to specify the relative coefficients $(1/q_i)(\partial q_i / \partial q_{\mathrm{d}n})$.) The coefficients $\partial q_i / \partial q_{\mathrm{d}n}$ may be independent of q_{m}, so that the influence errors are additive (as e.g., in inductive pick-up) resulting, typically, in a zero error with no change of gain. In other cases the coefficients are functions of q_{m}, making the gain a function of the influence variables. A typical example would be the change of gain with temperature of a load cell employing an elastic deformation element.

The above considerations apply to reversible effects of influence variables. Irreversible effects are not influence variables, but secular intrinsic errors.

(*c*) *Time variation of influence and intrinsic errors.* The time variations of the effects of influence variables are of importance. We distinguish between

(i) short-term variation, in which the influence variables change appreciably over the period of a measurement process; and

(ii) long-term variation, in which the influence variables change so slowly that they are effectively constant over the period of a measurement process.

Similar considerations apply to secular changes of intrinsic errors.

3. Systematic and Random Errors

We distinguish between systematic errors, which are the same in repeated measurements of the same value of a measured quantity, and random errors, which differ in repeated measurements of the same value of a measured quantity.

Systematic errors are due to

(a) intrinsic errors arising from a determinate false law; and

(b) influence errors arising from long-term influence variations or secular intrinsic instrument changes, which although basically random are effectively systematic for a particular measurement.

Random errors are due to

(a) intrinsic errors arising from an indeterminate response law; and

(b) influence errors arising from short-term variations of an influence variable.

With respect to systematic errors, we may distinguish between certain and uncertain systematic errors. Certain systematic errors arise when we have a known real instrument law, but where the nominal response law is different, or when we known the deviation of influence quantities from reference conditions, and the law relating the influence variables to the instrument response. Certain systematic variables can be compensated by correction. Uncertain systematic errors arise when the real response law is determinate but unknown, for the reasons explained earlier.

4. Uncertainty and its Specification

Certain systematic error is, by definition, a known quantity at every value of the measured quantity. Random errors and uncertain systematic errors are unknown quantities, and make it impossible to assign with certainty a value of the measured quantity to the result of a measurement.

Given a particular value q_i, indicated by an instrument under reference conditions, the true value of the measured quantity q_{m} is related to q_i by a conditional probability density function (pdf):

$$p(q_{\mathrm{m}} \mid q_i)$$

A measure of the dispersion of this pdf is a measure of the inherent instrumental uncertainty at an indicated value of q_i. The pdf can in theory be estimated by calibration.

In practice, it is assumed that the random errors for any measured value are distributed according to a normal distribution with a zero mean error:

$$p(\varepsilon) = \frac{1}{\sigma \sqrt{2\pi}} \exp\left(-\frac{\varepsilon^2}{2\sigma^2} \right)$$

where

$$\varepsilon = q_i - q_{\mathrm{m}}$$

and σ, the standard deviation of the error distribution, is a measure of the uncertainty of the measurement given a normal distribution. In practice σ is estimated experimentally by replicating the measurement of a particular value of the measured quantity.

The reason for assuming zero mean error is that an intrinsic random error with a nonzero mean can be taken to be an uncertain systematic error.

The assumption of a normal distribution is based on the supposition that, in a well-designed measurement, random intrinsic errors will be the result of a large number of small, uncorrelated component errors. Such a process will, according to the central limit theorem, result in a normal error distribution. This assumption does not hold for many less precise measurements, for which the random error is the result of a small number of substantial error components, possibly correlated. Nevertheless, the assumption of a normal error distribution is adequate for most practical purposes.

Given a set of n indicated values $q_{i1}, q_{i2}, \ldots, q_{in}$, for a particular value of q_m, the best estimate of q_m is

$$\hat{q}_m = \bar{q}_i = \frac{1}{n} \sum_{k=1}^{n} q_{ik}$$

The best estimate of σ^2 is given by

$$\hat{\sigma}^2 = \frac{1}{n-1} \sum_{k=1}^{n} (q_{ik} - \bar{q}_i)^2$$

Because the estimation of $\hat{q}_m$ by $\bar{q}_i$ was based on a limited number of observations, repeated determination of $\bar{q}_i$ would produce different values. For large values of n these values will be distributed according to a normal distribution. The best estimate of the variance of the latter distribution is given by

$$\hat{s}^2 = \hat{\sigma}^2/n$$

where s, known as the standard error of the mean, is a measure of the uncertainty in q_m.

Given a knowledge of $\hat{s}$, it is possible to calculate values $q_m + \delta_1$ and $q_m - \delta_2$, such that the true value of q_m lies between them with a chosen probability P; $q_m + \delta_1$ and $q_m - \delta_2$ are known as the confidence limits, and P is known as the confidence level. For a normal distribution $\delta_1 = \delta_2 = \delta$, and we have that

$$\delta = t\hat{s}$$

The value t (called Student's t) is available in tables which give t for various values of P and the degrees of freedom $(n-1)$ used in the calculation of s. Thus, for intrinsic errors, the uncertainty will be specified as $\hat{s}$ or as $\pm\delta$ at some specified confidence level. For influence errors it is usual to specify for an instrument influence error coefficients $\partial q_i/\partial q_{dn}$ (or, more usually, $(1/q_i)(\partial q_i/\partial q_{dn})$). The actual influence errors, however, depend on the extent to which the influence quantities deviate from reference conditions. If ε_{dn} is the error due to the influence quantity q_{dk} deviating by Δq_{dn} from the reference conditions, then, as shown previously (see Sect. 2.2),

$$\varepsilon_{dk} = \frac{\partial q_i}{\partial q_{dn}} \Delta q_{dn}$$

If the actual value of Δq_{dn} is unknown, we may know the pdf of Δq_{dn}, $p_{q_{dn}}(\Delta q_{dn})$. The pdf of the influence error due to q_{dk} is then

$$p_{\varepsilon_{dn}}(\varepsilon_{dn}) = \frac{p_{q_{dn}}(\Delta q_{dn})}{\left| \dfrac{\partial q_i}{\partial q_d} \right|}$$

The mean of $p_{\varepsilon d}(\varepsilon_d)$ is the best estimate of the

influence error ε_d, and the variance (or standard deviation) of $p_{\varepsilon d}(\varepsilon_d)$ can then be used as a measure of the uncertainty due to influence effects.

Systematic uncertainties are best specified in terms of limits $q_i + \delta_1$ and $q_i - \delta_2$ between which, for a given value of q_i, lies the true value of q_m. The limits of systematic uncertainty must be estimated by a mathematical analysis of errors.

In accurate measurement it is recommended that random and systematic uncertainties should be specified separately, as described earlier. In many industrial measurements systematic uncertainties are much larger than random ones, and the latter may be neglected.

5. Accuracy, Repeatability and Reproducibility

The accuracy of a measuring instrument is a measure of how close an instrument's reading is to the true value of the quantity measured.

The quantitative characterization of accuracy should be given in terms of random and systematic uncertainty. Since the uncertainties will be different at different values in the range of the instrument, the uncertainty should be specified as a function of the quantity measured. In practice it is common to specify the maximum uncertainty in the range as a fraction of the upper limit of the range. The actual accuracy of the instrument in use depends also on the magnitude of the influence error effects.

In many cases what is of interest is not the accuracy, but the repeatability of an instrument: its ability to give identical indications or responses for repeated applications of the same value of the measured quantity under stated conditions of use. Repeatability is essentially freedom from random intrinsic error; it is expressed quantitatively by a measure of dispersion of the pdf $p(q_i/q_m)$, or generally $\hat{\sigma}$. As with accuracy, it should be stated as a function of the measured value.

The accuracy of a measurement is how closely the results of that measurement correspond to the true value of the measured quantity (see Sect. 2). Accuracy should be stated in terms of the systematic and unsystematic uncertainties.

The repeatability of a measurement is a quantitative expression of the closeness of agreement between successive measurements of the same value of the same quantity carried out by the same method, by the same observer, with the same measuring instruments, at the same location and at appropriately short intervals of time.

The reproducibility of a measurement is an expression of the closeness of the agreement between the results of measurements of the same value of the same quantity, after correction for errors due to individual measurements having been made, for example, under different conditions, by different methods or with different measuring instruments.

6. Indirect Measurements

In indirect measurement the value of a quantity is calculated from measurements, made by direct methods, on other quantities having a known relationship with the quantity to be measured.

If q_m denotes an indirectly measured quantity, determined from direct measurements of the quantities q_1, q_2, ..., by the use of the relation

$$q_m = f(q_1, q_2, \ldots)$$

then the estimate of q_m is taken as

$$\hat{q}_m = f(\hat{q}_1, \hat{q}_2, \ldots)$$

where $\hat{q}_1$, $\hat{q}_2$, ..., are the estimates of the means of q_1, q_2, ...

Then, if for random uncertainties the estimated variances of q_1, q_2, ..., are $\hat{\sigma}^2(q_1)$, $\hat{\sigma}^2(q_2)$, ..., we have

$$\hat{\sigma}^2(q_m) = \left(\frac{\partial q_m}{\partial q_1}\right)^2 \hat{\sigma}^2(q_1) + \left(\frac{\partial q_m}{\partial q_2}\right)\hat{\sigma}^2(q_2) + \ldots$$

The variance of the mean is given by

$$\hat{s}^2(\bar{q}_m) = \left(\frac{\partial q_m}{\partial q_1}\right)^2 s^2(\bar{q}_i) + \left(\frac{\partial q_m}{\partial q_2}\right)^2 s^2(\bar{q}_2) + \ldots$$

In the equations above higher order terms are neglected and all components are assumed to be independent of one another.

For systematic uncertainties, if $\Delta(q_m)_1$ is the component of systematic uncertainty of q_m due to the systematic uncertainty Δq_1 in q_1, then

$$\Delta(q_m)_1 = \left|\frac{\partial q_m}{\partial q_1}\right| \Delta q_1$$

There is no rigorous way of combining the components of systematic uncertainty to give the total systematic uncertainty $\Delta(q_m)$. Two relations are used in practice. In one the components are simply added:

$$\Delta(q_m) = \left|\frac{\partial q_m}{\partial q_1}\right| \Delta q_1 + \left|\frac{\partial q_m}{\partial q_2}\right| \Delta q_2 + \ldots$$

This is likely to be an overestimate, and represents an estimate of the maximum uncertainty. In the other method the systematic uncertainties are combined in quadrature:

$$\Delta(q_m) = \left[\left(\frac{\partial q_m}{\partial q_1}\right)^2 (\Delta q_1)^2 + \left(\frac{\partial q_m}{\partial q_2}\right)^2 (\Delta q_2)^2 + \ldots\right]^{1/2}$$

This method tends to underestimate the uncertainty.

There may exist also an uncertainty in the relation $f(\ )$. This may be dealt with by introducing an uncertain parameter in the above equations.

7. Active Measurement

So far we have considered measurements in which the measured variable acts directly on the measuring instrument, as in the measurement of quantities associated with a power flow—for example, force, current or temperature.

Quantities which characterize the storage, transformation or transport of energy must be measured by active procedures. In active measurements the magnitude of the measured quantity is determined by interrogating the physical system by a physical input and observing the physical output. The measured quantity is obtained from values of the input and output, using a model of the physical system which relates the measured quantity to the input and output variables. (This is indirect measurement, as discussed in Sect. 6.) There are four main sources of error: in the determination of the interrogating input quantity, in the determination of the observed output quantity, the inadequacies of the system model, and the effects of noise.

See also: Errors: Avoidance and Compensation

Bibliography

Campion P J, Burns J C, Williams A 1973 *A Code of Practice for the Detailed Statement of Accuracy.* HMSO, London

Dietrich C F 1991 *Uncertainty, Calibration and Probability*, 2nd edn. Hilger, London

Hofman D 1982 Measurement errors, probability and information theory. In: Sydenham P H (ed.) *Theoretical Fundamentals*, Handbook of Measurement Science, Vol. 1. Wiley, Chichester, UK pp. 241–75

Jaech J L 1985 *Statistical Analysis of Measurement Errors.* Wiley, New York

Topping J 1972 *Errors of Observation and their Treatment.* Chapman and Hall, London

L. Finkelstein
[City University, London, UK]

Errors: Avoidance and Compensation

The essence of the functioning of instruments or instrument systems is to realize a known, nominal transformation of the input signal into an output signal. Errors are the departure of the actual transformation realized from the required nominal one.

Practical systems always involve errors. The nature and sources of errors are considered in the article *Errors and Uncertainty*. Error avoidance and compensation is thus an essential consideration in the design, manufacture and operation of instruments. The general principles of such error avoidance and compensation are the subject of this article.

1. Intrinsic and Influence Errors

Intrinsic errors are those departures from the nominal transformation which arise under reference conditions, those conditions of the environment under which the nominal transformation is intended to hold.

Influence errors are the effect on the output of the instrument or instrument system of departures of the environment from reference conditions.

The methods of error avoidance and compensation are different with respect to intrinsic and influence errors.

2. Calibration

Calibration is the determination of the nominal transformation characteristic of an instrument or instrument system. An instrument characeristic thus determined is accurate up to the errors inherent in the calibration process.

Intrinsic errors are thus avoided by a calibration which is accurate. Changes of instrument characteristics which occur with time are avoided by regular recalibration.

In many instrument systems facilities may be provided for calibration when they are actually *in situ*, that is, embodied in the measurement system. Intelligent or "smart" systems frequently incorporate facilities for such recalibration to take place automatically, with automatic correction of the nominal characteristic.

3. Correction of Instrument Transformation Characteristics

An instrument response law or input–output transformation is commonly desired to be linear and to have negligible dynamic error.

A response law which is stable with time but exhibits, for instance, significant nonlinearity or dynamic lag may be corrected by processing the output signal. If G_r is the real transformation and G the desired one, then the required output signal transformation is (G/G_r). Correction of static characteristics in this manner is satisfactorily realizable by digital signal processing.

Correction or compensation of dynamic characteristics in this way presents two problems. The high-pass filtering process required to compensate for dynamic lag enhances noise. In any case the undesirable dynamic characteristics are not generally known to an adequate accuracy or stable with time.

4. Construction for Avoidance of Intrinsic Errors

Essentially intrinsic errors are avoided by building instruments or instrument systems with stable characteristics.

Instrument working principles which are not dependent on material properties are preferred. Incomplete elements are avoided, and stable materials are used.

Where necessary, it is possible to use negative feedback. Components which do not have satisfactory characteristics are placed in the forward path of a high-gain feedback loop, and may carry the operating power, while the instrument characteristics are maintained by low-power-carrying precision components in the feedback path (see *Instrument Systems: Functional Architectures*).

5. Influence Error Model

To explain the avoidance and compensation of influence errors it is convenient to construct a model of instrument errors. This is shown in Fig. 1.

q_m is the measurand input, q_i is the indicated value or output, and q_d is the influence variable or disturbance or, more strictly, its deviation from reference conditions. The influence error is ε_d.

Then

$$q_i = Gq_m + G_d q_d$$
$$\varepsilon_d = G_d q_d$$

where G is the nominal instrument transformation or law and G_d is the influence error coefficient.

It can be assumed, for simplicity of argument, that

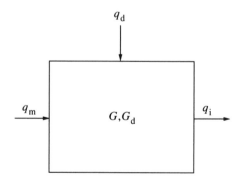

Figure 1
Influence error model

the relations are linear. For dynamic relations, these variables can be considered to be Laplace transforms.

The avoidance or compensation of influence errors requires that

$$\varepsilon_d = 0$$

that is,

$$G_d q_d = 0$$

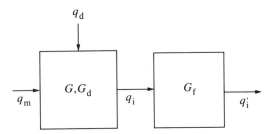

Figure 2
Disturbance feedback model

6. Elimination of Disturbances

The most effective method of avoidance of instrument error is to suppress deviations of influence variables from reference conditions, that is, to make $q_i = 0$.

Practical examples are the location of the instrument system in a controlled environment, or the suppression of major sources of disturbance, for example, fitting suppressors to sources of electromagnetic disturbance.

7. Elimination of Influence Sensitivity

If the presence of disturbances is unavoidable it may be possible to make instruments insensitive to them, that is, to make $G_d = 0$.

Typical practical methods are to place an instrument in a thermal insulating enclosure, to render it insensitive to temperature variations of the environment, or in a Faraday cage, to make it insensitive to electromagnetic disturbances.

Alternatively, it may be possible to avoid sensitivity to disturbances by appropriate construction. For example, if the cause of sensitivity to temperature is the thermal expansion of the dimension of components, it is possible to construct them of materials whose coefficients of thermal expansion are minimal. Another method is to follow a component in a chain with a positive disturbance sensitivity, by a component with an equal but negative sensitivity.

8. Disturbance Feedforward

If neither suppression of disturbance nor insensitivity to it are achievable it is possible to compensate for the effect of disturbances by disturbance feedforward. This is illustrated in Fig. 2. The disturbance q_d is sensed and an equal and opposite signal $-G_d q_d$ is fed to the output.

This may be accomplished off-line. If the temperature sensitivity of the device and the temperature deviation of the environment from reference conditions are known, it is poossible to calculate the compensation to be applied to the indicated value. This may be performed automatically by an intelligent or smart instrument.

Automatic disturbance feedforward may be built into an instrument in other ways than by intelligent compensation. One method is by the use of parallel or differential architectures. A typical example is an electrical resistance strain gauge instrument in which the strain gauges are connected in two opposite arms of a Wheatstone bridge, the other being formed by identical but unstrained resistors. The effect of temperature sensitivity on the strain gauges actively sensing strain is counteracted by the equal and opposite effect on the compensating gauges.

It is possible, in off-line use or intelligent instrumentation, to use a model of the disturbance q_d to estimate q_d as q_d' and to use a compensating quantity or signal $G_d q_d'$.

9. Feedback

The effect of influence variables may be avoided by a feedback architecture, discussed earlier. By constructing an instrument in a closed-loop architecture, in which components sensitive to influences are placed in the forward path of a high-gain feedback loop, and the required instrument transformation is maintained by insensitive components in the feedback path of the loop, the output is rendered insensitive to disturbances.

10. Insensitivity to Short-Term Influence Variations

Insensitivity to influence variations may be achievable by filtering. In Fig. 3 the block labelled G_f is a filter.

Thus,

$$q_i = G q_m + G_d q_d$$
$$q_i = G_f q_i$$
$$q_i' = G_f G q_m + G_f G_d q_d$$

If the frequency characteristics of q_m are substantially different from those of q_d, for example, if the measurand is constant, or varies slowly, and the disturbance

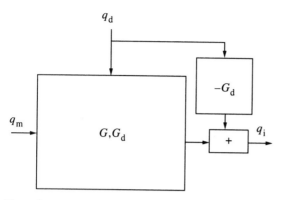

Figure 3
Filtering model

is high frequency, it is possible to make $G_f G_d q_d$, the effect of q_d negligible by an appropriate design of the filter G_f. It is necessary to achieve a balance between maximizing the suppression of the disturbance and minimizing the dynamic distortion of the measurand introduced by G_f (see *Signal Theory in Measurement and Instrumentation*).

One way of achieving this is to design the signal carrying the measurement information to have the appropriate frequency spectrum, by modulating a carrier signal of appropriate frequency spectrum by the measurand. A typical example is chopping an infrared beam in infrared intensity measurement.

See also: Errors and Uncertainty

Bibliography

Bentley J P 1983 *Principles of Measurement Systems*. Longman, London
Doebelin E O 1990 *Measurement Systems: Application and Design*, 4th edn. McGraw-Hill, New York

L. Finkelstein
[City University, London, UK]

F

Flow Measurement: Applications and Instrumentation Selection

The potential purchaser of a flowmeter, unskilled in the art of selection, can adopt one of several options before placing the order:

(a) engage an expert flow consultant to advise on selection (might be expensive);

(b) study the subject intensively and make one's own selection (there may not be the time or opportunity);

(c) invite quotations from flowmeter manufacturers and rely on their expertise to make a wise selection (vested interest might bias the recommendation); or

(d) prepare a tight and comprehensive specification of the flow metering application after due consideration of all the factors presented in this section and then invite competitive tenders from manufacturers.

If this last option is pursued, the intending purchaser might find the contents of this section to be useful as a form of checklist to ensure that flowmeter manufacturers are provided with all the essential information on which to make satisfactory competitive bids.

In the near future it is hoped that some of the hard work involved in selecting a flowmeter will be eased by the availability of intelligent knowledge based computer systems.

Selection cannot be considered until all the aspects of the flow application are fully understood. If the user decides to leave the selection to the manufacturer it is still the user's responsibility to provide the manufacturer with all the relevant facts concerning the application. This article guides the user to acquire all the information necessary for a successful choice.

The checklist factors logically fall into six study areas:

(a) measurement requirements,

(b) conditions external to the flow conduit,

(c) conditions within the flow conduit,

(d) properties of the fluid,

(e) accessories and installation, and

(f) economic factors.

1. Measurement Requirements

No selection procedure can give a good result unless there is a clear understanding of the functions to be performed by the flowmeter.

A number of basic questions have to be answered.

(a) Is it to be used in an open conduit or closed conduit system?

(b) What are the dimensions of the system?

(c) What are the chemical and physical characteristics of the fluid?

(d) What is the rate of flow or quantity flow, or both?

(e) What is the measuring fluid velocity or the flow rate?

(f) What is the volumetric flow, mass flow or reference flow?

(g) Is the flow batch or continuous? (Quantity meters are generally required for batch processes as they can indicate precisely the amount of an added component in such a process.)

(h) Are there any legal or industrial requirements? (If there is a legal requirement the meter needs to be certified by an appropriate body.)

(i) What are the safety requirements?

The following sections cover other requirements of a flowmeter.

1.1 Accountancy or Process Measurement

For accountancy purposes the meter should have a low and consistent uncertainty of measurement over its lifetime. In many process measurements, but not all, a good repeatability performance is sufficient and the uncertainty of measurement is rather less important.

1.2 Flow Rate Range (Maximum, Minimum, Normal)

If a wide range of flow rates is required, differential pressure type flowmeters are not very satisfactory because their range is limited to about 3:1 for best accuracy. Two flowmeters in tandem, one to measure the higher rates and the other to measure the lower rates, are sometimes used to extend the range to 9:1, but then an automatic changeover system on the output signal is required. The system then tends to become complicated and expensive.

Linear flowmeters such as the turbine, electromagnetic, ultrasonic, vortex and variable area, normally have a flow rate range of 10:1 or more, but this is dependent on the applicational conditions. The range is extendable when a lower level of accuracy is acceptable or when special techniques are applied such as automatic range changing or electronic signal processing.

A few nonlinear flowmeters having a flow characteristic given by the formula $Q = KY^n$, can also provide an extended range, where Q is the flow rate, K is the coefficient, Y is the measured flow variable, and n is greater than unity. A gate meter is a typical example of such a flowmeter.

1.3 Accuracy

The requirements need to be very clear as there is much confusion over the interpretation of this term. It is sometimes expressed as a percentage of full-range flow rate and sometimes as a percentage of actual flow rate. Further confusion can exist as sometimes accuracy is specified for a meter operating under reference conditions and at other times under conditions within specified limits of temperature, pressure, supply voltage, frequency, vibration, and so on. Accuracy specifications invariably refer to a new meter, and accuracy degradation with age is an unknown quantity but invariably occurs.

1.4 Linearity

Most meters exhibit small nonlinearities in their calibration curves. Sometimes these are systematic and a correction can be applied (a microprocessor application). In other cases the nonlinearities and hystereses are random by nature and correction is not possible. In general, users accept either linear or square root characteristics and regard departures from either as an inaccuracy.

A high degree of linearity simplifies computation where, for instance, two flow rates need to be ratioed or added, or where heat transfer is to be measured by multiplying a flow rate by a temperature difference. The microchip revolution has removed some of the problems arising from nonlinearities but adds to the cost and complication because of the hardware and software requirements.

1.5 Repeatability

The requirement for repeatability should specify whether it is for the short term or long term. Drift in calibration is a well-known phenomenon and is dependent on the type of flowmeter, its design and its applicational conditions—the latter being the main cause for calibration drift.

1.6 Life and Maintenance

Long life is a virtue in the sense that it implies slow deterioration in performance and reliability, but it is a virtue which can be overrated as the user can lose the advantage of more modern technology offering a better specification, greater reliability and lower maintenance.

Lower maintenance and diagnostic simplicity are very important considerations for today's economic conditions.

1.7 Pressure Loss

This can provide a problem for two reasons. For large-size meters (>500 mm), high-pressure losses can be responsible for high pumping or energy losses. For this reason venturis and low-loss devices are preferred to orifices. Better still, electromagnetic and ultrasonic meters can be employed with a negligible head loss.

There is another problem which arises when fluids flow by gravity in pipes or in open channels because any significant head loss would limit the throughput or cause back-up and flooding.

When referring to head loss associated with a flow device, the meaningful figure is that represented by unrecoverable losses and not necessarily by the head generated by the device, some of which can be recovered dynamically.

1.8 Hygienic Requirements

The measurement of fluid flow in the food industry can demand the installation of certain types of meters which meet specified hygienic requirements. Such requirements could include the use of special materials of construction, absence of cavities where stagnation could occur, ease of cleaning by simple disassembly, or the ability of the whole meter to withstand high temperature sterilization.

1.9 Dynamic Response

A meter which can perform perfectly well on a calibration rig may be less satisfactory on an industrial installation where the flow varies widely and rapidly. This aspect needs particular consideration when the dynamic responses to rising and falling flow rates are asymmetrical as in that situation serious errors can arise. Dynamic response can also be an important factor if the meter is part of an automatic control loop (see also effects of pulsating flow).

1.10 Signal Transmission

Industrial applications may involve transmission distances from a few meters to hundreds of kilometers, and the possibility of transmission errors needs to be considered, particularly if the transmission is over telephone lines and passes through several exchanges.

A flowmeter which has an inherent digital or quasidigital output signal can offer a potentially higher transmission accuracy than a comparable meter having an analog output signal. Also, the digital signal can have a higher degree of compatibility with digital systems and computers.

1.11 Power Supplies

Although the majority of flowmeters are designed to operate from industrial and domestic power supplies, there are applications where operating conditions make electrical power operation undesirable or where power supplies are nonexistent or inconvenient. A particular case occurs when legal metrology or custom transfer is involved, because problems would arise in the event of a power failure.

In the absence of domestic or industrial power supplies, consideration can be given to the use of batteries or solar power. Power from such sources can sometimes be conserved by intermittent measuring ("flash firing").

1.12 Adjustments to Calibration

Some flowmeters have user-accessible devices for adjusting the calibration range and zero. Such features are obviously unacceptable on meters involving custom transfer. They are also valueless to a user who has no facilities for calibration unless the adjusting means are provided with calilbration scales. In the latter case the flowmeter can be regarded as an adjustable range device which can provide a very important function to users who have to deal with any of the following problems:

(a) uncertainty concerning the maximum or minimum flows;

(b) large seasonal fluctuations in the flow rate;

(c) a foreseeable extension of the process involving larger flow rates; and

(d) the need to transfer the flowmeter to another application.

1.13 Pipeline or Insert Flowmeter

For large pipelines it is more economical but less accurate to use an insert flowmeter, that is, a small sized fluid velocity measuring device that can be positioned in the flow conduit where it can measure a fluid velocity that can be equated to flow rate.

2. Conditions External to the Flow Conduit

The performance and suitability of various flowmeters can depend on the environment in which they work. The following checklist refers to the various factors which need consideration.

2.1 Accessibility

For maintaining good long-term performance, accessibility is an important factor, facilitating ease of removal for recalibration, or convenience for in-line calibration, for maintenance and for diagnostics.

2.2 Temperature

Ambient temperature changes can cause errors due to:

(a) changes of density or viscosity of the measured fluid by heat transfer;

(b) differential temperature effects on the density of a fluid in pressure lines from a pressure difference device to its secondary measuring system; this occurrence can give rise to false heads;

(c) effects of temperature on electrical components;

(d) changes in elastic properties of membranes and springs; and

(e) damage can occur to components by exposure to both high and low ambient temperature. The possibility of pressure lines freezing or vapor lines condensing can be guarded against by tracing with steam pipes or electrical resistance cabling.

Modern flowmeters including the primary and secondary equipment can normally operate without special precautions over a temperature range of 0–80 °C. The primary measuring devices have a much wider range, some suitable for temperatures down to −250 °C and others up to 1500 °C.

High rates of change of temperature can sometimes be more troublesome than sustained high or low temperatures because the time constant of the compensating device may differ significantly from that of the flowmeter.

2.3 Humidity

High humidity can accelerate atmospheric and electrolytic corrosion and can damage electrical insulation. Low humidity can induce static electricity. Flow recorders using paper charts can be affected by extremes of humidity. Humidity problems can be induced by rapid environmental temperature changes.

2.4 Vibration

Vibration can cause wear and damage to pivots and bearings and can affect the function of unbalanced moving components. It can also cause malfunction of some vortex and ultrasonic flowmeters and differential pressure transmitters. Parasitic electrical signals can be generated in some high-impedance cables as a result of vibration.

2.5 Hostile Atmospheres

Hostile atmospheres are generally corrosive due to the presence of sulfur and nitrogen oxides, hydrogen sulfide or "marine" conditions. These can give rise to corrosive attack on vital parts of flowmeters, including electrical contacts, circuits and components if suitable protection is not provided.

2.6 Fire and Explosion Hazards

If fire or explosion hazards exist permanently or occur occasionally, or only in the event of some other mishap, then only certified instruments can be used which must be either intrinsically safe or housed in flameproof containments.

2.7 Electrical and Magnetic Fields

The performance of flowmeters incorporating electrical or electronic circuits may be subject to stray fields. These might occur naturally by lightning, in which case arrestors may be required, or the interference could be from electrical apparatus including power cables. Four specific dangers are:

(a) radio-frequency interference from walkie-talkies;

(b) the use of galvanic protection of pipelines (special danger to electromagnetic flowmeters);

(c) the existence of earth loops in the flowmeter circuitry; and

(d) the location of signal and power cables in the same duct.

2.8 Water Logging

In some applications, meters have to be installed below ground level with the consequent danger of water submersion. If an electromagnetic flowmeter or any other meter containing unprotected water-sensitive components or circuitry is to be used below ground level, a submersible design must be selected.

2.9 Waterproofing

The primary measuring device is generally designed to withstand outdoor weather conditions, but this is not always the case for the secondary equipment. In some industries flow measuring equipment may experience accidental or regular hosing down.

3. Conditions Within the Flow Conduit

3.1 Flow Profile

The flow profile (velocity variation across a section of the flow conduit) is probably one of the most important conditions affecting the uncertainty of flow rate measurement. Ideally, the flow profile at the point of measurement should be established by experiment or by modelling as a preclude to the installation of a flowmeter having a profile sensitivity (which applies to most types). However, this ideal approach is normally laborious and expensive and is rarely undertaken.

The presence of upstream bends, valves, obstructions and changes in cross-sectional shape or area of the conduit all affect the fluid velocity profile, and if this profile happens to be different from that which existed during the calibration of the meter, errors can arise which are difficult or impossible to predict. The subject is immensely complicated. Because the task of establishing the profile at different flow rates is difficult and time-consuming and generally not practicable, the conventional approach is to ensure that the lengths of straight upstream and downstream conduit are in accordance with published standards or manufacturers' recommendations. If these recommendations cannot be followed, the user might carry out an *in situ* calibration, but again this can be laborious and expensive. One practical solution is to fit a "flow straightener" upstream of the meter as this can reduce the asymmetry of flow profile and any rotational velocity components (see Sect. 3.2). Semipositive and positive

displacement-type quantity meters are not usually affected by asymmetric flow profiles. Coriolis-type flow rate meters are also relatively unaffected.

3.2 Swirl

Swirl frequently arises from the presence of two adjacent bends located in different planes. The effect on accuracy can be very serious, particularly for turbine meters, and to a lesser extent for electromagnetic meters and differential pressure devices. It can be minimized by long lengths of straight upstream pipe, but more compactly by a flow straightener, provided that the extra head loss caused by the straightener can be tolerated.

3.3 Protrusions

Accuracy can be prejudiced by the existence of protrusions in the pipeline. These can take the form of nodular deposits, burrs, a thermometer pocket or, more usually, accidental extrusion of gasket material at the upstream flange of the meter itself. An electromagnetic flowmeter with its corrosion-proof liner and straight-through bore appears to suffer less than other types of flowmeter from protrusion problems.

3.4 Pipe Bore: Size, Roundness, Roughness

Pipe bore characteristics apply more specifically to orifices which require the bores to be known and for the pipe to be round and internally smooth. The problem is less acute with flow devices employing electromagnetic, vortex and ultrasonic principles, for example, in which the critical section of the flow conduit is an integral part of the flowmeter. The problem is more acute with the simpler designs of insert meters or the shortened or wafer types which are clamped between pipe flanges.

3.5 Separation of Second Phase

Flowing fluids often contain one or more other phases varying from trace amounts (which can often be ignored) to a significant proportion of the volumetric or mass transport. These other phases can be immiscible liquids, gases, or solid particles. The problems of metering polyphase fluids are discussed later, but in this section attention is given to difficulties which arise when one phase separates out from the main stream or indeed is formed as scale at critical points in the metering system. Small quantities of transported second phase such as solid particles or gas bubbles can generally be tolerated either by the selection of a suitable meter or dealt with by vents, drains, probes for pressure tappings or by sumps. Orifice accuracy can be dramatically affected by deposited scale on the bore. Vortex meters can be affected by build up on the meter's internal bore or on the bluff body and electromagnetic flowmeters can be affected by build up on the electrodes. Ultrasonic clamp-on noninvasive flowmeters are least affected, but there are other inherent limitations to this type of flowmeter. In any case, any build up of scale on the bore of the meter will affect accuracy as in most techniques the measurement is derived from fluid velocity.

If the separated phase is a gas the problem can generally be solved if the flow conduit is given an appropriate inclination so that the gas is swept downstream by the fluid. Pressure difference devices need special care in selecting the location of pressure tappings and the run of pressure lines.

A particular example of phase separation can occur within a liquid flowmeter when there is a large pressure drop in the meter body which can give rise to the phenomenon of cavitation. Its effect has been experienced with turbine and vortex meters at the high end of their flow range and needs to be remembered where static fluid pressures are low and flow rates are high. The obvious solution is to select a meter of sufficient size to avoid cavitation.

3.6 Hydrodynamic Noise

The signal generated by a flow device invariably has a random noise content, which can impair accuracy of measurement. Some, if not all of this noise, has hydrodynamic origins caused by turbulence, vortices or even acoustic shock waves. As most users of control systems wish to avoid ragged records or noisy signals, a smoothing element is sometimes incorporated to act as a low-pass filter. In this case, pressure difference meters can suffer accuracy impairment if the square root of the mean value of the signal is measured instead of the mean of the square root. The problem does not exist to the same extent when a linear flowmeter is employed.

A peculiar problem can arise if the impedances to the fluctuating flows in the pressure lines to the flowmeter are not identical in opposing directions. The noise signal can then be partially rectified and, using an electrical analogy, the system behaves as a leaky diode tending to give a reading of the peak amplitude of the noise signal. It follows that pressure difference meters are more susceptible to hydrodynamic noise than linear flowmeters employing turbine, vortex, electromagnetic or ultrasonic principles.

3.7 Pulsations

Pulsating flow is a particular case of the problem dealt with in Sect. 3.6, but the noise signal is periodic and coherent instead of being random. Pressure difference meters read high on pulsating flows because the secondary measuring system provides a signal which is more nearly related to the square root of the mean value. Other errors can arise with pressure difference meters, due to inertia effects in the fluid. Linear flow rate meters and positive type meters are less susceptible to inaccuracies from pulsating flow, provided that the response time of the meter to flow rate change is at least an order of magnitude higher than the periodic pulse rate.

Electromagnetic and ultrasonic flowmeters are normally adequate for this purpose, but turbine meters and vortex meters are not always suitable depending upon pulsating frequency and wave form. In the event that the pulsating frequency approaches that theoretically due to vortex shedding from a bluff body of a vortex meter, then the meter becomes useless. The electromagnetic flowmeter would appear to be a good choice in these circumstances, but even so, severe problems could theoretically arise if the pulsating flow frequency approached the frequency of the magnetic field.

3.8 Bidirectional Flows

In some processes flow reversal may occur and it may be necessary to measure the reverse flows as accurately as the forward flows. This requirement would limit the choice of flowmeter as some types are unsuitable for this purpose. An electromagnetic flowmeter would be appropriate in this case if the measured fluid is electrically conducting.

3.9 High Differential Pressure

High differential pressure can be a problem for the manufacturer of orifice plates. Cases have been reported where the pressure difference is sufficient to cause buckling of the plate. A nozzle could be a better alternative.

4. Properties of the Fluid

When the fluid to be measured is specified, many of its properties can be found in the literature, but some properties important to accurate flow measurement, cannot be so easily established, particularly of two-phase or multiphase fluids. Also, many flow problems derive from impurities present in the fluid, the effect of which cannot always be quantitatively established.

4.1 Viscosity

Viscosity is a critical factor in the Reynolds number criterion, which means that for any specific flow, a change in viscosity can mean a change in the Reynolds number and a possibility of a change in the calibration curve of the flowmeter.

Viscosity effects vary according to the type of flowmeter used. Positive displacement meters can benefit at low flow rates from an increase in viscosity as the leakage past the pistons is reduced. Turbine meters tend to read higher than normal at some parts of the range and lower at others for a viscosity increase, but as the actual effect is dependent on turbine meter design, each manufacturer's product will behave differently.

The viscosity problem becomes even more complex when non-Newtonian fluids, such as slurries, are measured. Apart from the more obvious mechanical problems, such as meter chokage, the velocity profile can become profoundly changed and the flow pattern can exhibit the phenomenon known as "plug flow." Electromagnetic flowmeters and the more recently developed cross-correlation flowmeters are promising candidates for measuring slurries, but ultrasonic flowmeters can experience problems due to signal attenuation and dispersion.

For single-phase viscous liquids, special orifices have been developed which have a rounded or coned upstream edge, instead of the normal sharp edge. Coriolis-type meters are relatively uninfluenced by viscosity. The vortex meter performance is affected by viscosity, but not to the same extent as turbine meters or square-edged orifice and venturi tubes. Vortex meters should not be considered when the fluid kinematic viscosity materially exceeds $10\,\text{mm}^2\,\text{s}^{-1}$.

Doppler-type clamp-on ultrasonic flowmeters have been used to measure the flow of highly viscous fluids such as pitch, but the uncertainty of measurement is possibly 20–30%.

4.2 Two-Phase and Polyphase Fluids

Two-phase and polyphase fluids are, at best, difficult to measure but often impossible. In some cases the flowmeter signals are meaningless. A problem which is always present is the difficulty of calibrating the flowmeter due to the lack of test facilities covering a very wide range of polyphase fluids. Apart from the calibration difficulty there are other severe problems.

The consequences of phase separation can be serious. There are three typical problems: first, the phases may not be homogenized and the concentration ratios of the phases may continuously vary, for example, large gulps of air followed by slugs of liquid; second, solid separation can alter the profile of flow measuring devices and the effective cross-sectional areas of pipes and channels; third, the whole flow system can become choked with solids or become air locked. The possible alleviations to these problems include:

(a) flowmetering in a vertical pipe;

(b) use of mechanical or hydrodynamic agitation to improve homogenization upstream of the meter; and

(c) use of nonintrusive flowmeters, for example, electromagnetic type.

The selection of a suitable flowmeter for polyphase fluids must take into consideration whether the flow rate required is that of one phase only or the combined phases. For example, an electromagnetic flowmeter will measure the velocity of water in a sand and water slurry, and to convert this to water flow rate it is necessary to determine the water : sand volumetric ratio.

It cannot be assumed that all phases are moving with the same velocity, particularly in vertical pipes or when the density of one phase is significantly different from the carrier phase. In the case of the sand and water slurry referred to above, an electromagnetic flowmeter would measure the water velocity whilst a cross correlation meter would measure the sand velocity. If, instead of sand, the solid phase happened to be conductive, the electromagnetic flowmeter would measure the mean velocity of the two phases.

In the case of liquid and solid two-phase flows, the possibility of wear or erosion has to be kept in mind. Provided the liquid phase is conductive, an electromagnetic flowmeter offers the best solution, particularly as the bore of the flowmeter can be supplied with an abrasion resisting polyurethane liner.

4.3 Fluid Activity

Activity is used in the context of chemistry and radioactivity. Chemical activity has to be considered in relation to flowmeter corrosion and toxicity. Flowmeter manufacturers generally offer a selection of materials to withstand corrosive attack, but it is not always easy to specify suitable materials without some advice from the user. Electromagnetic and vortex flowmeters have been applied successfully on highly corrosive fluids. Although turbine meters have been used in certain corrosive applications, the existence of cavities between rotor shaft and support bearings, and the dependency on the bearings for reliable meter performance suggest caution in the selection of these meters for corrosive fluids.

Orifices and their associated differential pressure transmitters are available to operate with most of the commonly used corrosive fluids very satisfactorily, and provided the working flow range is not more than three or four to one, the orifice/transmitter is a very popular choice. In cases where very exotic materials are required to cope with very difficult and hostile situations, the transmitter can be protected by some means of isolation through fluid reaction systems or sealing chambers or separators.

Radioactivity presents special problems. For applications involving high levels of radiation, flowmeters offering long periods of reliable operation without maintenance are required, and this suggests noninvasive or nonintrusive types without moving parts and possibly remote electronics. Special procedures for thorough cleansing of the flowmeter before it is commissioned are generally mandatory.

Regardless of the type of activity involved, it is prudent to avoid using any meter that has glands or seals which could involve a leakage hazard.

4.4 Flammability

Fluids which are inflammable or react violently with other materials need flowmeters which are manufactured to a certified design, and this means that the user's choice of flowmeter is restricted. Turbine, vortex and orifice/transmitter flowmeters represent the popular choices of meters suitable for operations in hazardous conditions.

For the metering of high-pressure oxygen, the choice of material likely to be in contact with the gas is very restricted. Also, any sealing fluids which come into contact with the gas in the event of leaks or fractures have to be chosen from a very limited range. Flowmeters for this application also need to be completely clean and free from particulate matter, oil and grease.

4.5 Scaling and Other Deposits

Any fluid having a propensity to deposit scale should be regarded with suspicion. The effects can be dramatic:

(a) on electromagnetic flowmeters it can insulate and paralyse the electrodes;

(b) it can reduce the pipe diameter and clearances within the flowmeter;

(c) it can alter the profile of orifices, venturis, turbine meter rotors and the bluff bodies of vortex meters;

(d) it can block pressure lines and ducts; and

(e) nodules of scale can break away and become lodged in the vital parts of the meter.

The effect on calibration and performance of a meter due to any of these effects can be catastrophic. In many cases, the damaging effects of scaling cannot be avoided except by frequent stripping and cleaning, but in the case of the electromagnetic flowmeter, automatic electrode cleaning (by mechanical means, ac current or ultrasonic stripping) can overcome the major problem. Where scaling deposits occur, the choice could rest with the electromagnetic, ultrasonic or cross-correlation flowmeter, provided routine descaling is carried out at required intervals.

Deposits other than scale can cause trouble. A particular problem experienced with electromagnetic flowmeters is the build up of grease on the electrodes, but this can be cured by any methods used to remove scale.

4.6 Abrasiveness

Abrasiveness is a problem frequently encountered with a two-phase fluid, one phase being hard gritty particles. The effect on most types of flowmeters is initially a steadily increasing error and, ultimately, complete failure. Orifices lose their sharp edge which changes the coefficient, turbine meters suffer bearing damage, invasive-type ultrasonic flowmeters suffer damage to the piezoelectric transducers, vortex meters suffer damage to the bluff body and positive displacement meters either become jammed or become worn and inaccurate at the lower flow rates.

The electromagnetic flowmeter is less susceptible to damage of this kind due to its straight through design and the availability of special abrasion resisting linings. However, electromagnetic flowmeters have a relatively short life in some dredging applications, as occasionally large jagged rocks with mean diameters of up to 400 mm pass through the meter and tear away the insulating liner. In such cases there is evidence to suggest that cross-correlation flowmeters would offer a better solution to the problem, on the grounds that insulating liners are not required and replacement of the flow tube could be simpler and cheaper.

4.7 Compressibility

The compressibility of a fluid (if a gas or a vapor) needs to be considered in the following instances.

(a) Volumetric flow rate measured by pressure difference devices; here the flow equation relating differential pressure to flow rate contains a density term, which in the case of a gas is dependent upon line pressure.

(b) Differential pressure-type flowmeters: if the differential pressure is significant compared with the pipeline pressure, a correction may be required for the expansibility of the fluid across the measuring device.

(c) Mass flow rate: if the fluid is compressible it is necessary to choose a flowmeter which intrinsically measures the mass flow rate or to compensate for the effects of pressure.

(d) Supercompressibility: as the "perfect" gas laws may not apply to vapors or condensable gases, an additional correction would be required to allow for the supercompressibility of the fluid.

If a simple volumetric flow rate measurement is needed at the operating conditions existing at the point of measurement, a vortex meter provides a direct measurement independent of density and compressibility, but the majority of commercial vortex meters have an upper fluid temperature limit of 150 °C, although some designs can tolerate 400 °C. One manufacturer claims 600 °C.

4.8 Transparency

The transparency of a fluid is only important in those applications involving direct-reading variable-area meters, although in the future, transparency might be a requirement for some turbine and vortex meters using fiber-optic detectors.

4.9 Electrical Conductivity

Manufacturers of electromagnetic flowmeters define a minimum specific conductivity of the measured fluid below which errors in measurement can occur. However, manufacturers are trying to develop an electromagnetic flowmeter for nonelectrically conducting fluids.

4.10 Magnetic Properties

If the fluid passing through an electromagnetic flow-meter should happen to have a significant magnetic susceptibility (such as slurries containing iron or nickel ores) the effect of the fluid would be to alter the flux generated by the field windings and errors could result. Some manufacturers can supply designs which can compensate for this effect.

4.11 Lubricity

Flowmeters which operate with moving parts and bearing surfaces can be affected by the lubricating properties of the measured fluid. The effect is mostly experienced with turbine meters and semipositive or positive displacement meters. Curious results obtained from turbine meters can sometimes be explained by the combined effects of viscosity and lubricity.

5. Accessories and Installation

5.1 General

A flowmeter is normally calibrated under ideal conditions, in terms of stable temperatures, symmetrical velocity profiles, purity of fluid and the relatively short duration of the calibration process. Its performance over a period of time under industrial conditions is unlikely to reach the standard obtained in a calibration laboratory. The reasons underlying this deterioration are fourfold:

(a) the industrial installation differs from that in the calibration laboratory in terms of upstream and downstream conduit configurations (effect on velocity profile);

(b) physical conditions in the conduit and the external environment are likely to be more variable and more severe;

(c) deterioration occurs with time due to erosion, scaling, corrosion, and so on; and

(d) the absence of certain accessories to ameliorate the influence of undesirable industrial conditions.

5.2 Accessories

The purchaser of a flowmeter may make a wise selection but disappointing results could occur if the essential accessories are not installed at the same time. In the case of differential pressure devices the desirability of including the following accessories should be considered.

(a) Carriers and other means of fitting orifices in the pipe and bringing out tapping points (in some cases an interchangeable orifice carrier is justified so that orifices can be changed or inspected).

(b) Valves and manifolds for equalizing, draining, venting and isolation.

(c) Sumps, gas vents, probes (for pressure tappings) and drains.

(d) Cooling chambers when measuring condensable vapors (e.g., steam).

(e) Purges or separators for protection against corrosion, or to safeguard the measuring device against high fluid temperatures.

(f) Straighteners for improving the flow velocity profile; piezometer rings for averaging tapping point pressures.

(g) Steam or electric tracing of pressure lines to prevent freezing or condensation.

Other types of meters must also be provided with accessories in certain applications, for example.

(a) Strainers for turbine and positive displacement meters.

(b) Separators to prevent contaminating fluids such as water in oil from entering the flowmeter.

(c) Devices for maintaining electrode cleanliness in electromagnetic flowmeters (rotatable brushes, ultrasonic or electrolytic cleaners).

(d) Gas detectors to provide a warning if the flowmeter is not running full of liquid.

(e) In-line calibration systems: in some applications where the highest possible precision is required, meter provers may be required to enable the meters to be calibrated at frequent intervals.

(f) Diagnostic equipment: in order to simplify troubleshooting, specialized equipment which can check the operation of secondary flow measuring electronics is available from some manufacturers. This diagnostic equipment is sometimes called a simulator because it can generate a simulated flow signal for feeding into the secondary electronics and thereby check that all circuitry is functioning satisfactorily.

(g) Lightning arrestors: these are particularly important in some countries and where long transmission lines are involved.

5.3 Installation

Many of the difficulties experienced with flowmeters can be traced to poor installation.
  Some of the common faults include:

(a) orifices inserted the wrong way round;

(b) bad location of flowmeters where they experience a poor flow profile;

(c) presence of an unwanted phase (gas or liquid) in pressure lines connected to differential pressure devices;

(d) mechanical damage arising from locating the flowmeter in a vulnerable position;

(e) location of the flowmeter in a hostile environment or in an inaccessible position;

(f) incorrect orientation of the flowmeter, for example, an electromagnetic flowmeter with the electrodes in a vertical plane;

(g) incorrect gradients of pressure lines;

(h) siting of meter or its electrical transmission lines in a high electromagnetic field;

(i) siting flowmeters in undrained pits below ground level;

(j) fitting some flowmeters to vibrating pipes; and

(k) absence of essential accessories.

See also: Flow Measurement: Principles and Techniques; Flow Measurement: Economic Factors, Markets and Developments

R. S. Medlock[†]
[ABB Kent, Luton, UK]

Flow Measurement: Economic Factors, Markets and Developments

Most decisions on any subject tend to be based on optimal compromises, and the act of selecting a flowmeter is no exception to this rule.

The intending purchaser of a flowmeter generally has to make a selection from a number of various possible types offered by a number of manufacturers. Here it must be made clear that flowmeters of the same type and general specification supplied by a number of manufacturers do not necessarily function equally well in all applications. Two apparently identical turbine meters can work equally well on, say, flow rate measurement of water but will behave quite differently if a small percentage of air is present in the water. One flowmeter could give an error of, say, 1% while the other might give an error of 15%. Sometimes the anomalous results are not always explainable, which just adds to the difficulty of selection.

1. Technoeconomic Optimization

The industrial user is conservative in making a choice because of the dire results that can follow a bad decision—plant shutdown, loss of output, process inefficiency or hazardous conditions. The user therefore plays safe and orders equipment that has a good track record for the particular process requiring flow measurement. This approach is both popular and safe and accounts for the choice of an orifice and transmitter for about 50% of flowmeter sales.

By a process of evolution, a particular type of flowmeter may become the automatic choice for each individual application. It is the evolution of new processes and applications which call for an assessment of the price, performance, maintenance and reliability factors of a flowmeter. The user has to weigh up the characteristics of range, accuracy, repeatability, and sensitivity to process and ambient variables, and to balance these against cost of purchase and cost of maintenance. The cost of installation must not be forgotten as it can be a significant amount.

A list of desirable characteristics include:

(a) a wide operating temperature range;

(b) a wide measurement range;

(c) insensitivity to flow profile, swirl, viscosity and other physical properties of the fluid;

(d) noncorrodible and nondegradable materials of construction;

(e) small irrecoverable head loss;

(f) suitability for liquids, gases and multiphase fluids;

(g) availability in all practical sizes;

(h) safety in hazardous locations;

(i) immunity to vibration;

(j) fast response to flow changes;

(k) immunity to pulsating flow effects;

(l) accuracy;

(m) easily checked calibration, for example by dimensional metrology; and

(n) low purchase and maintenance costs.

There is no flowmeter available that meets all these requirements and it is unlikely that one will ever be developed. The list given above is a checklist to help the purchaser choose a technoeconomic compromise.

The economics of investing in a flowmeter is not confined to the purchase price. Other factors have to be assessed, such as those given in the "balance sheet" in Table 1.

The main problem is the difficulty of valuing each of the items in Table 1 in monetary terms but the probability is very high that the initial cost of purchase of the meter is one of the smaller values in the list. This leads to the conclusion that reliability with freedom from maintenance is the best criterion (almost regardless of initial purchase cost) provided that the accuracy attainable is adequate for the purpose.

Table 1
Economic factors involved in selecting a flowmeter

Liabilities	Assets
Cost of purchase	Higher operating
Cost of installation and	efficiencies
commissioning	Safety
Depreciation	Accurate costing
Cost of maintenance	Meeting statutory
Need for recalibration	requirements
Power requirements	Reduction of process costs
Energy losses across the	
meter	

2. Markets for Flowmeters

Ten main flowmeter market sectors can be identified:

(a) process (on line measurements) including utilities,

(b) other industrial,

(c) laboratory and scientific,

(d) domestic,

(e) automobile,

(f) aeronautics and aerospace,

(g) military,

(h) health care and biological,

(i) agriculture, and

(j) Environment and safety.

This classification is arbitrary as the areas cannot be precisely defined and because some flowmeters can serve more than one market. As the specifications and the applications are broad ranging it can be appreciated that the flowmeter design variants are enormous and the selection problems become complex. Apart from domestic water and gas meters, large scale manufacture to a standard design is not generally justified.

3. Future Developments

Flow measurement technology has made big advances since the 1960s, which have reduced the orifice and venturi tube market share. This is mainly because of the proliferation of new techniques of measurement, accelerated by the need to measure mass flow and the preference for noninvasive and nonintrusive designs. A noninvasive flowmeter is one in which the flow conduit is not penetrated, thus eliminating all leakage possibilities (e.g., an ultrasonic meter with clamp-on piezoelectric elements). A nonintrusive flowmeter implies penetration of the flow conduit but does not intrude beyond the inner surface of the conduit (e.g., an electromagnetic flowmeter).

There is a rapidly increasing demand for mass flow measurement particularly of the Coriolis type. Electromagnetic flowmeters have enjoyed a very successful period of development and could well become a market leader as demand increases and unit costs reduce. Vortex meters originally showed great promise as the main competitor to the orifice but their cost and initial teething troubles have made them less competitive. Nevertheless, the next generation product could restore the original promise. In the domestic field, gas and water meters could undergo considerable development as the demand increases for meters that can be interrogated and tariff changes are made remotely.

Electronic developments such as the microprocessor have had, and will have, a big impact on the performance of many flowmeters. By compensating for systematic errors the performance specification can be markedly improved. There are many other benefits which can be derived from incorporating microelectronics in flowmeter designs, for example: data storage and retrieval; application of signal processing to improve signal-to-noise ratios; multiplexing signals from several flowmeters; elimination of secondary parameter effects; improved maintenance and diagnosis; remote or auto range changing; operation of rule-based procedures; interrogation; and "smartness" or "intelligence."

See also: Flow Measurement: Applications and Instrumentation Selection; Flow Measurement: Principles and Techniques

R. S. Medlock[†]
[ABB Kent, Luton, UK]

Flow Measurement: Principles and Techniques

Historical interest in flow measurement can be traced back more than 1000 years, but it was less than 200 years ago that the first principles began to be understood. The initial incentive to measure flow was the need to charge consumers, and this led, in the early nineteenth century, to the development of water and gas meters to measure domestic consumption. Later, towards the end of the nineteenth century, measurement of flow rate became important for industry and the utilities. Pressure differential techniques held sway to satisfy this demand until about 1950, when alternate methods started to become available. Currently there are about 60 different basic designs of flowmeters, and these can be assembled into 11 groups which will be described in the following sections.

The world market requirement for flow measuring equipment required by industry has been assessed at US$ 1×10^9 annually.

Flow measurement is one of the most important of all process measurements. It is indispensable for costing purposes, as in the case of customer transfer, and it also plays an important role in many branches of science and technology. The term "flow measurement" is a generic one and embraces all the following meanings:

(a) flow rate measurement expressed in terms of volume per second;

(b) flow rate measurement expressed in terms of mass per second;

(c) flow quantity expressed in terms of volumetric transfer;

(d) flow quantity expressed in terms of mass transfer;

(e) flow rate or quantity expressed in "base" or "reference" volume terms which requires the measurement to be corrected for variations of temperature and pressure from reference values; and

(f) flow velocity expressed as the mean velocity of fluid in a pipe, or the point velocity in an undefined space (in the latter case the measurement is generally referred to as anemometry). If the area of the pipe is known, the volumetric flow rate is easily calculated from the product of area and mean velocity.

The technology of flow measurement needs to cope with an enormous range of flow rates, from drops to rivers, that is, about 12 orders of magnitude. The price of a flowmeter system can range over about six orders of magnitude. The variety of fluids which require to be measured is also extremely large, and includes single-phase gases, vapors, liquids and solids, and multiphase combinations. Since many flowmeters have a dynamic range of measurement of the order of ten to one, while the field of applications can involve flow rates over 12 orders of magnitude, it is clear that the hardware variety must be extensive. If the fluid variety is then considered, as well as other applicational requirements, it is apparent that the selection of a flowmeter is a complex task, calling for a considerable amount of skill and experience.

1. Flowmeter Configurations

A flowmeter is not necessarily a single item of hardware, although some of the simpler and cheaper designs, such as glass tube variable area meters, are just that. At the next level of complexity a flowmeter consists of two parts, a primary device which interfaces with the fluid and a secondary device which measures, processes and transmits the raw signal from the primary device. These two parts can be mechanically and electrically integrated at the point of measurement (as generally found with a vortex meter), or remote from each other (as generally found with a turbine meter). There are no rules to govern the type of configuration which is selected by the manufacturer according to market and technical requirements. The third level might be a combination of interconnected instruments, such as a flowmeter, temperature sensor and pressure sensor, in order to measure a mass flow rate or to correct for effects arising from temperature and/or pressure changes. At the second and third level of complexity the system needs to be associated with some form of receiving instrument, such as an indicator, recorder, controller, logger or computer. The ultimate level of complexity can be found in a flow metering station for the accurate measurement of oil flow, which is a system of numerous flow meters, accessories, control valves, and an automatic system for periodic calibration, which are all under computer control.

Finally, it must not be forgotten that in many applications certain accessories are required to enable the flowmeter to work reliably.

In this article it will not be possible to describe the total hardware requirements for flow, but only the primary measuring devices.

2. Flow Measurement Techniques

Various flowmeter classifications have been proposed, and will be discussed in the following sections; the more important commercially available types will be described.

2.1 Differential Pressure Type Flowmeters

Nearly all differential pressure flowmeters generate a differential pressure, which is a function of density and flow rate:

$$Q_v = K \left(\frac{\Delta h}{p} \right)^{1/2}$$

where Q_v is the volumetric flow rate, K is a constant, Δh is the differential pressure and p is the fluid density. As a result of this square root relationship, such flowmeters have a restricted flow range, typically 3:1, if high accuracy of measurement is required. The main types are orifices (see Fig. 1), venturis (see Fig. 2), nozzles (see Fig. 3), pitot tubes (see Fig. 4) and elbows (see Fig. 5).

The square-edged orifice (see Fig. 1) is the most frequently used device for industrial flow measurement. It has been developed and studied since the 1910s, and is simple to manufacture. Various profiles at the orifice bore have been developed to improve its performance with viscous liquids. The chord orifice,

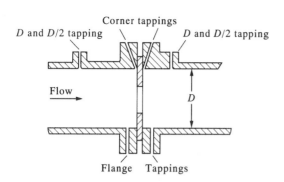

Figure 1
Square-edged orifice

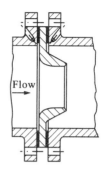

Figure 3
Nozzle

the eccentric orifice and the annular orifice have been developed for dealing with liquids and gases having a second-phase impurity which could build up as a secondary phase separation at the orifice face.

The venturi tube (see Fig. 2) has an advantage over the orifice because it is a lower energy absorber. The Dall tube and other similar low-loss devices are even better in this respect.

Other variants of the venturi tube exist and are used. The wedge flow tube can cope with second-phase impurities. The McCrometer design has an annular throat and is claimed to have a long range.

The nozzle (see Fig. 3) represents a halfway stage between the characteristics of an orifice and a venturi, and is frequently used for steam flow measurement. It is worth noting that for the same flow rate, the same pipe diameter and the same differential pressure, the orifice and the nozzle develop the same unrecoverable head loss. However, for the same β ratio (ratio of bore or throat to pipe diameter) and the same pipe, the nozzle has a lower unrecoverable head loss, but of course the differential pressure is then less from the nozzle than from the orifice. This distinction is not always clear from manufacturers' literature. Figure 6 illustrates the comparative net head loss of various pressure difference devices measured against their effective area ratio.

Pitot tubes (see Fig. 4) develop a pressure differential which is a function of the velocity of the fluid at the pitot tip. They are therefore used for fluid velocity measurement, as in ventilating ducts, or as flow rate metering devices, where the relationship between the average velocity in a pipe and the velocity at the pitot tip is known and constant. In cases where this relationship is not constant, multiple averaging pitot tubes are used to take account of the variable velocity profile across the pipe. The pitot venturi is used where the differential pressure needs to be enhanced to improve accuracy of measurement.

The elbow meter (see Fig. 5) is extremely simple and cheap but requires individual calibration.

Where the square root relationship needs to be avoided the linear resistance flowmeter can be employed. This operates on the principle of streamline flow through parallel channels, whereby the relationship between differential pressure and flow rate is linear.

2.2 Variable-Area Flowmeters

In variable-area flowmeters the differential pressure across a variable aperture is used to adjust the area of the aperture, which is then a measure of flow rate. In its simplest form (see Fig. 7) it consists of a tapered glass tube and a float which takes up a stable position,

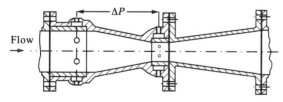

Figure 2
Venturi tube

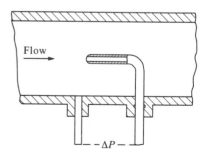

Figure 4
Single pitot tube

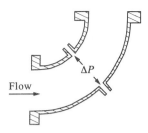

Figure 5
Elbow meter

Figure 7
Variable-area meter

where the upthrust due to the differential pressure across the float is exactly balanced by its submerged weight. The position of the float in the tapered tube is a measure of the effective annular area of the flow passage and therefore is an indicator of the flow rate. In another variable-area arrangement, the balancing force applied to a slideable cone-shaped body is provided by a spring. By suitably profiling the cone-shaped body, the pressure difference becomes a linear function of flow and provides a meter with an enhanced flow range. In some classifications this meter is considered to be a differential pressure flowmeter because the measurand is differential pressure. Gate meters are variations of the variable-area principle and rely on either gravity or a spring to create a balancing

force to oppose that developed by a differential pressure existing across the movable member.

2.3 Positive Displacement Flowmeters

Positive displacement flowmeters bear a relation to positive displacement pumps, but are used in the reverse mode. They are normally designed to measure only the volumetric quantity passed through the meter, but some are fitted with rate-of-flow attachments which compute the first derivative of the meter reading with respect to time. A parallel situation occurs with the rate flowmeter, which can have an attachment that measures the time integral of the reading to give quantity.

The operating principle of these meters can be seen in Fig. 8. They depend on the fluid powered mechanical cyclic movement associated with port openings and closings or the position of moving components to permit a fixed quantity of fluid to be discharged at each cycle.

Such devices are in common use for measuring domestic water consumption and are produced in millions of units every year. Other meters are more

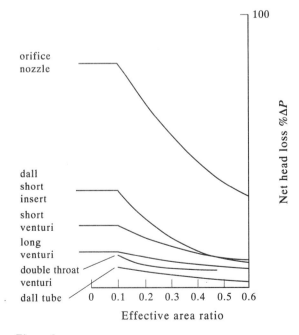

Figure 6
Comparative head loss of ΔP devices: effective area ratio equals discharge coefficient × area ratio × velocity approach factor

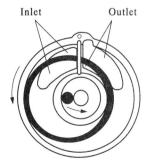

Figure 8
Rotary piston meter

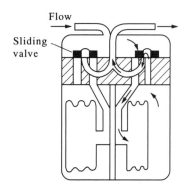

Figure 9
Domestic gas meter

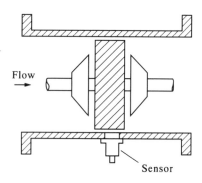

Figure 11
Turbine meter

frequently used for the measurement of hydrocarbon liquids.

For the accurate measurement of small gas flows, a simple device can be used which provides for a soap bubble film to be generated at the gas inlet port and then measures the time taken for the soap film to travel between two fixed markings which define the extent of a known fixed swept volume. The timing can be made automatically by employing optical sensors at the two fixed markings.

Figure 9 illustrates the domestic meter for measuring gas consumed in private houses. A wet gas meter is often used in laboratories for accurate measurement of gas flow.

Figure 10 shows the principle of the pipe prover, which is a form of positive displacement flowmeter of very high accuracy and repeatability which is used to calibrate other meters. Its main application is with liquid hydrocarbon flow, but attempts have been made to develop it for gas flow measurement. Basically it consists of a rubber sphere having an interference fit within a length of pipe in which the sphere acts as a piston. When the metering function is required, the sphere descends into the pipe through a valve, and its travel time between two detectors is electronically measured. The flow rate is then calculated from the

known volume between the detectors and the time of travel. The sphere is retrieved at the end of its travel and either returned to its original location or used in a return journey with the flow reversed.

There are two basic types of prover, unidirectional (see Fig. 10) and bidirectional, and both have a use according to application, although the bidirectional prover appears to enjoy a greater popularity. More recently compact provers have been developed which are very much smaller, although their operating principles remain the same. The designs are based on a precision bored and honed straight length of pipe.

2.4 Inferential Meters

Propeller or turbine type flowmeters exist in many commercial forms (see Fig. 11), using either impulse or reaction principles. Much ingenuity has been displayed in the detailed design of the rotors, housing and bearings to avoid metering limitations arising from effects of viscosity, bearing friction, signal detection and cavitation. In a recently developed design, bearing wear and friction have been eliminated by using a fully floating rotor supported by fluid dynamic forces. For the metering of domestic water consumption single-jet and multijet inferential meters are used in large numbers as the working parts can be produced from plastic mouldings very cheaply.

2.5 Tracer Techniques

Tracer techniques are based on the principle of adding a marker or tracer at one point or section of the flow path and detecting its presence at one or more points downstream.

For liquid flows (e.g., rivers and large pipelines) the tracer may be a brine solution which can be detected by electrolytic conductance, or a dye detected photometrically. For gases, the marker can be a safe radioactive isotope.

There are several tracer techniques which can be used for flow measurement.

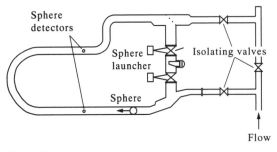

Figure 10
Unidirectional pipe prover

(a) *Transit time method.* A simple and rather crude method is simply to inject a tracer at a point A and measure the time t it takes to travel a known distance x downstream to a point B, and calculating the flow velocity by the ratio x/t. The flow rate can be calculated from the mean area of the conduit.

A more refined method is to inject the tracer at point A and then make two measurements at two downstream points, a distance x apart. The transit time between these two points is determined from the difference of the mean arrival times of the tracer.

(b) *Integration method.* This requires the injection of a known volume of tracer V, of concentration C_1, into the flow stream and, after thorough mixing, the concentration/time relationship is measured downstream. At the sampling section the concentration pattern follows a curved profile.

The flow rate Q can be calculated from the formula

$$Q = C_1 V \left[\int_0^\infty (C_2 - C_0)\, dt \right]^{-1} \qquad (1)$$

where C_1 is the original concentration of the tracer, V is the volume injected, C_2 is the concentration of the tracer at the sampling point, and C_0 is the background concentration of the tracer solution in the fluid before injection.

The integral in Eqn. (1) can be found by measuring the area of the concentration curve. It should be noted that this technique does not require a knowledge of distances.

(c) *Constant rate injection.* In this technique the tracer is added at a known rate and concentration. At a convenient downstream point after complete mixing the concentration of the diluted tracer is measured and the flow rate calculated from:

$$Q = q\left(\frac{C_1}{C_2}\right)$$

where q is the flow rate of the tracer, C_1 is the concentration of the undiluted tracer and C_2 is the concentration of the diluted tracer.

A more precise equation is

$$Q = q\left(\frac{C_1 - C_2}{C_2 - C_0}\right)$$

where C_0 is the initial concentration in the flow before injection. However, C_2 is very small compared with C_1, and C_0 is very small compared with C_2. Hence, the equation can be simplified.

There are other ways of "tagging" fluids in order to measure their velocity or flow rate. One sophisticated technique uses the phenomenon of nuclear magnetic resonance (NMR) (see Fig. 12). This technique is appropriate for fluids containing hydrogen or fluorine compounds having a nuclear magnetic moment. This

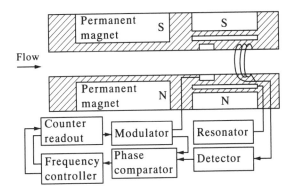

Figure 12
NMR flowmeter

requirement is met by all aqueous solutions and many organic liquids. The fluid first enters a permanent magnetic field and undergoes nuclear magnetization. It then enters a second magnetic field where resonance is accomplished by applying a radio-frequency (RF) signal. A modulating magnetic field creates demagnetized pockets and the modulating effects or tagging are detected by a final RF detector coil. The modulating and detected signals are fed into a phase comparator which adjusts the modulation frequency to maintain a fixed phase difference. The modulation frequency is then a measure of the flow rate.

The ionization flowmeter uses gaseous ions (produced by a radioactive source) as tracer particles. In its basic form an ion chamber located at a fixed distance downstream measures the ionization current. Recombination of ions takes place as the gas travels downstream; therefore, the rate of arrival of ions within the ion chamber, and hence the ionization current, is a function of the flow rate. In a more highly developed form, an additional ion collection chamber is interposed between the ionizing source and the final ion collection. The intermediate chamber is electrically pulsed at a fixed frequency and thereby periodically produces deionized pockets in the gas flow. The time delay intervals of these tagged pockets enable the gas velocity and thus the flow rate to be measured.

Cross-correlation flowmeters use the turbulent velocity or composition pattern in a flowing fluid as an inherent form of tagging (see Fig. 13). As the tagging patterns are very "noisy" and undergo fairly rapid dispersion, cross-correlation techniques are employed. In one version, two parallel ultrasonic beams are directed diametrically across the fluid and the amplitude-modulated signals at the receivers are demodulated and fed into a cross correlator. The upstream demodulated signal is fed through an adjustable time delay, the output is multiplied by the demodulated signal from the downstream receiver and the product is integrated to generate the cross-correlation function. The time delay for the tagged flow

119

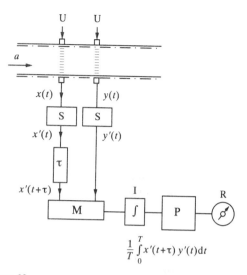

$$\frac{1}{T}\int_0^T x'(t+\tau)\,y'(t)\mathrm{d}t$$

Figure 13
Cross-correlation flowmeter: U, ultrasonic drive; S, signal processor; M, multiplier; I, integrator; P, peak measurement; R, readout; a, flow

disturbance pattern to traverse the distance between the two detectors is then obtained from the position of the peak of the cross-correlation curve. Electronic circuitry establishes this position and provides an output signal of the flow rate.

2.6 Oscillatory and Fluidic Flowmeters

Oscillatory and fluidic flowmeters rely on some fluid dynamic instability, which generates an oscillation whose frequency is proportional to the flow rate. One of the earliest commercial flowmeters of this type was the "Swirlmeter" (see Fig. 14), which has a set of fixed helical vanes to produce a swirl, followed by a venturi

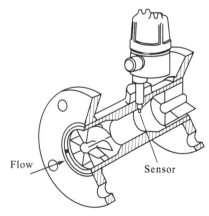

Figure 14
Swirlmeter

throat and expansion cone. This configuration results in the generation of a spiral-type vortex whose spiral frequency is a linear function of the flow rate. The frequency is measured by a heated thermistor detector.

The fluidic meter relies on the Coanda effect, whereby a fluid jet tends to attach itself to a fixed wall, provided the angle between the jet direction and the wall lies within certain limits. Detachment results from the routing of part of the flow back to the main jet, and thereby forcing the jet over to the alternative wall. The frequency of switching of this positive feedback system is a linear function of flow rate.

The generation of vortices when a fluid impacts a bluff body was observed by Leonardo da Vinci as far back as 1513, but it has only been exploited commercially for flow measurement since 1970 (see Fig. 15). In operation, vortices are shed from alternate sides of a bluff body mounted symmetrically and diametrically in a pipe. The frequency of shedding is a linear function of the flow rate, and is measured by a pressure or velocity detector. Some of the principal detectors used commercially to detect the presence of vortices are capacitors, thermistors, oscillators, pyroelectric elements, ultrasonic elements and oscillating disks.

A Japanese flowmeter has been produced in which the vortices cause physical vibration of the bluff body supported on pivots. A new flowmeter introduced by Badger Meter Inc. incorporates an oscillating vane. The liquid enters the meter and flows through an orifice in the main body. The differential pressure across the orifice causes a small proportion of the liquid to pass through the measuring chamber containing the oscillating vane. The hydrodynamic forces on the various surfaces of the vane cause it to oscillate at a frequency which is proportional to flow. This frequency is detected by an electronic proximity switch operated by a metal target fixed to the vane. A fluidic meter uses a jet of air or gas at a pressure of about 3 bar which is directed diametrically across the pipe, and its deflection (proportional to fluid velocity) creates a differential pressure across two symmetrically positioned receiving nozzles.

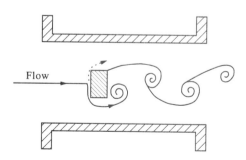

Figure 15
Vortex shedding flowmeter

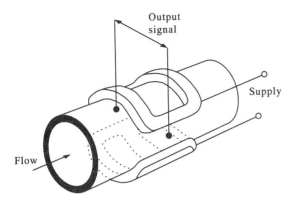

Figure 16
Electromagnetic flowmeter

2.7 Electromagnetic Flowmeters

In an electromagnetic flowmeter (see Fig. 16) the fluid acts as an electrical conductor moving in a magnetic field. A small voltage is thereby generated across an axis perpendicular to the axes of the flow and the magnetic field. The generated voltage is a linear function of flow. The noninvasive nature of this type of flowmeter has led to energetic development. In the original designs the magnetic field was alternating at the mains frequency to avoid measurement errors arising from spurious dc galvanic electromotive forces which developed at the electrodes. More recently, reversing or pulse-type dc fields have been employed to avoid the eddy current and quadrature effects of ac systems which may degrade the accuracy of measurement. Future developments are likely to permit the electromagnetic flow rate measurement of nonconducting fluids or to employ "buried" or insulated electrodes to measure conducting fluids. In the latter case the advantages would be freedom from electrode contamination and a lower sensitivity to nonsymmetrical flow profiles.

2.8 Ultrasonic Flowmeters

Ultrasonic flowmeters depend, for their operation, either on a change of frequency when a sonic beam is reflected from a moving body such as a dust particle or air bubble in the flow stream (Doppler effect) or on a change of sonic velocity in a flowing medium. A simple type of flowmeter can be constructed from two piezo crystals cemented to the outside wall of the pipe, one acting as a transmitter of ultrasonic energy and the other as a receiver. The receiver will detect reflections from moving particles in a liquid flow stream, from which the Doppler frequency can be separated and interpreted in terms of the flow rate.

Many commercial ultrasonic flowmeters have been developed which depend on the change of sonic velocity between two beams, one travelling in the direction of flow and the other in the opposite direction. There are many variations based on this principle. Most of the variations reside in the electronic circuitry, which attempts to overcome sources of error arising from changes in sound velocity due to temperature and composition changes, time delays in circuitry and transducers, occasional loss of signal and the difficulty of measuring the small difference in sonic velocity. The error arising from uncompensated temperature changes of the fluid is of the order of $0.4\% \; °C^{-1}$.

The physical design of transducers for transmission and reception of ultrasound has received considerable development, in order to maximize the transfer of sonic energy into the fluid. In some designs the transducers are "wetted," that is, they are in direct contact with the measured fluid, and therefore described as invasive but nonintrusive. Other designs clamp the transducers to the outside of the pipe wall with some acoustic bonding material. In this case the system is described as noninvasive. However, systems employing transducers clamped to the pipe cannot operate with gases, regardless of whether the Doppler or transit time technique is used.

Transit time flowmeters employ one of three possible techniques.

(a) Direct transit time measurement. This is sometimes referred to as a "leading edge" system. In this case an ultrasound pulse is fired across the path of the fluid in a diagonal direction, followed by a pulse in the opposite direction. If the travel times of the two pulses to reach their destination are T_1 and T_2, then the flow rate is proportional to $(T_1 - T_2)/T_1 T_2$, which is independent of the static sound velocity in the fluid. The time difference in small diameter pipes is very small and difficult to measure accurately, in which case a modified arrangement can be used in the configuration of the device.

(b) Phase difference measurement. This involves the measurement of phase difference between the two signals received from sinusoidal transmission in opposing directions. This technique requires compensation for the velocity of sound in the fluid.

(c) Sing around techniques. In this system a pulse is transmitted upstream from transducer A to transducer B. When B receives it another pulse is initiated at A, and so on. This sets up a frequency dependent on the time for the pulse to travel from A to B. The process is then repeated in the reverse direction by sending pulses from B to A in the downstream direction, which sets up a second and higher frequency. After allowing for delays in the electronics and transducers, the frequency difference is proportional to the flow rate, and is independent of static sound velocity in the fluid. The technique has a number of problems associated with it which need to be resolved electronically.

2.9 Mass Flowmeters

Mass flow measurements are relevant when mass or energy balances are required for industrial or custom transfer purposes. Many liquid fuels used for energy conversion are measured in mass units because the energy content per unit of mass has a high degree of constancy, regardless of composition, temperature and density variation.

The measurement can be made on a direct or an indirect basis. The term "direct" means that a technique is used which is directly influenced by the mass flow rate through the meter. The term "indirect" means that some other technique of flow measurement is used, such as volumetric flow rate, which is converted to mass terms by a secondary measurement such as density.

3. Direct Methods of Mass Flow Measurement

3.1 Critical Flow

A well-established method of measuring gas mass flow is to use a critical nozzle (see Fig. 17). In operation the nozzle is operated at an upstream pressure P_1 which in absolute units is of the order of at least twice the throat pressure P_2, the exact ratio being dependent on gas composition. This condition causes the gas in the throat to reach the speed of sound, a speed which cannot be exceeded, regardless of upstream or downstream pressures. When sonic velocity in the throat is established, the mass flow Q_m is given by the equation

$$Q_m = KP_1 T_a^{-1/2}$$

where T_a is the absolute temperature of the gas and K is a calibration constant. Although under critical flow conditions the downstream pressure P_2 does not affect the upstream pressure, it is necessary that the ratio $P_2:P_1$ does not exceed 0.8 for critical flow conditions to exist at the throat.

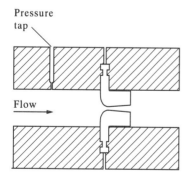

Figure 17
Critical flow nozzle

3.2 Flow Injection and Subtraction

If a fluid flow is split into two equal paths and one half has its rate increased by a fluid injection taken from the other half, then mass flow can be measured by a differential pressure device.

In one arrangement, four identical orifices are connected in the four arms of a bridge, together with a constant speed electric pump of the positive displacement type. The pump carries out the injection and extraction functions. If Q is the main flow, q is the injected or extracted flow, p is the density of the fluid, Δp is the differential pressure across the diagonal of the bridge, and K is a constant, then

$$P = K\left[\frac{(Q+q)^2}{2} - \frac{(Q-q)^2}{2}\right]p$$

$$P = (2Kq)Qp$$

$$P = \text{constant} \times \text{mass flow}$$

Mass flowmeters of this type are only suitable for liquids. They are particularly useful for very low flows from $0-5\ kg\ h^{-1}$, but are also available for flows up to $25\,000\ kg\ h^{-1}$. The technique has the major disadvantage of requiring a constant-speed positive displacement pump.

3.3 Fluid Damping in Oscillating Systems

In this section, the discussion will relate to oscillating techniques which measure mass flow rate by the damping effect on a vibrating element in a flow stream.

(*a*) *Vibrating vane.* A vibrating vane linear mass flowmeter was first described by Macdonald in 1983. The flowmeter consists of a thin metal or ceramic vane which is internally mounted axially from its upstream end, so that it is free to vibrate transversely.

It is excited at one of its flexural resonant frequencies by a piezoelectric device fixed to the center. Two other piezoelectric devices are attached near the two ends, and one of these is connected in a feedback loop with the one in the center. This feedback system maintains a flexural resonance in the vane. All three transducers are located at antinodal points. When the flow is zero the outputs of the sensors at the two ends are in phase with each other, but under flowing conditions a phase difference occurs. The wave travelling downstream is speeded up and the wave travelling upstream is slowed down. Mass flow is measured in terms of the time displacement of the signals generated by the two sensors at the ends of the vane.

(*b*) *Vibrating nozzle.* A mass flowmeter has been developed to measure gas–oil–water two- or three-phase fluids from oil wells. It consists of a vibrating flexible nozzle which is electromagnetically maintained at a constant amplitude by a feedback system. The flow passes through the nozzle which is surrounded by a gas pocket gratuitously trapped by virtue of the gas present in the fluid and the vertical disposition of the meter. The power required to maintain a constant amplitude of vibration is a function of the mass flow rate, and the resonant frequency is a function of the density. These two signals permit the mass flow rate of each phase to be computed.

3.4 Momentum Change

A momentum change flowmeter operates on the principle that a torque T is generated when a fluid flowing at a mass rate Q_m is subjected to a change in angular velocity $\Delta\omega$.

The equation describing the relationship between mass flow rate Q_m, the torque T and the angular velocity change of the fluid stream $\Delta\omega$ is

$$Q = KT/\Delta\omega$$

where K is a constant.
This equation can be realized in three practical ways.

(a) The torque required to drive an impeller at constant speed to generate angular momentum in the flow stream can be measured:

$$T \propto Q_m$$

where ω is constant.

(b) If a flow stream having an angular momentum generated by a constant speed impeller subsequently passes through a restrained straight bladed turbine in order to remove the angular momentum, then mass flow rate is measured by the torque generated by the turbine:

$$T \propto Q_m$$

where ω is constant.

The flow stream can be given angular momentum by a variable-speed motor driving an impeller, and the speed can be varied to maintain a constant torque. In this case, the mass flow rate can be measured in terms of motor speed $Q_m \propto \omega^{-1}$ (where T is constant). This inverse relationship can have the advantage of providing good resolution at low flows.

(a) *The angular momentum mass flowmeter.* One of the original meters of this type was manufactured by General Electric in the USA. This meter derives the mass flow rate from the torque required to remove a constant swirl velocity from the fluid. The swirl is created by passing the fluid through an impeller consisting of a series of radial vanes forming passages in line with the flow. The impeller is driven at constant speed. From the impeller the flow passes into a similar component or turbine which is constrained from rotating by a calibrated spring. The turbine removes the swirl, and the angular displacement resulting from the torque produced is measured by an electrical transducer. The principle is a simple one but the practical design is expensive and complicated.

(b) *The twin-rotor turbine meter.* This technique measures mass flow by having two in-line turbine rotors (whose blades have different helix angles) which are joined by a torsion spring coupling. As a result of the difference in helix angles and the presence of the coupling, the rotors cannot run at their natural speeds. One is underspeeding and the other overspeeding with the result that, as the mass flow rate increases, the relative angular displacement between them increases by an amount controlled by the torsion spring. Each rotor has a pick-off sensor as in conventional turbine meters. As the angular displacement between the rotors increases, the time interval between pulses from the sensors increases. This time interval is directly proportional to the mass flow rate.

The torque T exerted by the leading rotor on the trailing one will be:

$$T = KrpV^2(\tan\theta_1 - \tan\theta_2)$$

where K is a constant, r is the average effective radius of the rotor blade of each rotor, p is the density of the fluid, V is the velocity of the fluid, and θ_1 and θ_2 are the blade angles of the leading and trailing rotors, respectively. As r, θ_1 and θ_2 are constant the equation reduces to

$$T = K_1 p V^2$$

where K_1 is a new constant.
The angular displacement of the spring $\alpha = T/c$, where c is the torsion spring constant.

Thus, the angular displacement between the rotors is

$$\alpha = \frac{K_1 p V^2}{c} = K_2 p V^2, \qquad K_2 = \frac{K_1}{c}$$

In order to obtain mass flow rate pV, it is necessary to eliminate one power of V. This is easily done as the angular velocity of the turbine rotor assembly ω is proportional to the fluid velocity V, that is, $\omega = K_3 V$.

time between impulses = angular displacement/ angular velocity

$$= \frac{K_2 p V^2}{K_3 V}$$

$$= K_4 p V$$

$K_4 p V$ is proportional to the mass flow.

(c) *The GEC Avionics fuel flowmeter.* This flowmeter is similar in principle to the twin-turbine meter, and therefore does not require a constant-speed motor drive. Instead, the kinetic energy of the fuel flow provides the power to drive a bypass turbine, which is coupled by a spring to an impeller to impart angular momentum to the total flow. The speed of the turbine is controlled within the range of 70–400 revolutions per minute (rpm) by a spring-loaded bypass valve.

123

The torque provided by the turbine to drive the impeller is proportional to the product of its speed of rotation and the fluid mass flow rate, and is applied to the impeller through a helical spring. The impeller is surrounded by a drum fixed to the turbine shaft, in order to eliminate viscous drag between the impeller and stationary parts of the meter. Small magnets are fixed to both the drum and the impeller, and when rotation takes place these generate electrical pulses in their associated pick-off coils located on the outside of the meter body.

The time lapse between the pulses is proportional to the mass flow rate and is independent of the turbine speed. If, for example, at a constant flow rate the designed turbine speed was doubled, the angular momentum and, hence, the spring deflection would increase by a factor of two, but as the speed was doubled the time lapse between pulses would not alter.

3.5 Coriolis and Gyroscopic Techniques

Gustave Coriolis first described the Coriolis force in a paper published in 1835. A Coriolis force results from the acceleration of a body rotating about an axis when its radial distance from the axis undergoes change (see Fig. 18a)

In these circumstances a body of mass m undergoes two distinct components of acceleration and force:

(a) a radial acceleration $\omega^2 r$ towards the axis (centripetal); and

(b) an acceleration perpendicular to (a) equal to $2\omega v$, where ω is the angular velocity of rotation and v is the radial velocity.

The corresponding Coriolis force on a mass m is then $2m\omega v$. It should be noted that the centripetal acceleration does not change sign between clockwise and anticlockwise rotation, whereas the Coriolis acceleration changes sign if either ω or v changes sign.

(*a*) *The gyroscopic mass flowmeter.* This meter measures mass flow rate by a method which can be explained either in terms of a gyroscope or a modern vibrating Coriolis meter. Instead of having a solid mass rotating on bearings as with a conventional gyroscope, the gyroscopic effect is achieved by passing the fluid through a pipe bent into a ring about an imaginary axis AA. If the pipe is rotated about an imaginary axis BB passing through its inlet and outlet connections, precession occurs, and the flowing fluid tries to rotate about an axis CC. The presence of flexible couplings at the inlet and outlet allows limited motion to take place on two axes BB and CC. An oscillatory motion is imparted about axis BB of constant amplitude and frequency, and the measurement of mass flow is made by a sensor measuring the amplitude of oscillation about axis CC.

(*b*) *Modern Coriolis meters.* The first of these meters

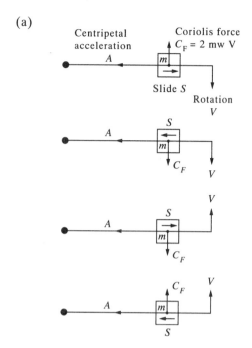

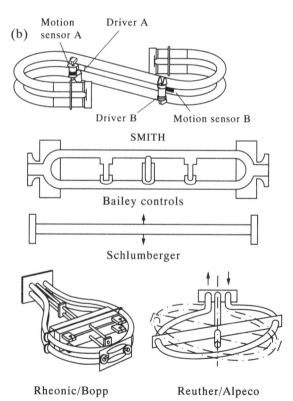

Figure 18
(a) Coriolis principle; (b) some commercial Coriolis flowmeters

was introduced in 1976 in the USA by Micromotion Inc. Until 1985 Micromotion had little competition, but since then several other manufacturers have entered the field.

The configurations developed by different manufacturers (see Fig. 18b) vary widely, but the operating principle is the same for each design. A relatively thin-walled tubular metal structure is anchored at its fluid inlet and outlet connections, and a vibrating motion is applied to one or more points in the structure by means of an electrodynamic actuator. The driven points on the system are chosen so that fluid travelling in a radial direction receives an angular oscillating momentum which creates the oscillating Coriolis force. The structure is maintained in an oscillating condition at its resonant frequency by a feedback system. The frequency is of the order of 75–100 Hz, dependent on the density of the fluid and the structural design, but harmonic frequencies up to 1000 Hz are also employed. In most designs the tubular structure is duplicated in a series or parallel arrangement so that the two sections cancel out the effects of external vibration.

The Coriolis components of force are detected by sensors at two points in the structure which give maximum sensitivity. The electronic signal processing of the signals from the sensors varies according to the manufacturer. The theory of operation shows that mass flow is directly proportional to the time lapse or interval between the outputs of the two sensors. Sometimes the phase shift between the two signals is measured, but in order to convert this measurement to mass flow the frequency needs to be taken into consideration. The resonant frequency is a function of the density of the fluid and this information can be extracted from the measuring circuitry.

The performance of Coriolis meters is remarkably good in terms of rangeability, accuracy, repeatability and stability, particularly when it is realized that the Coriolis forces produce sensor movements of the order of 25 μm full scale or 4° of phase shift from vibration amplitudes of about 2 mm. The performance of Coriolis flowmeters is very much a matter of compromise on the selection of the tube and its construction, for example, length, diameter, wall thickness, stress pattern, manufacturability, geometry, overall size, elastic constant, chemical inertness and cost. For maximum sensitivity, long, small-diameter, thin-walled tubes are required, but in this case the pressure drop is high and the static pressure is limited.

The advantages of Coriolis meters include their noninvasive, nonintrusive nature, and their lack of sensitivity to fluid parameters such as asymmetric velocity profile, viscosity, temperature, density, pulsating flow and two-phase flow. The disadvantages include high cost, high pressure drop, size limited to about 150 mm, sensitivity to stress corrosion, sensitivity to environmental vibration, and difficulty of manufacture.

While Fig. 18b shows some of the current variety of tube configurations, it is not possible in this article to analyze the advantages and disadvantages of each design.

3.6 Thermal Techniques

Thermal techniques rely either on a heated sensor being cooled by a flowing fluid, or on a measurement of temperature rise in a fluid as a result of a constant heat input.

Many designs exist and the manufacturers refer to them as mass flowmeters. As they show a cross sensitivity to secondary parameters such as specific heat, thermal conductivity, viscosity and fluid composition, they are not absolute mass flowmeters in the sense attributable to Coriolis meters.

However, there is a wide variety of designs from which to choose and, during the 1980s there were some significant improvements in their performance (due, in part, to electronic signal processing). Also, their cost is relatively low compared with many of the mass flowmeters so far described.

The application of thermal techniques is mainly to gas measurement, although they are occasionally applied to liquids.

(a) *Cooling effect.* The sensor can be a heated wire, thin film, thermistor, resistance temperature detector (RTD) or a thermocouple. The response of a heated sensor inserted in a flowing fluid is nonlinear with flow, the relationship being typically:

$$Q_m = K(H/\Delta T)^{1.67}$$

where Q_m = mass flow rate, K is a constant, H is the rate of heat loss and ΔT is the temperature difference between the sensor and the fluid. In order to measure ΔT a second unheated sensor is used and is inserted into the fluid stream to measure its temperature. Three modes of operation are possible. One mode provides constant power to the heated sensor (H = constant) and measures the temperature difference. The second mode provides the same power to both sensors—one in the flow stream and the other in a semistagnant bypass. The third mode maintains ΔT constant by a feedback control loop and measures the required heat input.

(b) *Thermal dispersion.* The FCI mass flowmeter also operates on the cooling effect but has an unusual mechanical construction (see Fig. 19). The heating power is constant and applied to a resistance heater inside a metal cylinder. A resistance temperature detector in a sheath is brazed to the metal cylinder to form an assembly. There is a duplicate assembly a short distance away, but here there is no heating element in the lower cylinder as this is used as a mass equalizer and balances the time constants of the two assemblies. The upper assembly is cooled at a rate which is dependent on the mass flow and the lower assembly measures the temperature of the incoming

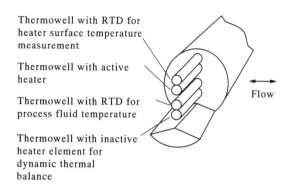

Figure 19
FCI thermal flowmeter

gas. The temperature difference is measured and converted to a linearized flow rate signal.

(*c*) *Temperature rise.* This technique involves a constant power source which heats the entire fluid stream. The heating element is generally inside the pipe. Two sensors are required; one to measure the temperature of the incoming flow prior to the heating element and the other to measure the temperature immediately downstream of the heater. The relationship between mass flow and temperature rise is an inverse one, according to the equation

$$Q_m = \frac{H}{C_p \Delta T}$$

where Q_m is the mass flow, H is the heat input, C_p is the specific heat and ΔT is the temperature rise.

In an alternative arrangement ΔT is maintained at a constant value by a dedicated control loop and the resulting power input H is a linear function of mass flow.

For measuring very low flow rates ($1 \, \text{cm}^3 \, \text{min}^{-1}$), the heater and the sensors can all be small thermistors supported in a capillary tube. For higher flow rates a bypass arrangement can be used.

(*d*) *Boundary layer.* A further variation of the temperature rise technique limits the heat transfer to the boundary layer of the fluid in the vicinity of the pipe wall. This requires the heating element and the sensors to be fixed to the exterior surface of the pipe. As only a small fraction of the fluid is heated it offers more economical operation. However, the performance is affected by the thickness of the boundary layer, which in turn is controlled by viscosity and the Reynolds number.

(*e*) *Temperature profile.* This technique, used in the Thermotube flowmeter, involves direct heating of the fluid flowing through a narrow-bore metal tube by passing a low voltage, high current through a length of the tube. At the points where the power leads are

connected to the tube, heat sinks are fitted. This arrangement creates a temperature profile and two thermocouples or equivalent sensors measure the temperature at two symmetrical points on the heated tube.

The temperature difference measured by the thermocouples will be zero under no-flow conditions. The temperature difference measured by the thermocouples can be interpreted in terms of mass flow rate. This technique is only suitable for low flows (up to $0.008 \, \text{kg h}^{-1}$ or $6 \, \text{l h}^{-1}$) through a capillary tube made of an electrically resistive alloy (e.g., constantan). For high flow rates, the device can be connected across a bypass.

3.7 Miscellaneous Mass Flow Measurement Techniques

(*a*) *Ionization.* Flowmeters based on the deployment of ion clouds have frequently been described in the literature. Figure 20 illustrates one example which can be used for the mass flow measurement of dry gases. The flowmeter consists of a hollow tube supporting internally two hollow coaxial cylindrical metal electrodes, spaced a short distance apart. A fine tensioned wire, supported by insulators, is located along the axis of the upstream electrode and is charged to a potential of 10–20 kV. This generates a cloud of positively charged ions, which at zero flow drifts to the upstream electrode. When flow occurs the ion drift is then towards the downstream electrode. By measuring the ion current to this electrode a signal is obtained which is proportional to mass flow. This is not an absolute mass flow meter as its calibration is dependent on gas composition, but it is claimed that it is independent of ambient temperature, pressure and velocity profile. It can be made bidirectional by adding an additional upstream electrode. It is claimed to have a linear

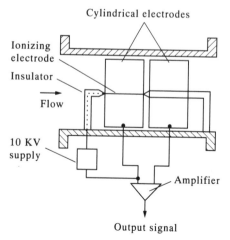

Figure 20
Ionization flowmeter

range of $\pm 150\,\mathrm{g\,s^{-1}}$, an accuracy of $\pm 3\%$ and a response time of less than 1 ms, which means that it should be able to cope with pulsating flow.

(b) *Solid flow.* Granular material can be measured in mass terms by allowing the material to fall from a fixed height onto an angled sensor plate. For accurate results, all material must strike the sensor plate at a constant speed and impact angle. The sensor plate deflects horizontally in direct proportion to the mass flow rate. This deflection is measured with a linear differential transformer.

(c) *Two-phase powder and air measurement.* There are many applications where it is necessary to measure solid rate of mass flow of a powder conveyed in a stream of air. The principle of operation involves the measurement of the dielectric constant of the stream with an insulated capacitor electrode fixed flush with the pipe wall. The flow can be regarded as a mixture of average flow and superimposed smaller irregular components which provide a "flow noise" effect. Experimentally it is found that there is a direct correlation between the "flow noise" and the mass flow.

Coriolis meters have been tested experimentally with some success to see if they can provide an accurate measurement of powder and air flows.

4. Indirect Methods of Mass Flow Measurement

The term "indirect" implies that mass flow rate is measured or computed in terms of two or more parameters. For example, measurement can be carried out by a conventional volumetric flowmeter and its output combined with other measurements such as density, pressure and temperature. Before accurate and reliable gas and liquid density sensors became available, the conventional technique for gases was to measure flow rate with an orifice or other differential pressure device and to combine its output signal with those from pressure and temperature sensors. Mass flow was then computed from the following formula

$$Q_{\mathrm{m}} = C(\Delta P \cdot P_{\mathrm{a}} S T_{\mathrm{a}}^{-1})^{1/2}$$

where C is a constant, ΔP is the differential pressure, P_{a} is the absolute pressure, S is the specific gravity and T_{a} is the absolute temperature. For further refinement of the calculation, corrections could be made for the departure of the gas from the ideal gas laws and for the effect of gas expansion on the downstream side of the orifice ($\Delta P/P_{\mathrm{a}}$). Alternatively, the gas volumetric flow could be measured by any appropriate technique and mass flow computed by combining the signal with that from a density meter

$$Q_{\mathrm{m}} = pQ_{\mathrm{v}}$$

where the suffixes m and v represent mass and volume measurements, respectively, and p is the gas density.

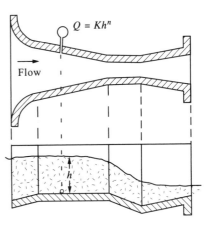

Figure 21
Parshall flume

With the advent of low-cost, reliable electronic computing, the indirect methods became commercially viable. It also encouraged other indirect methods to be examined, such as solids flow.

In solids flow, conveyor belts can be provided with load cells to continuously weigh the solid material carried on the belt. If a tachogenerator measures the speed of the belt, a simple mass flow can be computed from the equation $Q_{\mathrm{m}} = mv/s$, where m is the output of the load cell, v is the velocity of the belt and s is the length of the belt being weighed. An alternative technique is to measure the weight per unit length of belt with a nucleonic gamma-emitting isotope (e.g., ^{137}Cs), and compute the mass flow from a measurement of the belt speed.

5. Open Channel Flowmeters

In many applications involving irrigation, drainage, industrial effluents and sewage treatment, it is convenient to constrict fluid flow to open channels. Flow rates can then be measured either by fitting weirs, formed by shaped plates fixed in the channel, or by flumes, which are formed by shaping the section of the channel. An example of a design of a venturi flume is shown in Fig. 21. In all cases of open channel flowmetering the flow rate is measured in terms of liquid level and by applying the appropriate formula relating head or level to flow rate. The head or level of liquid may be measured using a mechanical float, an ultrasonic depth gauge or an air reaction bubbler system.

6. Anemometers

In some applications the total flow of a fluid is not required; only its velocity at some chosen point. An

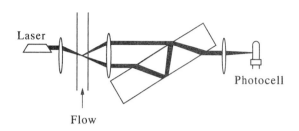

Figure 22
Laser Doppler anemometer

anemometer is then used and the techniques employed are similar to those for flow rate measurement.

Typical examples of anemometers are:

(a) pitot tubes for gases and liquids;

(b) hot-wire elements, which depend on the cooling effect on an electrically heated wire or thermistor when placed in flowing fluid;

(c) turbine or vane type instruments; and

(d) tracer types.

Other point measuring flow devices based on vortex shedding, electromagnetic induction and ultrasound can usefully be adapted for anemometry.

A particularly valuable anemometer for research work employs the laser Doppler technique (see Fig. 22), which is similar to that employed in ultrasound flow measurement (see Sect. 2.8), but uses laser light instead. If offers a unique opportunity for measuring a fluid velocity vector at a selected point in the fluid without interference to the flow pattern. The optical system is arranged to measure the frequency difference between the direct beam and that reflected from particles in the fluid. Velocities down to $10^{-5}\,\mathrm{m\,s}^{-1}$ and up to $800\,\mathrm{m\,s}^{-1}$ can be measured with this technique.

See also: Flow Measurement: Applications and Instrumentation Selection; Flow Measurement: Economic Factors, Markets and Developments; Pressure Measurement

Bibliography

Addison H 1949 *Hydraulic Measurements—A Manual for Engineers*. Chapman and Hall, London
Baker R C 1989 *An Introductory Guide to Flow Measurement*. Mechanical Engineering Publications, Bury St Edmunds, UK
Benard C J 1988 *Handbook of Fluid Flowmetering*. Trade and Technical Press
British Standards Institution 1991 *Guide to the Selection and Application of Flowmeters*, BS 7405. BSI, London
Cascetta F, Viga P 1988 *Flowmeters—a Comprehensive Survey and Guide to Selection*. Instrument Society of America, Research Triangle Park, NC
De Carlo J P 1984 *Fundamentals of Flow Measurement*. Instrument Society of America, Research Triangle Park, NC
Hayward A T J 1979 *Flowmeters*. Macmillan, London
Industrial Measurement Series—Flow A film produced by the Instrument Society of America, Research Triangle Park, NC
Instrument Society of America, Calculations and documentation on flow devices and fluids. Computer Software. FLOWEL 2.0 (IBM PC/XT/AT). ISA, Research Triangle Park, NC.
Linford A 1949 *Flow Measurement and Meters*. Spon
Miller R W 1989 *Flow Measurement Engineering Handbook*. McGraw-Hill, New York
Shell 1985 *Shell Flow Meter Engineering Handbook*. McGraw-Hill, New York
Spitzer D W 1984 *Industrial Flow Measurement*. Instrument Society of America, Research Triangle Park, NC

R. S. Medlock[†]
[ABB Kent, Luton, UK]

Force and Dimensions: Tactile Sensors

This article will deal only with the state of the art of research and development of sensors, the applications of which are important in handling processes, in which the term "sensor" does not only mean the measuring instrument itself, but also the electronic equipment for information processing as well as the interface of the handling device. In automated handling activities, for example, using industrial robots, the main aim is the substitution of manpower by appropriate, efficient and flexible automatic devices.

This kind of automation is used as an aid when human handling is:

(a) disagreeable,

(b) uneconomical, or

(c) inferior, because humans cannot reach the accuracy and loading capacity of automated handling systems.

1. Sensor Tasks

A handling device having capabilities comparable with those of man, must be equipped with two different sensor systems:

(a) an internal sensor system for observing the internal conditions of the automatic device (disposition, speed, torque of moving axes, voltage, currents of drives and control); and

(b) an external sensor system for supervision of the surroundings of the device (presence or absence, position, orientation, speed and classification of handling objects; comprehension of special object marks and the conditions of objects).

Without an appropriate external sensor system, handling devices are able to repeat a preprogrammed sequence of motions innumerable times within the device's accuracy; however, deviations of the peripheral conditions of time and position which require a correction of the movements or a change in the program sequence are not considered.

The technological aims inherent in the development and the consequent application of handling sensors are:

(a) an extension of the possibilities for application of handling systems by increasing the flexibility;

(b) an improvement of handling reliability by supervision;

(c) an improvement of the quality of working results obtained with handling devices, applied by understanding working conditions and states; and

(d) a building up of the preconditions for fully automatic handling systems by automatic checkup and control.

2. Sensor Problems

The main task of the sensor is to selectively acquire comprehensive information from objects as a whole, and to reduce and adapt this information for a few data which are significant for handling.

The parameters of the handling device which can be determined are mainly the path of motion, the speed of motion and the sequence of a series of motions. The amount of data required to determine these parameters is relatively small compared with the amount of data received by the sensor.

3. Steps in Signal Processing

The required data acquisition, concentration and computing is carried out in several steps, checked one after the other:

(a) selection of the sensor principle;

(b) selection of the space of installation and of the direction of the measuring instrument;

(c) determination of the mode of signal processing;

(d) analysis, interpretation and evaluation of the signal processed;

(e) selection of the information required; and

(f) adaptation of the interface between the sensor and the handling device.

4. Sensor Types

In order to develop the sensorial abilities experienced by man (seeing, hearing and touching) in the technological field, various measuring and processing principles are to be determined, tested and applied.

With regard to data acquisition, sensors can be roughly divided into noncontacting and tactile sensors. Noncontacting sensors are mainly optical, acoustic, inductive, capacitive and pneumatic sensors. Tactile sensors are connected with mechanical switches, mechanical path and speed receivers, and force–torque receivers equipped with piezoelectric crystals or strain gauges.

5. Contacting, Tactile Sensors

Contacting sensors require contact with the object to be measured. They can be used for the perception of the contact (presence or absence of objects), of the forces and torques occurring at the point of contact, and slips occurring during contact.

5.1 Tactile Single Sensors

The simplest form of contacting sensors are switches which, according to their arrangement and their position, can perform very different tasks. By means of switches, a searching function or a two-step action can be realized.

5.2 Tactile Multiple Sensors

By the arrangement of several switching elements, for example, in a line or in matrix form, tasks of position and pattern recognition can be performed. At the IPA, Institute for Manufacturing Engineering and Automation, Stuttgart, several sensors of this kind have been constructed and tested for practical applications (Abele 1981).

A sensor equipped with 256 contacting points on a surface of 66×66 mm has been applied for the control of machining results of the deburring of flat workpieces (see Fig. 1). The number of the contacted points increases when the number of burrs decreases. A 0.5 N force is necessary for each contact point. A similarly constructed sensor has been used for position recognition of simply shaped workpieces in an orienting system.

The problems to be solved which are connected with sensors of this type are:

(a) the realization of the conducting points as individual elements;

(b) high costs per conducting point;

(c) low density of conducting points;

(d) low reliability;

(e) large cabling expenditure;

(f) high conducting power per point required; and

(g) rigid arrangement.

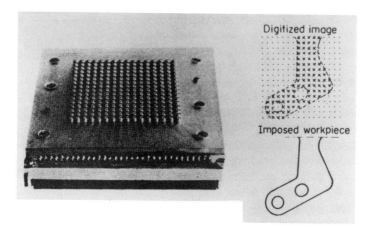

Figure 1
Tactile area sensor

5.3 Tactile Flat Sensors

Flat sensors, for which the position and number of conducting points on the surface are unimportant, can be installed directly on the surface of industrial robots as a security means for the prevention of collisions. If the flat sensor is exposed to a certain pressure, the machine can be switched off by an emergency/off circuit. The actual conducting element consists of a plastic sheet, embedded between two conducting electrodes; it considerably reduces the electrical resistance of transmission when pressure is exerted vertically on the surface.

5.4 Tactile Analog Sensors

By means of contact sensors, distances can be measured accurately. Linear potentiometers and inductive receivers which work according to the principle of differential transformation are useful in this field. Difficulties in practical applications arise with clamping. To date, it has not been possible to satisfactorily solve the mechanical problems.

On the other hand, these sensors can be used for highly accurate distance measurements. Distance measurements by contact sensors can be carried out without continuous contact with the object.

5.5 Slip Sensors

Sensitive parts should be gripped by using a power which is sufficient to handle the gripped object. This minimum handling power value is achieved the moment when no relative movement between the object and the gripper jaw occurs. In order to recognize such movements several solutions have been proposed and realized (Schweizer 1978). For example, the gripper jaws were equipped with rotatable balls, the movement of which could be detected optically. An interesting proposal was a sound receiver system integrated into the gripper jaw which gave an output

signal with each relative movement between the gripper jaw and the gripped workpiece. In another case a change of the movement of the workpiece was sensed by plastic sheets which changed their resistance when exposed to pressure.

5.6 Force–Torque Sensors

Among the tactile sensors the force–torque sensors form the most important group. These sensors measure reacting forces and torques occurring between an industrial robot and the surrounding object with which it comes into contact.

Four different methods have been developed to measure forces and torques occurring at the gripper of an industrial robot; these methods are as follows:

(a) a multidimensional force–torque sensor at the gripper flange (between the robot arm and the gripping tool);

(b) a force–torque sensor directly at the gripper jaw;

(c) a one-dimensional sensor on each robot axis; and

(d) a multidimensional force–torque sensor at the bottom of the solid workpiece or tool support, which measures reaction forces and torques during machining or joining.

Method (a) has been tested in different constructions. The IPA, Stuttgart, constructed several sensors equipped with extending strips and performed practical tests. The aims of this development were: (a) mechanical decoupling of single components; (b) overload security; (c) small dimensions; and (d) simple construction. A sensor of this kind, constructed on two spoked rings, is shown in Fig. 2a. The wire strain gauges are interconnected in the form of semibridges and fixed to the spokes of the rings. Decoupling is carried out by an analog circuit. The Istituto di Metrologia G. Colonnetti, Turin, developed a multi-

Figure 2
(a) Multidegree force momentum sensor (b) force–torque dynamometer (wrist type)

component force–torque dynamometer for robotics (wrist type). The sensor, shown in Fig. 2b, is a double parallelogram structure folded over itself, a configuration which permits a definite space saving (Barbato *et al.* 1990).

6. Sensor Interface

The connection between a sensor and the handling device controls demands different interfaces, depending on the kind of sensor used and on their control or adjustment applications as follows:

(a) sensors with conducting output—binary inputs (interconnecting inputs) of handling devices;

(b) analog sensors (one-dimensional)—analog input with rigid coordinate effects; and

(c) multidimensional sensors—parallel inputs of data (software oriented).

The amount of data to be transmitted by the interface is small with sensors used for control; by contrast, it is large and time critical with adjusting sensors.

Through the sensor, interface positioning and speed data of the handling device for changes of movement are transmitted. Besides these data, which are given as absolute or correction values, further information for control of the program sequence of the handling device is to be transmitted. On the other hand, control information or actual speeds must be transmitted to the sensor.

Between the sensor and the handling device, a coordinate transformation must be performed for the parameters which are dependent on geometry. The sensor works in a coordinate system which refers to the workpiece or sensor space or to the robot gripper, whereas the handling devices work in a coordinate system referred to as device axes.

7. Sensor Software

Hardware-performed sensor tasks form the basis for many application possibilities of a particular physical principle. However, in the end, they do not decide whether a project leads to the required result. Complex software, the means by which received information is filtered, classified, compared, evaluated, combined and computed, is dependent on both the receiver and the goal, and it is highly affected by variable boundary conditions. Evaluations and decisions must be made in a multidimensional range of values. These tasks are typical of artificial machine intelligence.

See also: Mass, Force and Weight Measurement; Mechanical and Robotics Applications: Multicomponent Force Sensors

Bibliography

Abele E 1981 Adaptive controls for fettling of castings with IR. *Proc. 1st RoViSec.* IFS Publications, Bedford, UK, pp. 81–90
Barbato G, Desogus S, Germak A 1990 Multicomponent force sensors for robotics. *Proc. ISMCR IMEKO Tc-17.* IMEKO, Budapest, pp. I 2.1–2.9
Bauzil G, Briot M, Robes P 1981 A navigation subsystem using ultrasonic sensors for the mobile robot HILAIRE. *Proc. 1st RoViSec.* IFS Publications, Bedford, UK, pp. 47–58
Espiau B, Andre G 1981 Sequential algorithms related to optical proximity sensors. *Proc. 1st RoViSec.* IFS Publications, Bedford, UK, pp. 223–31
Jablonowsky P 1981 Robots that assemble. *American Machinist.* Special Report 739. American Instrument Society
Jurevich E I 1978 Second generation robots of Leningrad Polytechnical Institute. *Ind. Rob.* **5**, 127–30
Kinncunan P 1981 How smart robots are becoming smarter. *High Technol.* **9**, 32–40
Larcombe M H E 1981 Carbon fibre tactile sensors. *Proc. 1st RoViSec.* IFS Publications, Bedford, UK, pp. 273–6
Mettin F 1984 Aufgabe von Sensoren. *AWF Seminar Sensoren fuer alle fertigungstechnik. Ausschusz fuer*

wirtschafliche fertigung. Internal Report, University of Stuttgart, Germany

Nevins J L, Whitney D E 1980 Assembly research. *Ind. Rob.* **7**, 27–43

Purbrick J A 1981 A force transducer employing conductive silicon rubber. *Proc. 1st RoViSec.* IFS Publications, Bedford, UK, pp. 73–80

Schweizer M 1978 Tactile sensoren für programmierbare Handhabungsgeräte. Ph.D. dissertation, University of Stuttgart

Veda M, Matsuda F, Matsugamoto S 1979 A simple distance sensor and a new minicomputer system, *Proc. 9th Int. Symp. Industrial Robots.* IFS Publications, Bedford, UK, pp. 477–88

Wang S M, Will P 1978 Sensors for computer controlled mechanical assembly. *Ind. Rob.* **5**, 9–18

Weissmantel H, Gairola A 1981 Die verwendbarkeit von leitenden kunstoffen fuer taktile. *Sensoren Feinwerktechnik und Messtechnik* **89**(2), 79–84

C. Ferrero
[Istituto di Metrologia G. Colonnetti, Turin, Italy]

Frequency and Time Measurement

The measurement of frequency is an important aspect of the work of the electrical engineer and is often encountered in the field of servo and allied systems. The advent of precise voltage to frequency converters enables other quantities to be measured more rapidly and accurately than hitherto. This article defines terms, describes the available reference sources, outlines some frequency measurement techniques and discusses errors.

1. Basic Definitions

1.1 Definition of Frequency

A simple definition of frequency f is the number of repetitive events per unit of time. Figure 1 depicts a series of repetitive events, each having a periodic time t. We therefore define

$$F \equiv 1/t \qquad (1)$$

In the International System of Units (SI)

The second is the duration of 9 192 613 770 periods of the radiation corresponding to the transition between the hyperfine levels of the ground state of the caesium 133 atom. [13th General Conference of Weights and Measures (CGPM) 1967]

Following Eqn. (1) the unit of frequency is then defined in absolute terms as "hertz" (Hz) implying the repetitive occurrence of f events per second.

The frequency difference between two sources can be expressed as

$$\Delta f = (f_a - f_b) \text{ (Hz)} \qquad (2)$$

where f_a and f_b are the measured (not nominal) frequencies. (Nominal value is a specified or intended

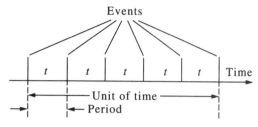

Figure 1
Definition of frequency

value independent of any uncertainty in its realization.)

The normalized frequency departure from nominal is derived by dividing Eqn. (2) by the nominal frequency f. Hence

$$\frac{\Delta f}{f} = \frac{(f_a - f_b)}{f} \text{ (dimensionless)}$$

1.2 Definitions of Accuracy, Stability and Spectral Purity

Frequency sources are usually specified in terms of accuracy, stability (or more correctly instability), and spectral performance. These terms may be defined as follows.

(a) *Accuracy* is the degree of conformity of a measured and/or calculated value to some specified value or definition. For example, the error of a source after corrections have been applied with respect to the National Standard. Accuracy is usually expressed as parts in 10^n, where n is commonly 6, 9 or 12.

(b) *Stability* is defined in terms of the frequency or time domain behavior of a device. In the time domain (i.e., where the sampling time is the varied quantity), a frequently used measure of the stability is the two sample Allan Variance or its square root (see Sect. 4).

(c) *Spectral purity* is the ratio of single sideband phase noise to carrier and is usually expressed in dB (in a 1 Hz bandwidth, at x kHz from the carrier).

2. Standards

2.1 Reference Sources

All measurements must be traceable to a reference source or standard. For general work, the integral crystal of the basic counter/timer may be adequate. For the highest accuracy measurements, it may be necessary to trace a frequency or time measurement from the measuring instrument via a standard frequency transmission to the National Standard.

Table 1
Characteristics of commercial frequency standards

Characteristic	Cesium	Rubidium	Crystal	
$\Delta f/f$ Accuracy	5×10^{-12}	5×10^{-10a}	High grade[b]	Low grade[b]
$\Delta f/f$ Stability				
1s	5×10^{-11}	5×10^{-12}	1×10^{-12}	1×10^{-7}
1h	1×10^{-12}	5×10^{-13}	1×10^{-12}	1×10^{-7}
1d	5×10^{-13}	5×10^{-13}	1×10^{-11}	1×10^{-6}
10d	5×10^{-13}	3×10^{-12}	1×10^{-10}	1×10^{-6}

a Applicable for 1 year after calibration   b Not applicable, calibration needed

Typical values for the accuracy and stability of the main reference standards likely to be encountered are to be found in Table 1. Cesium beam standards embody the SI definition of the second and can be used without calibration if an uncertainty of greater than parts in 10^9 is adequate. Nonetheless, an occasional check against a standard-frequency transmission is desirable to guard against unsuspected malfunctions. For an uncertainty of less than 10 parts in 10^{12}, checking (often continuously) against a reference frequency such as a reliable standard-frequency transmission is necessary. For very high accuracy, it may be necessary to apply corrections for the deviations of the standard-frequency transmission from the national standard of frequency and, at the very highest level, also for the deviations of the national standard from the international value, as determined by the International Bureau of Weights and Measures (BIPM).

2.2 Standard-Frequency Transmissions

Many countries maintain a system of standard-frequency transmission which radiate reference frequencies and time information. By means of a suitable tracking or phase-lock receiver, the user has access to several reference frequencies, each having an accuracy related to the controlling source of the transmissions, which in turn, is traceable to the National Standard. The user must, however, make appropriate allowance for any propagation errors. On the low-frequency (lf) and very-low-frequency (vlf) bands, this is likely to be negligible in most cases.

Tracking and phase-lock receivers cause the frequency of the local standard to be coherent with the instantaneous frequency of the received signal, usually averaged over several minutes. Depending on the radio propagation path, this may result in the local standard varying from hour to hour by a few parts in 10^{12}.

A number of transmissions radiate a "time code." A different form of receiver is required for this feature which will provide the user with year, month, day of month, day of week, hour, minute and second information. By international agreement, all transmissions are radiated within 100 µs of Coordinated Universal Time (UTC). This helps provide a reference date or to synchronize measurements at different locations.

3. Measuring Instruments

The availability of a plethora of electronic counter–timers has made the measurement of time interval and its reciprocal, frequency, a relatively simple procedure. Instruments are available with internal reference standards, and in many cases external sources can be connected. In the following applications it should be appreciated that many configurations are available within the same instrument, either by switching or inserting plug-in modules. Typical devices measure directly from dc to several gigahertz. The use of prescalers and heterodyne systems extend the range to the microwave region. More sophisticated instruments are controlled by microprocessor which is also available for conversion from the measured frequency to a related quantity such as velocity or revolutions per minute (rpm). Mathematical operations and statistics are included in the specification of some devices. Instruments may be operated by external computer via an appropriate control bus. Of these, the General-Purpose Interface Bus (GPIB) is most frequently encountered (BS 6146: Pts. 1 and 2; International Electrotechnical Commission Publication 625, Pts. 1 and 2, and Institute of Electrical and Electronics Engineers 488). Bus operation enables full setting of the arming, trigger and counting circuits, together with instrument status and output information.

3.1 Frequency Measurement

The basic frequency measurement instrument consists of an input signal processing stage, electronic gate, counter and time base. In operation, the gate is opened for a known period defined by the gate control circuits, referenced to either the internal or external time base. Typically, the gate would be opened for 1 s during which period the number of pulses allowed through would be counted and displayed. Resolution may be improved by increasing the gate time. The count registered represents the average frequency over the measurement time.

By introducing a prescaler in the input, the high-frequency (hf) limit of the device can be extended to the GHz range. A prescaler is a divider which reduces the input frequency by $1/k$ where k may be either a factor of two or ten. Naturally the instrument indication must be multiplied by k in order to obtain the true result. The alternative is to increase the gate time by the same factor. This is a cost-effective method of extending the frequency range of an instrument. Although the technique requires an increase in measurement time, the accuracy capability is identical to the basic instrument.

Heterodyne converters reduce the signal by frequency conversion, to a value within the range of the counter section. The indicated result must be added to the harmonic number derived from the heterodyne section.

3.2 Period Measurement

It has already been shown that the frequency of a waveform can be determined from its periodic time. However, it is sometimes advantageous to measure period as opposed to frequency and this is especially true when low frequencies are encountered. The disadvantage of period over direct frequency measurement is that it is necessary to calculate the final result. More sophisticated instruments will perform this operation automatically.

A counter configured to measure period will have the gate controlled by the unknown frequency applied to the input. The display will indicate the number of reference cycles counted in the time interval defined by the period of the input frequency.

As in direct frequency measurements, accuracy can be improved by increasing the measurement time. This is known as multiple period measurement and involves directing the unknown frequency through decade dividers before reaching the gate control circuit. The main gate is thus opened for 10^n periods, where n is in the range 1–5. The greater the value of n, the greater the resolution, although as with direct frequency measurement techniques, the time for a complete measurement is also increased.

It has been mentioned that frequencies at the lower end of the spectrum can be measured more accurately by the period method. This is true where the unknown is lower than the time-base frequency. At this point, the gating error (see below) is identical in both methods.

3.3 Instrument Accuracy

There are three sources of potential error inherent in the typical counter–timer: gating (sometimes referred to as ±1 count), trigger, and time-base errors. The overall measurement error will be the algebraic sum of the individual errors.

(a) *Gating error.* Due to the (general) absence of synchronization between the signal to be measured and the source defining the gate time, it is possible for the indication to be in error by ±1 count. By increasing the measurement time, the error from this source is minimized. To reduce this error without increasing the measurement time, some commercial instruments apply a low-frequency modulation to the time-base frequency—sometimes described as a jittered time base.

(b) *Trigger error.* For direct frequency measurements, trigger error is usually negligible. However, in the period mode, the input signal is used to control the gate. If this signal is degraded by noise or modulation, triggering may be erratic and the measurement degraded.

For period and time interval measurements, trigger error may be expressed as the ratio of gate time duration t to the total period T. It is also proportional to the relative amplitude of peak V_n noise voltage to the peak signal voltage V_s. Hence trigger error may be expressed as

$$\frac{\Delta t}{T} = \frac{kV_n}{V_s}$$

where k is a constant of proportion depending on the nature of the modulation. Although input filters can be used to reduce the effects of noise, it is essential to ensure that the trigger amplitude and slope controls on the instrument are correctly set. Clean, rapid rise-time signals should be used wherever possible for triggering.

As with gating error in frequency measurements, trigger error is reduced proportionally as the number of periods averaged is increased.

(c) *Time-base errors.* The crystal oscillator forming the time-base reference is subject to both short- and long-term variations which may affect the accuracy of the measurement. Many oscillators are temperature controlled, and sufficient time must be allowed for the system to stabilize. Where this is a problem, fast warm-up crystals are available which stabilize in a few minutes.

Short-term variations may affect short-period measurements and reference should be made to the manufacturer's specification to ascertain their possible magnitude. Long-term error is usually the result of crystal aging and, depending on the quality of the crystal, may amount to between 10^{-11} and 10^{-6} per day. The error is cumulative and in the more precise measurements it should be verified with respect to a known reference or standard frequency transmission.

For regular accurate measurements where time-base error may be significant, the external reference feature found on many instruments should be connected to a known source.

4. Very-High-Accuracy Measurements

The measurement techniques described so far are general in nature and applicable to a range of frequencies. However, it is often necessary to determine certain qualities associated with frequency standards themselves. One example is to establish quantitatively, fluctuations in frequency of a source. The following sections describe possible techniques.

4.1 Short-Term Frequency Instability Measurement

Measurements over sampling times less than about 1000 s are generally classed as short term. To obtain adequate resolution at short measurement intervals, it may be necessary to use some form of frequency multiplication. A typical commercial instrument will multiply frequency differences by up to a factor of 10 000. This is equivalent to comparing 1 MHz signals after multiplication to 10 GHz. The multiplied signal is applied to a frequency measurement system in the normal way. Frequency stability measures are generally denoted by σ, and are measured by taking a reasonably large number of successive readings y of the frequency of the device under test. As it is not possible to measure instantaneous frequency, any frequency measurement relates to a specific sampling time τ which must be stated. It can typically range from 10^{-3}–10^6 s and is frequently determined by the gate time of the counter involved in the measurement.

The two sample Allan Variance $y(\sigma)$ can then be computed from the relationship:

$$y(\sigma) = \left(\frac{\substack{\text{Sum of squares of differences} \\ \text{between successive readings}}}{2 \times \text{number of differences used}} \right)^{1/2}$$

To obtain reliable data, it is necessary for the instrumentation to have a bandwidth greater than $1/\tau$, where τ is the minimum desired sampling time. A short counter "deadtime" between measurements is also required. Generally, σ will be given as a normalized value, that is, the value obtained for the instability is divided by the frequency at which the measurement was performed.

For example, if $\Delta f = 10$ Hz was measured at a frequency of 5 MHz, then the normalized frequency instability is calculated:

$$\frac{\Delta f}{f} = \frac{10}{5 \times 10^6} = 2 \times 10^{-6}$$

Note that the result is dimensionless, and is still applicable if the original signal is multiplied or divided (synthesized) to a different value.

Usually, the system is arranged so that the reference source has superior stability to the device under test. Where similar devices are intercompared, the instability measured is apportioned between the two sources in the ratio of $1/\sqrt{2}$. Short-term stability assessment is a specialized aspect of frequency measurement and reference should be made to the literature for alternative methods and computation procedures.

4.2 Long-Term Frequency Stability Measurement

It is not convenient to measure long-term stability by any of the previous systems. A suitable method to investigate the daily performance of two high-grade frequency sources is to measure the phase change over a known time period. If the two sources are mixed and an analog voltage derived from the phase difference, this can be applied to a chart recorder and the phase change with elapsed time noted. A 1 MHz comparison frequency is convenient as it gives 1 μs full-scale on the recorder. For example, if a phase change of 2.30 μs was noted as the daily rate between two sources, then

$$\frac{\Delta f}{f} = \frac{\Delta t}{T} = \frac{2.30 \times 10^{-6}}{24 \times 60 \times 60} = 2.67 \times 10^{-11}$$

5. Time

Time can be considered to be of two related but somewhat different natures—time lapse and time of day. Time lapse is determined by counting the cycles of a known frequency that occur between the start and end of the period in question. For example, in exactly one day there will be 4.32×10^{11} cycles of a 5 MHz reference frequency, so that any instant can be designated by a particular cycle number later than the starting reference time.

Time of day requires a designated reference instant: thereafter an exact description of any subsequent instant requires counting the number of time intervals (microseconds, seconds, days, years, etc.) to achieve an exact description of the instant.

Time of day is commonly made available by time signals and, more coarsely, by the calender.

6. Precautions

Mention has already been made of the need to apply "clean" signals to the measuring system. Modulation, both AM or FM, is likely to affect adversely the input–trigger circuitry of a counter and produce erroneous readings. Similarly, 50 Hz is detrimental and care should be taken with earthing and earth loops if problems from this cause are to be avoided; 50 Hz chokes can be included in the signal lead in some cases.

In a high-accuracy measuring system comprising several instruments, it is always desirable to supply the time bases from the same source, preferably an external standard whose characteristics are known.

Measurements in standards laboratories often involve establishing the frequency difference between several sources of the same nominal frequency. In this type of measurement, a useful check can be made by measuring in pairs, that is, A–B, B–C, C–A; the algebraic sum of the readings should then be zero.

Bibliography

Allan D W 1966 Statistics of atomic frequency standards. *Proc. IEEE* **54**, 221–30

Audoin C, Vanier J 1976 Atomic frequency standards and clocks. *J. Phys. E.* **9**, 697–720

Barnes J A 1966 Atomic timekeeping and the statistics of precision signal generators. *Proc. IEEE.* **54**, 207–20

Barnes J A *et al.* 1971 Characterization of frequency stability. *IEEE Trans. Instrum. Meas.* **20**, 105–20

International Electrotechnical Commission 1979 *An Interface System for Programmable Measuring Instruments (Byte Serial, Bit Parallel); Part 1, Functional Specifications, Electrical Specifications, Mechanical Specifications, System Application and Requirements for Designer and User.* IEC 625: Pt 1. IEC, Geneva

International Radio Consultative Committee (CCIR) 1992 *Standard Frequencies and Time Signals.* Recommendations and Reports of the CCIR. International Telecommunications Union, Geneva

Kartaschoff P 1978 *Frequency and Time.* Academic Press, London

Ramsey N F 1956 *Molecular Beams.* Clarendon Press, Oxford

D. S. Sutcliffe
[National Physical Laboratory, Teddington, UK]

P. M. Clifford
[City University, London, UK]

I

Identification in Measurement and Instrumentation

Identification is the determination of a mathematical model of a dynamical system or process, based on input and output measurement data of that process. Thus, depending on the point of view, identification can be considered as:

(a) a type of straightforward measurement technique, when focusing on the data gathering;

(b) a type of instrumentation when focusing on the information processing technique or the data handling; or

(c) a type of model building, when focusing on the overall goal of the operation.

That goal, in relation to the prospective use of the model, implies that one may already have a particular class of models in mind. The slightly altered classical definition of the field as given by Zadeh (1962) is brought to mind.

> "Identification is the determination, on the basis of input and output, of a model within a class of models, to which the process under test is equivalent."

The scheme of identification (Eykhoff 1974), illustrated in Fig. 1, is also classical. This scheme is divided into two activities:

(a) modelling—deriving structural knowledge of the process under test by using, for example, physical laws that hold for the process or system under consideration; and

(b) estimation—deriving knowledge about parameters (and perhaps the order and states of that process) by using measurement data.

Identification is the combination of both these activities.

Measurement can be defined as the representation of attributes of objects and events by numbers or by other symbols. Hence, identification can be interpreted as an extended type of measurement theory and practice. It extends the scope of assignments to indirect measurements. Therefore, the following may be recognized:

(a) measuring unmeasurable things such as parameters (e.g., coefficients of a differential equation) that specify the dynamical behavior of a process and that cannot be determined in a direct way; and

(b) determining, estimating and reconstructing quantities that are difficult or expensive to measure

directly (e.g., specific state variables of a chemical process) by making use of model knowledge that relates such quantities to signals that can be measured in a better, cheaper, simpler and more reliable way.

Such indirect measurements are becoming more and more important for both technical and economic reasons.

1. Uses of Identification

Identification is being used for a variety of purposes.

(a) *Diagnosis*—for determining quantities that are not directly measurable; for example, in physical or engineering research.

(b) *Monitoring and fault detection*—for judging the quality, reliability and efficiency of processes in order to provide an adequate base for making decisions on overhaul, replacement, cleaning, and so on, of those (or part of those) processes; for example, the wear and tear of machines (Isermann 1984).

(c) *Design*—for attaining adequate models of a process (or parts of a process) in order that they can be used for design; for example, in designing an optimal and robust controller for that particular process.

(d) *Control (adaptive or self-optimizing)*—for continuously adjusting the control scheme in relation to changes that take place in the process (or the dynamics of the process) under control; for example, for adjustments to be made in order to obtain some specifically desired behavior.

(e) *Prediction*—for estimating future behavior; for example, weather, economic and managerial forecasts.

By definition, identification has to result in one of several models. For the dynamic behavior of processes, many different types of models present themselves as candidates (e.g., transfer function, difference equation, impulse response and state-space models). The choice from that wide variety depends on:

(a) the requirement of the specific application that one has in mind (goal oriented); and

(b) the properties of the estimation schemes that one would like to use (method oriented).

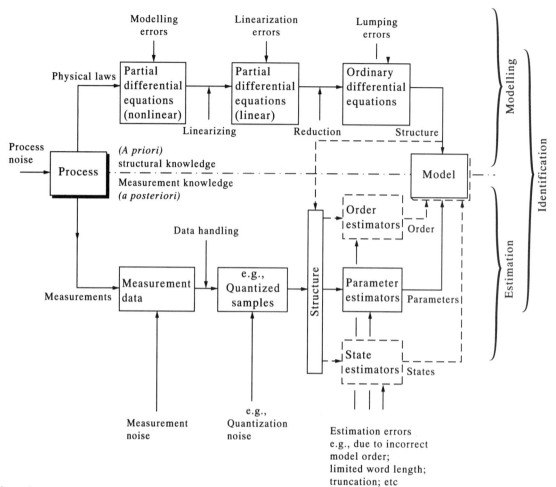

Figure 1
Scheme of identification

2. The Identification Protocol

As indicated earlier, when applying identification, a particular purpose for the use of the model to be developed is necessarily in mind. This implies a protocol of steps to be taken from the outset, the problem definition, until a satisfactory end result is reached, that is, a usable model has been generated. The essential elements of this identification protocol are given in Fig. 2 (Eykhoff 1984).

A division is made between modelling and estimation; which is illustrated in the following sections.

2.1 Modelling: Engineering Insight

In application-oriented work it is essential that the model builder, identifier and experimenter have an open mind for, and a clear view on, the real *a priori* knowledge that is available, and on the tacit assumptions that are embedded in his or her task.

The first decision that requires careful consideration is the choice of the demarcations of the process under study:

(a) what is considered part of the process and what is nonprocess or environment?

(b) what are (measurable) inputs and what have to be recognized as nonmeasurable inputs, that is, as disturbances?

Clearly, in real engineering situations such decisions require a knowledge, insight and intuition that, as yet, defy complete scientific argumentation. In such situations it is frequently unavoidable that prejudices creep into the modelling procedure.

2.2 Modelling: Physical Laws

The *a priori* knowledge associated with the modelling depends on the artful combination of techniques from

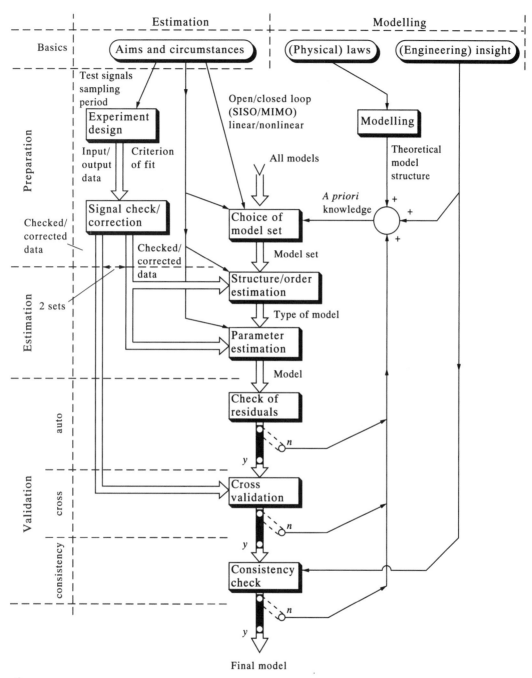

Figure 2
Identification protocol

many fields, for example, mechanics, thermodynamics, and so on. Judicious simplifications such as linearizing, lumping and reducing play an important role (see Fig. 1). Particularly crucial is the judgement on the acceptability of the various types of errors introduced by those simplifications: modelling, linearization and lumping errors.

For linear dynamic behavior there is a coherent and comprehensive theory of models. For nonlinear dynamics however, the situation is still rather poor.

Volterra series provide a coherent approach, but their application is limited due to the excessive number of parameters needed. Also, a technique called group method of data handling (GMDH) has been used in such cases (Ivakhnenko 1970). Apart from these general approaches, mostly incidental and application-specific models are being used.

2.3 Estimation: Aims and Circumstances

As indicated earlier, the goals of the identification procedure are of paramount importance. To a high degree they determine the answers to essential questions such as:

(a) for the particular type of application being considered, what kind of model would be adequate (explanatory, representation or prediction model)?

(b) can the process be treated as a set of single-input/single-output (SISO) models, or does a more complex multi-input/multi-output (MIMO) model have to be determined?

(c) are the nonlinearities that are present in the process essential, or can the identification be restricted to a set of models for the process while operating at a number of working points?

(d) next to process dynamics, should the disturbances also be characterized?

(e) what complexity of the model (e.g., number of parameters) would be adequate?

(f) are there closed loops to be considered, either directly recognizable or perhaps hidden, for example, by human intervention on the process during the measurements?

(g) can test signals be applied to the process? If so, what would be an optimal choice?

In particular, the reader should keep in mind the distinction between "science" and "art" aspects; that is, that some decisions defy complete scientific argumentation and have to rely on engineering insight.

2.4 Experiment Design

As measurements are indispensable, the sensor location problem also has to be considered if there is some freedom at all in the instrumentation.

Input signal design has been recognized as a useful tool for improvement of the accuracy of parameter estimates. In the literature, a number of aspects of input signal design for system identification has been discussed for various classes of models. An input signal, often used for system parameter estimation, is the pseudorandom binary signal (PRBS); this is simple to generate, has suitable properties, has the best power-to-maximum-amplitude ratio and is simple to apply.

2.5 Signal Check and Correction

In practice, the collected process data have to be visually checked through plotting and, in most cases, preprocessing has to be done before they can be used for the data processing/estimation task. Such preprocessing operations may consist of the following (Backx and Damen 1989a, b):

(a) *peak shaving*—the removal of spikes, which are mostly caused by measurement errors and by induction in sensor leads;

(b) *trend correction*—the removal of trends, which are often caused by all kinds of, mostly unknown, low-frequency disturbances acting on the process;

(c) *delay time compensation*—the compensation for the pure time delay by a time shift between the collected process input and output data;

(d) *offset correction*—the subtraction of the average signal values, to enable linearization around the selected working point; and

(e) *filtering*—the prevention of frequency aliasing that might occur by the sampling operation and for the improvement of signal-to-noise ratios.

2.6 Choice of a Model Set

The choice of a model set is based on the modelling activities described earlier, before the type of model, or at least its structure, is selected.

After these preparations the estimation phase is entered. Order estimation and parameter estimation, either implemented separately or combined into one computer program, depend on the *a priori* knowledge available; this knowledge ought to be decisive for the choice of estimation method (e.g., least squares, instrumental variable, Markov, maximum likelihood and Bayes). This will be briefly indicated in Sect. 3. Important considerations imply:

(a) the uniqueness of the error criterion;

(b) the unbiasedness of the estimator;

(c) the transformability of the model, obtained from estimation, to the type of model(s) needed for the particular application;

(d) the error propagation under transformation of the model;

(e) the computation time needed; and

(f) the numerical accuracy and computational stability.

The final, validation phase will disclose whether the model that results from the estimation, is adequate, according to the quality measures chosen, and is suitable for the intended purpose.

The autovalidation or check of residuals gives a measure of how well the model explains the process data that have been used for the estimation. The residuals, that is, the differences between the outputs of the process and the model, when their inputs are the same, should not contain any components that are correlated with the input; if that were the case then the "explaining power" of the model could still be improved.

Cross validation gives the same measure of model performance as autovalidation; now measurement data that have not been used during the estimation cycle will be dealt with.

The consistency check again leads to a confrontation with the physical and engineering insight of the experimenter. The model may fail for each of these validation steps. In such a case one must return to the selection of the model set and repeat the steps indicated.

3. *The Basic Approaches to Parameter Estimation*

The modelling path (see Fig. 1) is very much dependent on the type of process under consideration (which might be a cat cracker, an electric power generating unit, a human aorta, some econometric relation, etc.). The laws and relations to be used are quite specific to such particular processes. Consequently, this part of the identification procedure does not lend itself to a coherent presentation.

The structure of the process to be identified is conceptually illustrated in Fig. 3a, where u and x are the input and output of the dynamics under study; the observations on those signals may be contaminated by additive disturbances and noises, m and n, respectively. The dynamics may be influenced by unmeasurable inputs w that act as disturbances.

Many identification schemes are based on a comparison between the process and a model, using the measurable process signals. Figure 3 indicates the ways in which this can be done. In Fig. 3b the output of the process is compared with the output of the model M; in this case, e is the output error. In Fig. 3c the process input and output are applied to a generalized model, consisting of M_b and M_a; now e is the generalized model error. In Fig. 3d the process input and output, except the most recent output sample, are being used by a prediction model M_p, to forecast by $\hat{y}$ the present value of y; in this case e is the prediction error.

These errors may be minimized in some (e.g., quadratic) sense, with respect to the values of the model parameters $\hat{v}$; the corresponding values represent the estimates for the unknown parameters.

Another way of looking at estimation is based on projection of the process input and output signals on suitably chosen templates or time functions (Eykhoff *et al.* 1981). A specific technique based on this is the so-called instrumental variables (IV) method.

For the estimation path, a step towards order can be found in the recognition of the *a priori* knowledge,

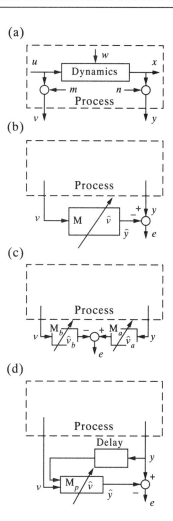

(a)

(b)

(c)

(d)

Figure 3
Parameter estimation process

that is available in particular identification/parameter estimation situations. This is illustrated by Fig. 4, which indicates the simplest nontrivial parameter estimation situation, and by Table 1. Here, u and y are the process input and output, respectively; n is the disturbance or noise; Θ is the unknown parameter to be estimated; N is the covariance matrix of the disturbances/noise n; $p(n)$ indicates the probability density function of these disturbances; $q(\Theta)$ indicates the *a priori* probability function of the parameters to be estimated; and $c(\hat{v}, \Theta)$ represents the cost function related to the estimated and the true parameter values $\hat{v}$ and Θ, respectively.

Based on the *a priori* knowledge available and on the preparedness to actually employ that knowledge, one has to choose from the methods indicated.

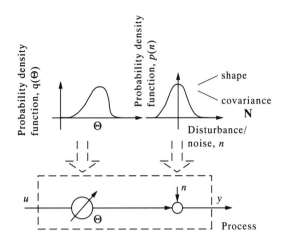

Figure 4
Estimation path

4. Conclusions

A thorough treatment of system identification would imply an in depth discussion of the topics indicated, as well as of a wide variety of other topics such as model structure determination, estimation of model order, measurements under closed-loop conditions and specific types of applications. System identification is characterized by a marked ambiguity. On the one hand a number of aspects clearly belongs to the realm of science; for example, the theory available for the many parameter estimation schemes that have been proposed and advocated in literature, as well as their coherence. On the other hand, when applied to practical problems, identification shows traits that refer to it being an art; for example, essential features that are related to practical applications as well as limitations of identification procedures, that can still be characterized as the artistic intuition of the experimenter or model builder.

This duality is partly responsible for the wide variety and large number of publications, differing in scope, depth, applicability and appeal.

Table 1
A priori knowledge versus estimation schemes applicable

Estimation techniques	*A priori* knowledge			
	$p(n)$ N	$q(\Theta)$ shape	$C(\hat{v}, \Theta)$	
Least squares (LS)	N	N	N	N
Markov (generalized LS)	Y	N	N	N
Maximum likelihood (ML)	Y	Y	N	N
Bayes	Y	Y	Y	N
Minimum risk/costs	Y	Y	Y	Y

Y = Yes; N = No

Bibliography

Backx T, Damen A 1989a Identification of industrial MIMO processes for fixed controllers; part 1, general theory and practice. *J. A* **30**(1), 3–12

Backx T, Damen A 1989b Identification of industrial MIMO processes for fixed controllers; part 2, case studies. *J. A* **30**(2), 33–43

Eykhoff P 1974 *System Identification, Parameter and State Estimation*. Wiley, London

Eykhoff P 1984 Identification theory; practical implications and limitation. *Measurement* **2**(2), 75–85

Eykhoff P, van den Boom A J W, van Rede A A 1981 System identification methods—unification and information development using template functions. *Proc. 8th IFAC World Congress*, Kyoto, Japan, August 1981

Isermann R 1984 Process fault detection based on modelling and estimation methods—a survey. *Automatica* **20**, 387–404

Ivakhnenko A G 1970 Heuristic self-organization in problems of engineering cybernetics. *Automatica* **6**, 206–19

Ljung L 1987 Identification: basic problem. In: Singh M G (ed.) *Systems and Control Encyclopedia*. Pergamon, Oxford, pp. 2239–45

Ljung L 1987 *System Identification: Theory for the User*. Prentice-Hall, Englewood Cliffs, NJ

Schoukens J, Pintelon R 1991 *Identification of Linear Systems: a Practical Guideline to Accurate Modeling*. Pergamon, Oxford

Vansteenkiste G C, Spriet J A 1987 Structure/parameter identification by simulation experimentation. In: Singh M G (ed.) *Systems and Control Encyclopedia*. Pergamon, Oxford, pp. 4690–2

Zadeh, L A 1962 From circuit theory to system theory. *Proc. IRE* **50**, 856–65

Zhu Y, Bachx T 1993 *Identification of Multivariable Industrial Processes (for Simulation, Diagnosis and Control)*. Springer, Berlin

P. Eykhoff
[Eindhoven University of Technology, Eindhoven, The Netherlands]

Image Processing and its Industrial Applications

Figure 1 shows the organization of an archetypal machine vision system, used for industrial applications such as robot control and inspection. A most important feature of this diagram is that it emphasizes the interdisciplinary nature of the subject. It is imperative that industrial machine systems are designed and installed by a team having a broad range of skills.

(a) *Mechanical engineering*. It is important that the object to be inspected is presented properly to the camera.

(b) *Illumination engineering*. The quality of the images "seen" by the camera is critically dependent on the provision of good lighting.

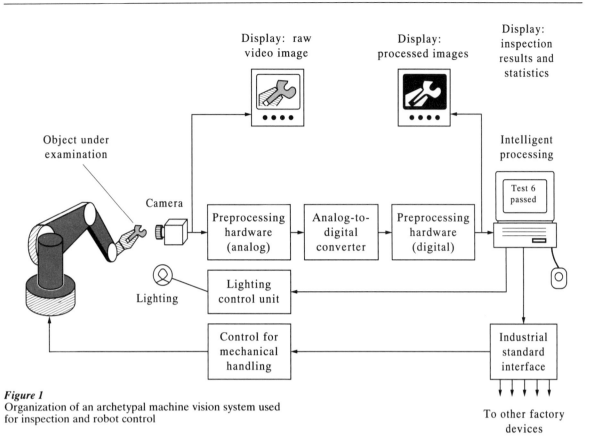

Figure 1
Organization of an archetypal machine vision system used for inspection and robot control

(c) *Optical engineering.* It is often possible to improve image quality considerably, using straightforward optical techniques. A clear distinction must be drawn between illumination and image formation; two different engineers might be needed.

(d) *Sensor (camera) technology.* Video sensors vary a great deal in weight, physical size, optical sensitivity, spectral and dynamic response, image burn in, long-term stability, electrical noise and their sensitivity to optical overload, vibration, magnetic fields, radiation, and so on.

(e) *Electronic signal processing hardware technology.* It is important to realize that a video camera generates data at a far higher rate than a conventional computer can process. Specialized processing hardware is therefore needed for many applications. The boxes labelled "Preprocessing" in Fig. 1 incorporate both analog and digital hardware.

(f) *Software engineering.* Some inspection systems require more "intelligence" than can be provided by dedicated signal processing hardware. It is important that the system can communicate in-

spection statistics to the factory personnel. Some inspection systems require *in situ* reprogramming.

(g) *Mathematics.* The algorithmic basis of the image processing is critical to the proper operation of the system. Where possible, the whole system, or its components, should be modelled mathematically.

(h) *Ergonomics.* Not visible in Fig. 1. The system has to work well with a wide range of skills found in the factory.

(i) *Industrial engineering.* Industrial inspection systems must continue working, even when neglected and abused, in a very dirty, hostile environment. The inspection equipment must synchronize with and work in harmony with existing machines in the factory. An automated visual inspection system must be viewed as part of a larger quality assurance activity.

1. Representations of Images

The representation of monochrome (gray scale) images will be considered. Let i and j denote two integers where $1 \leqslant i \leqslant m$ and $1 \leqslant j \leqslant n$. In addition, let

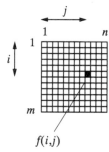

Figure 2
A digital image consisting of an array of $m \times n$ pixels. The pixel ith row and the jth column has an intensity equal to $f(i, j)$

$f(i, j)$ denote an integer function such that $0 \leqslant f(i, j) \leqslant W$. An array $\mathbf{F}$, defined as

$$\mathbf{F} = \begin{vmatrix} f(1,1), & f(1,2), & \ldots, & f(1,n) \\ f(2,1), & f(2,2) & \ldots, & f(2,n) \\ \cdot & \cdot & \cdots & \cdot \\ \cdot & \cdot & \cdots & \cdot \\ \cdot & \cdot & \cdots & \cdot \\ f(m,1), & f(m,2), & \ldots, & f(m,n) \end{vmatrix}$$

will be called a *digital image*. An address (i, j) defines a position in $\mathbf{F}$, called a pixel, pel or picture element. The elements of $\mathbf{F}$ denote the intensities within a number of small rectangular regions within a real (i.e., optical) image (see Fig. 2). Strictly speaking, $f(i, j)$ measures the intensity at a single point but, if the corresponding rectangular region is small enough, the approximation will be accurate enough for most purposes. The array $\mathbf{F}$ contains a total of $m.n$ elements; this product is called the spatial resolution of $\mathbf{F}$. We may arbitrarily assign intensities according to the following scheme:

$f(i, j) = 0$	black
$0 < f(i, j) \leqslant 0.33 \, W$	dark gray
$0.33 \, W < f(i, j) \leqslant 0.67 \, W$	mid-gray
$0.67 \, W < f(i, j) < W$	light gray
$f(i, j) = W$	white

Consider how much data is required to represent a gray-scale image in this form. Each pixel requires the storage of $\log_2(1 + W)$ bits. This assumes that $(1 + W)$ is an integer power of two. If it is not, then $\log_2(1 + W)$ must be rounded up to the next integer. This can be represented using the ceiling function [...]. Thus, a gray-scale image requires the storage of $\lceil \log_2(1 + W) \rceil$ bits. Since there are $m.n$ pixels, the total data storage for the entire digital image F is equal to

$$m.n \lceil \log_2(1 + W) \rceil \text{ bits}$$

If $m = n \geqslant 128$, and $W \geqslant 64$, a good image of a human face can be obtained. Many of the industrial image processing systems currently in use manipulate images

in which $m = n = 512$ and $W = 255$. This leads to a storage requirement of 256 kbytes image^{-1}.

A binary image is one in which only two intensity levels are permitted; W is equal to 1. This requires the storage of $m.n$ bits image.$^{-1}$

An impression of color can be conveyed to the eye by superimposing four separate imprints (black, cyan, magenta and yellow inks are often used in printing). Ciné film operates in a similar way, except that when different colors of light, rather than ink, are added together, three components (red, green and blue) suffice. Television operates in a similar way to film; the signal from a color television camera may be represented using three components:

$$\mathbf{R} = \{r(i,j)\}, \quad \mathbf{G} = \{g(i,j)\}, \quad \mathbf{B} = \{b(i,j)\}$$

where $\mathbf{R}$, $\mathbf{G}$ and $\mathbf{B}$ are defined in a similar way to $\mathbf{F}$, The vector $\{r(i,j), g(i,j), b(i,j)\}$ defines the intensity and color at the point (i, j) in the color image. Multispectral images can be represented using several monochrome images. The total amount of data required to code a color image with r components is equal to

$$m.n.r \lceil \log_2(1 + W) \rceil \text{ bits}$$

here, W is the maximum signal level on each of the channels.

Ciné film and television will be referred to, in order to explain how moving scenes may be represented in digital form. A ciné film is, in effect, a time-sampled representation of the original moving scene. Each frame in the film is a standard color or monochrome image and can be coded as such. Thus, a monochrome ciné film may be represented digitally as a sequence of two-dimensional arrays:

$$[\mathbf{F}_1, \mathbf{F}_2, \mathbf{F}_3, \mathbf{F}_4, \ldots]$$

Each $\mathbf{F}_i$ is an $m.n$ array of integers, as defined above, when discussing the coding of gray-scale images. If the film is in color, then each of the $\mathbf{F}_i$ has three components. In the general case, when there is a sequence of r-component color images to code, we require

$$m.n.p.r \lceil \log_2(1 + W) \rceil \text{ bits/image sequence}$$

where

(a) the spatial resolution is $m.n$ pixels;

(b) each spectral channel permits $(1 + W)$ intensity levels;

(c) there are r spectral channels; and

(d) p is the total number of stills in the image sequence.

Only those image representations which are relevant to the understanding of simple image processing functions have been considered. Many alternative methods of coding images are possible but they cannot be discussed in the space available here.

2. Image Processing Functions

The following notation will be used throughout this article, which will concentrate on gray-scale images, unless otherwise stated.

(a) i and j are address variables and lie within the ranges

$$1 \leqslant i \leqslant m$$

$$1 \leqslant j \leqslant n$$

Notice that i will be used to define the vertical position and j the horizontal position in the images (see Fig. 2).

(b) $\mathbf{A} = \{a(i,j)\}$, $\mathbf{B} = \{b(i,j)]$ and $\mathbf{C} = \{c(i,j)\}$.

(c) W denotes the white level.

(d) $g(X)$ is a function of a single independent variable X.

(e) $h(X, Y)$ is a function of two independent variables X and Y.

(f) The assignment operator "←" will be used to define an operation that is performed on one data element. In order to indicate that an operation is to be performed on all pixels within an image, the assignment operator "⇐" will be used.

(g) k, $k1$, $k2$ and $k3$ are constants.

(h) $N(i,j)$ is that set of pixels arranged around the pixel (i,j) in the following way:

$(i-1, j-1)$	$(i-1, j)$	$(i-1, j+1)$
$(i, j-1)$	(i,j)	$(i, j+1)$
$(i+1, j-1)$	$(i+1, j)$	$(i+1, j+1)$

Note that $N(i,j)$ forms a 3×3 set of pixels and is referred to as the 3×3 neighborhood of (i,j). In order to simplify some of the definitions, the intensities of these pixels shall be referred to, using the following notation:

A	B	C
D	E	F
G	H	I

Ambiguities over the dual use of A, B and C, should not be troublesome, as the context will make it clear which meaning is intended. The points $\{(i-1, j-1),$ $(i-1, j), (i-1, j+1), (i, j-1), (i, j+1), (i+1, j-1),$ $(i+1, j), (i+1, j+1)\}$ are called the 8-neighbors of (i,j) and are also said to be 8-connected to (i,j). The points $\{(i-1, j), (i, j-1), (i, j+1), (i+1, j)\}$ are called the 4-neighbors of (i,j) and are said to be 4-connected to (i,j).

3. Monadic, Point-By-Point Operators

Point-by-point operators have a characteristic equation of the form:

$$c(i,j) \Leftarrow g(a(i,j))$$

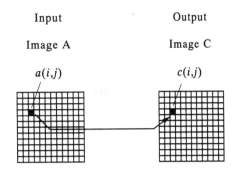

Input   Output

Image A   Image C

$a(i,j)$ $c(i,j)$

Figure 3
Monadic point-by-point operator. The (i, j)th pixel in the input image has intensity $a(i, j)$. This value is used to calculate $c(i, j)$, the intensity of the corresponding pixel in the output image

or

$$E \Leftarrow g(E)$$

Such an operation is performed for all (i, j) in the range $[1, m].[1, n]$ (see Fig. 3). Several examples will now be described.

Intensity shift

$$c(i,j) \Leftarrow \begin{vmatrix} 0 & 0 > a(i,j) + k \\ a(i,j) + k & 0 \leqslant a(i,j) + k \leqslant W \\ W & W < a(i,j) + k \end{vmatrix}$$

where k is a constant, set by the system user. Notice that this definition was carefully designed to maintain $c(i,j)$ within the same range as the input $[0, W]$. This is an example of a process called intensity normalization. Normalization is important because it permits iterative processing by this and other operators, in a machine having a limited precision for arithmetic (e.g., 8 bits). Normalization is implicit on those functions defined below, that are indicated by an asterisk (*).

Intensity multiply

$$c(i,j) \Leftarrow \begin{vmatrix} 0 & a(i,j).k < 0 \\ a(i,j).k & 0 \leqslant a(i,j).k \leqslant W \\ W & W < a(i,j).k \end{vmatrix}$$

Again, note the presence of normalization within the mapping formula.

Logarithm*

$$c(i,j) \Leftarrow \begin{vmatrix} 0 & a(i,j) = 0 \\ W \dfrac{\log(a(i,j))}{\log(W)} & \text{otherwise} \end{vmatrix}$$

This definition arbitrarily replaces the infinite value of log(0) by zero, and thereby avoids an impossible scaling problem.

Negate

$$c(i,j) \Leftarrow W - a(i,j)$$

Threshold

$$c(i,j) \Leftarrow \begin{vmatrix} W & k1 \leq a(i,j) \leq k2 \\ 0 & \text{otherwise} \end{vmatrix}$$

This is an important function, which converts a gray-scale image to binary format. Unfortunately, it is often difficult, or even impossible, to find satisfactory values for the parameters $k1$ and $k2$.

Highlight

$$c(i,j) \Leftarrow \begin{vmatrix} k3 & k1 \leq a(i,j) \leq k2 \\ a(i,j) & \text{otherwise} \end{vmatrix}$$

Squaring*

$$c(i,j) \Leftarrow \frac{[a(i,j)]^2}{W}$$

Modulus*

$$c(i,j) \Leftarrow 2 \left| \frac{a(i,j) - W}{2} \right|$$

4. Dyadic, Point-by-Point Operators

Dyadic operators (see Fig. 4) have a characteristic equation of the form:

$$c(i,j) \Leftarrow h(a(i,j), b(i,j))$$

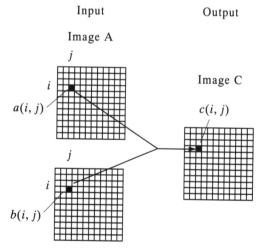

Input   Output

Image A

Image C

$a(i,j)$

$c(i,j)$

$b(i,j)$

Figure 4
Dyadic point-by-point operator. The intensities of the (i,j)th pixels in the two input images (i.e., $a(i,j)$ and $b(i,j)$ are combined to calculate the intensity $c(i,j)$ at the corresponding address in the output image)

There are two input images: $\mathbf{A} = \{a(i,j)\}$ and $\mathbf{B} = \{b(i,j)\}$, while the output image is $\mathbf{C} = \{c(i,j)\}$. It is important to realize that $c(i,j)$ depends only on $a(i,j)$ and $b(i,j)$. Here are some examples of dyadic operators.

Add* $c(i,j) \Leftarrow \dfrac{[a(i,j) + b(i,j)]}{2}$

Subtract* $c(i,j) \Leftarrow \dfrac{[a(i,j) - b(i,j)]/ + W}{2}$

Multiply* $c(i,j) \Leftarrow \dfrac{a(i,j).b(i,j)}{W}$

Maximum $c(i,j) \Leftarrow \text{MAX}(a(i,j), b(i,j))$

Minimum $c(i,j) \Leftarrow \text{MIN}(a(i,j), b(i,j))$

5. Local Operators

Figure 5 illustrates the principle of operation of the so-called local operators. The intensities of several pixels are combined together in order to calculate the intensity of just one pixel. Among the simplest of the local operators are those which use a set of nine pixels arranged in a $3*3$ square. These have a characteristic equation of the following form:

$$c(i,j) \Leftarrow g[a(i-1,j-1), a(i-1,j), a(i-1,j+1),$$
$$a(i,j-1), a(i,j), a(i,j+1), a(i+1,j-1),$$
$$a(i+1,j), a(i+1,j+1)]$$

where $g(.)$ is a function of nine varibles. This is an example of a local operator which uses a $3*3$ processing window, that is, it computes the value of one pixel on the basis of the intensities within a region containing $3*3$ pixels. Other local operators employ larger windows; these shall be discussed briefly later.

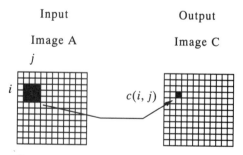

Input   Output

Image A   Image C

$c(i,j)$

Figure 5
Local operator. In this instance the intensities of nine pixels arranged in a $3*3$ window are combined. Local operators may be defined which use other possibly larger windows. The window may, or may not, be square and the calculation may involve linear or nonlinear processes

In the simplified notation which was introduced earlier, the above definition reduces to:

$$E \Leftarrow g(A, B, C, D, E, F, G, H, I)$$

6. Linear Local Operators

An important subset of the local operators is that group which performs a liner weighted sum, which are therefore known as linear local operators. For this group, the characteristic equation is

$$E \Leftarrow k1.(A.W1 + B.W2 + C.W3 + D.W4 + E.W5 + F.W6 + G.W7 + H.W8 + I.W9 + k2$$

where $W1$, $W2$, ..., $W9$ are weights, which may be positive, negative or zero. Values for the normalization constants, $k1$ and $k2$ are given later. The matrix

$$\begin{bmatrix} W1 & W2 & W3 \\ W4 & W5 & W6 \\ W7 & W8 & W9 \end{bmatrix}$$

is termed the weight matrix, and is important, because it determines the properties of the linear local operator. The following rules summarize the behavior of this type of operator.

(a) If all weights are either positive or zero, the operator will blur the input image. Blurring is referred to as low-pass filtering. Subtracting a blurred image from the original results in a highlighting of those points where the intensity is changing rapidly and is termed high-pass filtering (applying an electrical filter to a video signal produces similar effects to those obtained from the digital image).

(b) If $W1 = W2 = W3 = W7 = W8 = W9 = 0$, and $W4$, $W5$, $W6 \geqslant 0$, then the operator blurs along the rows of the image; horizontal features, such as edges and streaks are not affected.

(c) If $W1 = W4 = W7 = W3 = W6 = W9 = 0$, and $W2$, $W5$, $W8 \geqslant 0$, then the operator blurs along the columns of the image; vertical features are not affected.

(d) If $W2 = W3 = W4 = W6 = W7 = W8 = 0$, and $W1$, $W5$, $W9 \geqslant 0$, then the operator blurs along the diagonal (top-left to bottom-right); there is no smearing along the orthogonal diagonal.

(e) If the weight matrix can be reduced to a matrix product of the form $\mathbf{P.Q}$, where

$$\mathbf{P} = \begin{bmatrix} 0 & 0 & 0 \\ V4 & V5 & V6 \\ 0 & 0 & 0 \end{bmatrix}$$

and

$$\mathbf{Q} = \begin{bmatrix} 0 & V1 & 0 \\ 0 & V2 & 0 \\ 0 & V3 & 0 \end{bmatrix}$$

the operator is said to be of the separable type. The importance of this is that it is possible to apply two simpler operators in succession, with weight matrices $\mathbf{P}$ and $\mathbf{Q}$, in order to obtain the same effect as that produced by the separable operator.

(f) The successive application of linear local operators which use windows containing 3×3 pixels produces the same results as linear local operators with larger windows. For example, applying that operator which uses the following weight matrix

$$\begin{bmatrix} 1 & 1 & 1 \\ 1 & 1 & 1 \\ 1 & 1 & 1 \end{bmatrix}$$

twice in succession results in the same image as that obtained from the 5×5 operator with the following weight matrix:

$$\begin{bmatrix} 1 & 2 & 3 & 2 & 1 \\ 2 & 4 & 6 & 4 & 2 \\ 3 & 6 & 9 & 6 & 3 \\ 2 & 4 & 6 & 4 & 2 \\ 1 & 2 & 3 & 2 & 1 \end{bmatrix}$$

For the sake of simplicity, normalization has been ignored here. Applying the same 3×3 operator three times is equivalent to using the following 7×7 operator:

$$\begin{bmatrix} 1 & 3 & 6 & 7 & 6 & 3 & 1 \\ 3 & 9 & 18 & 21 & 18 & 9 & 3 \\ 6 & 18 & 36 & 42 & 36 & 18 & 6 \\ 7 & 21 & 42 & 49 & 42 & 21 & 7 \\ 6 & 18 & 36 & 42 & 36 & 18 & 6 \\ 3 & 9 & 18 & 21 & 18 & 9 & 3 \\ 1 & 3 & 6 & 7 & 6 & 3 & 1 \end{bmatrix}$$

All of these operators are also separable. Hence, it would be possible to replace the last-mentioned 7×7 operator with four simpler operators: 3×1, 3×1, 1×3 and 1×3, applied in any order. It is not always possible to replace a large-window operator with a succession of 3×3 operators. This becomes obvious when one considers, for example, that a 7×7 operator uses 49 weights and that three 3×3 operators provide only 27 degrees of freedom. Separation is often possible, however, when the larger operator has a weight

147

matrix with some redundancy, for example, when it is separable or symmetrical.

(g) In order to perform normalization, the following values are used for $k1$ and $k2$:

$$k1 \leftarrow \frac{1}{\sum\limits_{p,q} |W_{p,q}|}$$

$$k2 \leftarrow \left[\frac{1 - \sum\limits_{p,q} W_{p,q}}{\sum\limits_{p,q} |W_{p,q}|} \right] \frac{W}{2}$$

(h) A filter using the following weight matrix performs a local averaging function over an $11 * 11$ window:

$$\begin{bmatrix}
1, 1, 1, 1, 1, 1, 1, 1, 1, 1, 1 \\
1, 1, 1, 1, 1, 1, 1, 1, 1, 1, 1 \\
1, 1, 1, 1, 1, 1, 1, 1, 1, 1, 1 \\
1, 1, 1, 1, 1, 1, 1, 1, 1, 1, 1 \\
1, 1, 1, 1, 1, 1, 1, 1, 1, 1, 1 \\
1, 1, 1, 1, 1, 1, 1, 1, 1, 1, 1 \\
1, 1, 1, 1, 1, 1, 1, 1, 1, 1, 1 \\
1, 1, 1, 1, 1, 1, 1, 1, 1, 1, 1 \\
1, 1, 1, 1, 1, 1, 1, 1, 1, 1, 1 \\
1, 1, 1, 1, 1, 1, 1, 1, 1, 1, 1 \\
1, 1, 1, 1, 1, 1, 1, 1, 1, 1, 1
\end{bmatrix}$$

This produces quite a severe two-directional blurring effect. Subtracting the effects of a blurring operation from the original image generates a picture in which spots, streaks and intensity steps are all emphasized. On the other hand, large areas of constant or slowly changing intensity become uniformly gray. This process is called high-pass filtering, and produces an effect similar to unsharp masking, which is familiar to photographers.

7. Nonlinear Local Operators

Largest intensity neighbourhood function.

$$E \Leftarrow \mathrm{MAX}(A, B, C, D, E, F, G, H, I)$$

This operator has the effect of spreading bright regions and contracting dark ones.

Edge detector.

$$E \Leftarrow \mathrm{MAX}(A, B, C, D, E, F, G, H, I) - E$$

This operator is able to highlight edges (i.e., points where the intensity is changing rapidly).

Median filter.

$$E \Leftarrow \mathrm{FIFTH\ LARGEST}(A, B, C, D, E, F, G, H, I)$$

This filter is particularly useful for reducing the level of noise in an image. (Noise arises in all types of camera and can be a nuisance if it is not eliminated by hardware or software filtering.)

Sobel edge detector. *

$$c(i,j) \Leftarrow \frac{\{|(A + 2.B + C) - (G + 2.H + I)| + |(A + 2.D + G) - (C + 2.F + I)|\}}{6}$$

This popular operator highlights the edges of an image; points where the intensity gradient is high are indicated by bright pixels in the output image.

8. N-tuple and Morphological Operators

The *N*-tuple operators are closely related to the local operators and have a large number of (linear and nonlinear) variations. *N*-tuple operators may be regarded as generalized versions of local operators. In order to understand the *N*-tuple operators, let a linear local operator be considered, which uses a large processing window, (say $r \times s$ pixels) with most of its weights equal to zero. Only *N* of the weights are nonzero, where $N \ll r.s$. This is an *N*-tuple filter (see Fig. 6). The *N*-tuple filters are usually designed to

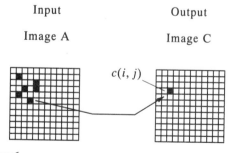

Input   Output

Image A   Image C

Figure 6
An *N*-tuple filter operates much like a local operator. The only difference is that the pixels whose intensities are combined do not form a compact set. A linear *N*-tuple filter can be regarded as being equivalent to a local operator which uses a large window and in which many of the weights are zero

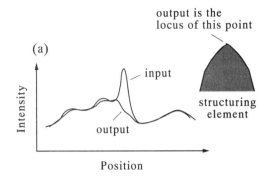

(a)

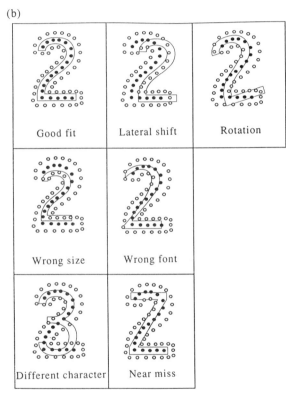

(b)

Figure 7
N-tuple and morphological operators

Figure 7a represents a one-dimensional morphological filter, operating on an analog signal (equivalent to a gray-scale image). The input signal is represented by the gray curve and the output by the black curve. In this simple example, the structuring element has an approximately parabolic form. In order to calculate a value for the output signal, the structuring element is pushed upwards, from below the input curve. The height of the top of the structuring element is noted. This process is then repeated, by sliding the structuring element sideways. This particular operator smooths the intensity peak but follows the input signal quite accurately everywhere else. Subtracting the output signal from the input would produce a result in which the intensity peak is emphasized and all other variations would be eliminated. A similar two-dimensional morphological operator might use a structuring element which has a conical or bell-shaped form.

Figure 7b represents the recognition of a numeral 2 using an *N*-tuple. The goodness of fit varies with the shift, rotation, size and font. Another character (Z in this case) may give a score that is close to that obtained from a 2 thus making these two characters difficult to distinguish reliably.

9. Edge Effects

All local operators and *N*-tuple filters are susceptible to producing peculiar effects around the edges of an image. The reason is, simply, that in order to calculate the intensity of a point near the edge of an image, information is required about pixels outside the image which, of course, are not present. In order to make some attempt at calculating values for the edge pixels, it is necessary to make some assumptions, for example, that all points outside the image are black, or have the same values as the border pixels. This strategy, or whatever one is adopted, is perfectly arbitrary, and there will be occasions when the edge effects are so pronounced that there is no alternative except to remove them by masking. Edge effects are important because they require that special provisions are made for them when several low-resolution images are patched together.

10. Global Image Transforms

An important class of image processing operators is characterized by an equation of the form $B \Leftarrow f(A)$, where $A = \{a(i, j)\}$ and $B = \{b(p, q)\}$. Each element in the output picture B, is calculated using all, or at least a large proportion, of the pixels in A. The output image B, may well look quite different from the input image A, or if it does not, the resemblance will probably be small. Examples of this class of operators are: lateral shift, rotation, warping, cartesian to polar coordinate conversion. Fourier and Walsh transforms.

detect specific patterns. In this rôle, they are able to locate a simple feature, such as a corner, annulus or numeral "2," in any position. However, they are sensitive to changes of orientation and scale. The *N*-tuple can be regarded as a sloppy template, which is convolved with the input image.

Nonlinear *N*-tuple operators may be defined in a fairly obvious way. For example, operators which compute the maximum, minimum or median values of the intensities of the *N* pixels covered by the *N*-tuple can be defined. An important class of such functions is the so-called morphological operators (see Fig. 7).

11. Intensity Histogram

The intensity histogram is defined in the following way:

(a) Let

$$s(p,i,j) \rightarrow \begin{vmatrix} 1 & a(i,j) = p \\ 0 & \text{otherwise} \end{vmatrix}$$

(b) let $h(p)$ be defined thus:

$$h(p) \leftarrow \sum_{i,j} s(p,i,j)$$

It is not necessary to store each of the $s(p,i,j)$, since the calculation of the histogram can be performed as a serial process in which the estimate of $h(p)$ is updated iteratively, as the input image is scanned through. The cumulative histogram $H(p)$, can be calculated using the following recursive relation:

$$H(p) = H(p-1) + h(p)$$

and

$$H(0) = h(0)$$

Both the cumulative and the standard histograms have a great many uses, as will become apparent later.

It is possible to calculate various intensity levels which indicate the occupancy of the intensity range. For example, it is a simple matter to determine that intensity level $p(k)$, which, when used as a threshold parameter, ensures that a proportion k of the output image is black. The mean intensity is equal to

$$\sum_{p} \frac{(h(p).p)}{m.n}$$

while the maximum intensity is equal to

$$\text{MAX}(p \mid h(p)) > 0)$$

One of the principal uses of the histogram is in the selection of threshold parameters. It is useful to plot $h(p)$ as a function of p. It is often found that a suitable position for the threshold can be related directly to the position of the "foot of the hill" or to a "valley" in the histogram.

An important operator for image enhancement is given by the transformation:

$$c(i,j) \leftarrow \frac{W.H(a(i,j))}{m.n}$$

This has the interesting property that the histogram of the output image $\{c(i,j)\}$ is flat, giving rise to the name histogram equalization for this operation. Histogram equalization is a data-dependent monadic, point-by-point operator.

An operation known as local-area histogram equalization is a local operator, which relies on the application of histogram equalization within a small window. This is a powerful filtering technique, which is particularly useful for texture analysis.

12. Binary Images

It will be convenient to assume that $a(i,j)$, $b(i,j)$ can assume only two values: 0 (black) and 1 (white). The operator $+$ denotes OR and $\bullet$ represents AND. Let $\#(i,j)$ denote the number of white points addressed by $N(i,j)$, including (i,j).

Inverse

$$c(i,j) \Longleftarrow \text{NOT}(a(i,j))$$

AND white regions

$$c(i,j) \Longleftarrow a(i,j) \bullet b(i,j)$$

OR

$$c(i,j) \Longleftarrow a(i,j) + b(i,j)$$

Exclusive OR

$$c(i,j) \Longleftarrow a(i,j) \sim b(i,j)$$

where $\sim$ denotes *Exclusive OR*.

Expand white areas

$$c(i,j) \Longleftarrow a(i-1,j-1) + a(i,j-1) + a(i+1,j-1)$$
$$+ a(i-1,j) + a(i,j) + a(i+1,j) + a(i-1,$$
$$j+1) + a(i,j+1) + a(i+1,j+1)$$

Note that this is closely related to the local operators defined earlier. This equation may be expressed in the simplified notation;

$$c(i,j) \Longleftarrow A + B + C + D + E + F + G + H + I$$

Edge detector

$$c(i,j) \Longleftarrow E \bullet \text{NOT}(A \bullet B \bullet C \bullet D \bullet F \bullet G \bullet H \bullet I)$$

Remove isolated points

$$c(i,j) \Longleftarrow \begin{vmatrix} 1 & a(i,j) \bullet (\#(i,j) > 1) \\ 0 & \text{otherwise} \end{vmatrix}$$

Count neighbors

$$c(i,j) \quad \#(a(i,j) = 1)$$

where $c(i,j)$ is a gray-scale image.

Region labelling. Imagine an image containing a number of separate blob-like figures. A region-labelling operator will shade the output image so that each blob is given a separate intensity value. The blobs could be shaded according to the order in which they are found during a conventional raster scan of the input image. Alternatively, the blobs could be shaded

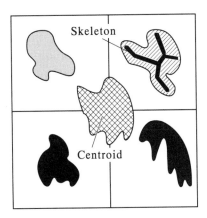

Figure 8
Shading blobs in a binary image according to their areas. The skeleton of the largest blob and its centroid are both shown

according to their areas; the biggest blobs become the brightest. Small blobs can be eliminated from an image (see Fig. 8).

Skeleton. The skeleton of a blob in a binary image is a "matchstick" figure, one pixel wide (see Fig. 8).

Centroid. The centroid of a blob determines its position within the image and is calculated using the formulas:

$$I \leftarrow \sum_j \sum_i \frac{(a(i,j).i)}{\mu_{i,j}}$$

$$J \leftarrow \sum_j \sum_i \frac{(a(i,j).j)}{\mu_{i,j}}$$

where

$$\mu_{i,j} \leftarrow \sum_j \sum_i a(i,j)$$

Although images are being considered in which $a(i,j)$ are equal to 0 (black) or 1 (white), it is convenient to use $a(i,j)$ as ordinary arithmetic variable as well. This results in simpler equations for the centroid (see Fig. 8).

13. Shape Descriptors

The following are just a few of the numerous shape descriptors that have been proposed:

(a) distribution of mass, as a function of distance measured from the centroid;

(b) the distance of the furthest point in the blob from the centroid;

(c) the number of protuberances as defined by that circle whose radius is equal to the average of the parameters measured in (a) and (b);

(d) the distances of points on the edge of the blob from the centroid, as a function of angular position;

(e) area/perimeter^2, (note that this ratio $\leqslant 0.25\,\pi$);

(f) the number of holes (i.e., black areas surrounded by white);

(g) the number of bays (consider a blob as being like an island);

(h) the ratio of the areas of the original blob to that of its circumcircle;

(j) the ratio of the area of the blob to that of its skeleton; and

(k) distances between joints and limb ends of the skeleton.

14. Implementation Considerations

All of the image processing operators that have been mentioned above can be implemented using a conventional programming language, such as C or Pascal. However, it is important to realize that many of the algorithms are time consuming when realized in this way. The monadic and dyadic operators can be implemented in a time $K.m.n$ s, where K is a constant, as can the local operators. However, some of the global operators require $O(m^2.n^2)$ time. With this in mind, it can be seen that a low-cost, slow but very versatile image processing system can be assembled, simply by plugging a *frame store* into a conventional desktop computer. A frame store is a device for digitizing video images and displaying computer processed/generated gray-scale and/or color images on a monitor.

The monadic operators can be implemented using a look-up table, which can be realized simply in a ROM or RAM. The dyadic operators can be implemented using a straightforward arithmetic and logic unit (ALU), which is a standard item of digital electronic hardware. The linear local operators can currently be implemented, using specialized integrated circuits. One manufacturer now sells a circuit board which can implement an $8*8$ linear local operator. Several companies market a broad range of image processing modules that can be plugged together, to form a very fast image processing system that can be tailored to the needs of the application.

Specialized architectures have been devised for image processing. Among the most successful are parallel processors, which may process the whole of one row of an image at a time (vector processor), or the whole image (array processor). Competing with these are systolic array, pipeline processors and

151

transputer networks. The relative merits of these are too complicated to explain here.

15. Applications

It is impossible to list more than a very small proportion of the applications that have been studied to date:

(a) detecting broken glass in food and pharmaceutical products (uses x rays);

(b) identifying and sorting pieces of leather in a shoe factory;

(c) measuring the shapes of loaves;

(d) inspecting the decoration patterns on cakes;

(e) measuring aircraft turbine blades;

(f) monitoring an aerosol spray cone (fuel jet);

(g) examining the inside surface of a car brake hydraulic cylinder;

(h) detecting faulty wet razor assemblies;

(i) inspecting bare and populated printed circuit boards;

(j) examining the cutting edges of surgical instruments;

(k) monitoring the fire in a wood-burning furnace;

(l) measuring the shape of the sails on a racing yacht;

(m) dissecting very small plants;

(n) locating the teats on a dairy cow, prior to placing a milking cup;

(o) inspecting the wafers used in the Roman Catholic Mass;

(p) locating air bubbles in television screens;

(q) inspecting knitwear; and

(r) placing motifs on garments.

These items were selected in order to demonstrate the enormous variety of potential applications that have been proposed for machine vision.

16. Conclusion

The image processing operators described in this article have all found widespread use in industrial vision systems. Other areas of application may well use additional algorithms for special effects. Two key features of industrial image processing systems are the cost and speed of the target system, that is, the one installed in a factory. It is common practice to use a more versatile and slower system for problem analysis and prototyping. The target system must continue to operate in an extremely hostile environment, which may be hot, greasy, wet and/or dusty. It must be tolerant of abuse and neglect. As far as possible, the target system should be self-calibrating and able to verify that it is "seeing" appropriate images. It should provide enough information (but not too much), to ensure that the factory personnel are able to trust it; no machine system should be built that is viewed by the workers as a mysterious black box. Consideration of these factors is as much a part of the design process as writing the software.

See also: Artificial Intelligence in Measurement and Instrumentation; Mechanical and Robotics Applications: Multicomponent Force Sensors

Bibliography

Ahlers R-J 1991 *Industrielle Bildverarbeitung*. Addison-Wesley, Bonn, Germany
Ballard D H, Brown C M 1982 *Computer Vision*. Prentice-Hall, Englewood Cliffs, NJ
Batchelor B G 1991 *Intelligent Image Processing in Prolog*. Springer, London
Batchelor B G, Hill D A, Hodgson D C (eds.) 1985 *Automated Visual Inspection*. North-Holland, Amsterdam
Batchelor B G, Waltz F M 1993 *Interactive Image Processing for Machine Vision*. Springer, Berlin
Breschi J 1979 *Automated Inspection Systems for Industry*. Verlag Chemie, Weinheim, Germany
Browne A, Norton-Wayne L. 1986 *Vision and Information Processing for Automation*. Plenum, New York
Davies E R 1990 *Machine Vision, Theory, Algorithms, Practicalities*. Academic Press, London
Dodd G C, Rossol L 1979 *Computer Vision and Sensor-Based Robots*. Plenum, New York
Gonzales R C, Woods R E 1992 *Digital Image Processing*. Addison-Wesley, Reading, MA
Hollingam J 1984 *Machine Vision, the Eyes of Automation*. IFS Publications, Bedford, UK
Horn, B K P 1986 *Robot Vision*. McGraw-Hill, New York
Marshall A D, Martin R R *Computer Vision, Models and Inspection*. World Scientific, Singapore
Parks J R 1978 Industrial sensory devices. In: Batchelor B G (ed.) *Pattern Recognition, Ideas in Practice*. Plenum, London
Pratt W K 1978 *Digital Image Processing*. Wiley, New York
Pugh A 1983 *Robot Vision*. Springer, Berlin
Rosenfeld A, Kak A C 1982 *Digital Picture Processing*, 2nd edn. Academic Press, New York
Russ J C 1992 *The Image Processing Handbook*. CRC Press, Boca Raton, FL
Schalkoff, R J 1989 *Digital Image Processing and Computer Vision*. Wiley, New York
Torras C (ed.), *Computer Vision, Theory and Industrial Applications*. Springer, Berlin

B. G. Batchelor
[University of Wales College of
Cardiff, Cardiff, UK]

Information Theory in Measurement and Instrumentation

The term information is currently used in many fields of science and engineering. Information theory was developed in the 1940s, especially for applications in communication (Shannon 1948). This theory—based on statistics—gives a good survey of the general aspects of information processing and, therefore, it is also applicable to measurement and instrumentation for solving problems of analysis and optimization.

1. Content of Information, Entropy and Channel Capacity

In general, in all fields of information processing and transmission including measurement, the problem as shown in Fig. 1 is to be solved. The output of the information source may be a digital or continuous signal, and is generated by a random mechanism having a probabilistic nature, otherwise the signal would be completely known and there would be no need to obtain it by means of measurement. The signal coming from the information source is the input signal of the encoder, modulator and transmitter, especially in measurement of the transducer.

If all possible symbols m that the information source is able to deliver have the same probability $p_1 = p_2 = p_i = 1/m$, then the content of the information in a bit (binary digit) of one message i is $\text{lb} = \log_2 = $ binary logarithm (Woschni 1982, 1988) is

$$s = \text{lb}\left(\frac{1}{p_i}\right)$$

The average information content over all source symbols is given by the source entropy H:

$$H = - \sum_{i=1}^{m} p_i \, \text{lb} \, p_i$$

H is a maximum when all symbols are equally likely; the maximum possible value of entropy H_0 in this case, with

$$\sum_{i=1}^{m} p_i = 1,$$

is given by

$$H_0 = \text{lb} \, m \ \text{bit/symbol}$$

This average information content per symbol and the entropy H give the number of binary decisions which are necessary, on average, to distinguish one state from all possible states.

These definitions are valid for discrete sources. For

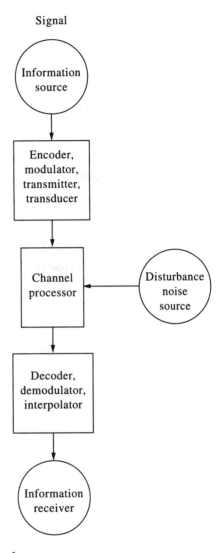

Figure 1
General problem of information processing and transmission

analog signals a differential entropy of a continuous distribution is defined (Feinstein 1958)

$$H(x) = - \int_{-\infty}^{+\infty} w(x) \, \text{lb} \, [w(x)] \, dx$$

The entropy of a continuous distribution is maximum for systems with amplitude limitation if $p(x) = $ constant and for systems with power limitation if $p(x)$ is a Gaussian distribution (Shannon 1948, Feinstein 1958, Woschni 1988). A measure of the difference between the maximum value H_0 and the real value H is the redundancy ΔH, where

$$\Delta H = (H_0 - H) \ \text{bit/value}$$

Table 1
Entropies for two signals x and y

Definition	Entropy equations	
	Discrete signals	Analog signals
Entropy of input signal x	$H(x) = -\sum_i p(x_i)\,\operatorname{lb}[p(x_i)]$	$H(x) = -\displaystyle\int_{-\infty}^{+\infty} w(x)\,\operatorname{lb}[w(x)]\,dx$
Entropy of output signal y	$H(y) = -\sum_j p(y_j)\,\operatorname{lb}[p(y_j)]$	$H(y) = -\displaystyle\int_{-\infty}^{+\infty} w(y)\,\operatorname{lb}[w(y)]\,dy$
Joint entropy of signals x and y	$H(x,y) = -\sum_i\sum_j p(x_i,y_j)\,\operatorname{lb}[p(x_i,y_j)]$	$H(x,y) = -\displaystyle\int_{-\infty}^{+\infty}\int_{-\infty}^{+\infty} w(x,y)\,\operatorname{lb}[w(x,y)]\,dx\,dy$
Conditional entropy	$H(x/y) = -\sum_j p(y_j)\sum_i p(x_i/y_j)\,\operatorname{lb}[p(x_i/y_j)]$ $= -\sum_i\sum_j p(x_i,y_j)\,\operatorname{lb}[p(x_i/y_j)]$	$H(x/y) = -\displaystyle\int_{-\infty}^{+\infty}\int_{-\infty}^{+\infty} w(x,y)\,\operatorname{lb}[w(x/y)]\,dx\,dy$
Transformation	$H(x;y) = -\sum_i\sum_j p(x_i,y_j)\,\operatorname{lb}\left[\dfrac{p(x_i)p(y_j)}{p(x_i,y_j)}\right]$	$H(x;y) = -\displaystyle\int_{-\infty}^{+\infty}\int_{-\infty}^{+\infty} w(x,y)\,\operatorname{lb}\left[\dfrac{w(x)w(y)}{w(x,y)}\right]\,dx\,dy$
Relations between the entropies	$H(x,y) = H(x) + H(y/x) = H(y) + H(x/y) = H(x/y) + H(y/x) + H(x;y) = H(x) + H(y) - H(x;y)$ $H(x/y) = H(x,y) - H(y) = H(x) - H(x;y), \quad H(y/x) = H(x,y) - H(x) = H(y) - H(x;y),$ $H(x;y) = H(x) + H(y) - H(x,y)$	

In general, the entropy for more than one variable $x_1, x_2, \ldots, x_r, \ldots, x_n$, results in entropies of higher order. This plays an important role in analyzing digital signals where a correlation exists between several bits in the signal sequence. It is also of value in relating signals at the input and output of measuring systems. The joint probability is $p(x_1, x_2, \ldots, x_r, \ldots, x_n)$, and the conditional probability is $p(x_1 | x_2, x_3, \ldots, x_r, \ldots, x_n)$.

For entropies it follows that the joint entropy is given by

$$H(x_1, x_2, \ldots, x_r, \ldots, x_n)$$

$$= -\sum_{x_1} \sum_{x_2} \cdots \sum_{x_n} p(x_1, x_2, \ldots, x_n)$$

$$\times \text{lb}[p(x_1, x_2, \ldots, x_n)]$$

and the conditional entropy is

$$H(x_1 | x_2, x_3, \ldots, x_n)$$

$$= -\sum_{x_1} \sum_{x_2} \cdots \sum_{x_n} p(x_1, x_2, \ldots, x_n)$$

$$\times \text{lb}[p(x_1 | x_2, \ldots, x_n)]$$

In Table 1 the entropy equations are placed together for the important measurement case of two fields of variables. In particular, this problem leads to the concepts embodied in Fig. 2, when expressed in terms of the entropies contained in Table 1.

If a channel with Gaussian signals with power P_s and noise P_n is given, the transformation given in Eqn. (1) follows (Shannon 1948, Goldman 1953, Fano 1961, Woschni 1988):

$$H(x; y) = \frac{1}{2} \text{lb} \left(1 + \frac{P_s}{P_n} \right) ; \text{ bit/value} \qquad (1)$$

The transformation $H(x; y)$ states that the information content is referred to as one transmitted value. If the transient time of the channel is given as $t_{tr} = 1/(2f_c)$ and f_c is the critical frequency, the channel transmits

$$I = \frac{H(x; y)}{t_{tr}} = 2f_c H(x; y)$$

bits per second, known as the information flow (Woschni 1988). The optimum value of this information flow for optimal coding is the channel capacity C_t (Shannon 1948), given by:

$$C_t = \frac{H(x; y)_{\max}}{t_{tr}} = 2f_c H(x; y)_{\max} \qquad (2)$$

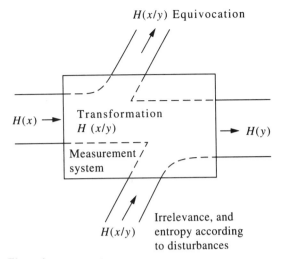

Figure 2
Entropies in measurement systems

From Eqn. (1) the channel capacity of a channel with noise is given as:

$$C_t = \frac{1}{2t_{tr}} \text{lb} \left(1 + \frac{P_s}{P_n} \right)$$

$$= f_c \, \text{lb} \left(1 + \frac{P_s}{P_n} \right) \approx 0.33 f_c \left. \frac{P_s}{P_n} \right| \text{dB}$$

This very important classical relation was first published by Shannon (1948). The relation shows that it is possible to "trade" signal-to-noise P_s/P_n for bandwidth f_c, and *vice versa*.

2. Examples of Applications of Measurement and Instrumentation

An important problem is that of the interface at the interconnection between two systems. Here the condition must be fulfilled that the channel capacity of the second system has to be the same as that of the first system, otherwise information would be lost.

Another example of the application of information theory to measurement is the comparison of digital and analog measurement (Woschni 1988). Figure 3 shows that the maximum number of distinguishable amplitude steps $\hat{m}_a$ or power steps $\hat{m}_p$ is obtained:

$$\hat{m}_a = 1 + \frac{\hat{Y}}{\Delta y} \qquad \hat{m}_p = \sqrt{\hat{m}_a} \qquad (3)$$

taking into account that zero is also a possible measured value. When the transient time $t_{tr} = 1/(2f_c)$, the channel capacity of the analog measuring system is

$$C_{\tan} = 2f_c \, \text{lb} \, \hat{m}_a$$

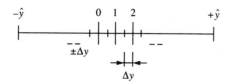

Figure 3
Maximum number of amplitude steps

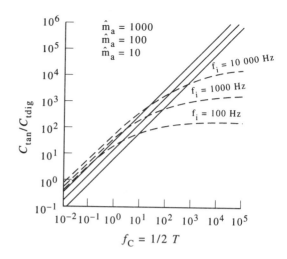

$$f_C = 1/2\,T$$

Figure 4
Channel capacity of analog and digital measuring systems

With a pulse frequency f_i the number of amplitude steps

$$\hat{m}_a = 1 + \frac{f_i}{2f_c}$$

leads to the channel capacity of the digital measuring system

$$C_{tdig} = 2f_c\,\mathrm{lb}\left(1 + \frac{f_i}{2f_c}\right)$$

The results are shown in Fig. 4. The limit between analog and digital systems due to

$$\cdot\frac{f_i}{2f_c} = \hat{m}_a - 1$$

is shifting towards higher frequencies due to increasing pulse frequencies of about one order of magnitude every seven years (Diebold 1976).

3. Limitations of Application: A Generalized Quality Criterion

The problem of obtaining a generalized quality criterion is treated in this section. This generalized quality criterion (QC) may be assumed to depend on a large number of parameters c_r, with weighting functions λ_r

$$QC = \sum \lambda_r(c_r) \qquad (4)$$

In the use of measurement techniques, typical parameters are: (a) the maximum number of amplitude steps $\hat{m}_a$ or power steps $\hat{m}_p$ as a measure of the static error Δy or the mean-square error $\bar{\varepsilon}^2$ as described by Woschni (1988) and given in Eqn. (3); and (b) the critical frequency f_c as a measure of the dynamic behavior. Assuming

$$\lambda_1 = \log(2f_c) \qquad \lambda_2 = \log(\mathrm{lb}\,\hat{m}_a) = \log(\tfrac{1}{2}\mathrm{lb}\,\hat{m}_p)$$

one gains the channel capacity C_t of Shannon's information theory, Eqn. (2), as a criterion, taking into account both the static and dynamic behavior

$$C_t = 10^{QC} = 2f_c\,\mathrm{lb}\,\hat{m}_a = f_c\,\mathrm{lb}\,\hat{m}_p$$

An interpretation by means of physical perception uses the memory capacity to store $\hat{m}_a$ equally probable measuring values s or the signal-to-noise ratio P_s/P_n

$$s/\mathrm{bit} = \mathrm{lb}\,\hat{m}_a = \mathrm{lb}\,\frac{y + \Delta y}{\Delta y}$$

$$= \mathrm{lb}(\hat{m}_p)^{1/2} = \frac{1}{2}\,\mathrm{lb}\,\frac{P_s + P_n}{P_n}$$

leading to $1/t_{tr} = 2f_c$ values per second to the channel capacity, Eqn. (2).

Similarly, the implication of two technical parameters by means of the classical information theory, a more generalized criterion, has to take into consideration not only statistical aspects as is done by the classical information theory, but also semantic and pragmatic aspects. However, this widening of information theory has not yet been developed in detail, and therefore Eqn. (4) may be used in the following form:

$$QC = C_t + \sum_{r=1}^{n} \gamma_r c_r^{\alpha} = 2f_c\,\mathrm{lb}\,\hat{m}_a + \gamma C^{\alpha} \qquad (5)$$

The problem of optimization, taking into account economic aspects, is a very important example.

The decreasing cost and rapid rise in the power and capacity of computers are currently shifting the optimum towards the broad application of computers in many aspects of measurement.

See also: Signal Processing; Signal Theory in Measurement and Instrumentation; Signal Transmission

Bibliography

Diebold 1976 Research Program. Europa Report No. E 146.

Goldman S 1953 *Information Theory*. Prentice-Hall, New York

Fano R M 1961 *Transmission of Information*. Wiley, Chichester, UK

Feinstein A 1958 *Foundations of Information Theory*. McGraw-Hill, New York

Shannon C E 1948 A mathematical theory of communication. *Bell Syst. Tech. J.* **27**, 379–423

Woschni E-G 1982 Signals and systems in the time and frequency domains. In: Sydenham P H (ed.) *Handbook of Measurement Science*, Vol. 1. Wiley, Chichester, UK

Woschni E-G 1988 *Informationstechnik*. Verlag, Berlin

E.-G. Woschni
[Technische Universität, Chemnitz-Zwickau, Germany]

Installation and Commissioning

The issues that must be considered in the installation and commissioning of instrument systems must be considered at the very inception of the idea. These considerations must continue to be addressed throughout the design and manufacture of products, their application, maintenance, and usage of the total system. Diagnostics for verification of system functionality must be an integral part of the instrument system and its components wherever possible.

It is essential that the physics of measurement and control be engineered into the particular instrument, because this will dictate the requirements in the real world. The information used in the design and manufacture of the instrument must be carried forward in the documentation accompanying the finished product, and conveyed to the parties responsible for preparing the details of the physical arrangement, installation drawings, and so on, to ensure the realization of a viable operating system that will meet the intended performance criteria.

1. Installation

Instruments may be divided into two broad groups, local mounted or remote. The first, local mounted, group contains those which display the measurand directly, transducers that convert the direct measurand to another form of energy for further handling, and devices for process manipulation such as control valves. The second, remote, group encompasses instruments which are remotely mounted from the direct measurand, but which are an integral part of the instrument system. These include digital distributed systems and devices such as receivers (both indicating and recording), controllers (individual or integrated), and some transducers and auxiliary devices, for instance, switches for alarm and control.

Installation considerations are governed by the purpose of the instrument and in general are under three major categories: mechanical, utility, and environmental. These categories are subdivided as follows.

1.1 Mechanical

Reliable measurement of the process variable is highly dependent on a proper mechanical installation. The following considerations must be addressed.

(a) Location of instrument:

    (i) minimization of measurement errors,

    (ii) accessibility for maintenance,

    (iii) freedom from vibration, and

    (iv) code requirements and regulations—national and local.

(b) Support:

    (i) structural,

    (ii) process (line), and

    (iii) code requirements and regulations—national and local.

1.2 Utility (Supply and Signal)

The quality of the supply medium and transmitted signals is highly dependent on minimizing or eliminating the effects of detrimental influences. The following must be taken into consideration.

(a) Electrical:

    (i) grounding,

    (ii) isolation,

    (iii) interference—transient and stability,

    (iv) voltage—transient and stability, and

    (v) code requirements and regulations—national and local.

(b) Pneumatic or hydraulic:

    (i) state,

    (ii) pressure, and

    (iii) code requirements and regulations—national and local.

1.3 Environment

Environmental conditions may adversely affect the performance of the instrument and the overall system. The following must be considered:

(a) temperature—process and ambient,

(b) metallurgy,

(c) safety, and

(d) code requirements and regulations—national and local.

For a proper installation these items must be addressed before proceeding with commissioning.

2. Commissioning

The first step in commissioning requires evaluation of the instrument (system) installation to ensure it conforms to the aforementioned considerations, as applicable. Only when these considerations are catered for can one proceed with a functional check of the instrument. Calibration must be performed in accordance with the respective manufacturer's documented procedures. In the case of the instrument system, the receiver (operator display, indicator, or recorder) should be calibrated against a primary standard with all other components in the system calibrated to the receiver. This technique of calibrating the system as a whole against the receiver minimizes the errors and uncertainties which might be additive when each system's components are individually calibrated against the primary standard.

Final commissioning must be carried out as the plant comes on stream. Attention must be directed toward elimination of foreign matter accumulated during plant startup, controllers must be tuned, and any malfunctions must be identified and corrected.

Diagnostics, including self-checking features which are an integral part of the system and of its components, ensure continued proper operation and minimization of maintenance.

See also: Life Cycle

W. A. Bajek and D. G. Plackmann
[UOP, Des Plaines, Illinois, USA]

Instrument Elements: Models and Characteristics

Instrument elements are the simplest components of instrument systems. They embrace, above all, transducers and sensors. This article, which should be read in conjunction with *Instruments: Models and Characteristics*, discusses special aspects of their models.

1. Power Flow Functions

The transfer of information-carrying signals in instruments takes place by power flow, so the behavior of instrument elements in particular sensors is described best in terms of power flow. Power flows between points or regions termed ports: for example, the pair

Table 1
Effort and flow variables

Form of energy	Effort e	Flow fl
Mechanical translation	Force f	Velocity $\dot{x}$
Mechanical rotation	Torque T	Angular velocity $\dot{\phi}$
Electrical	Voltage σ	Current i
Fluid	Pressure p	Flow rate $\dot{g}$
Thermal	Temperature θ	Heat flow rate $\dot{Q}$

of terminals of an electrical instrument, or the point at which a force or pressure is applied. The flow is effected by a pair of conjugate physical variables, the product of which is power; one of the variables is an effort, the other is a flow. (Some effort and flow variables are given in Table 1.) At each port one of the power variables is independent, constituting a signal input; its conjugate variable is then a dependent output.

There are several kinds of elementary operation on power flow as follows.

(a) Storage: energy is stored without loss.

(b) Conversion or transformation: energy is converted from one form to another, or else transformation takes place, causing an increase in effort or flow, and an accompanying decrease of the conjugate variable. Energy conversion may be bilateral, reversible between one form and another, or unilateral, converting input power unilaterally to heat.

(c) Transport: power flow takes place without storage or transformation between two ports, or is split or merged at junctions.

(d) Flow in and out: sources and sinks of energy are strictly always either stores or converters of energy.

(e) Control: a physical action at one port affects (controls) power flow at another port, without energy interchange between the ports.

2. Transducers

Transducers are instrument elements which either convert input signal-carrying power into output signal-carrying power, or in which the input signal controls output power. They may be sensors or effectors. In a physical element with a single port the relation between effort and flow is best represented by a generalized impedance or admittance. The generalized impedance is the ratio of effort to flow in Laplace transform form, that is

$$Z(s) = e(s)/fl(s)$$

$$
\begin{pmatrix}
e_1(s) \\
\vdots \\
\vdots \\
e_k(s) \\
fl_{k+1}(s) \\
\vdots \\
fl_n(s)
\end{pmatrix}
=
\begin{pmatrix}
Z_{11}(s) & \cdots & Z_{1k}(s) & A_{1(k+1)} & \cdots & A_{1n} \\
\vdots & & \vdots & \vdots & & \vdots \\
\vdots & & \vdots & \vdots & & \vdots \\
Z_{k1}(s) & \cdots & Z_{kk}(s) & A_{k(k+1)} & \cdots & A_{kn} \\
B_{(k+1)1} & \cdots & B_{(k+1)k} & Y_{(k+1)(k+1)}(s) & \cdots & Y_{(k+1)n}(s) \\
\vdots & & \vdots & \vdots & & \vdots \\
B_{n1} & \cdots & B_{nk} & Y_{n(k+1)}(s) & \cdots & Y_{nn}(s)
\end{pmatrix}
\begin{pmatrix}
fl_1(s) \\
\vdots \\
\vdots \\
fl_k(s) \\
e_{k+1}(s) \\
\vdots \\
e_n(s)
\end{pmatrix}
$$

Figure 1
Linearized model

The generalized admittance is then

$$Y(s) = fl(s)/e(s)$$

Consider a physical system such as an instrument with n ports: let the flow variable be independent at k ports, and let the effort be independent at the other $n - k$ ports. A linearized model for the system in Laplace transform form is given in Fig. 1, where the matrix elements are transfer functions representing the relation between a particular output variable and a particular input variable, other inputs being zero. The elements $Z(s)$ and $Y(s)$ are generalized impedances and admittances, as defined earlier, while the elements A and B are numerical effort or flow ratios, respectively. The diagonal elements represent impedances or admittances at a particular part, and the off-diagonal elements represent transduction—the effects of a signal at one port on the signal at another port.

A complete account of this matrix equation is given by Finkelstein and Watts (1983).

3. Sensors

It is the function of sensing which, in particular, distinguishes instruments from other forms of informa-

tion machine. In a sensor the measurand is a signal acting at an input port, and determines the output signal at another port, so that this signal is functionally related to the measurand. Sensors are based on one of three types of energy process: energy conversion, power flow control and coupled energy storage.

A typical sensor based on energy conversion is a coil moving in the field of a permanent magnet (see Fig. 2). There is a bilateral conversion of energy between the mechanical and electrical ports. For an ideal linear element with no storage or loss, the simple model can be used:

$$
\begin{pmatrix} V(s) \\ f(s) \end{pmatrix} = \begin{pmatrix} 0 & A \\ B & 0 \end{pmatrix} \begin{pmatrix} i(s) \\ \dot{x}(s) \end{pmatrix}
$$

where, from energy conservation, $A = -B$. This model shows that a moving-coil sensor can either produce a voltage which is a function of velocity and act as a velocity sensor, or produce a force which is a function of current and act as a current sensor.

A typical sensor based on power-flow control is an electrical-resistance strain gauge (see Fig. 3), in which the mechanical displacement (strain) controls the power at the electrical port. The equations describing the element are

$$V = R(x)i, \qquad f = kx$$

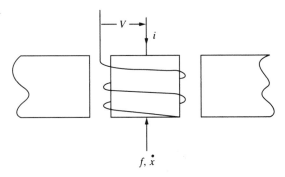

Figure 2
Electrodynamic element

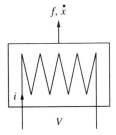

Figure 3
Electrical-resistance strain gauge

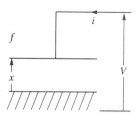

Figure 4
Capacitative electromechanical element

where R and k are the resistance and spring constants, respectively, and x is the displacement. The linearized incremental model of the element is

$$\begin{pmatrix} \delta V(s) \\ \delta f(s) \end{pmatrix} = \begin{pmatrix} R(\overline{x}) & A\overline{i}(s^{-1}) \\ 0 & k(s^{-1}) \end{pmatrix} \begin{pmatrix} \delta i(s) \\ \delta \dot{x}(s) \end{pmatrix}$$

where the variables δV, δf, δi and $\delta \dot{x}$ are incremental variations about a steady-state operating point $\overline{x}$, $\overline{i}$, and A is a constant of the element. The model shows that the variations in voltage output are functions of displacement. (The model is written in terms of x and not n to preserve a systematic approach to modelling, so that all matrix elements are impedances.) It can be seen that the element relating δV to $\delta \dot{x}$ depends on the element being energized by a current $\overline{i}$. The 2, 1 element of the matrix is zero, showing that signal flow is unilateral, the variables at the electrical port having no effect at the mechanical port.

A typical element based on coupled energy storage is a capacitive electromechanical element, such as a moving-plate capacitor (see Fig. 4). Energy can flow into and out of the element through either the electrical or the mechanical port, and is stored in a coupled form. The capacitance of the element is $C = \varepsilon A/x$, where ε is the permittivity, A the area of each plate and x the separation between them.

The equations of the element are then

$$V = (x/\varepsilon A)q, \qquad f = q^2/2\varepsilon A$$

The linearized model of the element, giving the incremental variations δV, and so on, about the operating point $\overline{q}$, $\overline{i}$ is

$$\begin{pmatrix} \delta V(s) \\ \delta f(s) \end{pmatrix} = \begin{pmatrix} \overline{x}/\varepsilon As & \overline{q}/\varepsilon As \\ \overline{q}/\varepsilon As & 0 \end{pmatrix} \begin{pmatrix} \delta i(s) \\ \delta x(s) \end{pmatrix}$$

It can be seen from the off-diagonal element that this sensor can operate effectively in an energy conversion mode at high displacement frequencies, producing either a voltage as a function of velocity or a force as a function of current. The 1, 1 matrix element shows that at low displacement frequencies the device operates effectively as a capacitive store, voltage being controlled by displacement.

In order to function the element must be energized; thus energy conversion elements are passive sensors deriving the whole of their output energy from the input. Controlled elements or coupled storage elements are active sensors requiring an energizing supply.

A sensor must take only a small amount of energy from the system under measurement; in other words, the signal conjugate to the input signal must be small. If the input is an effort the input impedance must be high, and if the input is a flow the input impedance must be low. In passive sensors the effective input impedance or admittance depends upon the energy taken out of the sensor, and all that energy comes from the system under measurement. In active sensors the energy at the output comes mainly from the energizing supply, so that it is possible to have a high-system-power output with low-signal-power input. This is signal power amplification.

See also: Transducers: An Introduction

Bibliography

Finkelstein L, Watts R D 1983 Fundamentals of transducers–description by mathematical models. In: Sydenham P H (ed.) *Handbook of Measurement Science*, Vol. 2. Wiley, Chichester, UK, pp. 747–95

L. Finkelstein
[City University, London, UK]

Instrument Systems: Functional Architectures

It is helpful to have a systematic organization and analysis of instrumentation. An instrument may be defined as a machine or system which is designed to maintain functional relationships between prescribed properties of physical variables, and must include a means of communicating information to a human observer. By concentrating on these functional elements and the various physical devices available for accomplishing them, one can develop an ability to analyze an instrument and see the reasons for its particular performance, and an ability to synthesize new combinations of elements leading to new and useful instruments.

1. Architecture

Information–machine systems such as measuring systems are formed by joining blocks with suitable connections, the structure of which is referred to as the architecture (Sydenham *et al.* 1989). A physical system may have many components or elements, and it is convenient to consider each such element as a block with its own input–output relationship, that is, its own transfer function. The blocks representing the various elements of a system are connected to use their functional relationship within the system, thus producing a block diagram for the system. This diagram may well be helpful in analyzing a measuring system, and considering each element as a block helps in the design of new systems.

Four symbols are used in block diagrams: (a) the "arrow" designates a signal which is a quantified physical variable acting only in the direction indicated; (b) the "block" designates the functional relationship between signals where, for example, *r* may be an input or cause and *c* is the output or effect (the transfer function is often written inside the block); (c) the "circle" signifies a summation point or comparator; and (d) the "branch" point describes the case when one signal simultaneously causes two separate effects.

2. Functional Description

The maintenance of functional relationships is mainly associated with the constancy of static calibration, although performance is defined in terms of the static and dynamic performance characteristics. The operation of a measuring system can be described in terms of the functional elements of the system. One can generalize a measuring system into such elements, and an example of this is shown in Fig. 1 (Jones 1977). Every instrument and measuring system is composed of one or more of these functional elements.

A physical variable may convey information through a change in one or more of its properties, for example, in amplitude, phase, frequency, duration, delay or state. The physical quantity of a particular property that is being measured, which is the input to the measuring system, is known as the measurand. The primary element or group of elements is that which first receives energy from the measured medium and produces an output depending in some way on the measurand. The measurand is always disturbed by the act of measurement, but good instruments are designed to minimize this effect. Primary sensing elements can have a nonelectrical input and output, for example elements such as a spring, manometer, Bourdon tube, orifice and diaphragm, or they may have an electrical input and output, for example, elements such as a filter and rectifier. Often the nonelectrical signal will be converted to an electrical signal by means of a sensor, for example, a strain gauge, thermistor or thermocouple, and the sensor may have associated signal conditioning and conversion equipment such as a bridge, amplifier and analog-to-digital converter. When the functional elements are actually separated, it becomes necessary to transmit the data, using a telemetry system, from one group of elements to another. The data may be processed, recorded and compared with the input (via an inverse transducer), before being communicated to a human observer in some form of display.

3. Input–Output Configuration

A generalized configuration which brings out the significant input–output relationships present in all measuring apparatus is shown in Fig. 2. Input quantities are classified into three categories: desired inputs, interfering inputs and modifying inputs (Doebelin 1990). Desired inputs represent the quantities that the instrument is specifically intended to measure. Interfering inputs represent quantities to which the instrument is unintentionally sensitive. Modifying inputs cause changes to the effects on the output of

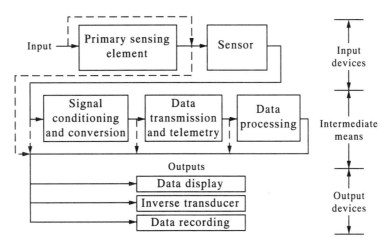

Figure 1
General functional diagram of a measuring system

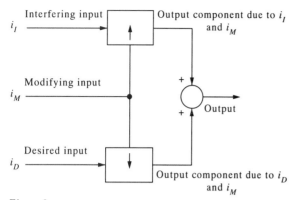

Figure 2
Generalized input–output configuration: i_M, modifying input; i_D, desired input; i_I, interfering input

desired and interfering inputs. The output of the measuring apparatus is the instantaneous algebraic sum of the output components due to the desired, interfering and modifying inputs. A measuring system may have several inputs of each of the three types and also several outputs.

Typical interfering inputs are electrostatic and electromagnetic pickup, and temperature effects causing drift such that an instrument gives an output for no desired input. Typical modifying inputs are temperature and battery voltage as, for example, in the case of a strain gauge in a Wheatstone bridge. The bridge output is proportional to gauge factor (which is sensitive to temperature) and to bridge battery voltage, therefore both temperature and battery voltage change the proportionality factor between the desired input (strain) and the bridge output voltage. In this case temperature is also an interfering input, as it causes gauge resistance change and bridge output voltage even when the strain is zero.

4. Transducers

In order to characterize transducers in the instrument field the following six signal domains can be considered: radiant, mechanical, thermal, electrical,

magnetic and chemical (Middelhoek and Audet 1989). Based on this division, measurement systems can be represented by the block diagram in Fig. 3. As an example, an electronic thermometer is indicated: thermal energy is converted into electrical energy, the electrical signal is amplified and converted into digital form, and finally the digital signal is converted into a radiant (optical) form, such that an observer can read the temperature visually.

A simple functional diagram of a modern measuring transducer is shown in Fig. 4 (Jones 1987). The sensor is seen as only a part of the transducer, with microelectronics and intelligence being incorporated into the device. Overall transducer performance is still highly dependent on that of the front-end, primary sensing element. It is now possible to modify on-line operational characteristics of the various elements in the transducer. It should be noted that an instrument user does not need to know about this functional grouping in the sensor.

5. Data Processing

In many situations it is necessary to process signals obtained from measuring transducers. Corrections may need to be made to the measurements of physical variables, to compensate for scaling, nonlinearity, zero offset, temperature errors, and so on. Calculations may involve addition, subtraction, multiplication or division of two or more physical variables and their associated constants. Averaging, square root extraction, squaring and logarithms are other simple arithmetic calculations which are sometimes required.

An example of arithmetic functions to be performed in an instrument system is shown in Fig. 5. Here to establish the mass flow rate of a gas, the gas volumetric flow rate is measured using an orifice plate, and the gas density ρ is determined by measurement of gas static pressure P_s and static temperature T_s (ρ is proportional to P_s/T_s). Now mass flow rate $= \rho Q$, where Q is volumetric flow rate. The differential pressure across the orifice $\Delta p \propto \rho Q^2$, and therefore, to obtain mass flow rate it is necessary to multiply by ρ and take the square foot of the result. This assumes a

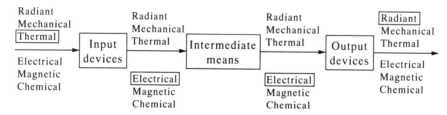

Figure 3
General block diagram of an instrumentation system indicating the six different forms of signal-carrying energies (an electronic thermometer is indicated by a rectangle)

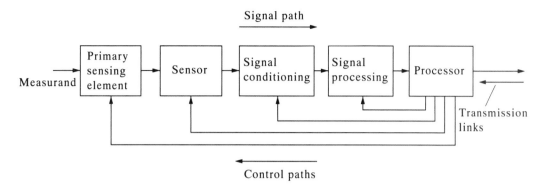

Figure 4
Functional diagram of a modern measuring transducer

constant coefficient of discharge and expansion ratio for the orifice plate.

6. Feedback Measuring Systems

There are only a few structural schemes employed in the construction of instruments and instrument systems, and one such scheme is the use of feedback. A feedback system can be thought of as a system which tends to automatically maintain a prescribed relationship of one system variable to another by comparing functions of these variables and using the difference as a means of control. The main characteristic of a feedback system is its closed-loop structure. A measuring system in which feedback is the basic structural arrangement is called a feedback measuring system (Jones 1977, 1982).

A general block diagram of such a system is shown in Fig. 6. Here the output signal (usually electrical) is converted to a form (usually nonelectrical, e.g., force) suitable for comparison with the quantity to be measured (e.g., force). The resultant error is usually transduced into electrical form and amplified to give the output indication. Normally a sensor and associated circuit has a nonelectrical input and an electrical output, for example, a thermistor, strain gauge and photodiode, whereas a so-called inverse transducer or precision actuator has an electrical input and a low-power nonelectrical output: for example, a piezoelectric crystal, translational and angular moving-coil elements can be used as inverse transducers. The sensor, inverse transducer and usually the amplifier must be close to the point of measurement, whereas the indicator may be some distance away. The feedback loop must have sufficient negative gain and the system must be stable. The system is driving fairly low-power devices at its output, and the inverse transducer essentially determines the characteristics of

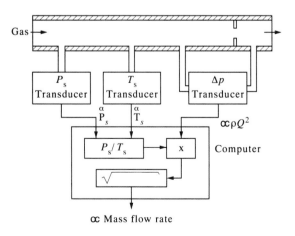

Figure 5
System to measure mass flow rate

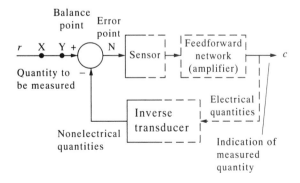

Figure 6
Diagram of a feedback measuring system: ---, electrical units or signals; ——, mechanical units or signals

the system, although noise connected with the sensor and amplifier input stage may well be important. In practice, the measurand may not be directly connected to the null or balance point, and there may well be one or more primary sensing elements inserted between points X and Y in Fig. 6.

Systems incorporating feedback involve comparison of two physically similar variables and production of a minimum or null (at point N in Fig. 6). In feedback measuring systems, pervariable (or "through" variable) balance occurs with force, torque, current and heat flow, while in the case of transvariables (or "across" variables), voltage, temperature and displacement balances are common.

7. *Examples of Functional Architecture*

A simple system for measuring small weights is shown in Fig. 7a. The weights W are placed on a plate fixed to the end of a pivoted beam such as to produce a torque in an anticlockwise direction. A spiral spring fixed about the pivot produces a torque in a clockwise direction proportional to beam angular deflection G. At some point the torques will balance, and G is an indication of W. The block diagram for this system shows that there is inherent feedback, in that the mechanical indication G produces a torque via the spring constant (feedback path) which subtracts from the torque produced by the weights on the beam (feedin path) to give an error torque. Linearity

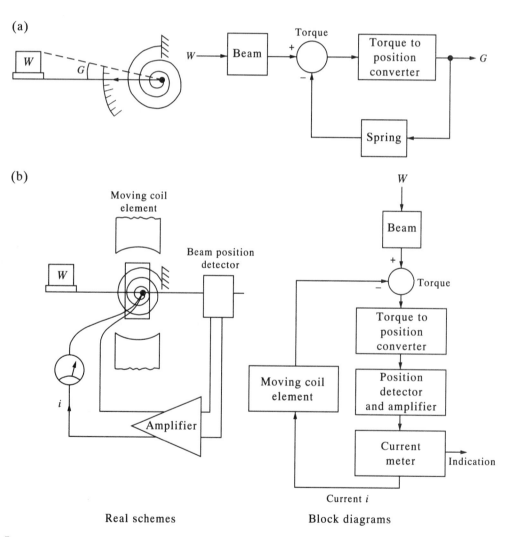

Real schemes   Block diagrams

Figure 7
Measurement of small weights: (a) simple system, (b) more complex system

between G and W is very dependent on the linearity between G and the spring torque.

A torque balance which results in very little beam movement and, therefore, greater accuracy is shown in Fig. 7b. Here the clockwise torque is produced by the current i in a moving-coil element fixed about the pivot point. Without weights the beam remains stationary and i is zero. When weights are applied the beam tends to move in an anticlockwise direction, and this movement is detected by sensors, for example, photodiodes, which produce an electrical signal; this is amplified to give current i, which, in turn, produces the counter torque to almost maintain beam balance. As there is little angular movement of the moving-coil element, the current–torque relationship is linear and the electrical output i is proportional to W. The output meter can be some distance from the point of measurement. Because there is little mechanical movement using this method of measurement, it can be used to measure more rapid changes in W than the fully mechanical scheme of Fig. 7a.

See also: Instrument Elements: Models and Characteristics

Bibliography

Doebelin E O 1990 *Measurement Systems—Application and Design*, 4th edn. McGraw-Hill, New York
Jones B E 1977 *Instrumentation, Measurement and Feedback*. McGraw-Hill, Maidenhead, UK
Jones B E (ed.) 1982 *Instrument Science and Technology*, Vol. 1. Hilger, Bristol, UK
Jones B E (ed.) 1987 *Current Advances in Sensors*. Hilger, Bristol, UK
Middelhoek S, Audet S A 1989 *Silicon Sensors*. Academic Press, London
Sydenham P H, Hancock N H, Thorn R 1989 *Introduction to Measurement Science and Engineering*. Wiley, Chichester, UK

B. E. Jones
[Brunel University, Uxbridge, UK]

Instrument Systems: General Requirements

The specification of the requirements for instruments and instrument systems is, as in the case of more general equipment and systems, a statement of the attributes they must have to satisfy the objectives of their use. This article considers the main features of the requirements which instruments and instrument systems must generally satisfy. It incorporates some of the aspects which must be considered in a design specification: i.e. the attributes which a design of equipment and systems must aim to realize and which extend beyond the requirements to incorporate, for example, aspects of manufacture, implementation and the like.

1. Requirements

Requirements for an instrument or instrument system should be the result of a requirement analysis which models the supersystem into which it is to be incorporated and its environment.

Instruments or instrument systems are never used in isolation, but always in a larger information processing and effectuation system. It is the objectives of the whole of that supersystem which must be considered.

Requirements should be determined not only with respect to the operational stage of the equipment or system but also with respect to the complete life cycle.

The formulation of requirements should as far as possible be abstract, that is, they should omit irrelevant details, and be implementation independent, thus ensuring that the widest possible range of solutions which meet the essential requirements can be considered.

Finally, the specification of requirements should be, as far as possible, precise and quantitative, so that the degree to which a particular concept satisfies it should be capable of verification. The specification of required attributes should, where possible, be in the form of a utility function, that is, in the form of a function which relates the value of an attribute to the degree to which the value satisfies the requirements. The specification will generally include constraints, and limits below or above which the values of the attribute must not lie.

In the general case of a requirement involving many attributes, the requirement specification should determine trade-offs.

2. Functional Requirements

The functional requirements are the attributes which the equipment or system must have in operation. They concern the inputs and outputs and the transformation between them, as well as the power supplies and the human–machine interface. They also involve a specification of the input and output signal power and the error characteristics. These requirements constitute the interface specification of the instrument.

For further discussion of functional requirements, see *Errors and Uncertainty, Instruments: Performance Characteristics*.

3. Constructional Features

The constructional features of instrument equipment must commonly be specified. They generally include outer envelope dimensions, there being commonly a restriction of size. Fixings and connections are commonly determined by the requirement to incorporate the equipment in an assembly. The weight of equipment is usually required to be specified, low weights being generally desirable.

Appearance is an important feature of equipment: good appearance is generally a consideration in its acquisition, and contributes to good operation and care of the equipment. It generally also contributes to the corporate image of the manufacturer and the user.

4. Manufacture

Equipment and systems must be as easy as possible to manufacture and assemble. In the case of instrumentation, testing and calibration are essential processes of manufacture. A significant part of instrumentation manufacture is software engineering (see *Construction and Manufacture of Instrument Systems*).

5. Distribution and Storage

Equipment needs to be transported from the place of manufacture to the place of use. This commonly involves their intermediate storage. Transportability and storability requirements, resulting from an analysis of the distribution and storage system, are an important part of instrument design. For some equipment this should include the requirement of portability.

6. Installation and Commissioning

Instrumentation must be installed and integrated in the system in which it is to operate, and commissioned, that is, made to operate in the working situation. Ease of installation and commissioning is an important requirement for any engineering equipment. In the case of instrumentation, *in situ* testing and possibly calibration facilities need to be considered as requirements (see *Installation and Commissioning*).

7. Operation Requirements

The principal requirements with respect to operation are those discussed above as functional requirements in Sect. 2.

It should be added here that the functional characteristics required must be maintained in the operational environment. In particular, the equipment must be protected against input overload and effects of power supply failure. It must be adequately robust against mechanical shock and vibration. Thermal effects of the environment must also be considered, and electromagnetic compatibility is an increasingly

important consideration. Equipment is commonly required to be protected against ingress of dust and against the effects of humidity. Resistance against chemical attack is sometimes required. Conversely, in some classes of application, sensors must not contaminate the measured object (see *Operational Environments*).

8. Economic, Logistic and Cognate Requirements

8.1 Reliability, Maintenance, Safety and Support

Reliability, maintenance, safety and support are important considerations (see *Reliability and Maintenance*; *Safety*).

8.2 Modifiability

It is important to consider and specify the ability of equipment to be modified: that is, its repairability, and its ability to be adapted to new requirements and to be upgraded.

8.3 Disposal Characteristics

All equipment ultimately comes to the end of its useful life and must be disposed of. This is not generally a problem with instrumentation, but it should not be ignored.

8.4 Economic Characteristics

For instrumentation as for all engineering equipment and systems, economic consideration lies at the core of the tasks of supply and acquisition. The manufacturer must consider the cost of supply: design, manufacture (including materials and components), distribution, installation, commissioning, support and sale. Overhead costs are part of the total cost. The price obtainable is generally determined by a market. The objective is generally profit: more resources are to be generated by the making of the product than are consumed. The user must generally consider the life cycle cost of the system: that is, the overall cost to acquire, operate, maintain, and withdraw it, considering the distribution of these costs over time. These costs must be balanced against gain, expressed as the effectiveness of the instrumentation arising from its use. Costs and gains need not always be financial. In many applications (e.g., scientific research, defence and health care) they are not. Nevertheless they should be quantified as far as possible.

8.5 Standards

It is commonly required that equipment and systems should conform to relevant industrial, national and international standards, which frequently determine design solutions.

8.6 Documentation

A significant requirement for instrumentation systems is that documentation should be provided which is adequate for operation, installation, commissioning, maintenance and modification.

See also: Design Principles for Instrument Systems; Life Cycle

Bibliography

Cook S 1990 Automatic generation of measuring instrument specifications. *Measurement* **6**(4), 155–60

L. Finkelstein
[City University, London, UK]

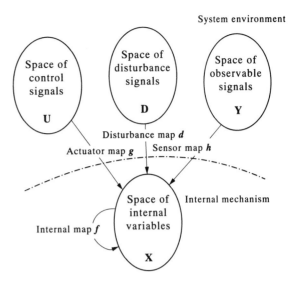

Figure 1
Final system: from interaction of the internal mechanism and the instrumentation scheme

Instrumentation in Systems Design and Control

The instrumentation of a process and selection of measurement variables (outputs), sensors and actuators (inputs) has a micro (local) as well as macro (global) aspect. The micro role of instrumentation has been well developed and deals with the problem of measurement, or implementation of action upon given physical variables. Instrumentation theory and practice deals almost exclusively with the latter problems. The macro aspects of instrumentation stem from the fact that designing an instrumentation scheme for a given process expresses the attempt of the observer (designer) to build bridges with the internal mechanism of the process, in order to observe and/or act upon it. What is considered as the final system is the object obtained by the interaction of the internal mechanism and the instrumentation scheme, as illustrated in Fig. 1.

In simple terms, the internal mechanism of the process is the set of all independent internal variables, attributes and signals, referred to as internal space **X** (irrespective of whether they can be measured or acted upon), and the relationships between them. These relationships express the physical laws describing the phenomena associated with the process and the coupling conditions arising because of the interconnections, and manifest themselves in the internal map f of the system. The process environment refers to the three sets: spaces of signals **U**, **Y**, **D**, which are referred to as the input space, output space, and disturbance space, respectively. **U** is the space of all external, arbitrarily assignable signals which may be applied to the system; **Y** is the space of all possible signals that may be measured; and **D** is the space of all disturbances that may affect the system. The result of instrumentation is the construction of two maps or functions. The first map, denoted by g, expresses the

coupling of input space to the internal variables space and is called the input or actuator map, since it is the result of selecting actuators. The second map, denoted by h, expresses the coupling of the internal variables to the environment and is called the output or sensor map, since it is the result of selecting sensors. The coupling of disturbances to the internal mechanism is expressed by the map d; note that disturbances may be measurable, unmeasurable, deterministic or stochastic, with corresponding implications on d, which is referred to as the disturbance map. For the sake of simplicity, the effect of **D**, will be ignored in the following sections.

What the observer understands as a system is the set $\Sigma = (\mathbf{X}, \mathbf{U}, \mathbf{Y}; f, g, h)$ of signal spaces and interrelationships between them. Clearly, the properties of Σ express the cumulative effect of the composition of the f, g, h maps, and thus their formation is affected by the selection of the actuator and sensor maps constructed by the observer.

The maps g and h may thus be considered as design parameters in the shaping of the characteristics of Σ, viewed as an information processing device or object to be controlled. The macro aspects of instrumentation are associated with the design of maps g and h, using global criteria and techniques, steming from the information processing and control capabilities of the resulting system. This body of rules, theory and techniques will be referred to as global instrumentation. Clearly, the macro aspects of instrumentation have to develop within the framework of constraints imposed by traditional instrumentation practice.

1. The Model Environment of Global Instrumentation

The characteristics and nature of global instrumentation depend on the type of possible available models used to describe the system; this is referred to as the model environment of the problem. Depending on the nature of the process and the modelling approach which is used, the following two classes of models can be distinguished: (a) internal models (IMs) and (b) external models (EMs).

1.1. Internal Models

These are described in terms of ordinary nonlinear equations and, if they are first order, they are called state-space models (SSMs). With such models we associate the n-, l-, m-dimensional, state, input and output vectors x, u, y correspondingly, and the respective spaces X, U, Y. The vector x is a function of the modelling exercise (i.e., assumptions, accuracy, etc.) and expresses a level of understanding about the internal process variables (Zadeh and Desoer 1963). The vectors u, y, express the process variables, which are used as inputs and outputs; if all physically possible variables, that may be acted upon and measured are included, then the model is referred to as extended SSM (ESSM). The system dynamics of SSMs are represented by

$$\dot{x} = r(x, u) \qquad y = h(x) \qquad (1)$$

where h, r, are vector-valued functions representing the sensor, composite internal and actuator maps, respectively. In the case of linear systems Eqn. (1) becomes

$$S(A, B, C): \dot{x} = Ax + Bu, \qquad y = Cx \qquad (2)$$

and the $n \times n$, $n \times l$, $m \times n$ matrices A, B, C represent the internal, actuator and sensor maps, respectively.

1.2 External models

If V, Z denote the spaces of all potential inputs and measurements, referred to as extended input–output spaces, respectively, and v, z are the corresponding p, q-dimensional vectors, then the internal map f is a vector-valued function $f: V \rightarrow Z$, where $z = f(u)$. For the case of linear, time-invariant systems f is a convolution function (Zadeh and Desoer 1963) or in the Laplace domain f is represented by the $q \times p$ rational transfer function matrix $F(s)$, for which

$$z(s) = F(s)\,v(s) \qquad (3)$$

Note that V and Z denote the potential input and output spaces, and the effective spaces are denoted by U and Y, and have corresponding dimensions l and m. If u and y, are the effective input and output vectors, then the corresponding model is illustrated in Fig. 2.

The final system is defined by the composite input–output map $w: U \rightarrow Y$, where $w = h * f * g$ and $*$ denotes map composition. In the case of linear,

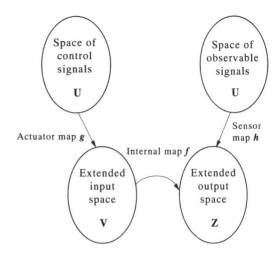

Figure 2
Model of potential and effective input–output spaces

time-invariant systems we have the transfer function matrix representation

$$y(s) = W(s)u(s), \qquad W(s) = H(s)F(s)G(s) \qquad (4)$$

where $H(s)$, $G(s)$ are the transfer function matrix representations of the sensor and actuator maps, respectively.

The linear systems defined by $S(A)$: $\dot{x} = Ax$, $F(s)$ will be named the internal and external-progenitor models, respectively. Progenitor models express the cumulative effect of our understanding of the basic subprocess dynamics and their interconnection. Alternative models to those defined earlier may also be used, such as the steady-state and the structural, graph models (Reinschke 1988). Such models are of special interest in the early design stages when accurate models are not available.

2. Fundamental Issues and Problems in Global Instrumentation

In general, instrumentation may be viewed as the body of knowledge of techniques that allows the selection of input and output schemes for a given process, with certain objectives and criteria. In view of the crucial role which input/output (I/O) structures play in control and signal processing, a refined classification of the issues involved in I/O selection is essential. We may classify the issues and problems as (a) model orientation problems (MOP); (b) model expansion problems (MEP); and (c) model projection problems (MPP). This classification is based primarily on the nature of issues, and secondarily, on the type of background concepts and techniques.

2.1 Model Orientation Problems

The classification of system variables as inputs and outputs is referred to as model orientation. In many systems, the orientation is not known, or the orientation changes, depending on the use of the system. Questions such as, when is a set of variables implied, or not anticipated by another, or when is it free, have to be answered, if model orientation criteria based on the nature of the process are to be derived. The specific use of the system may provide additional model orientation criteria.

2.2 Model Enhancement Problems

Defining input test signals and corresponding output measurements is an integral part of the identification and modelling exercise. Defining input/output schemes, with the aim of identifying (or improving) a system model, or reconstructing an unmeasured internal variable is what is meant by model expansion problems. Some distinct problem areas are as follows.

(*a*) *Additional measurements for estimation of variables*. Frequently, in process control, some important variables are not available for measurement. Secondary measurements have to be selected and used in conjunction with estimators to infer the value of unmeasurable variables. The proper selection of secondary measurements is a task of paramount importance for the synthesis of control schemes.

(*b*) *Input/output schemes for system identification*. The selection of input test signals and output measurements is an integral part of setting up model identification experiments. The identified model is a function of the way the system is excited and observed, that is, of the way the system is embedded in its experimental environment.

2.3 Model Projection Problems

The number of potential control variables p and potential measurements q, which ideally may be used, is quite large in many engineering designs. In an ideal design, unconstrained by resources and effort, all possible inputs and outputs should be used; however, economic and technical reasons force us to frequently select a subset of the potential inputs and outputs as effective, operational inputs and outputs. Developing criteria and techniques for selection of an effective input/output scheme, as projections of the extended input and output vectors, respectively, are called model projection problems (MPP). The two basic ingredients of a control-oriented MPP are the specific problem and criteria contexts; these two lead to a general philosophy for Global instrumentation.

(*a*) *The problem context of MPP*. Global instrumentation (GI) deals with the appraisal of alternative instrumentation schemes, represented by the maps g, h, and their final design using control and operability based criteria. Thus GI should provide concepts, tools and techniques, which may be used to give answers to the following problems.

(i) Define the lowest bounds for the number of effective inputs and outputs, which are needed for certain control schemes, or families of alternative control schemes.

(ii) Define the best location of effective inputs and outputs, as well as the structure of actuator and sensor maps, which may guarantee structural controllability and observability.

(iii) Evaluation of the degree of dependence and independence of given input/output instrumentation schemes, and its implications for process controllability and observability.

(iv) Evaluation of the effect of a selected input/output scheme on the control quality and characteristics of the final system, and selection of best schemes for easy and reliable control.

This list of problems is by no means complete and the tasks are rather general and vague. These problems may be transformed to concrete tasks, when assumptions about the system (stage of the design, nature of the model, engineering constraints, etc.) and the nature of the evaluation and selection criteria are specified.

(*b*) *The criteria context of MPP*. The control/operability quality criteria determine the difficulty of the control problem; they are defined in terms of the characteristics of the property indicators and the values of the system invariants, briefly described in the following paragraphs.

(i) *Property indicators*. These are functions of the given model parameters which describe the presence and/or degree of presence of one, or more than one, system property. Examples of such functions are the Nyquist diagrams, singular values (Maciejowski 1989), controllability observability grammians (Skelton 1989), and so on. The characteristics of property indicators (PIs) vary under system compensation, but within certain limits. In general, these limits are not well understood, but the values of certain invariants are actively involved in determining what can be achieved under compensation.

(ii) *System invariants*. These are functions of the given model parameters, the values of which remain the same under certain types of transformation/compensation schemes. As such, system invariants (SIs) characterize a family of systems obtained under compensation/transformation from a given system, and thus their values are involved in setting up what can be achieved under compensation. Examples of such functions are the system zeros (MacFarlane and Karcanias 1976), controllability, Forney indices (Kailath 1980), and so on.

(*c*) *A general design philosophy for global instrumentation.* The formation of the final structural characteristics of the system model is a dominant feature of MPP (and thus of GI), and it is reminiscent of an evolutionary process. The internal mechanism model *f* acts as the parent gene which predetermines a possible range of central characteristics of the final process. Selection of the *g* and *h* instrumentation maps corresponds to the final mutations which determine the system characteristics. The role of GI is thus to provide the concepts, tools and techniques that may direct the overall instrumentation design along favorable branches of the model evolutionary tree. This new role for instrumentation relies heavily on tools and concepts from control theory and design.

3. The Role, Content and Tasks of Control Theory and Design in the Development of Global Instrumentation

Control theory (CT) is the backbone to control systems design (CSD), since it provides the conceptual framework (concepts and tools) as well as the algorithms on which CSD philosophies, strategies and techniques are based. CT and CSD are well developed, especially in the context of linear systems; however, the development of CT has been almost entirely based on the assumption that the final model is given. There are a few examples where the fixed structure of the model is disputed, such as the zero assignment problem (selection of output matrix, or squaring down compensator with zero assignment criteria) (Rosenbrock and Rowe 1970, Kouvaritakis and MacFarlane 1976, Karcanias and Giannakopoulos 1989). Despite the fact that the formation of structure has not been properly addressed within CT, the basic concepts, tools and techniques needed for GI originate within CT. The need for development of GI defines new tasks, and emphasizes the role of existing areas in CT. These are given in the following sections.

3.1 Control Quality Criteria

These involve characterization of shapes or values of PIs and SIs, which may ease, or make difficult the control synthesis design problem. Integral parts of this task are: (a) establishment of links between the limits in the compensation of various PIs and relevant SIs and (b) further development of the solvability conditions of control synthesis problems, in terms of SIs and PIs. These problems are expected to lead to a classification of desirable or undesirable values of SIs and PIs, and are essential inputs for the following task.

3.2 Structure Synthesis Problems

These involve development of a methodology for shaping the instrumentation maps *g*, *h* with control-based criteria. Important aspects of this task are as follows.

(a) Understanding the mechanisms of formation of values of PIs and SIs as functions of the model parameters.

(b) Derivation of techniques for designing *g* and *h* maps, such that the formation of undesirable properties in the model are avoided and, if possible, desirable model characteristics are assigned. Clearly, such techniques should operate within the constraints, specifications and traditional instrumentation practices.

CT is the body of concepts, tools and techniques that deals with the qualitative and quantitative properties of a system model and the methods for solving a variety of control problems. The nature of CT clearly depends on the type of model which is used to describe the system. Linear system theory is the most developed and thus the initial effort to develop GI must be based on it. Issues related to nonlinearities may be handled, in the first instance, by linearization and then by posing problems of simultaneous design; that is, a common design for *g* and *h* maps for many operating points.

Bibliography

Chen C T 1984 *Linear Systems Theory and Design*, 2nd edn. Holt, Rinehart and Winston, New York

ESPRIT II Project 2090, EPIC 1989–92 Early process design integrated with control. Deliverables, D 1.1–1.3

Kailath T 1980 *Linear Systems*. Prentice-Hall, Englewood Cliffs, NJ

Karcanias N, Giannakopoulos C 1989 Necessary and sufficient condition for zero assignment by constant squaring down. *Linear Algebra & Appl.* **122–4**, 415–46

Kouvaritakis B, MacFarlane A G J 1976 Geometric approach to analysis and synthesis of system zeros. Part II: non-square systems. *Int. J. Control* **23**, 167–81

MacFarlane A G J, Karcanias N 1976. Poles and zeros of linear multivarible systems: a survey of the algebraic, geometric and complex variable theory. *Int. J. Control* **24**, 33–74

Maciejowski J M 1989 *Multivariable Feedback Design*. Addison-Wesley, Wokingham, UK

Morari M, Stephanopoulos G 1980 Studies in the synthesis of control structures for chemical processes: Part II. Structural aspects and the synthesis of alternative feasible control schemes. *AIChE J.* **26**, 232–46

Reinschke K J 1988 *Multivariable Control: A Graph-Theoretic Approach*, Lecture Notes in Control and Information Science, Vol. 108. Springer, Berlin

Rosenbrock H H, Rowe B A 1970 Allocation of poles and zeros. *Proc. IEE* **117**, 1079–83

Skelton R E 1988 *Dynamic Systems Control: Linear Systems Analysis and Synthesis*. Wiley, New York

Willems J C 1987 Models for dynamics. In: *Dynamics Reported*, Vol. 2, pp. 171–269
Zadeh L A, Desoer C A 1963 *Linear Systems Theory*. McGraw-Hill, New York

N. Karcanias
[City University, London, UK]

Instruments: Models and Characteristics

A model of a system is the description of the system in a formal language, such that relations between symbols in statements in the language imply and are implied by relations between the objects and attributes of the system and its components. Models of instruments in appropriate languages, or schemes of representation form the basis of the description, analysis and design of instruments. Characteristics of instruments, that is, relevant attributes and functional relations, are represented in models (see *Instruments: Performance Characteristics*; *Measurement Science*).

1. Instrument System

To consider instrument models it is essential to do so in terms of the architecture of a measuring instrument system. Such a system is shown in Fig. 1, which is discussed in more detail in the article *Measurement Science*. It is composed of a number of subsystems. First, there is the system under measurement and a sensor system connected to it which converts the measurand into a signal to be processed by the system. This is usually followed by a signal conditioning block which transmits the information to an information processing block. The output is passed to an effector block. The operation of the instrument system is controlled by a control block. A human–machine interface is used by a human operator to manage the system; the interface also includes the output to the operator.

2. Models of Instruments: General Considerations

Instruments are complex systems, the functioning of which is not apparent from an informal description, in natural language, of their physical embodiment, or a functional description of their components. Models with an appropriate scheme of representation are used to provide a description of the system, such that the functioning of the complete system is analyzable from a description, physical or functional, of their construction. In design, models are used to determine the function of a design concept from its construction or embodiment, and to determine an embodiment that can realize a particular function.

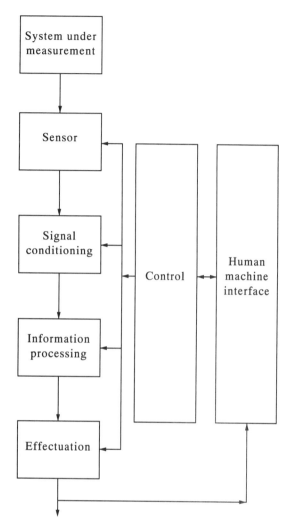

Figure 1
Architecture of an instrument system

Models involve abstraction, that is, the omission of irrelevant detail. This is required to reduce complexity, ensure tractability and give access to analytical tools. In the case of instruments a number of levels of abstraction can be distinguished in increasing levels of abstraction: embodiment models in terms of the physical variables, and the shape, dimensions and material properties of the system components and their physical interconnection; models of flow of power and power transformation; models of signals and their transformation; models of flow of symbol and symbol transformation; models of knowledge representation; and transformation by the system. Models may be functional, representing input–output relations in terms of mathematical or similar functions only, or embodiment models which represent input–output relations in terms of the attributes of the components of the system and of its structure.

To assist the tractability of the task, instrument models are based on the decomposition of the instrument modelled into simpler constructional or functional components. This is assisted by the fact that most instruments are built up from a small number of basic types of component, joined by a few basic types of interconnection structure. The components are modelled separately with the separate models, then integrated into a total model.

With reference to the earlier description of an instrument system, it is noted that in modern instrumentation the subsystems performing signal conditioning, information processing and control, as well as the human–computer interface are based on standard information technology. Their analysis and design as components does not form, in general, a task for instrument analysis and design. They are incorporated into instrument systems as blocks with known functional characteristics. The core aspects of instrument technology are twofold. First, there is the analysis and design of total systems performing a particular information or knowledge input–output transformation in terms of functional models of their components. Second, there is the analysis and design of sensors and their interaction with the system under measurement and the environment in terms of physical models.

Models of modern instrument systems, as those of other complex systems, are generally computer based. This is due to the high performance of modern computers, their high cost-effectiveness, the distribution of computer power, the friendliness of the human–computer interface and the use of high-quality packaged software.

Finally, it is necessary to state that the core feature of models of instruments is the representation scheme, or formal language, employed. This must be appropriate to the purposes for which the model is constructed, to the level of abstraction adopted and the analytical tools to be used. It is convenient to organize further discussion of the models of instruments in terms of representation schemes.

3. Linguistic Models

The description of instruments in natural language, accompanied by drawings, was the original form of models. It is still commonly used in textbooks and the like. However, while this is simple on the surface, it is imprecise and unsuitable for the analysis of complex devices.

With the advent of the use of artificial intelligence, a methodology of qualitative models is being developed. For example, the function of a voltmeter can be expressed by the statement: indication = function

(voltage). A system can be described by a functional model consisting of a set of such statements. It is possible to reason about them by appropriate rules of inference. The reasoning can be carried out in a computer using languages such as Prolog.

4. Pictorial Models

Embodiment models can be conveniently described by pictorial models. They range from artistic impressions to full engineering drawings, which enable a machine to be constructed. The production of such engineering drawings are the target at which the design of sensors aims.

Schematic drawings are a convenient way of representing a general embodiment concept. Electrical circuit diagrams are a typical example.

5. Mathematical Models

Mathematical models are a powerful means of analysis and design.

In the mathematical models of instruments symbols, signals or the physical variables describing power flow, depending on the level of abstraction adapted, are treated as variables. The relations between the variables, maintained by the system and its components, are described by algebraic equations, ordinary differential equations, partial differential equations, integral equations and the like. Using Laplace transforms reduces differential equations to algebraic ones.

There exists a variety of computer analytical tools for the analysis of such mathetical models, and for the design of systems described by them.

6. Diagrammatic Representation of Functional Models

To improve the transparency and ease of understanding of complex models of instrument systems, as well as control and other similar information systems, a diagrammatic representation such as that shown in Fig. 1 is generally employed. A functional component of the system is represented by a box with the inputs shown as arrowed lines coming into the box and the outputs shown as arrowed lines coming out of the box. The transformation between input and output is written in the box. This may be a linguistic description or a mathematical relation. Commonly, when input and output are represented by their Laplace transform the relation between them is termed the transfer function. It is written in the box. The output is the input multiplied by the transfer function.

7. Signal Flow Models

Models representing the flow of signals in systems are generally mathematical, and use both time-domain and frequency-domain models, and frequently employ diagrammatic representation. Computer tools for analysis and design are highly developed.

8. Power Flow Models

The transfer of information in instruments takes place by power flow, so the behavior of instrument elements in particular sensors is often conveniently considered in terms of power flow. A discussion of power flow models is given in the article *Instruments: Performance Characteristics*.

9. Detailed Physical Models

Detailed mathematical models of instrument elements representing the relation between input physical variables and output physical variables, in terms of the shape, dimensions and material properties of the element are generally complex, involving partial differential equations not uncommonly nonlinear and involving transcendental functions.

The approach to them is commonly by abstraction, reducing the representation of the element to a combination of simpler elements of the kind discussed in the article *Instruments: Performance Characteristics*. For more accurate models computer tools, such as those based on the finite-element method, form a powerful means of analysis.

See also: Instrument Elements: Models and Characteristics; Instruments: Performance Characteristics; Transducers: An Introdcution

Bibliography

Abdullah F, Finkelstein L 1983 Computer aided design of instruments sensors. In: Syrbe M, Thoma M (eds.) Fortschritte durch digitale Mess und Automatiesierungstechnik. *Proc. INTERKAMA Congr.* Springer, Berlin, pp. 551–62.
Finkelstein L, Abdullah F, Hill W J 1992 State and prospects of the computer aided design of instruments. In: Steussloff H, Polke M (eds.) Integration of design implementation and application in measurement automation and control. *Proc. INTERKAMA Congr.* Oldenbourg, Munich, Germany.
Mirza M K, Finkelstein L 1991 Mathematical modelling of instruments—application and design. In: Gardner J W, Hingle H T (eds.) *From Instrumentation to Nanotechnology*. Gordon and Breach, Philadelphia, PA, pp 55–71

L. Finkelstein
[City University, London, UK]

Instruments: Performance Characteristics

Instruments and instrument elements operate by maintaining a functional relation between an information-carrying input and an information-carrying output. In a complete measuring instrument the input is the measurand and the information output is termed the indication. The relation between the input and output of an instrument or instrument element, such as the relation between measurand and indication, and any parameters of such a functional relation are termed its performance characteristics.

1. Models

The performance of an instrument is described most completely by a complete mathematical model. Such models may be functional, representing the relation between input and output as a functional relation, or embodiment, describing the relation between input and output in terms of the instrument physical embodiment. Instrument models in general are considered in the article *Instruments: Models and Characteristics*.

This article is concerned with the principal features of the functional description of instruments.

2. Static and Dynamic Modes

Instruments may be used in two modes: static and dynamic. In the static mode a constant input or measurand is sensed by the instrument and the ultimate steady state is taken as the indication. In the dynamic mode a varying input is sensed by the instrument and the indication is given by the instantaneous output signal.

3. Response Law

The functional relation between the instrument information-carrying output or indication is termed its response law.

This response law is normally determined for the static mode of response, representing the ultimate steady-state indication for a constant measurand (see Fig. 1).

The minimum and maximum values of the input, or measurand, for which the instrument is effective, that is, gives an output compatible with further processing and with an acceptable error, are termed the lower and upper range limits, respectively. The interval between these limits is termed the range of the instrument and is specified by the limits. The algebraic difference between the upper and lower range limits of an instrument is termed its span.

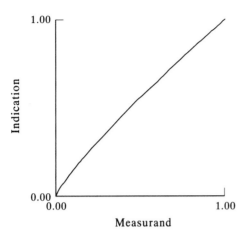

Figure 1
Typical static response law

The relation between the change of value of the instrument indication and the change of value of the input or measurand is termed its sensitivity. The response law of an instrument is usually ideally linear and, hence, the sensitivity is the slope of the response law. For nonlinear response laws the sensitivity is measured as the ratio of the finite change of the output for a finite change of the input at a specified point of the range, or the gradient of the response law at such a point.

The response threshold of an instrument, also sometimes known as the sensitiveness, is the minimum input or measurand level required to produce a response exceeding by a specified amount the indication of the instrument already present due to noise and other sources.

4. Errors

Real instruments do not, in practice, give their intended or nominal response to a particular measurand. The deviation of performance of a real instrument from the ideal is discussed in the article *Errors and Uncertainty*, which discusses the specification of errors. Such a specification of errors forms part of the performance characteristics of instruments.

5. Determination of Response Law

The response law of an instrument is generally determined by calibration against a standard, that is, by comparison with it. Such a standard is in turn calibrated against others of higher accuracy, right up to primary standards (see *Measurement Science*). The calibration results in a nominal response law together with a specification of the uncertainty involved in it.

For some classes of instrument element, for example, orifice plates, the response law can be determined from standard calibration curves, which have been determined for the complete class.

For a limited class of instruments, known as "calculable" instruments, the response law can be determined from a mathematical model. A simple example is a U-tube manometer in which the input pressure is balanced by a column of liquid. The height of the balancing column is the instrument indication. The relation between the indication and the measurand pressure can be calculated from the density of the fluid and the acceleration due to gravity.

6. Conditions of Use

The response law of an instrument is intended to be valid for what are known as particular conditions of use. Specifically, an instrument is intended to be valid for a defined set of reference conditions—that is, a range of influence quantities and variables, other than the measurand, which affect the instrument indication. The specification of these reference conditions is an essential part of the performance characteristics of instruments.

7. Dynamic Response of Instruments

Ideally, the instrument output or indication should respond instantaneously to changes of the input or measurand. Real instruments do not do so, showing a lag of output after input. This dynamic response is part of the specification of the performance characteristics of the instrument.

A complete specification of the dynamic response of an instrument with linear behavior is given by a measurand–indication transfer function $G(s)$, which also specifies the static response. In the case of nonlinear instruments a state variable model of the input–output relation is the most complete. In the case of sampled data systems a z transform transfer function is the most complete representation.

In the dynamic mode of use the instrument output or indication takes a finite time to reach the steady state following a change of the measurand. A typical dynamic response is shown in Fig. 2. The time taken for the instrument to reach the steady state or to approach it within a specified amount is termed the response time or, more properly, the settling time. If the indication exhibits an overshoot then the peak overshoot may also be specified as part of the performance characteristics.

In the case of instruments with first-order dynamics, the transfer function is given as:

$$G(s) = \frac{K}{1 + Ts}$$

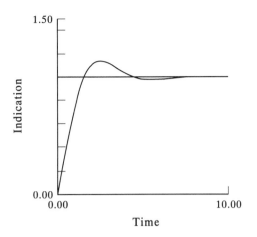

Figure 2
Typical dynamic response

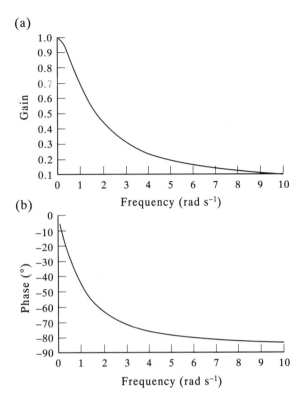

Figure 3
Typical frequency response

where the time constant T may be specified as a characteristic completely defining the dynamic response, in particular, the response time.

In the case of instruments with second-order dynamics, the transfer function is given as:

$$G(s) = \frac{K}{\omega_n^2 + 2\zeta\omega_n + 1}$$

It may be appropriate to specify the undamped natural frequency ω_n and the damping ratio ζ. These parameters again determine the dynamic response, including the response time, the peak time and the peak overshoot.

The transfer function of a linear instrument is completely specified by the appropriate frequency response. The frequency response is the response of the instrument to a harmonic input $q_m \sin(\omega t)$, resulting in a harmonic output $q_i \sin(\omega t + f)$. The response is typically presented as a plot of the ratio of amplitudes q_i/q_m against the frequency ω accompanied by a plot of the phase lag f against ω.

It is generally desired that the amplitude ratio should not vary with frequency and that the phase lag should be negligible over the range of frequencies in which the instrument is used in the dynamic mode.

While only a full frequency response constitutes a complete definition of the dynamic response of an instrument, it may be convenient to specify the instrument bandwidth, the range of frequencies over which the instrument frequency is flat, to within a specified amount (usually 3 dB or 0.707). The frequency response of a first-order instrument is shown in Fig. 3.

See also: Instruments: Models and Characteristics

Bibliography

Doebelin E O 1990 *Measurement Systems; Application and Design*, 4th edn. McGraw-Hill, New York
Bentley J P 1983 *Principles of Measurement Systems*. Longman, London

L. Finkelstein
[City University, London, UK]

Ionizing Radiation Measurements

Ionizing radiation loses energy when passing through any material medium due to ionization and excitation of the atoms in the medium. A single particle or photon can produce many free electrons both directly and indirectly, and hence, such radiation is very sensitive to detection. The wide diversity of possible interactions gives rise to very different attenuating effects with various absorbing materials, so that there is scope for the measurement of widely varying physical parameters. Suitable choice of radiation source and detection system, and optimization of instrument design with respect to geometry, shielding, and so on, are essential. The general principles of radiogauging have been established since about the early 1960s and are thoroughly discussed by Gardner and Ely (1967) and Cameron and Clayton (1971). A review of more recent applications is given by Palmer (1982). Modern developments in methods of detection and signal analysis have led to improvements in accuracy and response time with the use of weaker and therefore safer sources. Compact and versatile multichannel analyzers and microcomputers can be employed in on-line control systems, and can also be used in a variety of applications such as civil engineering and mining. Signal analysis and control systems are discussed in the article *Transducers, Ionizing Radiation*.

1. General Considerations for Optimum Design

As in any form of instrumentation, considerations of accuracy, sensitivity, stability, time of response, ruggedness, safety and expense may be conflicting, and instrument design must be adapted for the particular needs of any given situation.

1.1 Radiation Sources

Isotopes are readily available which cover a wide energy range of alpha and beta particles and photons. The characteristics of a number of commonly used isotopes are summarized in Table 1. For most purposes long life is an asset as the need for frequent recalibration is reduced. Measurements involving tracer techniques, however, require short-lived isotopes for safety reasons.

Electrically generated particles and x rays are sometimes necessary, for example, for proton production or for x-ray fluorescence measurements, but these usually involve greater expense in capital outlay and maintenance. Neutrons are also used in some measurements. They do not cause ionization directly but they interact with matter to produce ionizing particles, most commonly protons but sometimes heavier ions. These can also be detected either directly or indirectly as a result of further ionization, but the scope of neutron instrumentation is too wide to be covered in this article.

1.2 Radiation Detectors

A wide variety of sensors are available for the detection of radiation after its interaction with the test material. Careful consideration should be given to the choice of detector for any particular application, since this will strongly influence the signal-to-noise ratio and the stability of the instrument under working conditions.

If only the number of particles or photons is required Geiger counters can be used. Scintillation

Table 1
Some radioisotopes in common use

Source	Predominant energy (keV)	Approximate range
Alpha		in air
^{241}Am	5.5×10^3	
^{244}Cm	5.8×10^3	4.5 cm
^{252}Cf	6.1×10^3	
Beta		in Al
^{55}Fe	5	0.1 mg cm^{-2}
^{3}H	18	0.6 mg cm^{-2}
^{63}Ni	66	6.5 mg cm^{-2}
^{14}C	156	30 mg cm^{-2}
^{147}Pm	225	50 mg cm^{-2}
^{204}Tl	763	0.28 g cm^{-2}
^{90}Sr	550	1.1 g cm^{-2}
	2.3×10^3	
Gamma		
^{3}H/Ti	4.5	2.5 mg cm^{-2}
^{55}Fe	6	6 mg cm^{-2}
^{147}Pm/Zr	15.7	0.1 g cm^{-2}
^{210}Pb	46	1.6 g cm^{-2}
^{241}Am	60	2.5 g cm^{-2}
^{152}Eu/(^{151}Sm)	76	3.3 g cm^{-2}
^{147}Pm	121	4.6 g cm^{-2}
^{137}Cs	0.66×10^3	9.2 g cm^{-2}
^{60}Co	1.17×10^3	13.0 g cm^{-2}
	1.33×10^3	

counters, semiconductor detectors and proportional counters can give also the energy or spectral content of the radiation. In some systems, for example those using ionization chambers, electrical pulses resulting from individual events are collected and smoothed in an RC circuit and can be displayed in analog form on a ratemeter. Radiation detectors are discussed in the article *Transducers, Ionizing Radiation*.

1.3 Instrument Design

Owing to the statistical nature of the emission from radioisotopes, accuracy is limited by the count registered, N. The length of counting time available for any particular measurement is ultimately decided by convenience and stability considerations, and may be very short for on-line measurements. Hence, it is useful to refer to the standard deviation of the count rate, $\sigma_R = \sqrt{N}/t$. For maximum sensitivity in measuring a variable parameter x there must be a large change in the count rate R corresponding to a small change in x. The fractional change in x which has a 68% probability of detection is referred to as the resolving power of the instrument, and is given by the relation

$$\frac{\Delta x}{x} = \sqrt{2} \left(\frac{\sigma_R}{x} \right) \frac{dx}{dR}$$

The relative position of the source, the material under test and the detector must be carefully considered. In some circumstances collimation of the radiation is desirable so that it is incident on the specimen over a well-defined area. This reduces the effective source strength considerably, but it can also reduce shielding problems. If good collimation is not used, fixed or accurately reproducible geometry is essential since a small change in the relative position of the test material and the detector can lead to large errors. Measurements may involve the detection of transmitted, forward-scattered or backscattered radiation. The intensity of scattered radiation is always low compared with that of the incident beam, so it is desirable to collect the radiation through as wide an angle as possible. Careful shielding is necessary not only for safety purposes but also to prevent the detection of extraneous radiation which would reduce the signal-to-noise ratio.

2. Gamma Instrumentation

The interactions of gamma photons with matter are complex. In general, absorption is greatest at low energies, particularly in dense materials of high atomic number. At high energies scatter interactions predominate.

The fractional intensity of radiation transmitted by a uniform absorber of thickness x is given by:

$$I_x/I_0 = \sum P_i \exp(-\mu_i x) \quad \text{for all } i$$

where P_i is the proportional intensity of radiation of energy i which has an absorption coefficient μ_i in the absorber. Then the thickness of absorber τ required to reduce the intensity to half the incident value is given by $0.69/\mu$, where μ is the effective absorption coefficient of the beam. Due to large variations of μ_i with energy, transmission does not always undergo a simple exponential reduction over a very wide thickness range, the lower energy components being preferentially absorbed in the surface layers. In practice, these "softer" components are often removed by initial filtration to extend the useful range of measurement. It is often convenient to express the absorption in terms of the mass absorption coefficient $\mu_m = \mu/\rho$, where ρ is the density of the material, since μ_m does not change as rapidly as μ with atomic number.

2.1 Gamma Transmission Measurements

Gamma transmission has many applications in thickness and density measurements. Maximum sensitivity to thickness variation is achieved for values of μx within the limits $4 > \mu x > 1$, therefore the choice of isotope to be used is governed by the approximate value of thickness to be measured. NaI(Tl) scintillation detectors are most commonly used for such measurements, since they are highly gamma sensitive.

Gamma transmission techniques are used for measuring the thickness of rubber and metal strip and tube walls, for monitoring the loading on conveyer belts and for density measurements in fluids and slurries. They are also employed in liquid level gauges, in the assessment of interzone layers in extraction towers and mineral processing, in the measurement of water content in soil, and in sedimentation measurements. ^{60}Co, ^{137}Cs and ^{241}Am are the most commonly used sources.

2.2 Gamma Scattering Measurements

Backscatter gauges can be used to measure thickness when only one side of the test material is accessible. Such instruments usually consist of a ^{60}Co or ^{137}Cs source with a NaI(Tl) detector included in the same housing, but shielded from the source. Radiation incident on a homogeneous material such as a metal strip will be scattered back into the detector at a degraded energy. The response increases linearly with thickness for a fairly thin sheet, that is, up to a few millimeters for steel. At greater thicknesses it levels off to a steady value as the radiation scattered at depth in the material becomes self-absorbed before reaching the detector. Frevert (1986) describes a backscatter system for the precise gauging of the wall thickness of steel tubes, using either two or four accurately positioned ^{137}Cs sources.

Backscatter from a large volume of material increases with the density of the scattering material, but the response is nonlinear, first rising to a maximum and then decreasing with density, since the lower-energy backscattered radiation is absorbed more strongly by dense materials. The absorption of this radiation is also strongly dependent on the atomic number of the material, which complicates the response interpretation. Density and composition changes in soil can be observed, although they are not easily separated without further information. Backscatter instrumentation is also used in mining, the difference in backscatter response from coal and the surrounding shale being used to control the steering in coal cutting machines.

Dual gauges have been designed which give information about the scattered component and the attenuation of the primary beam, and these have particular application to soil characterization. Neutron gauges are often used, especially for assessing moisture content, since thermal neutrons interact readily with the hydrogen nuclei present in water molecules.

A review and bibliography of nuclear techniques in mineral exploration and mining is given by Mathew (1986), which includes information concerning measurement of the natural gamma activity of soil in various aspects of geophysical exploration.

2.3 X-Ray Fluorescence

Lower-energy photons are used mainly in x-ray fluorescence techniques. The inner shell electrons of atoms readily absorb photons, which have exactly the energy required to release them from the atom. The subsequent readjustment of the outer electrons results in the emission of low-energy x rays with wavelengths characteristic of the type of atom involved. This process of x-ray fluorescence is used for the chemical analysis of surfaces and small structures, and can also be used for thickness measurement of thin films or deposited layers. Monochromatic radiation of the optimum energy must be used, that is, slightly higher than the energy of the characteristic x rays emitted by the test material. If a suitable isotope is not available an x-ray generator must be used. Detection is by proportional counters of high resolution or by solid-state detectors which can identify the characteristic radiation against the general background of scattered radiation. Layers ranging, in general, from about 50 nm to 10^5 nm in thickness have been measured, the precision varying very much with the conditions, such as the coating and substrate combination or degree of surface contamination.

Kaushik *et al.* (1980) describe the measurement of up to three different metal layers on a Mylar substrate. Dostal and Rossiger (1987) discuss a method for the control of brass coating on steel wires, which is based on x-ray fluorescence analysis.

3. Beta Instrumentation

The stopping of fast electrons in matter involves elastic and inelastic collisions with atomic nuclei and electrons. In the process, energy is exchanged in multiple collisions, and the scattering process has a dominant role in the final energy distribution within the medium. The process is further complicated by the fact that most isotopic beta sources have a wide spectral range, the beta energy quoted being the maximum energy of the electrons emitted from the source. Despite the theoretical complexity of the processes involved, it is found empirically that the transmitted intensity I_t falls off exponentially over a fairly wide thickness range, and a convenient parameter μ_β, analogous to the mass absorption coefficient defined for gamma transmission, is often used. μ_β, usually measured in $cm^2 g^{-1}$ is very dependent on the energy of the beta source employed, but is roughly independent of the chemical nature of the absorbing medium, increasing only slightly with atomic number. However, variation in chemical constitution, for example, the presence of clays and titanium oxide in paper, or variation in water content, can affect thickness monitoring, since these factors can influence the density of the absorber.

3.1 Beta Transmission Measurements

For maximum accuracy in thickness measurements the beta source must be carefully chosen. In the case of gamma transmission measurements, the value of $\mu_\beta m$,

where m is the mass per unit area of the foil, should lie between the values 1 and 4. As may be seen from Table 1, beta particles are less penetrating than most gamma photons. In addition, beta-induced x rays (bremsstrahlung) produced by interaction with the nuclei in the absorber or in a collimator can be a problem when higher-energy beta particles are used, especially when materials of high atomic number are involved. For this reason gamma attenuation or backscatter methods are preferred for thickness measurements exceeding $0.8 \, g \, cm^{-2}$. Plastic scintillators are normally used as detectors for beta particles since they are less sensitive to gamma photons than NaI scintillators.

3.2 Beta Scattering Measurements

The backscattering of electrons occurs as a result of Coulomb interactions with nuclei. It increases rapidly with the atomic number of the scattering material and it is greater at low beta energies. This process can be used for the measurement of sheet thickness and coating thickness of high atomic number materials on low atomic number substrates. It can also be used for the analysis of two-component materials, for example, metallic alloys, and has the advantage that access to only one side of the material is required. The detector response is very dependent on the geometry of the system and the beta particle energy, and is of the form:

$$R = R_s(1 - \exp(-\mu_\beta m)) + R_0$$

where R_s is the response for infinite thickness of the surface material and R_0 is the response for zero thickness. Backscatter gauges are more applicable to measurements of thinner layers than transmission gauges, the approximate range covered being 0.1–100 mg cm^{-2}. Basic conditions governing sensitivity and reliability are discussed by Latter (1975).

Very thin films can be measured by electron excitation of characteristic x rays in a manner analogous to x-ray fluorescence. Isotopic sources or electrically accelerated electrons can be used for this purpose. Beta particle excitation has also been used for the assessment of particle size in ores.

4. Heavily Ionizing Radiation

Heavy ions, alpha particles and protons lose energy through a large number of small-loss interactions as they pass through matter. As a result they have a clearly defined range in any material, which can be empirically related to their energy.

4.1 Alpha Particle Measurements

Isotopic sources of alpha particles are available either as monoenergetic, thin sources or as thick sources with a broad energy spectrum, the latter being obtainable in greater intensities. The rate of energy loss for alpha particles of velocity v is of the general form:

$$\frac{dE}{dx} \propto \left(\frac{Z}{v^2}\right) \ln(Kv^2)$$

in a medium of atomic number Z, where K is a factor dependent on Z. The range of alpha particles is less than 50 μm in condensed matter, but thinner foils will transmit particles of reduced energy which can be measured precisely in thin-window semiconductor detectors; from the energy loss the foil thickness can be deduced. This type of gauge can be used to measure gas pressure, the distance between the source and detector being adjusted to suit the range of pressure to be covered. Humidity can also be measured, provided that the effects of pressure and temperature variations can be eliminated. Gauges employing thick alpha sources can be used in a simple counting mode with either a Geiger counter or a semiconductor counter. The number of particles reaching the detector falls off with increasing thickness or density of the interposed material, as increasing numbers of the lowest energy particles are completely stopped, but the instrument requires calibration since the response is not theoretically predictable. This is a simple gauge, versatile in application, and it can be used to control gas pressure systems.

4.2 Measurements with Accelerated Particles

Very thin films or surface layers can be measured by heavy-ion bombardment. Accelerated protons or helium ions can be used for this purpose. Isotopic alpha sources are not suitable as the intensities are too weak. Thickness is determined either from the count rate of backscattered ions or by measurement of the particle-induced x-ray emission. X-ray measurements are preferred for materials of high atomic number, but backscatter measurements must be used for atomic numbers <15. Such methods are more sensitive than electron-induced x-ray emission and avoid problems due to electron scatter. Thickness measurements are usually in the range 10^{-5}–0.1 mg cm^{-2}, but can be made outside these limits depending on the conditions. A comparative survey of methods for measuring thin layers is given by Brunner and Rossiger (1986).

Continuing improvements in data processing are likely to extend the possibilities of using nucleonic techniques in various industrial situations. Instrument design must also take account of stringent safety regulations, and in this respect the possibility of using weaker sources should be pursued.

See also: Transducers, Ionizing Radiation

Bibliography

Brunner G, Rossiger V 1986 Non-destructive testing of layers by means of ionizing radiation, Pt 1. Survey. *Feingeratetechnik* **35**(12), 561–4

Cameron J F, Clayton C G 1971 *Radioisotope Instruments*. Pergamon, Oxford

Dostal K P, Rossiger V 1987 Non-destructive investigations of coatings by means of ionizing radiation, Pt 4. X-ray fluorescence method for the determination of coating thickness and composition of brass on steel wires. *Feingeratetechnik* **36**(7), 306–7

Frevert E 1986 Method of non-contacting gauging of the wall thickness of tubes by backscattering of ionizing radiation. *Atomkernenergie* **48**(2), 87–90

Gardner R P, Ely R L 1967 *Radioisotope Measurement Applications in Engineering*. Van Nostrand Reinhold, New York

Kaushik D K, Singh S P, Bhan C, Chattopadhyaha S K, Nath N 1980 X-ray fluorescence to determine the thickness of single, double and triple-layered films of Cu, Bi and Au on Mylar substrates. *Thin Solid Films* **67**, 353–6

Latter T D T 1975 Measuring coating thickness by beta backscatter techniques. *Br. J. Non-Destr. Test.* **17**, 145–52

Mathew P J 1986 Nuclear techniques and instruments in mineral exploration, mining and processing. In: Rao S M, Majali A B, Deshpande R G, Murthy T S (eds.) *Industrial Applications of Radioisotopes and Radiation*. Wiley Eastern, New Delhi, pp. 127–53

Palmer R B J 1982 Nucleonic instrumentation applied to the measurement of physical parameters by means of ionizing radiation. *J. Phys. E* **15**, 873–83

R. B. J. Palmer
[City University, London, UK]

L

Level Measurement

The terms level or contents measurement are often used, but they are frequently confused. The correct choice of instrument or system is only possible if the precise requirement of the user is defined at the start of the design process, a common feature of the specification of any transducer or sensor in a system.

Level measurement systems are those that define the position of the interface between the tank contents and its immediate "atmosphere" and, using this information, provide the user with an indication of the depth of the contents (which may be characterized by ancillary equipment to give an inference of the mass or volume of the contents as a secondary stage).

The determination of the interface between the process fluid and its immediate atmosphere is the prime requirement of this system, and normally depends on the existence of a clean, horizontal interface between the two media. The existence of an unclear interface may cause considerable difficulties in this measurement (e.g., the existence of foam between liquid and gaseous media, or emulsion at an interface between two liquids).

Contents measurement is the measurement of the tank contents in volumetric units (e.g., liters, cubic meters, gallons) or mass units (e.g., kilograms, pounds, tonnes) and, in practice, these are obtained in many systems by inference from a level measurement system. This is the prime commercial use of level measurement systems in a wide spectrum of industrial and domestic applications.

1. Measurement Techniques

1.1 Dipstick

The dipstick is the simplest measuring device available which gives good accuracy in a vented tankage where it is possible to wet, mark or stain the dipstick as it enters the process fluid. Care must be taken to ensure that the dipstick is used in the manner defined at calibration (normally vertically), as any deviation from this will result in inaccuracies.

When guide tubes are fitted to ensure that the dipstick is used in the required inclination, or where diptapes are used on sounding tubes (e.g., on marine installations), care must be taken to ensure that surge effects are not produced within the guide and sounding tubes by overrapid insertion of the dip device, otherwise abnormally high readings will be obtained. Therefore, it is obvious that the apparently simple dipstick may be capable of producing either very high accuracy or very inaccurate readings, depending on the skill of the operator: this leads to a requirement by plant operators for measurement systems that are less dependent on human skills.

An optical dipstick, based on fiber optics has recently been reported by Augousti *et al.* (1990). The device uses a fluorescent fiber, where the amount of fluorescence channelled along the fiber is used as a measure of the depth of the liquid into which the dipstick is placed. Natural or strip light illumination of the fiber is used to excite the fluorescence initially; absorption by the liquid causes diminution of the fluorescence signal as a function of depth. The device is simple and suitable, in particular, for applications where inflammable liquids are used and thus a nonelectrical system is preferred.

1.2 Sight Glasses

Sight glasses allow direct visual observation of the liquid level within the sight glass and, if it is correctly positioned with the isolating valves fully opened, it can give a visual indication of the liquid level within the process vessel. It is very commonly used on steam-raising boilers, but it does, of course, normally require easy access to a point where a sight glass can be viewed, and hence varying techniques of transmission are employed to provide remote indication of this information on boiler control consoles. Sight glasses can be fitted with internal float mechanisms that can either trip external mechanical followers, or actuate electrical analog devices to provide a remote indication facility. The bicolor visual sight gauge is frequently employed for boiler measurements (Chappel and Ballash 1983). A prism is used in such a way that if steam is in the gauge port, red light is observed. Water in the port causes the red light to be refracted, with only green light being transmitted. The device may be adapted with the use of high numerical aperture fibers to transmit the optical indication to a central monitoring area. Such a sensor is passive, with no moving parts and is nonelectrical. A schematic representation of the device is shown in Fig. 1.

1.3 Floats

Float followers can be used to indicate interface level and to provide some of the highest-accuracy level indication systems available. They are used, for instance, for custody transfer of petroleum products, where close physical and fiscal control is required of the quantities involved as they change ownership. The technique has also been used for many years to provide the simplest level indication devices, normally described as "cat-and-mouse systems." The main limitation to this technique is the process pressure applied to the vessel and, hence, the requirement for

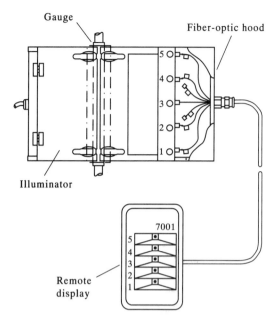

Figure 1
Fiber-optic water level gauge viewing system

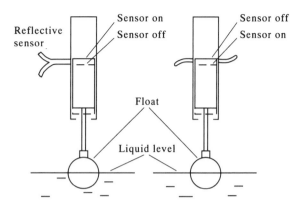

Figure 2
Liquid level fiber-optic switch, using a float arrangement:
(a) reflective, (b) transmissive

floats capable of withstanding high pressures. This involves increased wall thickness, and therefore increased mass and, as the device works on buoyancy, a restriction of the range of products to which the technique can be employed. The optical force or buoyancy liquid level sensor uses a float arrangement to track the level, using a reflective fiber-optic sensor (John 1982). The device is usually an interruption device to indicate the presence or absence of liquid. However, if the interruption device incorporates a transmissive binary code plate and multiple transmission-sensing fibers, the height of the liquid may be determined. An eight-bit code will give a resolution accuracy of approximately 0.5% of full scale (1:256). The simple form of the device is shown in Fig. 2.

1.4 Displacers

The displacer technique, normally involving a solid displacer supported by a spring suspension system, is widely employed as a level indication device for processes where high static pressures are typically in excess of 70 bar. The movement of the displacer as the interface moves up and down is transmitted via arms, torque tubes, magnetic coupling, and so on, to devices at low pressure that can be used for indication, switching or transmission.

1.5 Air Reaction Systems

Air reaction systems are among the simplest devices used for liquid level measurement and often involve the bubbler technique which, in its simplest form, is a

vented pipe inserted into the tank (see Fig. 3a). Air or inert gas is pumped through the pipe, and when its pressure is sufficient to clear the pipe of the tank contents, this pressure equals the hydrostatic head of the liquid. Thus, for a known density of tank contents, it is a simple process to deduce the level of product within the tank and, hence, the tank contents. This process can be applied to pressurized tanks by using a differential pressure technique with a reference connection to the tank top. For large distances between the tank and the indicator, twin impulse line systems are used to avoid system errors caused by the pressure drop in single impulse lines (see Fig. 3b).

The system has the advantage that it is widely understood and corrosion resistant materials are only needed for the pipe within the tank. Great care must be taken in the operation of the system to ensure that the tank contents are not forced into the instrumentation systems; a variety of auxiliary components are offered by manufacturers to minimize this risk.

1.6 Force Balance

This technique is a derivative of the air reaction system in which a 1:1 force balance transmitter maintains an instrument system pressure equal to the tank side pressure (see Fig. 3). It is one way of ensuring isolation in the instrument air supply and the tank contents and, hence, these devices are available in corrosion-resistant materials for chemical and hygienic applications.

Force balance transmitters that give a 3–15 psi (20–100 kN m^{-2}) pneumatic output signal are also available from manufacturers of pneumatic process control equipment and provide a very economical unit for short simple control loop requirements.

1.7 Hydrostatic Head

The pressure exerted by a column of liquid at a given level is proportional to the head and density of the

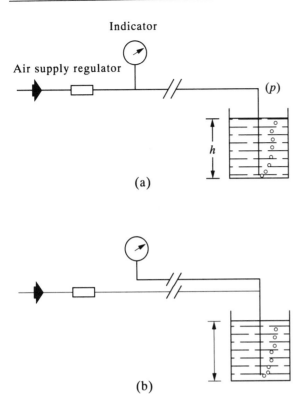

(a)

(b)

Figure 3
Air reaction sensors: (a) simple air purge system; (b) twin impulse line system

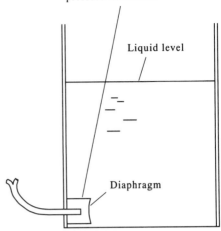

Figure 4
Liquid level sensor using a fiber-optic pressure transducer

fluid, the local value for *g* (acceleration due to gravity) and the atmospheric pressure above the liquid. The equation relating level height *H* and pressure *p*, is given by:

$$H = \frac{p \cdot SG}{d}$$

where *d* is the density of water and SG is its specific gravity. Due to higher pressure levels, high-density liquids offer a higher degree of measurement accuracy.

Hydrostatic pressure can be measured by a pressure sensor and indication system, which can be designed so as to remove the atmospheric pressure effects, and thus provide a very convenient liquid level indication system. The process forms the basis of the air reaction and the associated force balance devices described in Sect. 1.6, but it is more obviously employed in those installations where a simple pressure gauge is fitted at the base of a vented tank to indicate the tank contents. A more convenient variant uses isolating diaphragms and gas-filled capillary systems to provide remote indication of the tank contents. All hydrostatic contents gauging systems are affected by the density of the products being measured; this results in systems being calibrated, in general, for products at a given density, where volumetric indications are required. When tanks have a constant horizontal cross-sectional area, an indication of the mass of the contents is obtained that is independent of density.

Hydrostatic pressure measurements find wide application in all industries because of the reliability and proven nature of the sensor techniques employed. They are particularly suitable for very deep tankage, borehole, river level and similar applications, where the small size of the sensing unit is appropriate.

A reflective fiber-optic sensor used with a diaphragm has been employed in liquid level measurements, as shown schematically in Fig. 4 (Krohn 1988). Typical tank levels range from 6–20 m, with a pressure range from 0–2 bar. A resolution of 150 mm corresponds to a sensor accuracy of 1–2%. The measurement is complicated if the tank is pressurized. The preceding comments have referred, in general, to a conventional tankage with a freely vented space above the tank contents, or with a pressurized system using a differential pressure sensing technique to "back off" the so-called topping pressure. There are, however, other processes, such as one in which a tank is always full of liquid but the stored product (e.g., oil) floats above another fluid (e.g., water).

In this process the hydrostatic pressure instrument system can be designed in a number of ways; one is such that differential pressure is at its lowest when the tank is empty and at its highest when the tank is full, providing an indication of tank contents. However, the differential pressure obtained is very much lower than that obtained using an air-displaced tankage system, therefore great care is necessary in the choice of differential pressure sensor.

1.8 Conductivity Techniques

These have been used for many years for sensing the level of a conducting liquid, and depend on contact between one part of an electrical circuit (usually a probe) and the surface of a conducting liquid to complete a circuit to give an on/off or digital indication of a level being exceeded; a multiple of such devices can give a close approximation to an analog of the tank level, over a full working range. However, the technique is not suitable for nonconducting fluids, such as the majority of hydrocarbons, or "dirty" fluids where contamination may bridge electrodes.

1.9 Capacitance Techniques

The general method of operation is to measure the change in capacitance of a capacitor, one plate of which is the probe or sensor and the other is the wall of the tank, stilling well or other earthed component of the circuit. When the liquid is a conductor of electricity, it is necessary to use an insulated probe, but this is not necessary in the case of nonconducting fluids. The "capacitor" is built into a capacitor bridge circuit driven by an appropriate oscillator. The imbalance in the bridge gives an indication of the relative areas of the measuring capacitive plates that are coupled by the process liquid, hence the level of product in the tank can be determined if the characteristics of the system are known. As the dielectric constant of most liquids is highly dependent on temperature, some form of temperature correction is usually applied. When high accuracy of measurement is required, a reference system (comprising the probe and a tank wall reference that is always submerged) is used to standardize the system automatically for all changes of product.

Fail-safe capacitance level switches have been designed and supplied by European manufacturers and some have built-in self-checking procedures to monitor the measuring system from the sensor tip to the output relay, under the supervision of a microprocessor. Any fault in the complete circuit results in a fault-alarm light being illuminated. Normal operation is shown by the steady flash of a so-called heartbeat light-emitting diode.

Another capacitance device used for level alarms employs the "corner effect," which renders the device impervious to a wide change of dielectric constants or tank products. This, coupled with an angled head, which gives good drainage (and nontrapping of air), gives this an almost universal application, and offers increased safety.

1.10 Radio Frequency

The use of a device that vibrates in free air but is damped when covered by liquids or solids has been the basis of very reliable level switching units that have been available for many years. A more recent derivative of this technique is the use of a diaphragm or plate which is subjected to radio-frequency (rf) excitation; the damping of this plate detunes the circuit to give a discrete switching operation. These units can normally be tuned for varying degrees of sensitivity to make them suitable for a wide variety of applications as level switching devices for both liquids and solids.

1.11 Optical Techniques

A number of optical methods have already been discussed in previous sections. Optical methods, especially those involving optical fibers, have the distinct advantage of a passive mode of operation and ease of use for remote measurement, and they are nonelectrical and highly compatible with safety critical systems. In addition, they are often lightweight and operate when considerable levels of rf interference are present. Fiber-optic techniques often work best in clean and clear liquids, as certain opaque or dirty liquids tend to foul the optics and blind the sensor. In these cases, pressure-related measurement methods using optical fibers are best employed. Concepts that may be used in conjunction with fiber-optic level sensors include sight glasses, force or buoyancy, pressure or hydrostatic head, reflective surface methods or refractive index changes.

(*a*) *Refractive index change.* These devices function by transmitting light to a prism, typically quartz, into a medium of lower refractive index. In air (refractive index 1.0), the prism acts as a fiber optic with the air being the cladding. The prism is shaped to promote the back reflection. The essence of the mode of operation is that in the presence of the liquid, light is not totally internally reflected, but passes into the liquid. This is shown schematically in Fig. 5. With the use of proper electronic circuitry, discrimination can be achieved between liquid types, by the amount of light lost from the system.

For many years the total internal reflection technique has been used for level switching with a wide variety of low-viscosity liquids, from cryogenic liquid gas to low-viscosity hydrocarbons. Recent developments have included the use of multiple-prism arrays to give a near analog indication of level. These prism arrays can be mounted in such a way that they can discriminate levels to better than ±0.25 mm, but separation of sensors to a 1 mm spacing allows systems to ignore surface ripples and still provide an acceptable performance. Since refractive index change sensors require only small amounts of liquid to switch, leaks at lubrication seals can be easily detected and, if immersed in a liquid, small bubbles may be found. Further, the void fraction in a liquid may be determined using such methods.

(*b*) *Beam breaking.* The beam breaker technique is an elementary and long-established technique that may prove to be the solution to many level measurement problems, particularly where on/off control is required. The more sophisticated multicolored sight glass represents an elaboration of this technique.

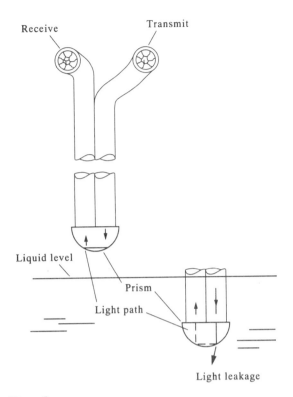

Figure 5
Fiber-optic liquid level switch

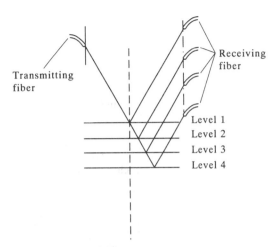

Figure 6
Surface reflectance liquid level measurement technique

(*c*) *Surface reflectance*. These devices use the reflectance of light from a surface to determine liquid level (King and Merchang 1982). A series of fibers may be employed in the configuration shown in Fig. 6. This approach is particularly attractive for monitoring levels which are inaccessible to contact methods, due to temperature or chemical composition. An example is the monitoring of the surface of molten glass. Higher viscosity fluids are preferred, with a smooth surface, giving a high surface reflectance. Low viscosity fluids, subject to surface vibration, are not candidates for this approach.

(*d*) *Optoelectronic devices for other physical parameters applied to level measurement*. Intrinsic fiber-optic physical parameter switches may be employed, in particular, using optoelectronics, for the reasons already discussed. A range of materials will change their properties with the environment, for example, pressure or temperature. Thus, surfaces may be detected by the different thermal properties experienced at an interface, for example, with liquid crystals. Further, polymeric materials can abruptly change their transmission or reflectivity at rather precise temperatures, and these may be used for liquid level sensing. Indeed, any material which undergoes such a physical state change is a candidate for this application.

2. Ultrasonic Techniques

One of the more valuable level measurement techniques is that of ultrasonics. Human hearing covers all or part of the range of 30 Hz to 15 kHz, depending on age and health. Any frequency above approximately 18 kHz is called ultrasonic and it is in this ultrasonic range that the units operate. The units normally work as sonar or pulse echo devices where a pulse of ultrasound is fired at the interface from either above or below the liquid surface. The time of flight of this pulse of sound to the surface, plus that of its reflection to a receiver (often in the same transducer housing), can be used to indicate the distance between transducer and surface, if the velocity of sound in that medium is known.

The velocity of sound in a liquid depends on a number of factors, the principal ones being temperature and density, and typically a temperature change of 5 °C can change the transit time by 1% of the range of the unit unless corrections are applied. These corrections can be determined by means of a temperature correction circuit or, alternatively, a second transducer system may be applied to a system with a "target" that is always submerged and is at a known distance from the reference transducer; this reference signal is then used to standardize the measuring unit.

Careful attention must be paid to the design of the transducer, since no transducer produces a pure, unidirectional pulse. It also needs control and attention during installation to eliminate the effects of divergence or side lobes. What must be taken into account are other factors that may increase the acoustic system loss, such as gas bubbles which reflect and disperse the beam. These may reach such a level that the signal can be lost. Solids in suspension can have the same effect.

A combined transmitter–receiver diaphragm in the transducer results in the need for a deadband of

185

typically 15 cm to allow any transmission resonance to die away before acceptance of incoming reflected signals; this results in loss of level measurement before that point. The successful application of ultrasonics to all systems therefore requires careful choice and installation to provide the level of performance of which these devices are potentially capable.

Mounting the transducer above the liquid interface overcomes the problems of gas or solid entrainment in the liquid, but does retain the requirement for standardizing the velocity of sound in the air or gas above the liquid. As gases and vapors often occur in layers of varying temperature, density and relative humidity above a process liquid, this can cause substantial standardizing problems. In this type of unit it is usual to have a deadband of approximately 400 mm to allow transmission resonances to die away.

The top-down "viewing" transducer can be used on hoppers containing solids for level indication, but care must be taken to minimize flying dust above any such solids, as these cause attenuation of the signal.

Overall, the ultrasonic system provides a very reliable level measurement system, typified by either noncontacting of, or minimum intrusion into, the process fluid, and is a viable instrument for level measurement and control as a direct result of the introduction of microprocessor-based techniques.

2.1 Observation

Although observation is not normally considered as a process control technique, it has been used for large hoppers containing wet solids that do not flow easily, or which may build up on hopper sides. In such cases the use of load cell techniques may not be acceptable to the user because of installation costs, and so on. The relatively low cost of weatherproof TV cameras and a centralized VDU monitor has enabled users to view hopper contents, and then manually control feed rates, vibrations, and so on, as required.

2.2 Radiation

The absorption of radioactive radiations by the contents of a vessel or process can be used for either switching or analog indication, and is particularly suitable for solids and aggressive liquids. The installation and application of these instruments should only be undertaken after very careful consideration of the application and installation, paying particular attention to the radiation source and the movement of plant operators.

3. Level Indication and Control for Bulk Liquid Storage

Petroleum, wine, beer, milk, chemicals and other liquids are frequently stored in large tanks. However, the majority of tanks exceeding 1000 m^3 in capacity are enclosed, vertical cylindrical containers for petroleum liquids stored at near atmospheric pressure. As many petroleum liquids are fairly volatile, a proportion of these tanks, including most of the very largest, have floating roofs to conserve vapor, and this in turn often influences the choice of level gauge and its installation.

Usually the petroleum storage tank is treated as a calibrated container, and to estimate the true quantity of the stored product, three measurements are needed: level, temperature and relative density. This is because petroleum has a high coefficient of thermal expansion which is a function of density. However, level is the primary measurement, being directly related to volume and the most accurate assessments of significant movements of the product into or out of the storage tank are obtained from the automatic level gauge readings referred to as the tank capacity tables.

The conventional automatic level gauges used for inventory or custody transfer purposes are very accurate instruments, and any major source of error in service can usually be attributed to an unsatisfactory mechanical installation, resulting from inadequate understanding of the structural problems associated with the tank.

The same principles apply to all vertical cylindrical tanks. When changes in the design codes were introduced to permit construction of very large tanks in higher quality steel, with more highly stressed shell plates, this became more apparent. It was subsequently realized that tank distortion caused by liquid head could have a serious effect on the installation of automatic gauges, and on gauging data or reference points for automatic and manual gauging. In some countries, level gauges may only be manufactured and installed in accordance with legally enforced standards which were, and in many cases still are, based on an incomplete appreciation of what actually happens to the tank as it is filled. Some years ago it was also felt that these methods of gauge mounting were probably structurally unsound.

Examination of extensive design data for numerous tanks, supported by actual physical measurements and calculations, revealed the exact nature of tank distortions as a tank was filled and emptied. An interactive computer program was developed, which, with the aid of the tank construction details and product density, would quickly carry out a complete stress analysis and accurately predict the tank deflections for different levels of product. By reference to much accumulated data, several simple principles were established, which have made it possible to recommend the best methods for gauge installation.

Much of the existing published information refers all level measurements to the bottom of the tank shell. This is not an ideal reference, and any attachment close to the bottom of the shell is probably less reliable, since maximum angular deflection from the vertical occurs at the bottom corner. Despite this, datum plates for manual gauging are still found cantilevered off the bottom of the shell, giving rise to very significant errors in measurement. The best reference or datum for the mounting of an automatic level gauge or for manual gauging is the actual bottom of the tank between 500 mm and 1000 mm radially inwards, where the bottom is only affected by the compressive load due to the head of liquid. This principle applies to all low-pressure vertical cylindrical tanks, but it is particularly relevant for tanks of 20 000 m^3 and greater capacity. Other known principles can be applied to avoid very significant errors, and many existing standards and codes of practice require continuous revision. The Institute of Petroleum has published a new standard for automatic tank gauging which embraces the latest ideas. It must principally cater for the types of level gauge in current use, but it does not preclude the introduction of new techniques. In the field of level gauging, however, extravagant claims are often made in good faith, but unfortunately, may sometimes be based on an inadequate understanding of the problems highlighted here.

Conclusions

A wide variety of level measurement techniques exist, and the choice of the most appropriate will depend upon a number of factors: cost, availability, environmental effects, ease of installation, level of training needed, and so on. Individual applications are numerous and both the fiscal and legislative needs for high-quality level measurement are clear. It is fortunate that such a wide variety of methods exists.

See also: Optical Measurements; Pressure Measurement

Bibliography

Augousti A T, Mason J, Grattan K T V 1990 A simple fiber optic level sensor using fluorescent fibers. *Rev. Sci. Instrum.* **61**, 3854–8

Chappel R. E, Ballash R T 1983 Fiber optics used for reliable drum water level indication. *Proc TAPPI Engineering Conf.* pp. 313–17

John R S Jr 1982 Fiber optic liquid level and flow sensor system. US Patent No. 4 320 394

King C, Merchang J 1982 Using electro-optics for non-contact level sensing. *InTech.* **29**, 39–40

Krohn D A 1988 *Fiber Optic Sensors—Fundamentals and Applications.* Instrument Society of America, Research Triangle Park, NC

K. T. V. Grattan
[City University, London, UK]

C. H. Goode
[KDG Mobery, Crawley, UK]

I. W. F. Paterson
[Whessoe Systems and Controls, Darlington, UK]

Life Cycle

Instrument equipment and instrument systems, like other engineering products and systems, have a life cycle, a series of stages which they undergo from their initiation to their ultimate withdrawal and disposal. In the design and management of such equipment and systems it is necessary to consider all stages of their life cycle and not only the operational stage of their life.

1. Life Cycle

A discussion of the life cycle of instrument equipment or systems is best undertaken with reference to Fig. 1, which illustrates the sequence of all the stages through which such equipment or systems pass.

It is useful to distinguish products, with respect to their genesis, into special and general products. Special products are designed and made to a commission, for example a special instrument for a particular experiment, or a control system for a particular plant. General products are designed and made by a manufacturer on a speculative basis and sold in the market, for example, standard oscilloscopes, pressure gauges, and so on.

Life cycles may be viewed from two different perspectives, that of the manufacturer and that of the user. These are usually different organizations or different bodies within the same organization.

The product initiation stage is the set of all the activities concerned with identifying a need, the possibility of satisfying it and the commitment to do so. It is part of product policy and management: the processes by which users plan and decide to acquire certain goods and services, or manufacturers decide what to make and supply.

Design is the process which takes the statement of need, originating from the product initiation stage, and translates it into a specification of a device, equipment or system to satisfy that need, such that it can be made or implemented.

The manufacturing stage takes its specification from the design stage and makes or implements the product. In the case of instrument equipment and systems, the manufacture of the product involves its calibration and testing as a most important stage.

The distribution of the product involves the transport of the product from the site of manufacture to the site of use, including any storage. The design of instrument equipment must therefore take into account the transportability of the equipment, its storage characteristics, and the like.

The processes of installation and commissioning are concerned with bringing the product into operation, including its incorporation into any larger system, or, in the case of an instrument system, the integration of components or subsystems into a single entity. Again,

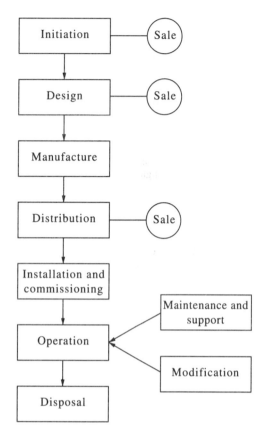

Figure 1
Life cycle

in the case of instrument equipment, the process of commissioning involves testing and possibly calibration in operation.

Operation is the basic process of use of the product. In parallel with the operation of the product, there are processes of maintenance and logistic support. The former is concerned with keeping the product in an operational state. It involves testing, repair and preventive maintenance. The latter is concerned with providing the supplies required, the provision of spares and the like. Also, in parallel with the process of operation, there is, in general, a process of modification.

It is useful to recognize the distinctions between: (a) corrective modification, which is concerned with remedying faults of design or manufacture; (b) adaptive modification, which is concerned with adapting (or tailoring) the product to new uses; and (c) enhancement modification, which is concerned with enhancing the performance of the product, for example, by retrofitting of new technology.

Finally, there is the disposal of the product at the end of its useful life. It takes place when either the

product has ceased to maintain an adequate level of performance, including having a low availability or, more generally, a product is withdrawn when this is cost-effective.

Disposal involves decommissioning and waste disposal. The product may be transferred to another user, in which case it begins its life again from the distribution stage.

The manufacturer divides the life cycle into a supply phase and a support phase. The supply phase involves all the stages of the life cycle from initiation to the transfer of ownership to the user. The support phase may involve distribution, installation and commissioning and, finally, aspects of maintenance, logistic support and modification.

The sale of an equipment or system is one of the most essential aspects of its life. In the case of special, that is, commissioned products, the sale may take place at, or following, the design stage. Alternatively, it may precede the design, the commission being placed on, for example, a design, make and commission basis. In the case of general, or speculative products, the sale takes place at the distribution stage.

The user divides the life cycle into an acquisition phase and a use phase. The acquisition phase may, in the case of a special commissioned product, start with the initiation stage, and extend as far as the installation and commissioning stage. It may start at the design stage and terminate after manufacture or distribution. In the case of standard products, the acquisition phase may consist of no more than activities at the distribution phase concerned with the choice of one product from among a number on offer. The use phase encompasses all the stages of the life cycle that follow the acquisition phase.

It is useful to examine briefly the financial aspects of the product life cycle from the point of view of the user. The illustrative case of a special, commissioned, capital equipment or system will be taken.

The flow of income to the manufacturer consists of an initial negative flow representing the expense of designing, manufacturing, delivering, installing and commissioning the equipment or system, followed by a positive income representing payment on completion.

The corresponding flow of income to the user consists of an initial expenditure, or negative income, representing the costs of acquisition, including the effort of the user as well as payment to the manufacturer. Thereafter there is income attributable to the capital equipment, net of the costs and support, and the like. This income will, in general, increase over an initial period, representing learning and running in, achieve a maximum, and thereafter decline with the age of the equipment, due to decreased effectiveness and increased support. The equipment is disposed of, and perhaps replaced, by the user somewhere during this decline. Disposal and replacement may be due to new equipment becoming available in which it is cost-effective to invest.

2. *Context and Literature*

A review of life cycles of engineering products is given in Finkelstein and Finkelstein (1991). The life cycle is generally considered from the point of view of systems engineering (see Kline and Lifson 1968 for a comprehensive analysis), terotechnology (see Husband 1976 for a review) or life cycle costs (see Blanchard 1978 for an authoritative analysis). These points of view are generally concerned with a single item of equipment as seen by a user, and have an origin in the literature of the design of military equipment. Another point of view is that of marketing, concerned with a general product as seen by a manufacturer (the text of Kotler and Armstrong 1989 is an excellent exposition of this perspective).

See also: Design Principles for Instrument Systems; Installation and Commissioning; Instrumentation in Systems Design and Control

Bibliography

Blanchard B C 1978 *Design and Management to Life Cycle Costs*. M/A Press, Portland, OR
Finkelstein, L, Finkelstein A C W 1991 The life cycle of engineering products—an analysis of concepts. *Eng. Manag.* **1**, 115–21
Husband T M 1976 *Terotechnology*. Saxon House, Farnborough, UK
Kline M B, Lifson M W 1968 Systems engineering. In: English J M (ed.) *Cost-Effectiveness*. Wiley, New York pp. 11–32
Kotler P, Armstrong G 1989 *Principles of Marketing*, 4th edn, Prentice-Hall, Englewood Cliffs, NJ

L. Finkelstein
[City University, London, UK]

A. C. W. Finkelstein
[Imperial College of Science, Technology and Medicine, London, UK]

M

Mass, Force and Weight Measurement

Mass and force are principal physical quantities essential to engineering measurements. Mass is the quantity of matter in a body; alternatively, it is a quantity which characterizes the inertia of that body. Force is the measure of attraction or repulsion between masses. The unit of mass is the kilogram, which is equal to the mass of the international prototype of the kilogram. The prototype kilogram is a solid cylinder of platinum–iridium kept by the International Bureau of Weights and Measures (BIPM) at Sevres, Paris.

Mass and force are related through Newton's second law of motion, $F = ma$, where a is the acceleration of the mass m. Although mass is a fundamental quantity, force is not, incorporating dimensions of length and time as well as mass. Its unit is the newton (N) within the International System of Units (SI), which is the force which, acting on a mass of 1 kg, results in an acceleration of $1 \, \mathrm{m \, s^{-2}}$.

Weight is defined as the force resulting from the gravitational attraction between the mass of a body and the mass of the earth. The resulting acceleration of this body is the gravitational acceleration, denoted by g. The actual value of g varies with location and the distance between the mass to be measured and the earth. The value of g in $\mathrm{m \, s^{-2}}$ at a geographical lattitude of ϕ degrees and an altitude of h meters above sea level can be computed from the expression

$$g = 9.780\,318\,4\,(1 + 0.005\,302\,4 \sin^2 \phi$$
$$- 0.000\,005\,9 \sin^2 2\phi) - 3.086 \times 10^{-6} \, h \quad (1)$$

The standard value of g is accepted as $9.806\,65 \, \mathrm{m \, s^{-2}}$; however, it varies by $\sim 0.1\%$ in the UK and by 0.5% over the surface of the earth between the equator and the poles (from $9.780 \, \mathrm{m \, s^{-2}}$ to $9.832 \, \mathrm{m \, s^{-2}}$), and decreases by $\sim 0.03\%$ for each 1000 m increase in altitude.

The weight of a mass also depends on the amount of gas it displaces, so for very accurate weight measurements in air, the ambient temperature, barometric pressure and humidity should be taken into consideration as well as the presence of large masses near the weighed object. The relative error caused by weighing iron or water of the same mass is $\sim 0.1\%$, due to this differing buoyancy effect.

When the value of g is determined accurately for a specific location, the gravitational forces acting upon accurately known masses may be computed to establish a standard of force. This forms the basis of deadweight calibration of force measuring transducers and systems.

1. Weight Measurement: General

The measurement of weight is a special case of force measurement and is highly developed, largely because it is the principal means of determining the quantity of goods exchanged or sold. It is also a critical operation in the efficient management of a modern industrial plant employing materials handling and storage.

The majority of industrial weighing systems incorporate a force transducer known as a load cell, which can be electrical, hydraulic or pneumatic. Although load cells are mainly used for weighing purposes, they can also be employed to measure other quantities derived from force: thrust, density, stress, torque, pressure, altitude, sound and acceleration.

Most electronic weighing systems utilize load cells based on strain gauges. A typical system comprises a mechanical arrangement for collecting and applying the force to be measured to the load cell, a load cell as the force transducer producing an electrical signal as a function of the force applied, and a signal processing system. These electronic weighing systems lend themselves readily to incorporation into computer based control and data processing systems for the efficient and economic use of raw and semiprocessed materials.

There are two basic methods of force and mass measurement: direct comparison and indirect comparison. The former uses an equal arm balance to compare known and unknown masses, utilizing the null balance technique. Indirect comparison employs calibrated transducers to measure the unknown mass.

2. Direct Comparison Method

This is possibly the simplest mass measuring method. The moment produced by the unknown force on one arm, called the load arm, is directly compared with the moment produced by the standard known mass on the other arm, called the power arm. When the beam is horizontal, a null position is obtained, indicating that the forces acting on each arm are equal, provided that the arms are equal. An optical micrometer is generally incorporated in these instruments to detect and measure small amounts of imbalance of the null position. For very high accuracy, measurements should be conducted in controlled environmental conditions.

This type of balance, also known as the analytical balance, is designed to measure up to 150 kg and can achieve a sensitivity of 1 ppb.

3. Indirect Comparison Method: Lever Systems

When large masses are to be measured, the equal arm balance becomes inadequate. In this case either a

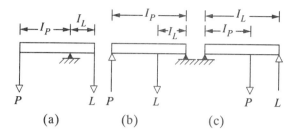

Figure 1
Classes of lever: (a) Class 1, $P = L(I_p/I_L)$; (b) Class 2,
$P = L(I_L/I_p) \lessgtr 1$; (c) Class 3, $P = L(I_L/I_p) \geqslant 1$ (after
Erdem 1982 © Institute of Physics, London. Reproduced
with permission)

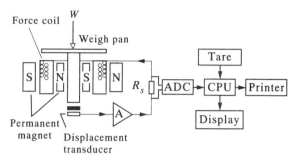

Figure 2
A typical force balance weighing system (after Erdem
1982 © Institute of Physics, London. Reproduced with
permission)

single beam with unequal arms or a system employing
a number of unequal arm beams or levers is used. The
unequal arm beam balance is commonly known as the
steelyard or the Roman steelyard; in which the fulcrum point is fixed and the balance is obtained by
moving the known weight or poise along the power
arm. The Roman steelyard is commonly incorporated
in multiple lever systems to achieve balance detection.

There are basically three types of lever. Class 1
(see Fig. 1a) has a fulcrum point between the load to
be measured and the counterbalancing force. Class 2
(see Fig. 1b) has the load application point between
the fulcrum and the power application point. Class 3
(see Fig. 1c) has the power application point between
the fulcrum and the load application point. This type
of lever is seldom employed in weighing machines,
since it needs counterbalancing forces greater than the
load being measured. It is, however, used in force
generators where a small but accurately known force is
multiplied by a precisely defined lever ratio and
applied to a force transducer under calibration.

4. Force Balance Technique

This technique is unique in its operation in that a
feedback loop is employed to compare the electrical
output with the force input. A typical force balance
system (see Fig. 2) uses a displacement transducer to
sense the deflection of the weighpan. This displacement signal is then fed through an amplifier to the
force coil to restore the weighpan to its original
position. The current flowing in this feedback loop is
converted to voltage across the sensing resistor R_s and
displayed in digital format by the use of an analog-to-
digital converter.

The construction of these devices is complex and
they tend to be more costly. However, force balance
systems have major advantages. They have superior
dynamic performance and they are independent of
zero instability, temperature effects and hysteresis.
Balance is achieved after only a very small deflection.

This type of device has a measuring range from <25 g
to >100 kg, with a resolution of 0.1 μg to 0.1 g. It is
mainly an alternative to the mechanical analytical
balance.

5. Transducing Force into Pressure

This is achieved by the use of hydraulic or pneumatic
load cells. The hydraulic load cell is filled with a fluid,
usually oil, and has preload pressure of 0.2 MPa.
Application of the load to the loading plate or button
of the device increases the fluid pressure, which is
displayed on a dial. It is also possible to use an
electrical pressure transducer to translate the fluid
pressure into an electrical signal. These load cells are
inherently very stiff, deflecting only 0.05 mm under
rated load conditions. The load-indicating pressure
gauge can be located several meters away from the
load cell by the use of a special fluid-filled hose.
Where more than one load cell is used, a specially
designed totalizer unit has to be incorporated to
combine the outputs.

Hydraulic load cells do not need external power and
are inherently suitable for hazardous area applications. An accuracy of up to 0.25% can be achieved
with careful design and favorable application conditions.

The pneumatic load cell has operating principles
similar to those of the hydraulic load cell, in that the
load is applied to one side of a piston or diaphragm of
a flexible material and balanced by the pneumatic
pressure on the other side. This counteracting pressure
is a measure of the load applied and may be displayed
on a dial or translated into an electrical signal by the
use of an electrical pressure transducer.

6. Gyroscopic Load Cell

This load cell utilizes the force sensitive property of a
gyroscope mounted in a gimbal system. Its rotor,
mounted in the inner gimbal, is rotated at constant

speed. The inner gimbal is mounted in the outer gimbal, which is suspended between two swivel joints. The load to be measured is applied in tension to the device through the swivel joints. This force causes the outer gimbal to precess about its vertical axis. The rate of this precession is a measure of the applied load, and can be measured by a disk having a large number of lines mounted on the outer frame: these lines are counted, or the time taken for each revolution of the outer frame is monitored by a sensor. The gyroscopic load cell can accommodate loads up to 25 kg and can have an accuracy of 10 ppm.

7. Vibrating Wire Transducer

This device exploits the principle that the resonant frequency of a vibrating wire changes when a tensile force is applied to it. A typical device consists of a pretensioned ferromagnetic wire placed in a permament magnetic field. The wire is driven with an alternating current from an oscillator to set up transverse vibrations. These vibrations are picked up by a detector coil magnetically coupled to the wire and its output signal is fed to the driving oscillator to form a self-oscillating system to maintain the vibrations of the wire at its resonant frequency. Each resonant frequency is a measure of the corresponding applied tensile load.

The relationship between the applied load and the resonant frequency is nonlinear, typically 0.5–1% of span. However, the use of two vibrating wires preloaded by a common reference mass can produce a linear relationship. The load is applied in such a way that the tension in one wire increases while the tension in the other decreases. The ratio of the two resonant frequencies is directly proportional to the applied force. A well-designed vibrating wire transducer incorporates some means of inherent temperature compensation. The load-bearing capacity of these devices is normally limited to 15–20 kg.

8. Piezoelectric Transducer

A piezoelectric crystal is a type of material which generates a potential difference between selected surfaces when subjected to a force. The magnitude and polarity of the induced surface charges are measures of the magnitude and the direction of the force applied.

The most commonly used material is natural quartz crystal, although other piezoelectric materials such as tourmaline, Rochelle salt and lithium sulfate are also used for various specific applications. However, a number of force transducers employ synthetic piezoelectric crystals such as lead zirconate titanate. A typical piezoelectric transducer, denoted as an active transducer, consists of a quartz disk cut along the axis to exploit the longitudinal piezo effect. It produces an output of $-4 \, \text{pC N}^{-1}$ and has a nonlinearity of ~1% of the span. The output of these transducers is monitored by very-high-input impedance amplifiers, such as electrometer amplifiers or charge amplifiers.

The use of piezoelectric transducers is generally limited to dynamic applications, such as vibrational analysis and measuring transient forces in machining operations.

9. Elastic Transducers

Many force transducers employ some form of elastic load-bearing element. Application of load to the elastic element causes it to deflect and this deflection is then sensed by another secondary transducer which converts it into an output. This may be in the form of an electrical signal in linear variable differential transformers (LVDTs) and strain gauge load cells or a mechanical indication as in the case of proving rings and spring balances.

The LVDT is a variable inductance displacement transducer. It is essentially a transformer which provides an ac output voltage as a function of the displacement of its movable magnetic core. The primary winding is energized with an ac reference signal and the two secondary windings, symmetrically spaced from the primary, are connected 180° out of phase and used as the output coils. The amplitude of the output signal depends on the coupling between the primary and secondary windings, which in turn depends on the position of the core. There is also a phase shift of 180° in the output signal as the core goes through the null position.

LVDT devices are used to sense the deflection of elastic load-bearing members such as proving rings, diaphragms and columns. This combination is commercially available as LVDT load cells with a typical nonlinearity of 0.05–0.1% of the span. The measurement chain employing these load cells may have to incorporate a phase sensor to distinguish the tension and compression loads.

A strain gauge load cell basically consists of an elastic load-bearing member, usually referred to as the stressed member or billet, on which a number of resistance strain gauges are bonded. This elastic member bears the load to be measured and produces a strain field whose amplitude is related to the load through its modulus of elasticity and its geometrical shape.

The strain gauge makes use of the principle that certain conductive materials change their electrical resistance linearly as a function of their dimensional deformation under mechanical load. The gauge factor k, constant for a given material, is the ratio of the change of electrical resistance R to the applied strain ε:

$$k = \frac{\Delta R/R}{\Delta l/l} = \frac{\Delta R/R}{\varepsilon} \qquad (2)$$

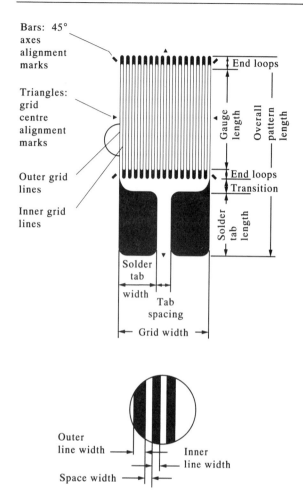

Figure 3
Details of metal foil strain gauge

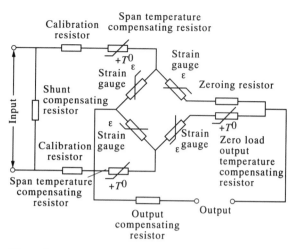

Figure 4
Circuit diagram of a typical industrial load cell (after Erdem 1982 © Institute of Physics, London. Reproduced with permission)

The material and shape of the load-bearing member depends on a number of factors, such as the range of load to be measured, dimensional limits, technical specifications and production costs. It is usual to select aluminum as material for the units measuring loads below 500 kg and steel or stainless steel above that. The shape selected is normally a double or single bending beam or ring for ranges up to 2000 kg, columns or shear beams for ranges up to 50 t, and single or multiple columns for ranges above 50 t. Figure 5 illustrates the cross-sectional view of an industrial load cell with a column load-bearing member.

The most common materials used for the manufacture of metallic strain gauges are copper–nickel and nickel–chromium alloys, which have a gauge factor of ~2. The most frequently used strain gauge types are wire, unbonded, semiconductor, thin film and metal foil, which is the most widely used and available type (see Fig. 3). It consists of metal foil mounted on a backing or carrier, photoetched to a selected pattern.

The strain gauges bonded to a load-bearing member of a load cell convert the strain produced to a resistance change. This resistance change is detected by connecting these strain gauges into a Wheatstone bridge configuration, exciting it with a reference voltage and then measuring the millivolt signal produced across the bridge. The circuit diagram of an industrial load cell given in Fig. 4 shows several correction and compensation components to realize the full capability of a strain gauge bridge and achieve the required specification.

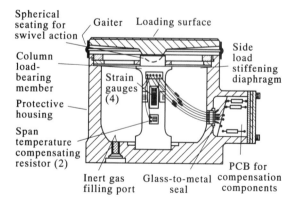

Figure 5
Hermetically sealed industrial load cell with a column type load bearing member (© Negretti Automation, Aylesbury. Reproduced with permission)

Figure 6
A selection of industrial load cells with their load application attachments (© Negretti Automation, Aylesbury. Reproduced with permission)

A typical precision industrial strain gauge load cell, when fully compensated and calibrated, has a non-linearity of ±0.02%, a hysteresis of 0.01% and a repeatability of 0.01%. Zero and span temperature coefficients would be ±5 ppm K^{-1}, in the typical operating range of −40 °C to +80 °C, and deflection will be 0.1–1 mm, depending on the design and range.

Figure 6 shows a selection of industrial strain gauge load cells covering a measurement range of 250 kg to 50 t.

10. Load Cell Systems

In most weighing applications, it is usual to employ multiples of load cells connected either in series or parallel to increase the load measuring range, and to support the structure being weighed. Such combinations can avoid the inaccuracy that would occur in a single load cell system when the center of gravity of the measured load shifts its position, for example, when weighing powder and granules. When the product being weighed is liquid and the accuracy of the system is not critical, then it is possible to use a combination of dummy load cells or pivots and active load cells.

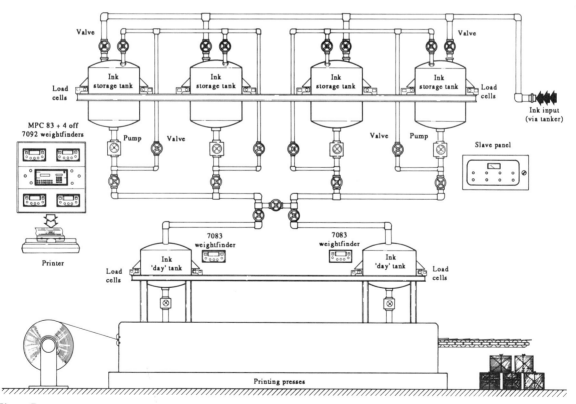

Figure 7
An industrial weighing system for the ink storage and distribution in a newspaper press (© Negretti Automation, Aylesbury. Reproduced with permission)

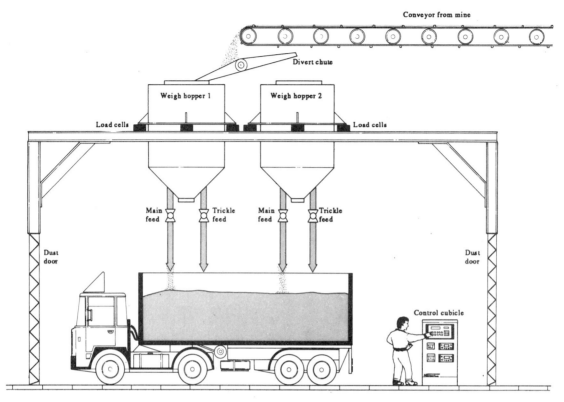

Figure 8
An industrial weighing system for batch discharge and replenishment in a mine (© Negretti Automation, Aylesbury. Reproduced with permission)

The load cells are normally connected in parallel and excited from the same source. Although series connection offers a higher output voltage for a given load, it is more complex in installation, requiring a separate excitation voltage source for each load cell.

Load cells are part of a weighing chain and their combined output signal needs to be amplified and conditioned before it can be meaningfully displayed or used in a control system. There are basically two methods of establishing this chain. The ac system excites the load cells with a periodically varying input which may be in the form of a sine or square wave, and the output is processed through an ac amplifier, synchronous demodulator, filter and dc amplifier. The dc system excites the load cell with a constant dc voltage and amplifies its output through an instrumentation amplifier.

In many industrial weighing applications, the distance between the load cells and the instrumentation may be considerable, causing a voltage drop along the cable. This voltage drop and its temperature dependance can contribute significantly to the total system error. This effect can be eliminated by the use of a six-wire excitation system. This technique involves sensing the excitation voltage at the transducer and of the cable, and maintaining it at a constant level by automatically adjusting the excitation voltage at the source.

A typical industrial weighing system, shown in Fig. 7, illustrates how the control of storage and distribution of printing ink is achieved in a newspaper plant. Such a system would be capable of controlling the supply of inks to the presses, alarms and automatic stop sequences as well as preventing spillages and providing management information on the instantaneous stock levels and ink usage rates.

Figure 8 shows another industrial weighing application where continuously mined raw material is transported by a conveyor and stored in either of the two weighed hoppers, then batch discharged into trucks. The replenishment of the hoppers and the discharged quantities are controlled by signals from the load cells.

When the weighing is to be carried out in an area classified as "hazardous area," for example, certain parts of petrochemical complexes where explosive gases may exist and may be ignited by a spark, then all electrical equipment installed in this area may need to be certified to appropriate safety standards.

See also: Force and Dimensions: Tactile Sensors; Transducers, Magnetostrictive; Transducers, Piezoelectric; Transducers, Strain Gauges

Bibliography

Bass H G 1984 *Intrinsic Safety Instrumentation for Flammable Atmospheres*. Quartermaine House, Sunbury, UK

Beckwith T G, Buck N L 1969 *Mechanical Measurements*. 2nd edn. Addison-Wesley, Reading, MA

Bolton W C 1980 *Engineering Instrumentation and Control*. Butterworth, London

Colijn H 1975 *Weighing and Proportioning of Bulk Solids*. Trans Tech, Bay Village, OH

Considine D M (ed.) 1971 *Encyclopedia of Instrumentation and Control*. McGraw-Hill, New York

Erdem U 1982 Force and weight measurement. *J. Phys. E* **15**, 857–72

Neubert H K 1975 *Instrument Transducers: An Introduction to Their Performance and Design*, 2nd edn. Clarendon, Oxford

Norden E 1984 *Electronic Weighing in Industrial Processes*. Granada, London

Norton H N 1982 *Sensor Analyser Handbook*. Prentice-Hall, Englewood Cliffs, NJ

Institute of Measurement and Control 1993 A procedure for the specification, calibration and testing of strain gauge load cells for industrial process weighing and force measurement, WGL 9301. IMC, London

Window A L, Hollister G S 1982 *Strain Gauge Technology*. Applied Science, London

U. Erdem
[Negretti Automation, Aylesbury, UK]

Measurement and Instrumentation: History

Measurement and instrumentation have always been part of human life, providing the means to gather knowledge about the physical world so that man can better understand and control his existence.

Instrumentation has steadily been improved through the adoption of scientific principles of engineering processes, applied in innovative ways. Availability of instruments has been an important factor both in scientific discovery and in bringing new and improved applications of science and technology into existence. The history of measurement is entwined with that of science and engineering (Ellis 1973, Klein 1975, Sydenham 1979).

1. Ancient Times to the Dark Ages (circa 30 000 BC to 1000 AD)

Earliest man, circa 30 000 BC, probably made crude measurements in response to daily needs. The earliest traceable records include a device used for weighing (limestone beam, Naqada, Egypt, about 4500 BC), sets of weights (that may predate that period) and a water clock (Karnak Temple, Egypt, 1400 BC). Contemporary writings often discuss simple instruments.

The first units of measurement were based on available physical objects, such as 5 handbreadths for the cubit (a unit of length), and the palm seed for the carat (a unit of mass). Unfair measurement practices and the need for more precise trade required a centrally defined standard artifact to be adopted for each unit as the basis against which working units were compared; this led to legal metrology. Legal standards of length, mass and volume exist from Babylonian times (circa 2000 BC).

Measurements were also made to assist astronomy, being motivated by need for knowledge about the seasons and by the requirements of religious practices. For example, emergent theoretical geometry was applied to measure the diameter of the Earth. Eratosthenes, circa 250 BC, used a deep water well and a small calibrated bowl to monitor angles cast by shadows of the sun (Sydenham 1979).

The first instruments were based on the application of intuitively obvious principles, and many used elementary mechanics. Greek and Roman writings provide statements of the basics of mechanical levers used to form weighing balances. Some instruments involved elementary optical, thermal or hydraulic concepts in their operation. Except in medical applications, electrical energy was not exploited for measurement purposes.

Measurement tools and practices in the ancient world have been extensively reported; Sydenham (1979) gives an introduction. Our understanding of the attitudes of these early instrument builders is, however, supported by limited primary evidence that is often incomplete, has been translated many times and rarely covers an event in detail. A small range of artifacts exists in museums. The technological history of this period is far from adequately known and, unfortunately, some historians have reconstructed quite fanciful instruments from the known record.

A small, fragmented, part of the knowledge developed in the ancient Greco–Roman and Hellenistic world was passed on through the Islamic, Eastern civilization, where significant scientific improvements were made to timekeeping mechanisms (evidence suggests that the Chinese had similar scientific ability at that time) and to weighing instrumentation, possibly promoted by the then developing interest in alchemy. The weighing balance (see Fig. 1) of Al-Khazini (1121 AD), for example, incorporated advanced concepts of air buoyancy correction (Harrow and Wilson 1976).

Studies on Islamic science (Harrow and Wilson 1976), on the technology of the ancient world (Singer et al., 1954) and on early China (Needham 1954) collectively suggest that more research into early global measurement science and technology is needed.

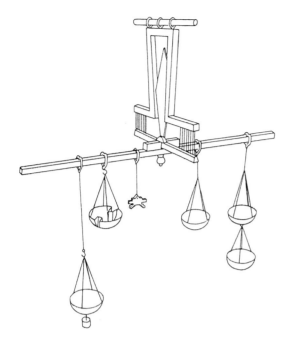

Figure 1
Balance of Al Khazini, 1121

The first 10 000 years of the history of settled man did not provide any great advances in measurement, but it did yield many instrument forms still in use—weighing balances, volume measures, simple surveying devices, geared mechanisms, the magnetic compass, the plumb bob, the straight edge, and water-level and escapement clocks.

2. Renaissance, Middle Ages and the Birth of the Scientific Method (1000 AD to 1800 AD)

As intellectual capability reawakened in the European world, measurement and instrumentation again began to progress, but now at a much faster rate. Weights and measures apparatus—length rods, volume measures, weighing balances and time-keeping instruments–again became part of the people's lives. Legal metrology developed in many countries, especially after such events as the declaration of the Magna Carta in 1215, which specifically mentioned the need for trading standards of measurement.

A new type of instrument capability occurred during this period. Instrument forms emerged that enhanced man's natural senses, enabling him to appreciate previously unobserved variables of his world. The microscope and the telescope, for example, both came into extensive use early in the seventeenth century, following gradual development of the underlying science and experimentation throughout the previous

century. Microscopic and far-distant objects could be studied, extending the known world and furthering scientific pursuits into the nature of life and the universe.

"New" science could not have advanced without the support of ever-improving measurement capability, for this provided the means to prove the many hypotheses generated by it. The importance of measuring instruments to the progress of science and technology was, however, generally not given due credit—as is still the case today. Often a new instrument clearly paved the way for new scientific achievement. Conversely, scientific pursuits often set the scene for instrument development, both in terms of making available new principles for design and in creating needs for improved and different measurements.

The sophistication of mechanical instruments continued to progress closely in hand with scientific understanding. For example, the use of materials in structures needed knowledge of material properties. Da Vinci conceived a tensile testing apparatus around 1500, but that of Musschenbroek (circa 1720), was far more sophisticated, closely resembling machines used today (see Fig. 2).

Fine mechanics allows sensitive magnetic compasses and dip needles to be made, aiding global navigation. Scale-making skills and dividing engines provided more precision in surveying and astronomical instruments. Clockwork advanced to provide chronometers that could keep time accurately to within seconds over a period of months by incorporation of compensation; this level of precision was needed for navigation between continents.

Optical theory and its application to instrumental science, industry and commerce became established in this period. Single-element optical telescopes and microscopes were improved by the use of multiple lenses (and mirrors) which were manufactured from special optical materials to give higher-gain systems with smaller limiting, aberrations. Optical instruments were added to other apparatus as sights and magnifiers in sextants and dividing engines and, later, as parts of new instruments that made use of the dispersion of radiation.

Extensive combination of mechanics, optics, magnetics and electrics (in the eighteenth century) took place to form quite complex instruments that needed specialist manufacturers to craft the apparatus devised by scientists. Supported by wealthy patrons, who also desired esteem functions be built into their apparatus, an instrument industry flourished.

Science developed in close connection with instrument capability. Thermometry, for example, became established with agreed (but numerous) scales, its apparatus being widely similar. Its underlying science (in which heat is regarded as a mode of energy, not a fluid), however, did not come about until the nineteenth century; this exemplifies the development of instruments without adequate scientific understand-

Figure 2
Musschenbroek's tensile testing apparatus of circa 1720

ing. In science and industry, instruments are the vital "eyes" and "ears" for observing processes.

In 1600, Gilbert wrote of both magnetic and electric forces, and from those times the science of each was gradually exposed. (The scientific coupling of electricity to magnetism was, however, only enunciated in the nineteenth century.) Fine mechanics featured prolifically in the instruments used for this work, because many experimental studies converted unobservable physical effects into observable mechanical deflections.

The use of electricity was delayed because the first observations and experiences were of the very-high-voltage, easily discharged, electrostatic form. Early attention to electrostatics may seem to have been an inefficient diversion that slowed electrical discovery. It was, however, laying the foundations for understanding atomic structure, electrons, and the basics of electronic methods that allowed instrumentation to blossom in the twentieth century.

3. The Nineteenth Century – The Emergence of Electrics

In the late eighteenth century interest in "animal electricity," thought to reside within physiological beings, led—by chance encounter, often the basis of discovery (Batten 1968)—to Galvani observing, in 1780, that electricity caused twitching in dissected frogs' legs. His work, and that of others, was soon applied, with Volta creating the first primary electrochemical cell in 1800.

This simple source of electricity was the key to developments in electrical science and instrumentation, for it could be manufactured with ease by any experimenter and possessed electrical properties more suited to research and application. Wide availability of the "battery" allowed the properties of "coil and current" electricity (Frith and Rawson 1896) to be discovered, yielding seminal knowledge such as Ohm's law of direct current circuits in 1826 and the basic laws of alternating current circuits by circa 1850.

The key instrument of this period was the sensitive galvanometer, based on the electromagnetic motor effect. Oersted's work in 1819 (on magnetic fields of a current-carrying wire causing a magnetized needle to deflect) started a progression of electrical indicating instruments that enabled increasingly smaller electric effects to be observed as they were continually improved.

Science thus obtained greatly improved ability to sense small physical effects: Boys detected a candle flame at a distance of 5 km in 1889; Callendar, in 1900, used a galvanometer coupled to a thermopile to detect the radiation of the moon. Radiation laws were revealed which, in turn, allowed more instruments to be invented.

The sensitivity of galvanometers was improved to the point where the molecular vibration of material—Brownian motion—became the limiting fundamental effect. Galvanometers also assisted development of intercontinental telegraphy and paved the way towards automated processes, such as temperature control in furnaces based on a galvanometer incorporated as part of a potentiometric recorder.

The derivative nature of instrumentation was, again, clearly apparent. Flowmeters, medical diagnosis bullet detectors, and crack detectors, for example, made use of telegraph and telephone components. The materials and principles used in instruments were almost always first developed for nonmeasurement applications; a notable exception was Invar, a temperature-independent material created for making improved long-length surveying tapes.

Around 1800 the French government considered adoption of a consistent, decimal-based, "metric" system of measurement units (Klein 1975). A system was implemented in Holland in 1809 and in France in 1837. The metric system was gradually adopted, but in many forms. It was not until the twentieth century, however, that a more truly unified (but still not fully adopted) agreement—the Système International Des Unités (SI)—was put into place.

By 1900 the existence of many laboratories for national measurement standards activity (Cochrane 1966), physical and life science research and astronomical observation, and of teaching organizations, collectively provided commercial incentives for instrument manufacturing companies (Cambridge Instrument Company 1955) to emerge. These suppliers subsequently carried scientific laboratory instruments into industrial applications by making the instruments robust and available at an attractive price.

Contributions to the philosophy of measurement can be found in the nineteenth century, but they did not lead to major interest within the scientific community. Helmholtz (1887) analyzed the knowledge of counting and measuring. Lord Kelvin, who was associated with many instrument developments, wrote a chapter on measurements in a physics text but this was only a statement of typical laboratory measurements.

Galton set up anthropometric measurement of the people of London, bringing quantification to the life sciences in 1884. Researchers of education applied scales to measures of education, Binet's 1900 work being well known.

The nineteenth century closed with measurement having a noticeable and vital role in scientific advance, being applied throughout a wide spectrum of living, and with industry beginning to learn of the advantages of improved measurement. Measurement systems were processing gathered data with mechanical computers. Signal levels from sensors needed to be large to find application outside the laboratory. It was also difficult to carry out computations or to process data.

4. The Twentieth Century — Application and Commercialization

Around 1880 Edison discovered, by chance, the principle of the electronic rectifier. He patented the effect in 1883 but did not exploit the idea. This event, along with work with high-voltage generating coils, cathode-ray discharge tubes, and electric indicating meters, formed the birthplace of electronics, a device-driven technology that subsequently had great influence on measurement and instrumentation.

The initial years of electronic development (from the 1910s to the 1940s) were resourced by the needs of communication. The first thermionic device applications (Fleming's diode of 1904 and De Forest's triode of 1907) were aimed at improved radio-wave detection. Electronic knowhow spread rapidly into both business and domestic fields with popularization of radio as a hobby, which made components and methodology freely available.

Application of electronic technique to measurement and instrumentation has its first evidence in the combined use of electronic amplifiers with the photoelectric cell in the late 1920s. Early electronic instrumentation applications included newspaper counting, talking-film sound track, facsimile transmission of radio pictures, and television.

Other instrument applications soon emerged as supporting items in larger systems. Telemetry of railroad signal lights, to convey signal information into the driver's cabin, was in operation in 1924. Laboratory instruments that were developed included ultraviolet dosage measurement, microphotometers, color matchers, the vacuum-tube voltmeter, and the cathode-ray oscilloscope, which was in use as a calibrated instrument by 1930.

Electronic methodology provided means to gather information using many sensor techniques that could now be operated without human presence. In the 1940s highly developed, but limited, mechanical methods of data storage and processing gave way to electronic counterparts. By 1955, valve-based equipment was giving way to solid-state transistor systems.

Integration of electronic circuitry led to today's use of purpose-made chip sets that are tailored to the task in hand using software tools. All of these have become part of instrumentation.

Until the 1940s, fine mechanical skills had allowed computing machines and recorders to progress (Eames and Eames 1973). Clockwork reached its pinnacle around 1970. Optical techniques were revitalized by combination with electronic sensing and processing and with the development of the laser radiation source; this allowed coherent radiation methods, such as holography, to be exploited in measurement. The need for process control led to many robust sensors that initially fed chart recorders and, later, control loops. Military needs created many new instruments—from gun rangers, periscopes, bombsights, navigational equipment, radar and gyroscopes through to laser guidance systems and missile sensors. An account of the scientific instrument development environment over this period has been given by Jones (1988).

Measurement and instrumentation is just beginning to be recognized as part of information technology. Ironically, the designer of an instrument system will often need to use knowledge and technology of all the components of information technology, and yet instrument technology is still not recognized as being as scholarly or intellectually demanding as its component parts.

The sophistication and complexity of measurement systems has risen to the level where collection of data from many sensors has required the concept of distributed data networks. Elements of intelligence are being included to make sensors more effective and self-sufficient. Instrumentation of the automotive industry has provided another collection of componentry that measurement can usefully adopt. New sensing principles continue to be discovered. Applications continue to multiply. The history of measurement and instrumentation reveals an ever-quickening pace of growth of capability, and need for it in every sphere of life. Measuring instruments can be expected to continue being tools of knowledge and control.

Bibliography

Batten M 1968 *Discovery by Chance—Science and the Unexpected.* Funk and Wagnells, New York
Cambridge Instrument Company 1955 75 years. Internal publication, Cambridge Instrument Company, Cambridge, UK
Cochrane R C 1966 *Measures for Progress—History of the National Bureau of Standards.* National Institute for Science and Technology (formerly National Bureau of Standards), Washington, DC.
Eames C, Eames R 1973 *A Computer Perspective.* Harvard University Press, Cambridge, MA
Ellis K 1973 *Man and Measurement.* Priory Press, London
Frith H, Rawson W S 1896 *Coil and Current, or the Triumphs of Electricity.* Ward Lock, London
Harrow L, Wilson P L 1976 *Science and Technology in Islam.* Crescent Moon Press, London
Helmholtz H V 1887 Zahlen und Messen erkenntnis—theoretisch betrachet. Philosophische Aufsaetze Eduard Zeller gewidmet, Leipzig. English translation by Bryan C L. 1930. *Counting and Measuring.* Van Nostrand, New York
Jones, R V 1988 *Instruments and Experiences.* Wiley, Chichester, UK
Klein H A 1975 *The World of Measurements.* Allen and Unwin, London
Needham J 1954 *Science and Civilisation in China.* Cambridge University Press, Cambridge
Singer, C, Holmyard E J, Hall A R 1954 *A History of Technology.* Oxford University Press, Oxford
Sydenham P H 1979 *Measuring Instruments: Tools of Knowledge and Control.* Peregrinus, Stevenage, UK

P. H. Sydenham
[University of South Australia,
Ingle Farm, Australia]

Measurement: Fundamental Principles

Measurement is the process of empirical, objective assignment of numbers to the attributes of objects and events of the real world, in such a way as to describe them.

This article gives a brief outline of the foundational concepts of measurement and their philosophical background, explaining the logical basis of the measurement process. It will, as far as possible, attempt to be simple in its presentation of concepts, while giving the essential aspects of formal measurement theory.

1. Definition of Measurement

Before embarking on a formal presentation of the logical basis of measurement it is useful to discuss its definition.

Measurement is the assignment of numbers to properties of objects and events, and is thus the description of properties of objects or events, and not of the objects or events. Measurement is based on a clear concept of a property, as an abstract aspect of a whole class of objects, of which individual instances or manifestations are the subject of measurement.

The definition states that the assignment of numbers in measurement is such that the numbers describe the property of the object or event. The meaning of this can be explained as follows. Consider that a number, or measure, is assigned, by measurement, to the property of an object, and other numbers are assigned by the same process to other manifestations of the property. Then the numerical relations between the numbers or measures imply and are implied by empirical relations between the property manifestations.

It follows that measurement is a process of comparison of a manifestation of a property with other manifestations of the same property, a common feature of informal definitions of measurement.

The next aspect of the definition of measurement that requires discussion is the fact that measurement is an objective process. By this is meant that the numbers assigned to a property by measurement must, within the limits of error, be independent of the observer.

This definition of measurement stresses the fact that measurement is an empirical process. This means that it must be the result of observation and not, for example, of a thought experiment. Further, the concept of the property measured must be based on empirical relations.

There is a divergence of views as to whether any descriptive assignment of numbers is adequate for the process to qualify as measurement. At one extreme, broadly in the social and behavioral sciences, there is the view that any empirical, objective assignment of numbers which describes a property manifestation, can be termed measurement. At the other extreme is the view that only numbers which reflect in some way a ratio to a unit magnitude of a property are true measures. This is the classical view and that of most informal definitions of measurement in physics. Many other definitions imply that, to be true measurement, the assignment of numbers must imply at least an empirical order among the property manifestations, corresponding to a concept of ordering according to magnitude. This article assumes the first view, without denying the validity of the others.

The significance of measurement in science arises from its objectivity, the conciseness and, usually, objectivity of the descriptions it provides, and the fact that they can be reasoned in formal systems such as mathematics.

2. Formal Theory of Measurement

The concept of measurement presented in Sect. 1 will now be presented formally using the representational or model theory approach.

A representational theory of measurement has four parts:

(a) an empirical relational system corresponding to a quality;

(b) a number relational system;

(c) a representation condition; and

(d) a uniqueness condition.

These will now be considered.

2.1 Quality as an Empirical Relational System

Consider some property, and let q_i represent an individual manifestation of the property, so that a set of all possible manifestations can be defined as:

$$Q = \{q_1 \ldots\}$$

Let there be on Q a family R of empirical relations $\mathbf{R}_i$

$$R = \{\mathbf{R}_1, \ldots, \mathbf{R}_n\}$$

Then the property is represented by an empirical relational system Q, termed a quality:

$$Q = \langle Q, R \rangle$$

2.2 Numerical Relational System

Let N represent a class of numbers

$$N = \{n_1 \ldots\}$$

Let there be on N a family P of relations

$$P = \{P_1, \ldots, P_n\}$$

Then

$$N = \langle N, P \rangle$$

represents a numerical relational system. Commonly, N is the real number line.

2.3 Representation Condition

The representation condition requires that measurement be the establishment of a correspondence between quality manifestations and numbers, in such a way that the relations between the referent property manifestations imply and are implied by the relations between their images in the number set. Formally, measurement is defined as an objective empirical operation M

$$M: Q \rightarrow N, \text{ so that } z_i = M(q_i)$$

such that $Q = \langle Q, R \rangle$ is mapped homomorphically into (onto) $N = \langle N, P \rangle$. This homomorphism is the representation condition.

Measurement is a homomorphism because M is not one-to-one; it maps separate but indistinguishable property manifestations to the same number.

$$S = \langle Q, Z, M \rangle$$

constitutes a scale of measurement for Q.

$n_j = M(q_j)$, the image of q_j in N under M, is called the measure of q_j on scale S.

2.4 Uniqueness Condition

The representation condition may be valid for more than one mapping M. One may admit certain transformations from one scale of a property to another, without invalidating the representation conditions. The uniqueness condition defines the class of scale transformations to those for which the representation condition is valid. This is discussed further in Sect. 3.

3. Examples of Scales of Measurement and Their Formation

3.1 Extensive Measurement

Extensive measurement is the basis of physical measurement and will be considered first.

The extensive scales of physical measurement are based on establishing for the quality Q, for which a scale is to be determined, of an empirical ordering with respect to Q for all objects possessing elements of Q, together with an operation of combining the objects, elements of Q, which has, with respect to Q, the formal properties of addition, such as commutativity, associativity and the like.

With an empirical ordering operation and an additive combination thus established, one proceeds to the setting up of a scale. A single object is chosen as a standard and its quality manifestation is assigned the number 1. One then constructs, or seeks, another object with quality Q, equivalent to the first with respect to Q. The additive combination of the first and second object is then assigned the number 2. By repeating such combinations, one can generate an infinite series of standards. Fractional standards are generated by the following process. By constructing or finding two empirically equivalent standards, the additive combination of which is empirically equivalent to the standard assigned, we have a standard which can be assigned the number $\frac{1}{2}$. Measurement of a quality manifestation then consists of comparing an unknown manifestation to the series of standards, and assigning the number of the standard to which it is empirically equivalent.

As an example, in the measurement of mass, the equipoise balance offers the means of establishing empirical order. If the arm balances, the masses in the pans are equivalent. The tipping of the balance indicates that one mass is heavier than the other. Thus, a series of weights can be ranked in order of heaviness. The lumping together of two objects is, with respect to mass, an operation which has the properties of addition.

3.2 Matching Scale

A matching scale for a quality Q is based on the establishment of an empirical indifference relation on the set of quality manifestations Q having the properties of reflexivity, symmetry and transitivity.

A set of differing objects possessing the quality Q are selected to form a standard set. Numbers (or other symbols) are then assigned to each element of the standard set, the same number not being assigned to two differing elements. The fundamental measurement operation M of the scale consists of an empirical operation, in which measurands are compared with members of the standard. A measurand is assigned the number of the standard to which it is equivalent.

An example of this form of scale is a color code in which the relation 'matches' constitute the empirical indifference relation. In the establishment of a color code, the indistinguishability relation is established, based on a color match. The relation is an equivalence relation. A set of colored objects are selected as standards, each a distinct color manifestation. Each is assigned a different number or other symbol, such as a word label. Any unknown color manifestation is compared with the standards and if it matches one of them it is assigned the same number or symbol as the standard.

Matching scales are not generally considered measurement scales, since they are not quantitative in the sense that they do not establish, on Q, relations formally similar to the relations greater or less. Also, it is not generally practical to establish sufficient standard elements to ensure that every measurand can be matched with a standard and, hence, assigned a measure.

3.3 Ranking Scales

In ranking scales for a quality, an empirical order system is established on Q. A set of differing standard objects possessing the quality is then selected and arranged in an ordered standard series, according to the empirical order. Numerals are assigned to each standard, in such a way that the order of numerals corresponds to the order of standards. Any measurand can then be compared with the elements of the standard series in the same way as in matching scale measurement. If a measurand is equivalent to a standard it is assigned the numeral of the standard. If a measurand is not equivalent to any standard, one can determine between which two standard elements it lies in the empirical order system.

The best example of a ranking scale of measurement is the Mohs scale of hardness of minerals. Ten standard minerals are arranged in an ordered sequence, so that precedent ones in the sequence can be scratched by succeeding ones and cannot scratch them. The standards are assigned numbers one to ten. (The sequence is talc, 1; gypsum, 2; calcite, 3; fluorite, 4; apatite, 5; orthoclase, 6; quartz, 7; topaz, 8; corundum, 9; diamond, 10.) A mineral sample of unknown hardness which cannot be scratched by quartz and cannot scratch it, is assigned measure 7.

4. Formation of Scales Based on Models

Section 3 considered measurement scales formed by direct mapping from a quality relational system to a numerical relational system. Frequently, however, scales of measurement for qualities are constructed indirectly through a relation of the quality to be measured and other qualities, for which measurement scales have been defined. The reason may be, for example, the impossibility of setting up a satisfactory measurement scale directly. Thus, for example, an extensive measurement scale for viscosity or density cannot be set up, since there is no appropriate combination operation having the properties of addition. Another reason is the wish to set up a consistent set of measurement scales, in which a minimal set of qualities with direct scales is defined, and scales for other qualities are defined by derivation from them.

In its simplest form, consider a case when every object that manifests the quality to be measured exhibits a set of other qualities which are measurable. Then to each manifestation of the measurand quality q_i, there corresponds a vector of measures of the associated measurable quantities:

$$\mathbf{N}_i = [n_{i1}, n_{i2}, \ldots, n_{in}]$$

If these vectors can then be mapped by a function M() into the real number line so as to assign to $\mathbf{N}_i$ a number $M(\mathbf{N}_i)$, and if the numbers so assigned to the quality manifestations by this process, imply and are implied by empirical relations between the quality manifestations, then this sets up an indirect scale of measurement.

Consider, as an example, the scale of density measurement of homogeneous bodies. Each such body possesses mass m, and volume v (where m and v are assumed to be measures on previously defined scales). It is an empirically established law that objects of the same material, and hence, conceptually of the same density, have the same ratio (m/v). When different materials are ordered according to our concept of density they are also ordered according to the respective ratio m/v. Hence, a scale of density measurement is based on the ratio of mass to volume.

A few observations should be made here. The mapping M is not unique in its order of preserving properties with respect to density. For instance $(m/v)^2$ would be an equally valid derived measure of density. The form is chosen to result in the greatest simplicity of mathematical relations involving density. The properties of the function M, given earlier, are an idealization of real observations.

In general, once an indirect scale of measurement has been defined it is treated as a definition of Q.

The description of qualities by multidimensional arrays of measures, where the measures are not combined, is known as multidimensional measurement. An example might be a characterization of a shape by a set of geometrical measures.

Measurement by combining component measures, so that the resultant number characterizes the place of the measure and quality manifestation in an empirical order is termed conjoint measurement. It is based on the conjoint establishment of order on the measurand quality and on the component qualities.

In physical measurement, indirect measures of qualities are obtained as multiplicative monomial functions of the measures of component qualities. This is generally known as derived measurement (Krantz *et al.* 1971).

5. Uniqueness: Scale Types and Meaningfulness

The requirement that the fundamental measurement procedure of a scale should map the empirical relational system Q homomorphically into the numerical relational system N does not determine the mapping uniquely.

There is an element of arbitrary choice in the setting up of scales of measurement. For example, in the case of scales based on additive combination, the choice of the unit standard is arbitrary.

The requirement of homomorphism thus defines a class of scales which may be called equivalent. The class of transformations which transform one member of a class of equivalent scales into another is called the class of admissible transformations. The conditions which admissible transformations must satisfy are known as the uniqueness conditions. They specify that a scale is unique up to a specific class of transformations.

Scales can be classified by the classes of transformations admissible for them. Let n be numbers representing measures on a scale, and let $n' = F(n)$ be the corresponding numbers on the transformed scale. The generally accepted classification of scales is given in Table 1.

The problem of the meaningfulness of statements made about a quality, in terms of its measures, is important. Such a statement is meaningful if its truth of such a statement is unchanged by admissible transformations of the scales of measurement.

6. Measurement and Other Forms of Symbolic Representation

Measurement is only one form of representation of entities by symbols and it is closely related to other

Table 1
Classification of scales

Scale type	Admissible transformation $F(\ )$
Nominal	$n' = F(n)$, where $F(\ )$ is any 1-1 transformation
Ordinal	$n' = F(n)$, where $F(\ )$ is any monotonic increasing function
Interval	$n' = an + b$, where $a > 0$
Ratio	$n' = an$, where $a > 0$

forms of symbolization. This is discussed in the article *Measurement Science*, which also considers how this underlies the concept of information and information machines.

See also: Measurement Science; Measurement: System of Scales and Units

Bibliography

Finkelstein L 1982 Theory and philosophy of measurement. In: Sydenham P H (ed.) *Handbook of Measurement Science*. Wiley, Chichester, UK, pp. 1–30
Krantz D H, Luce R D, Suppes P, Tversky A 1971 *Additive and polynomial representations*, Foundations of Measurement, Vol. 1. Academic Press, New York
Pfanzagl J 1968 *Theory of measurement*. Verlag Chemie, Wurzburg, Germany
Roberts F S 1979 *Measurement Theory*. Addison-Wesley, Reading, MA

L. Finkelstein
[City University, London, UK]

Measurement Science

Measurement science is the science of measurement and instrumentation in the sense of a systematically organized body of knowledge, with general concepts and organized, generic, transferable principles, which underlie the technology of measurement and instrumentation. The science is widely and increasingly recognized as a discipline. It forms the basis of this Concise Encyclopedia. It is also described by other terms, such as instrument science, principles of instruments, measurement and instrumentation science and the like.

The science of measurement and instrumentation is based on the treatment of measurement as an information process, and of instruments as information machines. It is founded on the concepts and principles of the science of information and systems.

1. Measurement as an Information Process and Measuring Instruments as Information Machines

Measurement is the assignment of numbers or other symbols by an objective, empirical process to attributes of objects or events of the real world, in such a way as to describe them. The information may be viewed as what is carried by a symbol about a referent, by virtue of a defined relation the symbol bears to the referent. Measurement may thus be viewed as an information process (see *Measurement: Fundamental Principles*).

Instruments are information machines, which sense a power or material flow from an object under measurement at the input, assign to it a symbol and carry out operations on the symbol, providing at the output either a display symbol to a human operator (a symbol which is processed further by other processes or information machines), or finally effectuate the information by operating actuators or similar machines.

It is convenient to discuss the fundamental principles of instruments and the measurement process in terms of the architecture of a measuring instrument system. Such a system is shown in Fig. 1. The system consists of a number of subsystems. First, there is the system under measurement. This is connected to a sensor system and acts on the sensor by a flow of matter or energy. The sensor converts this flow into a

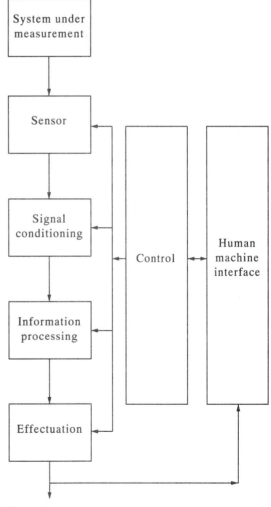

Figure 1
Architecture of an instrument system

signal maintaining a functional relation between the input flow and the information-carrying characteristics of the signal. There is usually a signal conditioning block which converts this signal into a symbol which may be conveniently handled by the following block, which performs any required functions of information transformation and communication. This system passes the information to the effector block for further processing or to the human operator. The measuring instrument system operates under the control of a control block. An important part of the system is the human–machine interface. Through this interface the operator effects supervisory control of the measurement process. The interface also embodies any displays.

In the great majority of modern systems, and to a rapidly increasing extent, once information has been acquired by a sensor and conditioned, it is processed and effectuated by standard computing equipment. Control of the measurement process and the display of information to the operator also takes place through standard human computer interfaces. Thus, much of instrumentation is implemented by modern computer equipment.

As measurement is an information process, and instruments are information machines, realized substantially by standard computer technology, it is argued that systematic principles of measurement science and its instrumentation form part of the wider science and technology of information and knowledge. There are, however, specific aspects and problems in measurement and instrumentation science. In terms of this architecture, the principles underlying measurement and instrumentation can now be reviewed in summary.

2. Symbolic Representation

It is now appropriate to define, more formally, the concepts of the representation of the attributes of objects or events of the real world by symbols.

An object, termed the symbol, may be said to represent or carry information about another object or event termed the referent, by bearing a known relation to it. Information about the referent consists of the symbol, together with the representation relation.

A referent relational system (a referent set of objects or events and a set of relations on them), may be represented by a symbolic relational system (a symbol set of objects or events and a set of relations on them), by a mapping which is a homomorphism or an isomorphism.

Measurement can thus be seen as the assignment, by an objective, empirical process, of symbols, as defined above (most commonly, but not necessarily numbers), to attributes of objects and phenomena of the real world in such a way as to characterize them.

In terms of the above definition of information, measurement is a symbolization, which maps an empirical relational system of a class of attributes of real objects or events into a symbolic system by a mapping which is a homomorphism and which is empirical and objective. It is not essential, though often required, that the symbols be real numbers.

The underlying principles are a key element of measurement and instrumentation science. They are based on the representational theory of measurement and its extensions, which are founded on model theory. Measurement and allied symbolic representation can now be seen as an aspect of the general principles of knowledge representation.

Since measurement is generally recognized as a knowledge acquisition process, it is important to mention here the concept of knowledge. In machine information and knowledge processing, knowledge has a technical meaning. It consists of symbolic formulas in a language (a symbolic representational system) which can be used to describe extralinguistic entities and their relations. Measurement yields such symbolic formulas.

Reasoning is the application, to formulae, of such a language of rules of transformation, such that, if the original sentence corresponds to a true relation in the world described so does the transformed sentence. Measurement systems may be said to perform such transformations.

The above study of the basic concepts of information and measurement analyzes the representation conditions underlying symbolization and the uniqueness of particular schemes. It is of particular significance to the symbolic description of aspects of the world not hitherto describable by formal symbol. The related theory can now be seen as an aspect of the general principles of knowledge representation.

3. Instruments as Information Machines

Having described instruments as information machines, it is appropriate to develop this concept further.

A machine is a contrivance which transforms a physical input into a physical output for a specified purpose. Information machines are machines or systems of machines which have, as their function, the acquisition, processing, outputting and effectuation of information. They operate by transforming input symbols into output symbols by defined transformations. The symbols are attributes of physical variables termed signals.

Information machines or systems of machines comprise computers, communication systems and what is termed instrumentation. The last is defined here as an information machine or system of machines which has, as its function, the acquisition, processing, outputting and effectuation of information from the real world. This function is commonly measurement.

4. Signal and Information Theory

In information machines information is carried by the magnitude or attributes of the time variation of a physical variable termed the signal.

The principles underlying the analysis and design of signals and of signal transformation processes are based on signal theory and the so-called information theory, which are, thus, fundamental aspects of measurement science and technology.

The use of digital computation for information processing has made the discrete representation of signals and signal processing the most significant area, and also the development of the theory and methods of advanced filtering and the like.

For further discussion see the articles *Information Theory in Measurement and Instrumentation* and *Signal Theory in Measurement and Instrumentation*.

5. Information Machines as Systems

Information machines are most effectively viewed using the methods of systems theory and technology. The essence of these methods is first, analysis and synthesis of such machines as systems of simpler blocks (complex and diverse systems are built up from a limited set of simpler components and connection architectures). Second, systems are viewed holistically, and their analysis and synthesis must consider any supersystem in which the information system is embedded, and also the environments in which it exists during its life cycle.

The substantial realization of instrument systems by computer hardware and software has reinforced the use of the systems approach.

6. Instrument Models

Information machines are most usefully described by abstract models. Such models make clear the isomorphisms that exist among systems of different physical nature.

There are a number of levels or perspectives of abstraction and a variety of representational schemes which may be employed. The models may be, at the lowest level of abstraction, of a physical configuration in terms of drawings, full or schematic. At a higher level, they may be power or matter flow models, using, among others, equivalent circuit, bond graph, structure graph or allied schemes of representation. At a further level of abstraction, they may be signal flow models which only consider the information-carrying variable. At the highest level of abstraction, information and knowledge flow models may be used. Graphical models, such as block diagrams and the like, are particularly convenient for human use, but the increasing application of computer tools for analy-

sis and design has advanced the use of symbolic formulae as models.

The effect of advances in the science underlying information technology has been to develop the understanding of the process of abstraction and to develop general methods of abstract modelling, particularly in relation to data and knowledge representation and processing.

The analysis and design of much of the modern instrument system uses models at a high level of abstraction. What must be noted is that the sensor component, the interaction between the sensor and measured object, and the measured object itself, must be modelled in terms of power flows and, ultimately, in terms of physical configuration, approaches that are not greatly considered in much of modern information technology, but which take a most significant place in the science of measurement and instrumentation.

For further discussion see the article *Instruments: Models and Characteristics*.

7. Architectures

There are a limited number of structures in which the elements and subsystems of instrumentation systems are configured, in particular, on simple systems: chain, parallel and feedback structures. The principles underlying the functioning of such structures forms an important part of measurement and instrumentation science. The current trend is towards larger and more complex structures, and the principles of computer architectures are acquiring significance to measurement systems.

For further discussion see the article *Instrument Systems: Functional Architectures*.

8. Errors and Uncertainty

The theory of errors and uncertainty in measurement forms a central part of measurement and instrumentation science, and is its most distinctive component.

The problems of error and uncertainty are in many ways general to all information machines and processes, being the problem of the distortion of signals by transformations, and the estimation of signals in the presence of noise. They are specially important to measurement processes and instruments presenting specific aspects, and demand special treatment in their context.

There is the well-known body of random error theory based on the assumption of Gaussian error distribution and the statistical methods derived from it. Further, however, there are the principles of treatment of systematic errors based on the analysis of instrument models, and the effects of sensitivity to parameter variations and external influences. Following from error theory, there is a systematic body of principles of error avoidance and compensation, based on the theory of invariance, systematically employing methods such as disturbance suppression, disturbance feedforward, feedback and filtering. The latter presents, as already mentioned, a link to signal theory.

For further discussion see the articles *Errors and Uncertainty* and *Errors: Avoidance and Compensation*.

9. Inferential Measurement

It can be observed that measuring systems can only sense active or directly observable quantities, that is, those which characterize the flow of energy and matter. Those quantities which are passive, that is, quantities which characterize storage, transformation or transmission of energy or matter, can only be observed by a process of inferential measurement. This consists of interrogating the system under observation by exciting it and estimating the quantities under observation as parameters of a model of the system. This problem of system identification is specifically a measurement problem, although it is mainly treated from the perspective of control theory, with some distortion of its measurement aspects.

For further discussion see the article *Identification in Measurement and Instrumentation*.

10. Design

The science and technology of measurement and instrumentation should be concerned, at its core, not with the description and analysis of measurement and instrumentation systems, but with their design. There exists an established methodology of engineering design which is at the core of measurement and instrumentation science. The essential aspects of this methodology are well known. It is not possible to review them here, or to discuss the very rapid development of this methodology under the impact of advances in knowledge engineering and lessons from software engineering.

The design of measurement and instrumentation systems is a special application of the general design methodology. Its distinctive feature is that it encompasses approaches used in hardware and software design in information technology as well as approaches to the search for physical realizations of required functions used in mechanical engineering. Systematic approaches to the generation of design concepts will become an ever more important aspect of measurement and instrumentation science.

For further discussion see *Design Principles for Instrument Systems*.

11. Classification

While the science of measurement and instrumentation is based on the assumption that it is not to be a collection of catalogs, a systematic approach to the design of measurement and instrumentation systems demands that knowledge about the working principles and archetypal forms of instrument elements and components must be systematically classified and organized so that it can be conveniently stored and retrieved. The development of appropriate classification and organization principles is a major new trend in measurement and instrumentation science.

12. Measurement and Instrumentation Science and Physics

This article is concerned with the systems and information principles of the measurement process and measuring instrumentation. These principles are, in general, abstract and are not concerned with physical realization.

However, the measurement process is substantially concerned with some aspects where such systems and information perspectives of the principles of measurement and instrumentation are inadequate. They are the system under observation and the sensor-measured system interaction. Further measurement is generally carried out in order to either describe or control physical systems. The treatment of these aspects represents special problems of measurement and instrumentation as a science. It is, in these areas, closely linked to the discipline of physics.

As stated earlier, measurement and instrumentation systems are considered in the science of measurement and instrumentation in terms of abstract mathematical models. Physics is equally covered by a set of formal theories, yielding models of physical systems and phenomena. The compatibility of modelling principles and knowledge representation schemes between the science of measurement and instrumentation and the science of physics is an essential aspect of the former.

13. Intelligent and Knowledge-Based Processing of Measurement Information

The processing of the information acquired by sensors in modern instrumentation is performed by computing subsystems of high information and knowledge processing power, which may be termed intelligent. It is not possible to review here the concepts of intelligent instrumentation, but some remarks should be made about the impact of the advent of such instrumentation on the general principles of measurement and instrumentation.

First, the development of artificial intelligence has reinforced the links between measurement and instrumentation science and the general principles underlying information technology. Second, it has promoted the study of the formalization and organization of the principles underlying information processing. Third, recognizing that instrumentation is commonly designed to emulate animal and, specifically, human senses and brains, it has stimulated research into the links between measurement and instrumentation and cognitive sciences, in particular such problems as sensing, perception, intelligence, reasoning, and the like.

For further discussion see the article *Artificial Intelligence in Measurement and Instrumentation*.

14. Measured Information and the Supersystem

The information and knowledge output of instrumentation is, in general, required, not for its own sake, but for the purposes of some supersystem.

Measurement information may be used in some information system, which interprets the information and reasons about it, for instance, for the purpose of diagnosis. The theory of measurement and instrumentation is thus linked with the general theory of information systems, which is rapidly developing.

Most commonly, measurement information is required for the control of some system. Measurement and instrumentation science shares with control science, a great number of concepts and principles, since they are both concerned with information processes. There is now a developing theory concerned with the measurement information required in large control systems, an important new trend in measurement and instrumentation science.

See also: Artificial Intelligence in Measurement and Instrumentation; Errors and Uncertainty; Identification in Measurement and Instrumentation; Information Theory in Measurement and Instrumentation; Instrumentation in Systems Design and Control; Instruments: Models and Characteristics; Instruments: Performance Characteristics; Measurement: Fundamental Principles; Signal Theory in Measurement and Instrumentation.

Bibliography

Finkelstein L 1977 Instrument science. *J. Phys. E* **10**, 566–72
Finkelstein L 1985 State and advances of general principles of measurement and instrumentation science. In: *Advances in Measurement and Instrumentation Education. Proc. 1984 International Measurement Confederation Conf.* IMEKO, Budapest
Finkelstein L 1990 Measurement and instrumentation as a discipline in the framework of information technology. Knowledge based measurement. VDI Report No. 856. VDI–Verlag, Dusseldorf, Germany, pp. 275–83
Sydenham P H (ed.) 1982–1983 *Handbook of Measurement Science*, Vols. 1–3. Wiley, Chichester, UK

L. Finkelstein
[City University, London, UK]

Measurement: System of Scales and Units

The importance of accurate and reliable measurement has been stressed throughout much of recorded history, from the earliest days of the use of measurement instruments in trade and commerce, and references to the use of "honest scales" and the avoidance of "dishonest standards of length, weight or quantity" go back to early biblical times. In order to ensure that such standards are met, and that definitions of units of measurement are respected (a practice not common throughout much of the history of the subject), there must be a method of defining each of the measurement units used, both nationally and internationally. This provides a standard of measurement which exists to enable a comparison of the unit(s). This has been historically defined by the physical representation of the unit of measurement. Thus, if, for example, the definition of a unit of mass such as the kilogram is required, it may be obtained with respect to an arbitrary material standard such as the mass of a well-defined substance. However, the early definition of the kilogram as the mass of a cubic decimeter of water at $4\,^\circ\text{C}$ (at its maximum density) showed the deficiencies of a standard based on a material such as water, a liquid at the defined temperature, subject to purity problems, evaporation and transport difficulties of the standard, and other factors, which yielded it impractical for an accurate standard.

1. Scales and Units

The material representation of the kilogram was defined by the International Prototype Kilogram, in the mid-nineteenth century, an artifact constructed of a carefully machined cylinder of platinum of high purity, and remains as such today. Thus, the long-term stability of the standard could be more easily ensured than by using a water standard, although the inertness of a platinum cylinder cannot simply be assumed.

The need for national and international standards has led to the establishment of national standards laboratories, initially to regulate standards for trade, but now fulfilling a wider standards function (the economic implications of which cannot be underestimated). Thus, laboratories such as the National Institute of Science and Technology (NIST, formerly the National Bureau of Standards) fulfill that function for the USA and in Europe, the National Physical Laboratory) (NPL) in the UK and the German National Metrology Laboratory (PTB) in Germany are among the major keepers of national and international standards. The international regulation of standards occurs through the dialog between laboratories, and a program of standards intercomparison, with mutual acceptance. The aim of this activity is the characterization of standard measurement parameters with the lowest acceptable level of uncertainty in their measurement, and the wide dissemination of such

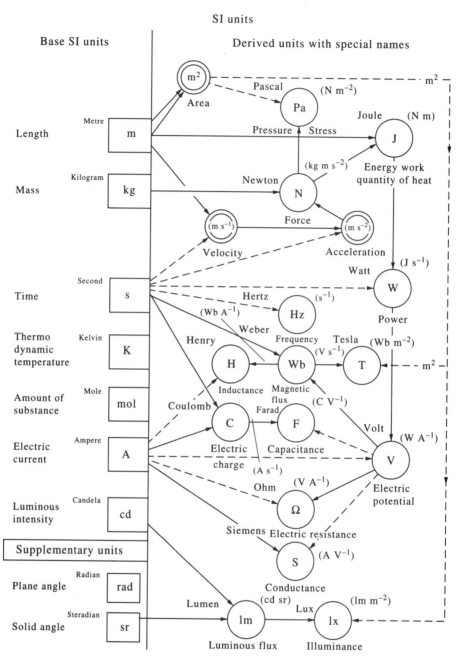

Figure 1
Chart showing relationship of SI measurement units to base units: – – –, division; ——, multiplication (reproduced from MHI 1972 Metric Conversion in Building and Construction © Standards Australia)

standards to manufacturing and calibration laboratories throughout their spheres of influence. Thus, while these national laboratories deal with international or primary standards, it is the lower accuracy (and lower cost) secondary standards, and their derivatives, which are used for regular calibration purposes.

The maintenance of accurate primary standards,

and research into their improvement and intercomparison, is an expensive exercise requiring highly specialist staff; not all countries find it economical to maintain all major standards. However, the most important, and those held by the major laboratories, are the base units of the now internationally accepted SI system of metrology. This is based on a series of units: the meter, kilogram, second, ampere, kelvin and candela; the hierarchy of major derived units is shown schematically in Fig. 1 (Sydenham *et al.* 1990). This hierarchy is one of the strengths of the SI system of units, in that permutation of the basic units can lead to a series of derived units, with special names, commonly employed in everyday measurement; for example, the unit of force (newton), the unit of power (watt), the unit of frequency (hertz), and so on, and other common derived units such as that of velocity (ms^{-1}) and area (m^2).

There are a number of detailed texts which deal with the methods of producing standards for the SI base units discussed and their maintenance. However, to illustrate the principle that the technology is evolving and not stationary, while the definition of the kilogram remains artifact based, the definition of the meter and second have changed. Initially, the meter was established as a unit of length equal to 10^{-7} of the polar quadrant of the earth. This was later refined using a platinum bar and, before 1960, the standard for the meter was based on a carefully preserved and more stable platinum–iridium bar kept in Paris. The early definition of the second was based on a fraction (1/86 400) of the mean solar day. These standards have moved to those based on optical radiation, for increased accuracy. The optical meter was redefined in 1960 in terms of 1 650 763.73 wavelengths of ^{86}Kr emission, in a vacuum. However, in 1983, the situation was improved with the definition of an exact quantity for the speed of light in a vacuum and the meter in terms of the radiation path in a fixed time, reflecting the quality of optical measurements that could be made. Thus, new definitions are now possible, to higher accuracies, using the most suitable physical phenomena. Work is advancing towards the possible replacement of the prototype kilogram as a fundamental standard, using a definition based on electrical parameters. The definition of standard units is shown in Table 1 (Morris 1988).

The measurement of temperature represents an interesting challenge to the metrologist. A temperature scale is applied to correlate numerical values to some defined temperatures. However, to produce a functional scale of temperatures, ideally requires one which is valid in any temperature range and is totally independent of the working substance. The thermodynamic Kelvin scale, based upon the efficiency of an ideal reversible Carnot cycle, is such a scale, as discussed by Herzfeld (1987). By means of such a scale, any chosen thermal state may be assigned a numerical value of temperature, and the scale defined upon a known temperature difference between two fixed points (e.g., boiling and freezing of water), or a defined value of one specific temperature. Thus, such fixed points, to which certain temperature values may be attributed, are required. A working substance is needed to establish the scale, and a correlating function is required to relate the temperature to the property of the substance. Complications arise because there is no direct method for the measurement of temperature, by comparison with the other measurements discussed, and although any material, in principle, may be used, certain stable substances are preferred. The temperature unit has been related to the properties of the common substance, water, and the kelvin is defined as 1/273.16 of the thermodynamic temperature of the triple point of water, a replacement of the pre-1954 definition of the scale based upon a 100-degree difference between the boiling point of water and the melting point of ice. The Carnot cycle is impossible to achieve in practice, but it may be shown that the thermodynamic temperature scale may be reproduced by a gas thermometer, with an ideal gas as the working substance, whereas a noble gas, at low temperature can serve as an adequate representation for the idealized medium. Thus, a temperature relationship given by:

$$T_1 = T_2(1 - \eta)$$

Table 1
Definitions of standard units

Length	The meter (m) is the length of path travelled by light in an interval of 1/299 792 458 s
Mass	The kilogram (kg) is equal to the mass of a platinum–iridium cylinder kept in the International Bureau of Standards, Sevres, France
Time	The second (s) is the duration of $9.192 631 77 \times 10^9$ cycles of radiation from vaporized ^{133}Cs (an accuracy of 1 in 10^{12} or 1 s in 36 000 years)
Temperature	The temperature difference between absolute zero and the triple point of water is defined as 273.16 kelvin (K)
Current	One ampere (A) is the current flowing through two infinitely long parallel conductors of negligible cross section placed one meter apart in vacuum and producing a force of $2 \times 10^{-7} N\,m^{-1}$ of conductor
Amount of substance	The mole is the number of atoms in a specified mass of ^{12}C

Table 2
The temperature fixed points which define IPTS-68 and ITS-90[a]

	Scale			
	IPTS-68		ITS-90	
Equilibrium state	T_{68} (K)	t_{68} (°C)	T_{90} (K)	t_{90} (°C)
Vapor pressure point of helium	not defined	3–5	270.15	−268.19
Triple point of equilibrium hydrogen	13.81	−259.34	13.8033[b]	−259.346[b]
Boiling point of hydrogen at a pressure 33 330.6 Pa	17.042	−256.108	17[b]	−256.15[b]
Boiling point of equilibrium hydrogen (see IPTS-68), or gas thermometer point of helium	20.28	−252.87	20.3[b]	−252.85[b]
Triple point of neon	not defined		24.5561	−248.5939
Boiling point of neon	27.102	−246.048	not defined	
Triple point of oxygen	54.361	−218.789	54.3584	−218.7916
Triple point of argon	83.798	−189.352	83.8058	−189.3442
Condensation point of oxygen	90.188	−182.362	not defined	
Triple point of mercury	not defined		234.3156	−38.8344
Triple point of water	273.16	0.01	273.16	0.01
Melting point of gallium	not defined		302.9146	29.7646
Boiling point of water	373.15	100	not defined	
Freezing point of indium	not defined		429.7485	156.5985
Freezing point of tin	505.1181	231.9681	505.078	231.928
Freezing point of zinc	692.73	419.58	692.677	419.527
Freezing point of aluminum	not defined		933.473	660.323
Freezing point of silver	1235.08	961.93	1234.93	961.78
Freezing point of gold	1337.58	1064.43	1337.33	1064.18
Freezing point of copper	not defined		1357.77	1084.62

a The values of the temperature fixed points with the exception of the triple points and the 17.042 K point are given at a pressure $p_0 = 101\,325$ Pa
b Given for $[H_2]_{cqm}$, which is the equilibrium concentration of the o- and p-molecular forms of hydrogen

may be represented between temperatures T_1 and T_2, with $T_1 > T_2$ and η being the efficiency of the reversible heat engine used. Either pressure difference at constant volume or volume difference at constant pressure can be used as the measure of temperature. Gas thermometers based on this principle can be used up to temperatures of 1350 K. The scale may be extended using radiation measurements from blackbody radiators, where the Planck Law gives the ratio of spectral radiant intensities, $W(T_1)$ and $W(T_2)$, as a wavelength λ, where

$$\frac{W(T_1)}{W(T_2)} = \frac{\exp(h/\lambda T_2 - 1)}{\exp(h/\lambda T_1 - 1)}$$

Here, h is Planck's constant (0.014 388 mK) and the temperatures T_1 and T_2 represent the reference and unknown temperatures, respectively. Due to the complexity of the gas thermometer for other than fundamental laboratory measurements, a more practical scale was needed. Over the twentieth century, several such scales were devised, culminating in the Interna-

tional Practical Temperature Scale of 1968 (IPTS-68), refined in 1990 as ITS-90. In these scales, the Celsius temperature is defined by:

$$t = T - T_0$$

where $T_0 = 273.15$ K. In these practical scales, every temperature measurement based on it closely approximates the true thermodynamic temperature. The basis of the scale is a series of fixed points, defined on the basis of suitable working substances, for the different regions of definition, as shown in Table 2 for both the IPTS-68 and ITS-90 scales. The latter scale gave better agreement with the corresponding thermodynamic temperatures, with improved continuity and precision, and a lower range was defined. Also, the elimination of boiling points in fixing the scale is an improvement in the accurate use of such a temperature scale. The Fahrenheit scale is still widely used, especially in the the USA, and its relationship to the Kelvin and Celsius scales is shown in Table 3. A detailed discussion of the scales, and their practical implementation, has been given by several authors (e.g., McGee 1992).

Table 3
Conversion of temperature scales

		To be determined in	
Scale	Given	(°C)	(°F)
Celsius	$X\,°C$	X	$1.8\,X + 32$
Fahrenheit	$X\,°F$	$0.5556(X - 32)$	X
Kelvin	$X\,K$	$X - 273.15$	$1.8(X - 273.15) + 32$

Source: McGee (1992)

2. Traceability and Calibration

It is essential that measurements made with everyday laboratory and industrial instruments can produce consistent results to achieve national and international confidence and trade using such standards. The concept of international projects, for example, in aerospace would be impossible without agreement on standards of metrology, as the results of measurements from different laboratories must be consistent for the work to proceed smoothly. The process of traceability is one that ensures that each level of calibration can be traced legally to the next, through a system of certification and the dissemination of standards. This links the standard held by the national laboratories to the instrument used on the shop floor, as shown schematically in Fig. 2 (Morris 1988). This underpins the process of calibration, by which the output of an individual instrument can be traced back to a national standard. Calibration is the act or result of a quantitative comparison between a known standard and the output of a measuring system, measuring the same quantity, to determine the scale for the measuring system. The process may be carried out with reference to the primary standard (primary calibration) but, more usually, it is performed against another device of lower accuracy, itself related to the primary calibration. Such secondary standards are widely used in laboratory practice. At each level, there is an increase in the degree of uncertainty (usually around three to ten times at each step), and thus, primary standards must be better than those used further down the tree. Thus, a base unit standard of length with an uncertainty of, say, 1 in 10^{11}, will result in, say, an uncertainty of 1 in 10^6 at the working level of engineering practice.

It is reasonable to assume that a new instrument obtained from a reputable manufacturer will be calibrated when delivered. However, its behavior will gradually diverge from the stated specification for a variety of reasons, due to the effects of mechanical wear, dirt, the environment, and frequency of, and care in, usage. The process of routine recalibration may then be performed, to readjust the instrument and, if necessary, replace faulty or worn components, as a procedure representing good practice in engineering, to restore the device to a calibrated state. Thus,

the maintenance of instruments in a calibrated state may well require their protection from the environment and careless use and, even if these precautions are taken, there may still be the need for recalibration due to long-term drift and mechanical or chemical changes to the instrument which are unavoidable, even with the most careful treatment of the device. Calibration is only valid if it is performed under the environmental conditions stated in the manufacturer's specifications, and the instrument is stored and used in the approved manner. The effects of pressure, humidity, temperature, and so on, are often significant for calibration, and corrections to the instrumental reading may be specified by the manufacturer.

In recent developments, the incorporation of intelligence into instruments can bring about the reduction in output error due to such changes, if the nature of the effects are well known and categorized, and suitable, reproducible correction factors are included in the associated intelligence (often through the use of a microcomputer) of the instrument. This will require

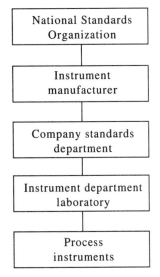

Figure 2
Instrument calibration chain schematic of the dissemination of standards from national organization to instrument user (after Morris 1988)

the use of accurate secondary transducers to measure these parameters, for input to the intelligent instrument. The achievement of such reliable inputs from transducers, themselves liable to drift, and environmental factors is not a trivial exercise.

3. Conclusion

The importance of accurate and reliable standards cannot be underestimated; considerable progress has been made and is continuing to improve and disseminate such standards through intercomparison and traceable calibration, as a vital element of successful international trade and commerce.

See also: Measurement: Fundamental Principles; Measurement Science

Bibliography

Herzfeld C M 1987 The thermodynamic temperature scale, its definition and realization. *Temperature: its Measurement and Control in Science and Industry*, Vol. 3, Pt 1. Van Nostrand Reinhold, New York, pp. 41–50.
McGee J 1992 *Temperature Measurement*. Wiley, New York.
Morris A S 1988 *Principles of Measurement & Instrumentation*. Prentice-Hall, Hemel Hempstead, UK.
Sydenham P H, Hancock N D, Thorn R, 1990 *Introduction to Measurement Science*. Wiley, New York.

K. T. V. Grattan
[City University, London, UK]

Mechanical and Robotics Applications: Multicomponent Force Sensors

The increasing interest in the study of techniques for the measurement of forces, however oriented in space, and resolvable into the main six components of the force tensor, namely, three forces and three moments, arises from definite requirements for:

(a) the development of adaptive-control, machine tools and robots;

(b) the advanced development, in the aircraft and automobile sectors, of prototype study in wind tunnels with the aid of multicomponent balances; and

(c) the reduction of the uncertainty of force standard deadweight machines.

In consideration of the aforementioned requirements, it has been found necessary to design and construct:

(a) multicomponent dynamometers suitable for different applications (Levi 1967, Broussaud 1968,

Dubois 1974, 1977, Cornut and Schulz 1983, Yoshida 1984, Hatamura *et al.* 1988, Barbato *et al.* 1990, Ferrero *et al.* 1986; and

(b) calibration systems supported by appropriate calibration methods (Dubois 1976, Takada *et al.* 1988, Ferrero 1990).

1. Calibration Procedures

Calibration of a multicomponent dynamometer essentially consists of determination of the sensitivity of the dynamometer output channels to the different components applied: axial and transversal components Z, X, Y; bending and twisting moments L, M and N. These components, known and defined as related to a reference axis, are generated by means of six-component calibration systems.

The output signal y_n (where $n = 1$ to 6) of a six-component dynamometer is a function of applied components x_i (where $i = 1$ to 6) and can be adequately approximated by a second-order polynomial of the type:

$$y_n = A_0 + \sum_{i=1}^{6} A_i x_i + \sum_{i=1}^{6} \sum_{j=1}^{6} A_{ij} x_i x_j \quad (1)$$

The first-order terms correspond to the linear effects of the six components; the second-order terms correspond to double interactions and can be interpreted as a variation in sensitivity to a component, caused by the deformation induced by another component and to individual quadratic effects, namely, to effects due to nonlinearity of the transfer function for a given component. With calibration data, the equations for dynamometer use are obtained in the following form:

$$x_n = a_0 + \sum_{i=1}^{6} a_i y_i + \sum_{i=1}^{6} \sum_{j=1}^{6} a_{ij} y_j y_j \quad (2)$$

The components applied to the dynamometer under different loading conditions can be determined by means of Eqn. (2).

The main calibration methods used can be schematically outlined as follows.

(a) Two-wedge method: transverse components (X, Y) are proportional to axial load Z and functions of inclination angle β and the φ azimuth angle; $L = M = N = 0$.

(b) One-wedge methods: components (X, Y) and bending moments L, M are proportional to Z and functions of angles β, φ and of the radius of loading cap; $N = 0$.

(c) Transverse load method: XL and YM are applied at the same time; $Z = N = 0$.

(d) Multiposition method: bending moments L, M are proportional to Z; $X = Y = M = 0$.

Table 1
Simple and cross combinations of the three orthogonal forces and the three moments

Forces			Moments		
X	Y	Z	L	M	N
	XY	XZ	XL	XM	XN
		YZ	YL	YM	YN
			ZL	ZM	ZN
				LM	LN
					MN

(e) Multicomponent method: independent or contemporary application of the six components of the force tensor is possible.

A complete calibration of a multicomponent dynamometer consists of a complete determination of six linear coefficients A_i, and 21 quadratic coefficients A_{ij} for each channel. The calibration requires separate and independent application of the six main components, that is, the three orthogonal forces (X, Y, Z), the three moments (L, M, N), and the 15 cross combinations listed in Table 1. The six simple cases allow the 36 linear coefficients A_i and the 36 quadratic coefficients A_{ij} to be defined. The 90 interaction coefficients A_{ij} ($i = j$) can be defined by 15 cross combinations.

Only with a complete six-component calibration system can all the 27 terms in Eqn. (1) be determined. Fortunately, not all the linear and quadratic coefficients prove significant; they depend on dynamometer structure and rigidity, on machining tolerances, strain gauge position, and so on.

2. General Evaluation of Multicomponent Dynamometers

According to the specific literature, an ideal multicomponent dynamometer must exhibit:

(a) high repeatability and reproducibility;

(b) independence from support and load application conditions;

(c) high sensitivity to different components of the applied force; and

(d) relative freedom from cross talk.

The main sources of cross talk are:

(a) variations in interface conditions with changes in the applied load;

(b) nonoptimal position of strain gauges; and

(c) incorrect dynamometer geometry.

The structure of a multicomponent dynamometer has to be very complex if one has to mechanically analyze the different components of the applied force to be measured, taking into account the inevitable limitations (dimensions, constraint types, etc.) and the different contradictory requirements, such as sensitivity and stiffness, in order to minimize the influence of disturbance quantities and factors (temperature, pressure, vibrations).

The materials to be used and the metallurgy treatments to be applied must ensure the best mechanical and elastic characteristics (fatigue, creep, hardness, tensile values).

According to their conditions of use and to the materials (steel, beryllium, aluminum, etc.) the stress to be applied must be limited, both in the measurement sections and in supports. Limit design values (with steel dynamometers) are from 150–250 Mpa for the measuring sections, and 600–700 MPa for the support and clamping parts.

Multicomponent dynamometers are essentially of two types, namely, single-block (or integral) and composite (assembled or built up) dynamometers.

2.1 Single-Block Dynamometers

It is well known (Dubois 1974, Ferrero *et al.* 1983) that complex integral structure of dynamometers involves thinning out several parts (elastic hinges, etc.) in order to obtain the mechanical decoupling necessary for the analysis of the components to be measured. However, mechanical decoupling cannot be pushed too far for reasons of stiffness and machining capability.

A single-block dynamometer therefore has certain intrinsic limitations that prevent cross talk effects (or interactions) from being reduced, but has the advantages of less weight, smaller dimensions and high stiffness.

2.2 Composite Dynamometer

The configuration of six load cells in a composite dynamometer makes it possible to analyze components more easily than with single-block dynamometers. The elastic flexures at the two ends of these dynamometers allow mechanical decoupling to be made very effectively. These elastic flexures produce decoupling whose efficiency is stronger, the higher the ratio of longitudinal to transverse stiffness. This ratio may vary between 200 and 5000 with most dynamometers. Interactions are therefore very weak, not to say negligible.

3. Comparison of the Characteristics of Multicomponent Dynamometers

3.1 General Evaluation

Tables 2 and 3 and Fig. 1 give the characteristics of a number of different multicomponent dynamometers to be used:

(a) as reference standards in force comparisons;

Table 2
Main data of a representative group of strain gauge multicomponent dynamometers

| | Capacity or maximum test load | | | | | |
| | Forces (kN) | | | Moments (kN m) | | |
Type	X	Y	Z	L	M	N
Slotted solid bar	20	20	300	0.5	0.5	0.45
Built-up, cradle type	2	2	500	1	1	0.15
Built-up, cradle type	1	1	15	0.1	0.1	0.15
Built-up, cradle type	0.2	0.2	100	0.16	0.16	0.05
Built-up, cradle type	6.5	22	22	19	3.5	3.5
Built-up, two pieces	5	5	5	0.5	0.5	0.5
Built-up, cradle type	60	15	17	6	6	7
Sting, Φ 19 mm	0.03	0.09	0.18	0.0017	0.01	0.007
Sting, Φ 40 mm	0.2	0.32	1.5	0.8	1.5	1

(b) in wind tunnel applications (sting or wall balances especially those produced by the Office National d'Etudes et de Recherches Aérospatiales (ONERA));

(c) in robotics; and

(d) in the field of machine tools.

A general evaluation can be given on the differences between the various dynamometers, by comparing the hypervolume of the components, which is the locus of the points of the six-dimensional space representing the maximum value of the components, which can be applied to the dynamometer without altering its metrological characteristics.

The ratio of the secondary components (the transverse forces and the bending moments) to the main component (see Table 2, Fig. 1), demonstrates the peculiarity of the multicomponent dynamometers specifically designed to check force standard machines with respect to those used in other fields; for the dynamometers to be used with the standard machines this ratio if $X/Z \approx Y/Z = 10^{-3}-10^{-4}$, while for those utilized in wind tunnels (see Fig. 2) and with machine tools these values of $X/Y = Y/Z$ range from 0.1 to 0.05. In robotic applications (see Fig. 3) the ratio between all the force components is usually one, while the ratio between the bending moments and the forces is about 70 mm (see Table 3), which is close to the physical dimensions of the dynamometer. In the transducer developed at the Istituto di Metrologia G. Colonnetti, Torino, Italy, for example, the ratio is about 60 mm, for all three bending moments (M_x/Z, M_y/Z, M_z/Z).

3.2 Multicomponent Dynamometers as Transfer Standards

As regards the multicomponent dynamometer to be used to check force standard machines, the main conclusion which can be drawn from the results of

recent international comparisons of force standard deadweight machines (Ferrero 1990, Ferrero and Zhong 1991), is that the measurement of parasitic components is a necessary step to maintain and/or improve the metrological characteristics of the primary force standards.

Table 4 summarizes the main characteristics of the four dynamometers specially designed to test force standards: the two ONERA four-component dynamometers of the integral type and the two IMGC six-component dynamometers of the composite type. The advantages of the composite type dynamometer

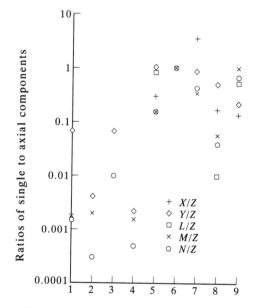

Figure 1
Component ratios for the different types of multicomponent dynamometers shown in Table 2

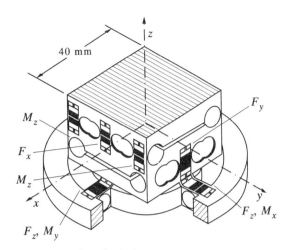

Figure 2
Six-component assembled dynamometer, utilized in a wind tunnel

Figure 3
Multicomponent force–torque dynamometer for robotics (wrist sensor)

are further enhanced in the composite "cradle" configuration adopted for the two IMGC dynamometers to be used to check force standard machines (Ferrero *et al.* 1983); the elements measuring the axial component and bending moments works in tension by means of double decoupling end flexures. Figure 4 shows the IMGC 100 kN six-component dynamo-

meter. It must be noted that with this configuration

(a) the path followed by the force flux lines between the load application point and the surface on which the dynamometer rests is such as to ensure remarkable insensitivity to interface conditions; and

Table 3
Characteristics of force–moment transducers used in robotics

Constructor	Force (N)	Moment Force (mm)	Diameter (mm)	Height (mm)	Weight (kg)
Lord Corporation	70–550	122	80–150	28–39	0.3–1.5
Ono (1) Tokyo University	10	100	80	80	
Ono (2) Tokyo University	30	50	80	80	
Hatamura Tokyo University	50	100	60	60	0.2
Mitsuishi Tokyo University	308	40	130		
IMGC	220	60	73	46	0.55

Table 4
Physical and structural characteristics of multicomponent dynamometers to check force standard machines

Dynamometer	BNM–ONERA integral type	ONERA 2 integral type	IMGC 1 built-up type	IMGC 2 built-up type
Number of components	4	6	6	6
Type	compression	tension	compression	compression
Capacity (FS) (kN)	300	300	100	500
Weight (kg)	34		150	450
Height (mm)	300	300	435	700
Diameter (mm)	150	150	310	400
Stiffness (mm)	0.125	0.27	1.63	1.6
$(kN \, mm^{-1})$	2400	111	61	300
$(kN \, m \, mrad^{-1})$	2.86	73.4	0.55	

(a)

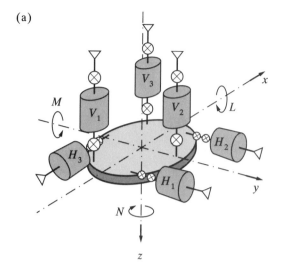

(b) instead of simple strain gauge bridges, it is possible to use load cells of high accuracy, which are intrinsically insensitive to side components. These load cells ensure a highly favorable sensitivity ratio between the components to be measured (S_M) and the parasitic (S_P) components ($S_M/S_P \gg 1$).

The design characteristics of this kind of dynamometer (X/Z, $Y/Z = 10^{-3}-10^{-4}$), and the simultaneous existence of other restraining and boundary conditions, mainly geometric and those concerning compatibility with the existing force standard machines (e.g., dimensions, shape), set narrow limits to the design of certain structural elements, such as vertical decoupling hinges and the upper support plate.

The composite six-component dynamometers have proved to be powerful diagnostic tools to detect possible contact points along the load transmission line. They have the additional advantage of low coefficients for component interaction, together with dynamometer insensitivity to support and load-application conditions. Table 5 gives the sensitivity matrix of the IMGC dynamometer. This means a high repeatability and reproducibility and, consequently, a low machine–dynamometer interaction (rotation effect). It is then possible to determine the real intrinsic characteristics of the force standards and the second-order effects, such as:

(a) the influence of the load transmission systems;

(b) the influence of different weight piece combinations; and

(c) dynamic effects, in connection with different methods of load application (overshoots).

In accordance with the foregoing considerations, and as regards the calibration of force standard machines (determination of their intrinsic metrological characteristics: repeatability, reproducibility, influence of load oscillation, etc.), it is pointed out that a built-up dynamometer is, at the moment, the only practical solution (Gosset and Nossent 1986, Ferrero 1988).

Instead, with a monoblock (integral) multicomponent dynamometer, variations in interface conditions modify the distribution of the force flux lines through the various strain gauge bridges; this can alter measurement results with a large dispersion of values observed with the dynamometer at the same angular position and at different positions relative to the machine axis.

See also: Force and Dimensions: Tactile Sensors

Bibliography

Barbato G, Desogus S, Germak A 1990 Multicomponent force sensors for robotics. *ISMCR IMEKO TC-17.* IMEKO, Budapest, pp. I1.2.1–2.9

Figure 4
IMGC six-component dynamometer: (a) schematic drawing; (b) photograph of a single component

Table 5
Matrix sensitivity coefficients with sequential or direct method of the 100 kN dynamometer

Terms	Units	Channel					
		V_1	V_2	V_3	H_1	H_2	H_3
Linear							
Z ⎫		−0.649 83	−0.648 05	−0.649 15	0.022	−0.044	−0.0149
X ⎬	N	1.48	0	−1.48	7.78	−7.95	−14.305
Y ⎭		0.86	−1.71	0.845	10.69	10.51	−0.011
L ⎫		4.425	−8.8	4.35	0.14	−0.15	−0.215
M ⎬	N m	−7.58	−0.04	7.62	0.15	0.080	0.070
N ⎭		0	0	0	−55.5	55.6	−38.5
Second-order							
Z^2 ⎫	10^8	−0.785	−0.143	0.0515	−4.50	1.30	−0.95
ZX ⎬	$\overline{N^2}$	3.25	0	−2.90	−16.0	16.0	25.8
ZY ⎭		1.25	−2.75	1.25	−10.8	−10.8	0
ZL ⎫	10^7	25	−52	29	105	115	0
ZM ⎬	$\overline{N^2 m}$	−50	0	50	−110	100	200
ZN ⎭		0	0	0	740	−720	340

Broussaud P 1968 Balances et dynamomètres utilisés au centre d'Essais de Modane – Avrieux. Note Technique ONERA. Office National d'Etudes et de Recherches Aérospatiales, Toulouse, France, p. 122

Cornut A, Schulz T 1983 Multicomponent force/moment transducer for industrial robot. *Proc. Weightech*, London

Dubois M 1974 Design and manufacture of high precision strain gauge dynamometers and balances at the ONERA. *Strain* **10**, 188–94

Dubois M 1976 Etalonnage de balances dynamométriques à six composantes. Office National d'Etudes et de Recherches Aérospatiales, Toulouse, France

Dubois M 1977 Multicomponent force transfer standard of 300 kN. *Proc. 6th IMEKO-TC3 Meeting*. IMEKO, Budapest

Ferrero C 1988 Evaluation of force standard machines with two multicomponent dynamometers. BCR—Applied Metrology. Commission of the European Community, Luxembourg

Ferrero C 1990 Multicomponent calibration systems to check force sensors. *Proc. 1st ISMCR – IMEKO TC-17 Meeting*. IMEKO, Budapest, pp. 5.1.1–1.16

Ferrero C 1990 The measurement of parasitic components in national force standard machines. *Measurement*, vol. 8(2) Measurement, Sunbury-on-Thames, UK, pp. 66–76

Ferrero C, Marinari C, Martino E 1983 Analysis and calibration of the IMGC six-component dynamometer. BCR Technical Report. Commission of the European Community, Luxembourg

Ferrero C, Marinari C, Martino E 1986 Main metrological characteristics of the IMGC six-component dynamometer. Report in Applied Measurements, Vol. 2. Hottinger Baldwin Messtechnick, Darmstadt, Germany

Ferrero C, Zhong L Q 1991 International comparisons of axial load in deadweight force standard machines by means of the IMGC 6-component dynamometer. *Proc. 7th Mondial Congr. IMEKO*. IMEKO, Budapest

Gosset A, Nossent P 1986 Mesure de plusieurs machines de force etalons à l'aide d'un dynamomètre multicomposantes de capacité 300 kN en compression. BCR Technical Report. Laboratoire National d'Essais, Paris

Hatamura Y, Matsumoto K, Morishita H 1988 A miniature 6-axis force sensor of multilayer parallel plate structure. *11th IMEKO World Conf. Sensors Sector*. IMEKO, Budapest, pp. 621–36

Levi R 1967 Drill press dynamometers. *Int. J. Mach. Tool Des. Res.* **7**, 269–87

Mitsuishi M, Hatamura Y, Nagao T Development of a sensor integrated manufacturing robot. *Proc. 20th ISIR*, pp. 111–18

Takada R, Ono K, Ogata K, Kusaki T 1988 An analysis of errors on 6-component force/moment calibration machines. *Proc. 11th IMEKO World Conf. Sensors sector*. IMEKO, Budapest, pp. 121–30

Yoshida T 1984 Six-component force transducer and its applications. *Proc. 10th IMEKO Conf. Measurement of Force and Mass*. IMEKO, Budapest

C. Ferrero
[Istituto di Metrologia G. Colonnetti, Turin, Italy]

Microwave Measurements

Microwave energy concerns extremely high frequencies or wavelengths of micrometer dimension; in descriptive nomenclature microwave transmission (mostly through guided structures) is a wave propagation phenomenon—thus, we have the appellation "microwaves." Microwave measurement covers the detection, evaluation and interpretation of energy so that microwaves of low power may be measured: a continuing and intimate process that constitutes most of the material of the following sections. Microwave frequencies occupy a considerable band of the electromagnetic (EM) spectrum, stretching from around

300 MHz up to 10 THz (1 THz = 1000 GHz). The corresponding wavelength range is from 1 m to 0.03 mm. At least two-thirds of this part of the EM spectrum has been intensively used as a scientific tool for the investigation of matter in its four states: solid, liquid, gaseous and plasma. This is because a great spectrum of resonance and relaxation processes in molecules and transport phenomena in solids and plasma are governed by energy transitions which equal the energy quantum of microwaves. For example, in crystals we can generate or convert energy by means of atomic oscillations in and around the avalanche condition.

In communications, microwave frequencies consume a proportionally smaller percentage of the total bandwidth of information channels where modulation rides on the main carrier wave; this inherently permits the use of more channels than at lower carrier frequencies. This is being fully explored, in particular for space communication and telemetry.

In navigation and radar, microwave antennas give extreme directivity because the short wavelengths allow focusing and directivity in much the same way as lenses focus and direct light rays.

For industrial processes, such as heating, drying and cooking, and in remote on-line measurements and control, automation and computer control of industrial processes is currently being introduced at a fast rate. These recent advances in manufacturing techniques are attributed to the development of state-of-the-art sensing systems in the optical, infrared and microwave regions. Here, microwave systems exhibit special advantages because of their unique features:

(a) the microwave sensing element can propagate through free space and often allow remote sensing;

(b) various insulating solid materials are opaque to light and infrared (ir), but transparent or semi-transparent to microwaves, hence allowing effective continuous and fast probing of objects noninvasively;

(c) some gases and water react specifically on selected microwave frequencies, allowing measurement of water and gas concentrations in complex mixtures; and

(d) it is now appreciated that when a particle beam generated by high-power microwave is passed through hazardous radioactive waste, it can reduce the half-life by an order of magnitude, and thus the high-powered microwave beam can provide valuable information about the half-lives of new radioactive waste materials.

In biomedical applications, microwaves have been used to produce hyperthermia, either invasively (interstitial probes) or noninvasively (radioactive applicators, endocavitary probes), and to detect and localize subcutaneous thermal gradients via radiometric techniques.

Around 1980, the first active microwave imaging techniques were applied to isolated dog kidneys. Their enhanced projections confirmed beyond any doubt the potential application of microwaves for providing a noninvasive access to any physiological or physical factors that depend on the dielectric properties of human tissues. Water content, temperature, blood flow rate and phase changes are some significant examples. More about this challenging aspect of microwave measurement is given in Sect. 3.1.

Much progress has been made recently in order to qualify the terahertz region of the EM spectrum for new and more exciting applications. This frequency band spanning the spectral range from 100 GHz to 10 THz is one of the last major windows in the EM spectrum to be explored. The primary applications for terahertz technology have been in basic scientific research, including astrophysics, atmospheric physics, plasma diagnostics and laboratory spectroscopy. The rotational emission lines of simpler molecules occur in this band. Measurements of these spectra allow the determinations of abundances, distributions and kinematic properties of the medium in which molecules are located. More recently, radar and communications systems have employed this spectral range to combine the frequency resolution and agility available in the microwave regime with the high spatial resolution using modest apertures typical of optical technology. Because the earth's atmosphere is opaque except for a few discrete windows in this frequency range, much of the technology development is being directed toward space qualifiable components. Achieving the full potential of the terahertz region, however, has been limited by the lack of suitable technology. For space applications, new solid-state approaches such as quantum well and Josephson junction oscillators, though less mature and of high technical risk, have the potential of revolutionizing remote sensing spectroscopy in the terahertz band.

In this article three main topics of measurement principles will be discussed and illustrated with some practical examples. First, wave propagation where the transmission medium does not affect the wave. Measurements of geometrical quantity such as distances and levels in containers for liquids and solids, control of movement of objects, surfaces or interfaces, velocity and vibration, and determinations of object dimensions are all performed using radar and surface imaging techniques. Second, interaction of microwaves with matter, in which the transmission medium changes the probing wave. Areas of application include measurement of complex frequency response of dielectric and magnetic materials, especially the determination of the water content in organics and inorganics, continuous measurement of gas concentrations in gaseous mixtures, and microwave active imaging and tomography. Finally, radiometry, in which the medium itself transmits the information at microwave frequencies. Examples include measurement of subsurface temperatures using the total or partial transparency of many materials.

1. Radar and Active Surface Imaging

In free space, electromagnetic waves propagate with the speed of light, one of the best-measured physical quantities:

$$C_{\text{vacuum}} = 2.9979 \times 10^8 \text{ m s}^{-1} \tag{1}$$

A fundamental measurement setup for propagation effects is shown in Fig. 1. An elementary form of radar consists of a transmitting antenna (T_x) emitting electromagnetic radiation generated by an oscillator of some sort, a receiving antenna (R_x) and an energy-detecting device, or receiver. The receiving antenna collects the returned energy and delivers it to a receiver. The time of flight τ can be calculated from the distance r and the velocity c:

$$\tau = \frac{2r}{c} \tag{2}$$

When a continuous wave (CW) at a fixed and constant frequency is transmitted, the delay time τ cannot be measured. The mixing product of the transmitted and the received signal yields a phase as a function $\phi(r)$ of distance but this distance information is ambiguous because of the 2π periodicity of the phase. Consequently, fixed-frequency CW signals can only be applied for the measurement of moving targets. When a target moves with a constant velocity v, then the resultant phase shifts linearly with time. This is the well-known Doppler effect. In the frequency domain, the moving target generates the Doppler frequency:

$$f_d = 2(v/c)f_0 \tag{3}$$

where f_0 denotes the signal frequency and f_d is a direct measure of the target velocity. Observing the Doppler signal in the time domain, a phase shift of π, a half sine wave, corresponds to a target displacement of a quarter wavelength. A laboratory model of a 35 GHz vibration meter yields a sensitivity of better than 1 cm V. Applying a narrow-band filtering at the output, periodic vibrations can be measured down to below 1 nm ($\sim 1 \text{ nV}$).

If the absolute distance has to be measured, the transmitted signal must be modulated. There are three principal methods, namely, single-frequency, pulse and frequency-modulated continuous-wave radar (FMCW).

1.1 Single-Frequency Radar Systems

In single-frequency radar systems the microwave is either amplitude modulated (AM) or frequency modulated (FM). The modulating frequency must satisfy the relation:

$$c/f_m = \lambda_m > r_{\text{max}} \tag{4}$$

where r_{max} is the maximum range to be measured. The phase difference of the modulation of the transmitted and received signal is given by

$$\Delta\phi = 2\pi \left(\frac{2r}{\lambda_m}\right) \tag{5}$$

and since $2r$ is always less than λ_m then $\Delta\phi$ does not exceed 2π and hence leads to an unambiguous measurement of range. This method finds applications in experimental car radars. It is not an accurate method, however.

1.2 Pulse Radar Systems

The pulse technique relies on transmitting very short pulses of duration T and repetition time T_s. The choice of T and T_s are governed by the following equations:

$$T_s > \frac{2r_{\text{max}}}{c} \tag{6}$$

$$T < \frac{2\Delta r}{c} = \Delta\tau \tag{7}$$

where Δr is the minimum distance between two objects which are to be resolved. Take, for example, two objects 10 cm apart and at $r_{\text{max}} = 30 \text{ m}$, then $T_s > 0.43 \text{ }\mu\text{s}$ and $T < 0.6 \text{ ns}$.

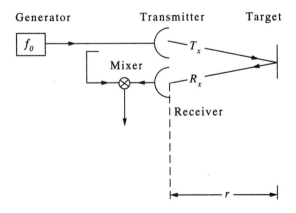

Figure 1
Elementary single-frequency CW radar response for target at rest and moving target. The output signal is proportional to $\sin(2\pi f_0/c)2r$ for a target at rest or $\sin(2\pi f_0(2v/c)t)$ for a target moving with velocity v

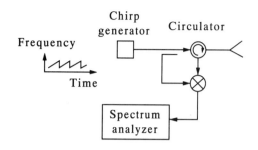

Figure 2
Block diagram of FMCW radar

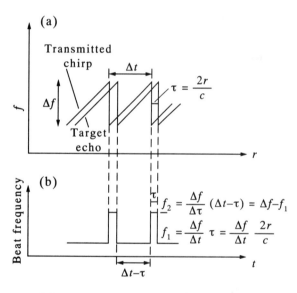

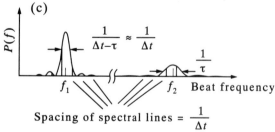

Figure 3
(a) The frequency difference f_1 between the transmitted chirp and the target echo is proportional to target range; (b) the beat frequency as a function of time, showing components both at f_1 and f_2; (c) the spectrum of the beat frequency, showing $((\sin x)/x)$ envelopes, modulating a line spectrum. The amplitude of the component at f_2 has been exaggerated for clarity.

If we restrict the product $T\Delta f$ to unity, as is usually the case, then the bandwidth of the pulse Δf in this case equals to $1/(0.6 \times 10^{-9}) = 1.7\,\mathrm{GHz}$, which is quite unacceptable as it exceeds by orders of magnitude the bandwidth of "normal" radars.

1.3 Frequency-Modulated Continuous-Wave Radar Systems

The principles of frequency-modulated continuous-wave (FMCV) radar systems are very simple, yet FMCW radars have some interesting and elegant properties that make them very well suited to a number of specialist applications. Consider the FMCW radar shown in Fig. 2. Here, the transmitted waveform has a constant amplitude, but a linear sawtooth variation of frequency with time and is often referred to as a chirp. A target echo will consist of a replica of this transmitted waveform, delayed by the two-way propagation delay $\tau = 2r/c$, with r and c as previously defined. The instantaneous frequency difference between the received signal and the transmitted signal (at the instant reception) is a constant, proportional to the target range, therefore, a measurement of this frequency difference will yield the target range.

Figures 3a and 3b show the transmitted and received signals in greater detail. The beat frequency will actually consist of two components f_1 and f_2 but, provided τ is small compared to the sweep period Δt, f_2 will be much greater than f_1 and the component at f_2 will have considerably less energy than that at f_1. Figure 3c shows the spectrum of these beat frequencies. Because of the repetitive nature of the signal, the spectrum of the beat frequency will consist of a set of spectral lines spaced at $1/\Delta t$. These are modulated by $((\sin x)/x)$ envelopes centered on f_1 and f_2, each of width equal to the inverse of their respective durations. The width of the f_1 envelope defines the resolution of the radar, that is, the ability of the radar to distinguish two targets closely spaced in range. Thus, the resolution in the frequency domain is $1/\Delta t$; in the time domain this is scaled by the sweep rate $\Delta f/\Delta t$ and, in turn, in the range domain by $c/2$ (time translates to two-way range by the factor c and to one-way range by the factor $c/2$), from which the range resolution Δr is $c/2\Delta f$.

1.4 Active Surface Imaging

The spatial resolution achieved with radars is determined by the directive properties of the antennas. In this section we discuss holographic images, the principle of which is illustrated in Fig. 4 for microwave imaging. For example, replacing the laser in Fig. 4 by a microwave source, and the photographic plate by thermally sensitive materials such as liquid crystal plates, will result in a microwave imaging system. In the quasiholographic system shown in Fig. 5, the amplitude and phase of the scattered field are measured at several points forming a square pattern

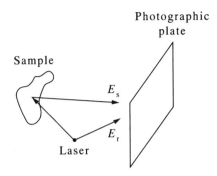

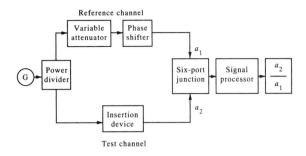

Figure 6
A schematic diagram of the automatic vector voltmeter system

Figure 4
An optical hologram

over a certain aperture R and the inverse Fourier transform is computed to produce a microwave image. The lens aperture is much greater than the wavelength—so much so that the radiated field can be described quite accurately as a single, fundamental Gaussian mode. Aperture R is scanned with power-detecting probes. An automatic vector voltmeter (Kamarei 1991) may be built using a six-port junction; the concept is simpler than might at first appear.

1.5 Automatic Vector Voltmeter in Microwave Imaging
A vector voltmeter is essentially a device that measures the complex voltage ratio of two input waves a_1 and a_2. In order to get the information related to an amplitude and phase signal, four power readings are required. An automatic vector voltmeter may therefore be realized using a six-port junction on the principle that a reference signal a_1 and a signal containing the useful information on the test channel a_2 constitute the two inputs of a six-port junction, as shown in Fig. 6. This six-port junction linearly combines the two input signals and, because of the square-law detectors on its four outputs, it delivers four dc voltages. The complex ratio a_2/a_1 may then be calculated. Image reconstruction is achieved with the

knowledge of amplitude and phase of the scattered field on an arbitrary surface R which can be made of up to 32×32 measurement points (Kamarei 1991). Two methods for calibration are available: the Taylor series expansion (Cletus *et al.* 1975) and the eigenvalue (Harry and Susman 1977). The latter is preferred and involves a matrix description of the problem leading to a formulation of the calibration constants as the solution to an eigenvalue problem.

2. Interaction of Electromagnetic Waves with Matter

2.1 Propagation Factor in Dielectric Medium
While propagating in nonconducting or poorly conducting media (dielectric materials or simple dielectrics), microwaves suffer some attenuation. This arises from the friction that results when the electromagnetic field of the wave acts with a force on the electric charges (e.g., ions, electrons, polar molecules) or magnetic dipoles in the medium, causing them to be slightly displaced. This effect can be taken into account by defining the complex propagating factor

$$k = k' - \mathrm{j}k'' \qquad (8)$$

where k'' is the loss factor, and inserting this propagation factor into the equation of a plane wave travelling in the direction of the x-axis, as given by

$$\mathbf{E} = \mathbf{E}_0 \exp(-\mathrm{j}kx) \qquad (9)$$

This is an oversimplified form of a wave equation with the time dependence term dropped. The real exponential function now describes exponential damping with propagated distance. The general definition for k is

$$k = \omega\sqrt{\mu\varepsilon} = \omega\sqrt{\mu_0\varepsilon_0}\,\sqrt{\mu_r\varepsilon_r}$$

$$= k_0\sqrt{\mu_r\varepsilon_r} = k_0\,\mathrm{Re}(\sqrt{\mu_r\varepsilon_r}) + \mathrm{j}k_0\,\mathrm{Im}(\mu_r\varepsilon_r) \qquad (10a)$$

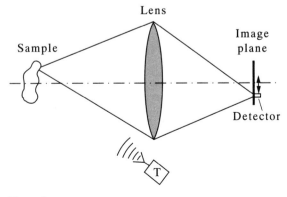

Figure 5
Microwave imaging with a lens-focused system

223

where

$$\varepsilon_r = \varepsilon_r' - j\varepsilon_r''$$

$$\varepsilon_r = \varepsilon_r'(1 - j\tan\delta) \qquad (10b)$$

$$\mu_r = \mu_r' - j\mu_r''$$

and where $\tan\delta$ is the loss tangent and is a measure of the rate of attenuation to a propagation wave. (The loss factor ε_r'' is always positive and usually much smaller than ε_r'. The former approaches zero for a lossless medium. The minus sign in the loss tangent equation (10b) is a direct consequence of the fact that the physical media attenuate waves rather than amplifying them.) This equation relates the complex propagation factor to the complex relative permittivity ε_r and the complex relative permeability μ_r of the media. ε_r describes the interaction with the electric field, whereas μ_r describes the interaction with the magnetic field. The relative permittivity ε_r is also a measure of the polarizing effect on the internal field of any physical materials subjected to an external electric field. In media where the constants ε_r, μ_r or both, are greater than one, the speed of propagation is lower than in vacuum. The real part of the permittivity ε_r affects the speed of propagation. In a nonmagnetic lossless medium, the speed of propagation is given by

$$c = \frac{c_{\mathrm{vacuum}}}{\sqrt{\varepsilon_r'}}$$

and $\qquad\qquad\qquad\qquad\qquad (10c)$

$$\lambda \simeq \frac{\lambda_0}{\sqrt{\varepsilon_r'}}$$

where λ_0 is the wavelength in free space $= 2\pi/k_0 = c_{\mathrm{vacuum}}/f$.

Because the permittivity and permeability are complex, there are four practically independent constants $(\varepsilon_r', \varepsilon_r'', \mu_r', \mu_r'')$ describing the electrical properties of the medium. These four constants depend on the other physical properties (moisture, composition, density, temperature, structure, etc.) of the medium and on the measurement frequency. If we knew the relationships between ε_r and μ_r this would allow us to measure any of the aforementioned physical properties. Such is the foundation for the majority of microwave sensors, described by Nyfors and Vainikainen (1989). Figure 7 shows how the dielectric medium slows down microwave propagation.

The loss factor ε_r'' is always positive and is usually much smaller than ε_r'; the former approaches zero for a lossless medium. The minus sign in the loss tangent equation (10b) is a direct consequence of the fact that physical media attenuate waves rather than amplify them.

2.2 Refraction and Reflection

The phenomenon of refraction and reflection at oblique incidence at boundaries between areas with

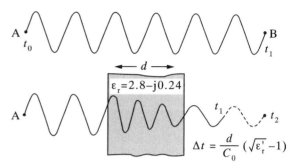

Figure 7
The effect of microwave propagation through a dielectric. The time arrival of t_2 is greater that that to free space t_1 by an interval $\Delta t = d/c(\sqrt{\varepsilon_r'} - 1)$

different ε_r or μ_r is similar to that in optics. Looking at the ray geometry in Fig. 8a, where a wave is incident at the interface from medium 1 to medium 2, Snell's law may be written:

$$\frac{\sin\theta_1}{\sin\theta_2} = \frac{\lambda_1}{\lambda_2} = \frac{k_2'}{k_2'} \qquad (11)$$

For the lossless nonmagnetic case,

$$\frac{\sin\theta_1}{\sin\theta_2} = \frac{\sqrt{\varepsilon_{r2}'}}{\sqrt{\varepsilon_{r1}'}} = \frac{n_2}{n_1} \qquad (12)$$

where n is the refractive index used in optics. The general consequence of refraction is that a wavefront will be bent in the case of oblique incidence at the interface between two media, as shown in Fig. 8a. The refraction also changes the normal and tangential components of the field strengths. Satisfying the boundary conditions therefore requires a third wave component, a reflected wave. The field Fresnel reflection coefficients for lossy media (see Fig. 8b) are given by

$$\Gamma_v = \frac{-(\mu_{r1}'/\varepsilon_{r1}')^{1/2}\varepsilon_{r2}'\cos\theta_1 + (\mu_{r2}'\varepsilon_{r2}' - \mu_{r1}'\varepsilon_{r1}'\sin^2\theta_1)^{1/2}}{(\mu_{r1}'/\varepsilon_{r1}')^{1/2}\varepsilon_{r2}'\cos\theta_1 + (\mu_{r2}'\varepsilon_{r2}' - \mu_{r1}'\varepsilon_{r1}'\sin^2\theta_1)^{1/2}}$$

$$(13)$$

$$\Gamma_h = \frac{(\varepsilon_{r1}'/\mu_{r1}')^{1/2}\mu_{r2}'\cos\theta_1 - (\mu_{r2}'\varepsilon_{r2}' - \mu_{r1}'\varepsilon_{r1}'\sin^2\theta_1)^{1/2}}{(\varepsilon_{r1}'/\mu_{r1}')^{1/2}\mu_{r2}'\cos\theta_1 + (\mu_{r2}'\varepsilon_{r2}' - \mu_{r1}'\varepsilon_{r1}'\sin^2\theta_1)^{1/2}}$$

$$(14)$$

where Γ_v and Γ_h refer to the vertically and horizontally polarized electric fields, respectively.

The general case of lossy media is slightly more complicated. The field reflection coefficients for lossy media are

$$\Gamma_H = \frac{-\varepsilon_2\mathbf{n}\cdot\mathbf{k}_1 + \varepsilon_1[\omega^2(\mu_2\varepsilon_2 - \mu_1\varepsilon_1) + (\mathbf{n}\cdot\mathbf{k}_1)^2]^{1/2}}{\varepsilon_2\mathbf{n}\cdot\mathbf{k}_1 + \varepsilon_1[\omega^2(\mu_2\varepsilon_2 - \mu_1\varepsilon_1) + (\mathbf{n}\cdot\mathbf{k}_1)^2]^{1/2}}$$

$$(15)$$

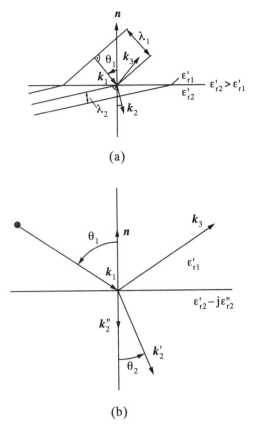

(a)

(b)

Figure 8
Refraction and reflection of waveforms at dielectric
boundary: (a) when medium 2 is lossless and
homogeneous; (b) when medium 2 is lossy

$$\Gamma_E = \frac{\mu_2 \mathbf{n} \cdot \mathbf{k}_1 - \mu_1 [\omega^2(\mu_2 \varepsilon_2 - \mu_1 \varepsilon_1) + (\mathbf{n} \cdot \mathbf{k}_1)^2]^{1/2}}{\mu_2 \mathbf{n} \cdot \mathbf{k}_1 + \mu_1 [\omega^2(\mu_2 \varepsilon_2 - \mu_1 \varepsilon_1) + (\mathbf{n} \cdot \mathbf{k}_1)^2]^{1/2}}$$

(16)

Figure 8b shows that the real (k') and imaginary (k'')
parts of the wave vector will be separate vectors
pointing in different directions. The wave is called an
inhomogeneous wave and consequently cannot be
divided into vertically and horizontally polarized components.

3. Applications

3.1 Medical Active Microwave Imaging

As mentioned earlier, the possible usefulness of active
microwave imaging techniques was realized around
1980, and encouraging results obtained around 1983
(Bolomey *et al.* 1989) have stimulated the development of a 2.45 GHz microwave camera for biomedical

applications. The clinical usefulness of an image
depends on both spatial resolution and contrast. The
mutual importance of these two criteria depends on
the specific domain of application. Concerning spatial
resolution, the diffraction effects constitute the main
limiting cause. As in any diffraction-limited optical
instrument, spatial resolution is of the order of the
wavelength in the considered medium. Furthermore,
due to diffraction effects, microwaves do not propagate according to linear paths and more complex
reconstruction algorithms are needed. With respect to
contrasts, microwaves are just as good as, if not better
than, other already well-established imaging modalities such as x rays or ultrasound. Air–water or
bounded water–free water transitions lead to drastic
changes in the complex permittivity of living tissues,
resulting in very sensitive microwave images.

3.2 Tomography

Tomography (x-ray, nuclear magnetic resonance and
microwave) normally means producing cross-sectional
images of the interior of a sample, whereas holography (light) means three-dimensional imaging of the
surface of a sample. Tomography methods are divided
into transmission tomography and diffraction tomography. The former is used with very weakly inhomogeneous materials, where the wavefront
propagates almost undisturbed. The image is constructed using probes to monitor any change of phase
or amplitude at the end of the propagation path,
which is assumed to be a slightly curved path due to
weak inhomogeneity of the medium. The sample may
be rotated in three axes and measurements recorded
to form a matrix description of the problem leading to
a formulation of a three-dimensional image. Diffraction tomography may be applied on samples with
slightly stronger inhomogeneities, but which are still
sufficiently homogeneous to fulfill the Born approximation at the measurement frequency (i.e., the
assumption that the field inside the inhomogeneities is
equal to the field outside). The field after transmission
is considered to contain two components, namely, the
attenuated incident field and the scattered field.

3.3 Microwave Radiometers

A radiometer is essentially a sensitive receiver designed to measure the radiated thermal noise power
emanating from a body. The radiometer overall gain
including that of the antenna must be high. Any
instability due to thermal drift or gain variation must
be included in the calibration strategy.

Microwave radiometry can be used in remote sensing to measure temperature profiles, that is, temperature versus depth, as well as the thickness of a
microwave semitransparent layer on a reflecting surface. This takes into account the physical relation
between reflection, transmission and absorption of
electromagnetic energy at an interface between two
media; the emissivity can be obtained from a reflection

measurement simultaneously with the radiation measurement.

In medical applications, microwave radiometry has been applied to noninvasive measurement of subcutaneous tissue temperatures. Imaging of subcutaneous temperature distributions has been realized at both millimeter and micrometer wavelengths. Results obtained for human muscle and skin temperature measurements using a radiometer and thermocouples attached to skin (the depth tested by the radiometer is of the order of several millimeters at 9 GHz) showed muscle temperature was more than 3 °C, higher than skin temperature (Lynch *et al.* 1985, Mamouni *et al.* 1987, Paulsen *et al.* 1988). Moreover, subcutaneous and muscle temperature increased during hard exercise while skin temperature correlatively decreased due to the rise in evaporated heat loss. The probes were of coaxial waveguide transition type and the waveguide section was fitted with a low-loss dielectric. The temperature probe was in flush contact with the skin.

In the steel continuous casting process, the measurement and control of slag (which is essentially a nonmetallic combination of various oxides in a solid-to-molten state on top of the mass of molten steel) thickness in both the ladle and tundish has long been an area of concern to steelmakers. An innovative approach to this problem has been proposed by Rizk (1993). This approach involves an on-line microprocessor-controlled microwave radiometer system for the measurement of thicknesses of hot nonmetallic material and involves multifrequency measurements.

See also: Optical Measurements

Bibliography

Bolomey J Ch, Gaboriaud G, Berthaud P, Cottard G 1989 2.45 GHz microwave camera for bio-medical applications. *PIERS 1989*, pp. 479–80

Cletus A, Hower C A, Keith C 1975 Using an arbitrary six-port junction to measure complex voltage ratios. *IEEE Trans. Microwave Theory Tech.* **23**(12), 978

Collin R E 1966 *Foundation of Microwave Engineering*. McGraw-Hill, New York

Harry M, Susman L 1977 A six-port automatic network analyzer. *IEEE Trans. Microwave Theory Tech.* **25**(12), 1086

Kamarei M 1991 Vector voltmeter applications in microwave imaging. *Microwave J.* **34**(11), 102–14

King J A (ed.) 1988 *Materials Handbook for Hybrid Microelectronics*. Artech House, Dedham, MA

Larsen L E, Jacobi J H 1986 *Medical Applications of Microwave Imaging*. Institute of Electrical and Electronics Engineers, New York

Laverghetta T S *Handbook of Microwave Testing*. Artech House, Dedham, MA

Lynch D R, Paulsen K D, Strohbehn J W 1985 Finite element solution of Maxwell's equations for hyperthermia treatment planning. *J. Comput. Phys.* **58**(2), 246–69

Mamouni A M *et al.* 1987 Passive subcutaneous temperature measurement for investigation. *Proc. 17th European Microwave Conf.*, Rome, pp. 381–5

Nyfors E, Vainikainen P 1989 *Industrial Microwave Sensors*. Artech House, Dedham, MA

Paulsen K D, Lynch D R, Strohbehn J W 1988 Three-dimensional finite, boundary and hybrid element solutions of the Maxwell equations for lossy dielectric media. *IEEE Trans. Microwave Theory Tech.* **36**(4), 682–93

Pichol Ch, Jofre L, Peronnet G, Bolomey J Ch 1985 Active microwave imaging of inhomogeneous bodies. *IEEE Trans. Antennas Propag.* **33**(4), 416–25

Rizk M S 1993 Proposed microwave radiometry technique to remotely measure the thickness of slag or molten powder layers over molten steel. *J. Microwave Power* (in press)

Skolnik M I 1970 *Radar Handbook*. McGraw-Hill, New York

M. S. Rizk
[City University, London, UK]

N

Networks of Instruments

The developments in very-large-scale integrated circuits (VLSI) technology and microprocessors has led to the concept of intelligent instruments. These incorporate processors with a sensor or display to form a single device. A communications network could be used to interconnect sensors and displays to form a distributed instrumentation and measurement system. The use of a generalized communications system facilitates the inclusion of actuators and controllers to form distributed computer control systems.

1. Motivation for a Communications Network

The advantages of using a communications network include the following.

(a) Reduction in cabling by replacing individual cables to each sensor/display by a bit serial, bus-type network. This can reduce wiring costs in process control applications.

(b) Digital rather than analog signalling techniques can be used. These are less susceptible to noise and are easier to regenerate.

(c) Communication networks typically include error detection and correction mechanisms, as well as diagnostic facilities which can improve system reliability and availability.

(d) Distributed systems based on a communications network permit incremental expansion by simply connecting new devices, improved reliability by replicating components, and flexibility. (Sloman and Kramer 1987.)

The main problem with using a communications network is that the interconnection interface must be specified in terms of a standard which defines a communication protocol, that is, message formats, rules and procedures for exchanging messages, as well as the electrical and physical characteristics of the physical interconnection interface. This is inherently more complicated than the traditional standards used for interconnecting instruments, based on simple 20 mA signalling, requiring internationally agreed standards.

2. Network Architecture

A communications network is the means by which components can exchange information by transferring messages over a transmission medium. If the components include local processing power and cooperate in order to achieve some overall goal, then the interconnected components form a distributed processing system. The network architecture defines the overall topology (i.e., interconnection structure) of the system as well as the structure of the communications software.

2.1 Network Topology

The bus topology shown in Fig. 1 is the favored topology in instrumentation systems because it results in minimum cable length and the transmission medium is inherently passive. A single component failure is less likely to affect communication between other components. However, large systems, such as those used for factory automation or process control, are likely to consist of multiple interconnected subnets as shown in Fig. 2. A subnet will interconnect the sensors, actuators, controllers or displays for a particular subsystem or automation cell. A site-wide bus is then used to interconnect the subnets to permit interaction between subsystems and communication with a central control room. Bridges are used to interconnect similar types of subnets and gateways interconnect dissimilar networks. A large organization may have sites in different parts of the country and these sites can be linked by a wide area, public network to form an integrated communications system.

2.2 International Organization for Standardization Reference Model

The International Organization for Standardization (ISO) have developed the Open Systems Interconnection (OSI) reference model as a framework for describing general communication systems and for the development of communication standards. This reflects the fact that most communication systems are layered with specific protocols to perform particular functions at each layer. The ISO reference model (see Fig. 3) identifies seven layers, as follows.

(a) *Application layer*. The application-specific protocols together with the software components which interact by using these protocols.

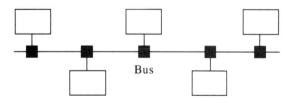

Figure 1
Bus local area network

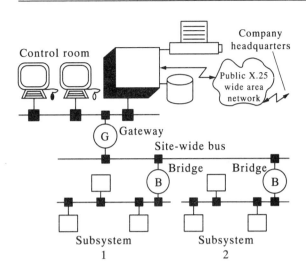

Figure 2
Interconnected subnets

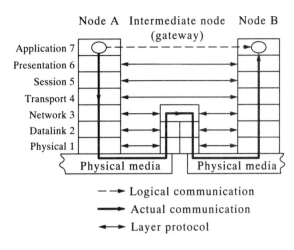

Figure 3
ISO reference model

(b) *Presentation layer.* Resolves differences in the way computers and programming languages represent information. For example, floating point numbers, integers and characters are represented by different bit patterns in different processors.

(c) *Session layer.* Establishes an association between two application entities by which they can communicate, for example, to exchange a file or to relate a particular actuator to a particular controller.

(d) *Transport layer.* Provides an end-to-end reliable communication path between the computers connected to the network, possibly spanning multiple subnetworks; responsible for segmenting large-application messages into smaller ones acceptable by the communication system and performs error control to recover from lost or corrupted messages.

(e) *Network layer.* The delivery service which transfers the messages across the network and performs routing if there is more than one path between a source and destination; it is also responsible for resolving differences between different types of subnets.

(f) *Datalink layer.* Responsible for transferring messages across a single physical transmission medium (i.e., a single subnet), which may involve error control and flow control if the transmission medium has high error rates.

(g) *Physical layer.* Concerned with the physical transmission medium, signalling and modulation.

Local area networks, of which a bus is an example, have another layer called the Media Access Layer (MAC), which corresponds to part of layers 1 and 2 of

the ISO model (see Fig. 3), and is responsible for controlling which of the stations connected to a shared transmission medium can transmit at any one time.

The reason for this layered structure is to permit flexibility for change. For example, the transport protocol can be independent of the type of network, whether it is a local-area or wide-area network. Similarly, dependencies on the type of local-area network (LAN) can be kept within layer 2.

3. A Communication Protocol

This section describes a simplified version of the Manufacturing Automation Protocol (MAP) datalink layer as an example of the elements of a protocol needed within a communication system. More detailed discussion of communication protocols can be found in Sloman and Kramer (1987), Halsall (1992) and Tanenbaum (1988).

3.1 Media Access Control

A bus is a shared transmission medium to which multiple nodes are connected. A media access control (MAC) mechanism is needed to prevent more than one node transmitting a message at the same time, otherwise these messages would interfere with each other and be corrupted. MAP specifies token passing. A token is a special control message which is passed round the nodes connected to the bus. The holder of the token is allowed to transmit an application message (and possibly receive a response) before passing the token on to another node. A node wishing to send a message must wait for the token to be received before it is permitted to transmit. The nodes form a logical ring and each node must know its predecessor from which it receives the token and its successor to

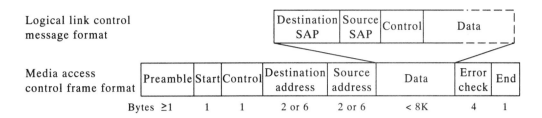

Figure 4
Token bus frame format

which it must pass on the token. The protocol specifies how nodes are added or deleted from the logical ring and how to regenerate a token if it gets lost, or on initialization.

A typical message has the format shown in Fig. 4. The media access control fields are described as follows and the logical link control (LLC) fields are described in Sect. 3.2.

(a) *Preamble*: To permit bit and byte synchronization to be achieved by receivers.

(b) *Start*: Start of message delimiter.

(c) *Control*. Contains control information for token management.

(d) *Destination address*. To identify the destination node to which the message is being sent. A message can be broadcast to all nodes on a subnet or a selected group of nodes.

(e) *Source address*. To identify the sending node in case a response is needed.

(f) *Data*. Control information plus data from higher layers.

(g) *Error check*. Redundant information, typically a cyclic redundancy check computed over the contents of the message, which can be used to detect a corrupted message due, for example, to noise on the transmission medium.

(h) *End*. End of message delimiter.

3.2 Logical Link Control

Error control is normally performed by means of positive acknowledgements and retransmissions. The source transmits a message and starts a timer. If the timer expires, the sender retransmits the message. This is repeated up to some maximum retry limit (e.g., three times) after which the source assumes the destination has failed and indicates a fault. If the message was correctly received by the destination it sends an acknowledgement which permits the source to cancel the timer and send a new message or, in the case of MAP, pass on the token to the next node.

3.3 MIL-STD 1553B

MIL 1553B is a 1 Mbit s^{-1} bus designed for use as a single bus consisting of a single controller and up to 30 terminal nodes. All control of the communication system resides with the controller, which can send a message to a terminal node, request a terminal node to respond with a message or nominate a sender and a receiver terminal node, in which case the sender transmits one message to the receiver. Messages are transmitted (bit serial) but consist of one or more 20 bit words. The controller always specifies the message length (MIL 1553B 1978).

The protocol defines the following.

(a) *Command word*. Sent to a terminal to set its mode of operation—e.g., listener, sender, shutdown, self test, transmit status, retransmit last message, etc.

(b) *Data words*. Contain 16 bits of data which can follow a command word, and a message can contain up to 32 data words.

(c) *Status word*. Returned by a terminal after receiving a message to indicate whether the message was correctly received, or whether the terminal is requesting the controller to perform some action.

The controller is responsible for initiating retransmissions, and so on, to recover from errors. The terminal nodes are assumed to have limited intelligence. The MIL 1553 bus is far less sophisticated than the MAP token bus, and hence it is more amenable to hardware implementation. Although it has been around for much longer, it has not been used much outside the avionics industry.

4. Conclusions

The trend in instrumentation networks is likely to be towards the use of standard communication systems such as MAP, as this will permit the integration of the instrumentation system with control to form a distributed computer control system. It is likely that in the near future standard communication systems such as the full seven layers of MAP will be available as a single VLSI component.

The methods to be used for designing distributed systems and the languages for programming the software are still subjects for research (Kramer *et al.* 1989).

See also: Digital Instruments; Signal Processing

Bibliography

Halsall F 1992 *Data Communications, Computer Networks and Open Systems*, 3rd edn. Addison-Wesley, Reading, MA

Kramer, J. Magee J, Sloman M 1989 Building distributed systems in conic. *IEEE Trans. Software Eng.* **15**(6), 663–75

MIL 1553B 1978 *Aircraft Internal Time Division Command Response Multiplex Data Bus*

Rodd M, Deravi F 1989 *Communication Systems for Industrial Automation.* Prentice-Hall, Englewood Cliffs, NJ

Sloman M, Kramer J 1987 *Distributed Systems and Computer Networks.* Prentice-Hall, Englewood Cliffs, NJ

Tanenbaum A 1988 *Computer Networks*, 2nd edn. Prentice-Hall, Englewood Cliffs, NJ

M. S. Sloman
[Imperial College of Science, Technology and Medicine, London, UK]

Noise: Physical Sources and Characteristics

The term "signal" is defined as a function, usually of time or space, capable of carrying information. Similarly, the term "noise" refers to any unwanted signal which interferes with the desired signal. It may therefore include many types of man-made disturbances, such as mains pickup or transients, as well as more fundamental processes such as Johnson noise or Brownian motion. The former are, in a sense, predictable, and therefore in principle removable, whereas the latter are inherently unpredictable and therefore cannot be removed (unless the bandwidth used is reduced to zero). Such fundamental noise processes occur in many different systems, such as mechanical, electrical, thermal, radiant and magnetic systems.

The most basic form is known as thermal noise, which arises from the principle of equipartition of energy. This states that each degree of freedom of a system in thermal equilibrium with its surroundings has a mean kinetic energy of $kT/2$, where k is Boltzmann's constant and T is the absolute temperature.

This gives rise to Brownian motion (in mechanical systems), Johnson noise (in electrical systems) and radiation noise (in thermal systems). Thermal noise is always associated with the dissipative element in a system, and its magnitude always involves the product kT.

A less basic but very common form of noise is known as shot noise. Shot noise processes depend on the discrete nature of matter. A stream of randomly emitted particles, such as electrons crossing a barrier, shows statistical fluctuations about a mean value. In the case of electrons, the type of noise is actually known as shot noise, and involves the electronic charge e. For photodetectors the effect is known as radiation noise and involves fluctuations in the photon streams to and from the detector.

Thermal noise and shot noise are usually "white" (i.e., they have a flat power spectrum over a wide range of frequencies), but the type of noise known as $1/f$ noise is so called because its power spectrum varies inversely with frequency. It is mostly found in semiconductors and appears to be associated with fluctuations in the number of carriers in the conduction band. The effect becomes important at low frequencies, when the observed noise may be many orders of magnitude greater than for white noise processes.

1. Noise Sources in Physical Systems

1.1 Electrical Systems

(*a*) *Johnson noise.* This is named after the person who first demonstrated it experimentally, and was predicted theoretically by Nyquist from purely thermodynamic considerations. However, the most instructive derivation is obtained by considering the motion of electrons in a conductor. The mean electronic velocity is determined by the principle of equipartition, and an electron moving at a given velocity v will produce an instantaneous current proportional to $e \times v$. Such a current will continue until the electron changes its velocity by colliding with an atom of the crystal lattice, or reaches an edge of the conductor, when a new current proportional to the new velocity is produced.

A single electron thus produces a succession of current pulses, and it can be shown that the resultant power spectrum for the current fluctuations in a conductance G is given by

$$P_i(f) = 4\,GkT\,\mathrm{sinc}^2 f\theta$$

where the sinc function is defined by

$$\mathrm{sinc}\,x = \frac{\sin \pi x}{\pi x}$$

and where θ is the mean time between collisions. The

corresponding expression for the voltage fluctuations in a resistance R is given by

$$P_v(f) = 4RkT \operatorname{sinc}^2 f\theta$$

Johnson noise is usually considered white, since the power spectrum is essentially flat up to several hundred megahertz.

The mean square voltage across a resistor R in bandwidth Δf is therefore given by

$$\bar{v}^2 = 4RkT\Delta f$$

which is perhaps the most usual expression.

This explanation of Johnson noise has the virtue that it immediately makes clear that "perfect" conductors, capacitors or inductors do not produce Johnson noise, since the electrical carriers do not collide with the lattice and there is no change in velocity. Strictly, however, such a lossless device cannot achieve thermal equilibrium as it has no mechanism with which to interact, therefore the matter is somewhat academic.

In practice, a capacitor C or inductor L will always have some stray resistance associated with it; it is this dissipative element that enables equipartition to be obtained. The mean square voltage $\bar{v}^2$ appearing across an RC combination is given by

$$\frac{1}{2} C\bar{v}^2 = \frac{1}{2} kT$$

so

$$\bar{v}^2 = \frac{kT}{C}$$

and similarly, for an LR combination the mean square current is given by

$$\bar{i}^2 = \frac{kT}{L}$$

These expressions do not involve the resistive element specifically, but the temperature T is actually that of the resistive component.

The magnitude of Johnson noise is fairly small at room temperature. A useful figure is the rms noise in a resistor of $1\,\mathrm{k}\Omega$ in a bandwidth of $1\,\mathrm{Hz}$, which is often quoted as $4\,\mathrm{nV}$ per root Hz.

(b) *Shot noise.* This was first noted in vacuum diodes, but also occurs in semiconductor diodes and in photodetectors. It arises from the statistical nature of the emission of electrons, or from the crossing of a potential barrier. The fluctuation in number of electrons emitted per second follows a Poisson distribution about the mean value, and has the interesting property that the variance σ^2 is exactly equal to the mean value. If the mean number of electrons emitted in time t is $\bar{n} = \sigma^2$ then $\bar{n} = It/e$, where I is the mean current.

If the effective bandwidth is taken as

$$\Delta f = \frac{1}{2t}$$

the current fluctuation becomes

$$\bar{i}^2 = 2eI\Delta f$$

which is the usual formula for shot noise. Alternatively, an emitted electron can be considered to produce a pulse of current of height A and width θ where

$$\int_0^\theta i(t)\,\mathrm{d}t = e = A\theta$$

Taking the Fourier transform gives

$$F(f) = A\theta \operatorname{sinc} f\theta$$

and the power spectrum for n electrons is given as

$$P_i(f) = 2eI \operatorname{sinc}^2 f\theta$$

Shot noise is thus white, up to frequencies of the order $1/\theta$, which may be as high as $100\,\mathrm{MHz}$ for a vacuum diode, though smaller in semiconductor diodes. Unlike Johnson noise, which is not dependent on the "granular" nature of electricity, shot noise specifically involves the electronic charge e, and its magnitude would be different if e were different.

The observed magnitude of shot noise is often rather larger than that of Johnson noise, though a strict comparison is inappropriate since shot noise is inherently a current fluctuation in a unidirectional mean current, and Johnson noise is independent of any mean current, depending only on resistance and temperature.

1.2 *Other Physical Systems*

The equivalent of Johnson noise occurs in several physical systems. Most systems have two passive storage elements and one passive dissipative element (e.g., capacitors, inductors and resistors in electrical systems). The basic noise generation process in the dissipative element produces an expression for the power spectrum of the form given in the following equation:

$$P(f) = 4RkT$$

The principle of equipartition leads to formulae such as

$$\bar{v}^2 = \frac{kT}{C}$$

and

$$\bar{i}^2 = \frac{kT}{L}$$

for the mean square values of the variables in the storage elements.

In mechanical systems the effect is known as Brownian motion, after the botanist who first observed it. Considering a simple system comprising a mass M suspended by a spring of compliance C_m, subject to viscous damping by the surrounding air, the fundamental noise process is due to collisions of air molecules with the mass. The forces produced cover a wide frequency range and have a white power spectrum given by

$$P_f(f) = 4R_m kT$$

where R_m is the mechanical resistance. The mean square velocity of the mass is given by

$$\bar{v}^2 = \frac{kT}{M}$$

and the mean square force in the spring by

$$\bar{f}^2 = \frac{kT}{C_m}$$

This corresponds to the expressions for Johnson noise. It is interesting to note that the same expressions would apply even if the system was in a vacuum, provided there was a mechanism (e.g., thermal conduction through the spring) for achieving thermal equilibrium.

Brownian noise is usually very small in most practical applications, since the masses involved are relatively large. However, the effect is important in sensitive galvanometers and seismometers, and the fundamental limit to the detection of earth motion by a seismometer is set by the Brownian motion of the suspended mass.

The expressions occur in somewhat modified form in thermal systems where, for example, the mean square fluctuation in temperature δT of a system of thermal capacitance C at temperature T is given by

$$\overline{\delta T^2} = \frac{kT^2}{C}$$

In other systems, such as fluid or magnetic, thermal noise is usually masked by larger effects specific to the system such as ripples or vortices in fluids or magnetostrictive noise in magnetic systems.

The equivalent of shot noise occurs only rarely. One example is the fluctuation in the stream of photons to and from a detector, leading to what is known as radiation noise. The results obtained differ only slightly from those predicted classically (essentially using $\delta T^2 = kT^2/C$), and are in any case usually masked by electrical Johnson or shot noise in practice.

2. Characteristics of Noise

2.1 Amplitude Distribution

Perhaps the most characteristic property of noise is that the probability density distribution of its instantaneous amplitude is Gaussian, that is:

$$p(v) = \frac{1}{\sigma\sqrt{2\pi}} \exp(-v^2/2\sigma^2)$$

In information theory this is the distribution which can "carry" most information and, similarly, the one that is most disruptive of a signal. All forms of noise can be thought of as the instantaneous sum of a large number of small effects (individual current pulses for Johnson noise), and such processes always produce a Gaussian distribution.

This applies to all forms of noise, including $1/f$ noise and band-limited noise. The envelope of random noise follows a Rayleigh distribution and its phase is uniformly distributed.

2.2 Power Spectrum

Randomness in the time domain implies flatness in the frequency domain, since this is the most disruptive or least predictable situation. All basic forms of noise (Johnson, Brownian, shot, etc.) are inherently white over a wide range, although the power spectrum always falls to zero at very high frequencies.

The power spectrum of a function is related to its autocorrelation function by a Fourier transform. The autocorrelation for ideal white noise (for which $P(f)$ is constant over all frequencies) is an impulse of strength equal to $P(f)$. If such noise is passed through a filter to produce band-limited noise, the autocorrelation function is broadened, being a sinc function for a rectangular filter.

2.3 Representation of Noise

A useful mathematical representation for noise can be obtained by slicing the power spectrum into sections of width Δf; the instantaneous noise $v(t)$ can then be written as:

$$v(t) = \Sigma \sqrt{2P\Delta f} \cos(\omega_i t + \phi_i)$$

where P is the (constant) power spectrum, and ω_i and ϕ_i are the angular frequency and phase of the ith slice. Further simplifications of this expression for narrow-band noise give the following equation:

$$v(t) = x(t) \cos \omega_0 t + y(t) \sin \omega_0 t$$

where $x(t)$ and $y(t)$ are slowly varying Gaussian parameters, and ω_0 is the central frequency.

Alternatively

$$v(t) = r(t) \cos(\omega_0 t + \phi(t))$$

where $r(t)$ is the envelope and $\phi(t)$ the phase.

2.4 Addition of Noise Generators

Unlike signal generators, for which the rms sum is simply the sum of the individual rms values, noise generators add by adding their mean square values, for example:

$$\bar{V}^2 \text{ sum} = \bar{v}_1^2 + \bar{v}_2^2 + \dots$$

This is because noise sources are usually independent and incoherent, so that a term of the form $(v_1 + v_2)^2$ leads to $\bar{v}_1^2 + \bar{v}_2^2$; the central term $2v_1 v_2$ has to be included if the generators are not totally independent.

It is therefore very easy and convenient to calculate the net effect of all the noise sources in a system, since each generator is usually specified by its mean square value.

2.5 Noise in Networks

In a passive network the total noise is determined by the noise in the dissipative elements. If the network presents an impedance

$$Z(f) = R(f) + jX(f)$$

the thermal noise is determined solely by the resistive component $R(f)$, and the power spectrum is given by

$$P(f) = 4R(f)kT$$

For example, a parallel RC combination has

$$Z(f) = \frac{R}{(1 + j\omega CR)}$$

so

$$R(f) = \frac{R}{(1 + \omega^2 C^2 R^2)}$$

and

$$P(f) = \frac{4RkT}{(1 + \omega^2 C^2 R^2)}$$

The total noise power is given by

$$\int_0^\infty P(f)\, df$$

This is easily shown to be $kT/2$, as expected from equipartition. Alternatively, the noise due to the resistor can be considered to be filtered by the transfer function $H(f)$ of the RC network, which acts as a low-pass filter with

$$H(f) = \frac{1}{(1 + j\omega CR)}$$

The noise bandwidth of a network is defined as

$$\Delta f_\text{n} = \frac{1}{H_0^2} \int_0^\infty H^2(f)\, df$$

where H_0^2 is the peak value, which evaluates to $1/(4RC)$ for the RC network. The total noise output is given by

$$4RkT \times \frac{1}{(4RC)} = \frac{kT}{C}$$

as before. Similar expressions apply in mechanical and other systems.

In an active network, such as an amplifier, the noise sources (thermal, shot, etc.) are referred to the input of an equivalent noiseless amplifier. They are often expressed in terms of their Johnson noise equivalent resistances (i.e., the value of resistance in which the Johnson noise would equal the value of the generator). The noise performance of the network can be given in terms of a noise figure F, if used with a resistive source, where

$$F = \frac{(\text{signal to noise ratio with ideal amplifier})^2}{(\text{signal to noise ratio with actual amplifier})^2}$$

F is unity for a perfect (noiseless) amplifier. Practical amplifiers show significant $1/f$ noise at frequencies below 100 Hz and the expressions for the noise equivalent resistances have to be modified appropriately.

See also: Errors and Uncertainty; Errors: Avoidance and Compensation

Bibliography

Fellgett P B, Usher M J 1980 Fluctuation phenomena instrument science. *J. Phys. E* **13**, 1041–6
Johnson J B 1928 Electrical noise. *Phys. Rev.* **32**, 97–109
Mottchenbacher C D, Fitchen F C 1973 *Low-Noise Electronic Design*. Wiley, New York
Nyquist H 1928 Thermal agitation of electric charge in conductors. *Phys. Rev.* **32**, 110–13
Shottky W 1928 Spontaneous current fluctuations in various conductors. *Ann. Phys.* **57**, 541
Usher M J 1974 Noise and bandwidth. *J. Phys. E* **7**, 957–61

M. J. Usher
[University of Reading, Reading, UK]

Nondestructive Testing, Electromagnetic

Photons of frequencies covering virtually a continuous electromagnetic spectrum from zero to hard x-ray and gamma-ray frequencies are used for nondestructive testing, but electromagnetic methods are commonly understood to be those encompassing the range from microwave frequencies downwards. They can be subdivided on a spectral basis into microwave methods (typically 1–15 GHz), eddy current testing (typically from a few to a few hundred kilohertz), magnetic induction, hysteresis and particle methods (usually

~50 Hz, i.e., mains frequency), potential drop techniques (typically from zero to a few kilohertz), and electrostatic methods (contact methods of resistance and capacitance measurements, thermoelectric, triboelectric and electric particle). The general principle behind electromagnetic testing is that field vectors are changed by variations in the permittivity, permeability or conductivity of the object under inspection and these changes are monitored by some kind of field probe, typically a coil. When the variations in permittivity, permeability and conductivity are local through the presence of a defect in the form of a crack or inclusion, lines of electric displacement and magnetic induction are distorted in the vicinity of the defect, to satisfy continuity requirements. The position of the field distortions, and thus the defect, can be detected with some precision by the use of sufficiently small probes.

1. Theoretical Principles

The electromagnetic field at any point in a medium is completely defined by the equations

$$\mathrm{div}\,\mathbf{D} = \rho \qquad (1) \qquad \mathbf{B} = \mu_0\mathbf{H} + \mathbf{M} \qquad (6)$$

$$\mathrm{div}\,\mathbf{B} = 0 \qquad (2) \qquad \mathbf{J} = \sigma\mathbf{E} \qquad (7)$$

$$\mathrm{curl}\,\mathbf{E} = \partial\mathbf{B}/\partial t \qquad (3) \quad \mathrm{div}\,\mathbf{J} = \partial\rho/\partial t \qquad (8)$$

$$\mathrm{curl}\,\mathbf{H} = \sigma\mathbf{E} + \partial\mathbf{D}/\partial t \qquad (4) \qquad \mathbf{D} = \varepsilon\mathbf{E} \qquad (9)$$

$$\mathbf{D} = \varepsilon_0\mathbf{E} + \mathbf{P} \qquad (5) \qquad \mathbf{B} = \mu\mathbf{H} \qquad (10)$$

where **B** is the electric field strength, **D** is the electric displacement, **B** is the magnetic induction, **H** is the magnetic field strength, ρ is the charge density, J is the current density, σ is the conductivity, **P** is the electric polarization, **M** is the magnetization, μ is the permeability, μ_0 is the permeability of free space, ε is the permittivity, and ε_0 is the permittivity of free space.

The specific properties of a medium are summarized in just three quantities: **M** (or equivalently μ), **P** (or equivalently ε) and σ. Generally μ, ε and σ could be tensors but for simplicity we will assume isotropic conditions at any point so that these quantities are scalars. For a linear medium, that is, one in which μ, ε and σ are independent of time and amplitude, and there is no free charge, Maxwell's equations, that is, Eqns. (1–4), become

$$\mathrm{div}\,\mathbf{D} = 0 \qquad (11)$$

$$\mathrm{div}\,\mathbf{B} = 0 \qquad (12)$$

$$\mathrm{curl}\,\mathbf{E} = -\,\partial\mathbf{B}/\partial t = \mu\,\partial\mathbf{H}/\partial t \qquad (13)$$

$$\mathrm{curl}\,\mathbf{H} = \sigma\mathbf{E} + \partial\mathbf{D}/\partial t = \sigma\mathbf{E} + \varepsilon\,\partial\mathbf{E}/\partial t \qquad (14)$$

The nondivergent quantities **D** and **B** are continuous at boundaries regardless of any spatial or temporal

variations in ε, μ or σ, while **E** and **H** are not. The vector identity curl curl = grad div $-\nabla^2$ can be used to yield the following equations of field propagation:

$$\nabla^2\mathbf{E} = \mu\varepsilon\,\partial^2\mathbf{E}/\partial t^2 + \mu\sigma\,\partial\mathbf{E}/\partial t \qquad (15)$$

$$\nabla^2\mathbf{B} = \mu\varepsilon\,\partial^2\mathbf{B}/\partial t^2 + \mu\sigma\,\partial\mathbf{B}/\partial t \qquad (16)$$

Identical equations obviously exist for **D** and **H**, but **E** and **B** are the most commonly used for the purposes of mathematical presentation: they are regarded as the most physically significant quantities through being directly defined (and measurable) in terms of the force exerted on a static charge and a moving charge (current), respectively, independently of material properties. From solutions of Eqns. (15, 16), **D** and **H** are readily generated when required, for example, for boundary condition determination, by the application of Eqns. (9, 10).

If the first term on the right-hand side only is present, these equations represent undamped wave motion at a constant speed:

$$c = (\mu\varepsilon)^{-1/2} \qquad (17)$$

But if the second term is finite, that is, σ is nonzero, the equations represent damped waves, associated with the generation of eddy currents and Joule heating losses, with a strong frequency-dependent phase velocity. As **E** and **B** vary spatially and temporally in the wave propagation direction, it is clear that all four field vectors must lie in a plane orthogonal to the direction of propagation to be consistent with the nondivergence of **B** and **D**.

1.1 Propagation in Dielectrics

In dielectrics the second term in Eqns. (15, 16) is zero, but in the first term we will now acknowledge the possibility of a frequency-dependent dielectric and magnetic hysteresis loss by denoting the permittivity and permeability as complex quantities ε^* and μ^*, in which case

$$\mathbf{E} = \mathbf{E}_0 e^{-\alpha x}\exp[\mathrm{j}(\omega t - k\omega)] \qquad (18)$$

$$\mathbf{B} = \mathbf{B}_0 e^{-\alpha x}\exp[\mathrm{j}(\omega t - k\omega)] \qquad (19)$$

where

$$\alpha + \mathrm{j}k = \mathrm{j}\omega(\varepsilon^*\mu^*)^{1/2} \qquad (20)$$

and the electromagnetic wave impedance of the dielectric is defined by

$$Z = \mathbf{E}/\mathbf{H} = \left(\frac{\mu^*}{\varepsilon^*}\right)^{1/2} \qquad (21)$$

1.2 Propagation in Electrical Conductors

A wave of angular frequency ω varies in time according to $e^{\mathrm{j}\omega t}$ so that the first term in Eqns. (15, 16) varies as ω^2, while the second varies as ω. Thus, at sufficiently low frequencies, if σ is nonzero, the second term becomes dominant and the field equations approximate to

$$\nabla^2 \mathbf{E} = \mu\sigma\,\partial\mathbf{E}/\partial t \qquad (22)$$

$$\nabla^2 \mathbf{B} = \mu\sigma\,\partial\mathbf{B}/\partial t \qquad (23)$$

Plane wave solutions in a homogeneous unbounded medium take the form

$$\mathbf{E} = \mathbf{E}_0\,e^{-x/\delta}\cos(\omega t - x/\delta) \qquad (24)$$

$$\mathbf{B} = \mathbf{B}_0\,e^{-x/\delta}\cos(\omega t - x/\delta) \qquad (25)$$

where the subscript refers to field amplitude values at $x = 0$; δ, the depth at which the field amplitudes have fallen to $1/e$ of the values $x = 0$, and known as the skin depth, is given by

$$\delta = (2/\sigma\omega\mu)^{1/2} \qquad (26)$$

and the intrinsic impedance of the medium $\mathbf{E/H}$, is given by

$$\mathbf{Z} = \mu/(\varepsilon - j\sigma/\omega)^{1/2} \qquad (27)$$

where, although ε is indeterminate for a metal if it is set equal to the highest known value for a dielectric, it would be negligible compared with σ/ω, even at visible light frequencies.

The speed of the propagation is

$$c = \omega\delta = (2\omega/\sigma\mu)^{1/2} \qquad (28)$$

indicating a strong frequency dependence (dispersion) and approaching zero as ω approaches zero.

This approximation is valid in good conductors, that is, metals and metal alloys, at all frequencies up to the visible spectrum.

1.3 Effect of Crack-Like Defects and Other Material Homogeneities on Field Lines

So far we have dealt only with solutions to Eqns. (15, 16) for propagation in infinite and spatially homogeneous media. Regardless of inhomogeneities, lines of **D** (in the absence of free charge) and **B** are continuous in space. In contrast, at a spatial discontinuity of $\varepsilon_1-\varepsilon_2$ in ε, some lines of **E** either terminate on negative charge induced in the surface of discontinuity ($\varepsilon_2 > \varepsilon_1$) or extra lines are generated from a positive induced charge (negative $\varepsilon_2 < \varepsilon_1$). Similarly, at a discontinuity $\mu_1 - \mu_2$, a discontinuity in the density of lines of **H** which is associated with an induced surface distribution of current, lines may be envisaged as being generated from or terminating on fictitious north or south poles, respectively. The boundary conditions on the field vectors at the discontinuities are that there must be continuity of the normal

components of **D** and **B** (i.e., lines of **B** and **D** form continuous closed loops in space in the absence of free charge and current), and continuity of the tangential components of **E** and **H**.

To satisfy these conditions, it follows that for oblique incidence, the lines of **D** and **B**, while remaining continuous, must be refracted when they cross the boundary, and generally there will be discontinuities in all vectors **D, E, B** and **H**. It will be helpful to consider the boundary conditions in more detail. Denoting normal and tangential components by subscripts n and t, respectively, we have

$$(\mathbf{D}_1)_n = (\mathbf{D}_2)_n, \qquad (\mathbf{B}_1)_n = (\mathbf{B}_2)_n \qquad (29)$$

$$(\mathbf{E}_1)_t = (\mathbf{E}_2)_t, \qquad (\mathbf{H}_1)_t = (\mathbf{H}_2)_t \qquad (30)$$

and combining with Eqns. (9, 10) we obtain

$$(\mathbf{E}_2)_n = (\mathbf{E}_1)_n\varepsilon_1/\varepsilon_2, \qquad (\mathbf{H}_2)_n = (\mathbf{H}_1)_n\mu_1/\mu_2 \qquad (31)$$

$$(\mathbf{D}_2)_t = (\mathbf{D}_1)_t\varepsilon_2/\varepsilon_1, \qquad (\mathbf{B}_2)_t = (\mathbf{B}_1)_t\mu_2/\mu_1 \qquad (32)$$

It is clear, from these boundary condition discussions alone, that any defect within which ε or μ values are significantly different from those in the surrounding material produces substantial perturbations in the electromagnetic field in the material, for example, air, within the volume of the defect and in the immediate surrounding volume of material into which the defect also reflects and scatters waves.

As an example, Fig. 1 illustrates the common case of a material surface containing an air-filled planar crack defect running normally, with radiation travelling normally and the **B** and **D** vectors being parallel to the surface. The lines of **D** and **B** in the material run parallel to the surface and follow continuous paths

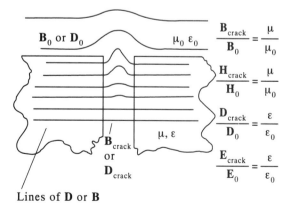

Lines of **D** or **B**

Figure 1
Distortion of lines of electric displacement and magnetic induction in the vicinity of a surface breaking crack

normally across both crack walls. However, the requirement of continuity of properties within a uniform medium means that **B** and **D** must fall off gradually outside the specimen surface, but within the region of the crack. Thus, the lines of **B** and **D** must be locally kinked within the volume of the crack, as illustrated, to produce locally different values of **B**, **H**, **D** and **E** outside the specimen, just over the crack. The values differ from corresponding ones elsewhere just outside the specimen surface by factors of up to μ/μ_0 (for **B** and **H**) and $\varepsilon/\varepsilon_0$ (for **D** and **E**). For ferromagnetic and ferrimagnetic materials, with μ much greater than 1, the predominant change is thus in **B** and **H**. However, even when the factors are only unity, the sharp change in direction of the external field vectors will be detectable in principle.

The remainder of this article is devoted to specific techniques which vary according to part of the frequency spectrum and/or type of instrumentation used.

2. Microwave Testing

Microwaves are applicable to bulk inspection of all dielectric materials and the penetration capability in the lowest frequency ranges (longest wavelengths) can be many meters (providing there is no excessive water content or other source of heavy dielectric loss), comparing favorably with kilohertz-frequency ultrasound, which has a similar wavelength range. Like kilohertz ultrasound, long-wavelength microwaves are particularly suitable for the inspection of coarsely inhomogeneous matter such as wall boards and concrete.

2.1 Testing by Reflection, Scattering and Refraction

Applying the boundary conditions given by Eqn. (28), it is easily shown that at a planar boundary, the power reflection and transmission coefficients R and T, respectively, at normal incidence, are given by

$$R = 1 - T = \frac{(Z_1 - Z_2)^2}{(Z_1 + Z_2)^2} \tag{33}$$

where the subscripts denote the two media and Z is related to ε and μ by Eqn. (26). For the detection of thin cracks, microwaves cannot compete with ultrasound of the same wavelength, as the following representative calculation using Eqn. (33) shows. Consider an air-filled crack in concrete, which (taking $\varepsilon/\varepsilon_0 = 9$) implies that $Z_1 = Z_3 = Z_2/3$ for microwaves and $Z_1 = Z_3 = 3 \times 10^4 Z_2$ for ultrasound. Taking a wavelength of 10 mm (corresponding to 10 GHz microwaves and 400 kHz ultrasound in concrete), for a crack width (L) of 0.5 mm, corresponding to $k_2 L = 0.314$, we obtain $R = 0.14$ for microwaves and $R = 0.999\,995$ for ultrasound. If the crack width falls to 0.1 mm, R for microwaves falls to 7×10^{-3}, and for ultrasound it differs from 1 by less than 2×10^{-7}. In

most common solids lower values of $\varepsilon/\varepsilon_0$ would be obtained so that the above estimates of R for microwaves would be still lower. In conclusion, microwaves cannot, in general, detect hairline cracks, but gross cracks such as areas of disbonding in fiberglass laminated boards and other insulating laminates have been detected with microwaves. Clusters of submillimeter pores in rubber (a material very difficult to penetrate by ultrasound even at low kilohertz frequencies) have been detected by the monitoring of high-angle scatter microwaves.

Broadband pulse–echo techniques (see Sect. 2.3) are rapidly becoming popular for the inspection of civil engineering structures such as old masonry-arch and concrete bridges, and highway foundations, partly under the pressure of impending EC legislation on increases in permitted axle weights for heavy goods vehicles. The method is providing serious competition from the acoustic impulse (hammer) technique in this area (Forde and McCavitt 1993). It has been proposed that some structures might be effectively imaged without depth distortion, by the use of velocity tomography (Bridge *et al.* 1990), analogous to attenuation tomography used in medical radiography.

2.2 Testing from Dielectric Loss Measurements

Many polar dielectric materials have resonance like relaxation losses (i.e., a large imaginary component to ε) in certain regions of the electromagnetic spectrum. For some materials, notably water, these peaks occur in the microwave region of the electromagnetic spectrum. Thus, minute quantities of water and also some other polar liquids and dielectrics can be detected using microwave through-transmission techniques. Like the real component of ε, dielectric loss can also vary with the control conditions operating in production processes, such as the curing of resins and vulcanization (during which cross-links replace freely vibrating groups including polar groups) and oxidation and so on (Bridge 1987). For these reasons microwaves are used widely for the on-line high-speed monitoring of the moisture content of paper, cardboard and paper-based laminate boards during production.

2.3 Instrumentation

Both continuous-wave and pulse techniques are used. Above a few gigahertz, transmission and reception are via horns terminating in waveguides; alternatively, a miniature diode detector can be employed. The transmitter and receiver can be located in parallel, for dielectric loss measurement by through transmission, or at oblique angles for scatter and reflection measurements. Velocities can be measured in through transmission by setting up standing waves from a metal reflector, and by using a diode detector (which gives minimal interruption to the field) to measure the spacing of nodes and antinodes. Current pulse techniques involve broadband transmission, that is, pulse lengths as short as half a wavelength are used to maximize spatial resolution in the discrimination of echoes. Such systems are popularly known as impulse radar or ground probing areas.

3. Eddy Current Methods: Metal Inspection

3.1 Probe Coil Impedance Measurements

Consider a coil carrying alternating current I with an impedance

$$Z = R_0 + j\omega L_0 \tag{34}$$

when located in free space. The flux of magnetic induction threading the coil, $N_0 = L_0 I$, is in phase with I. When the coil is placed near an electrical conductor the perturbation of the electromagnetic field, in general, causes the impedance to change to

$$Z = R + R_0 + j\omega L \tag{35}$$

The electric field generates eddy currents in the conductor and the joule heating loss associated with these currents must be supplied by the coil circuit; the loss rate can formally be represented by $I^2 R$, where R must be positive. The flux threading the coil in the presence of the conductor must generally have a different phase from I and N_0, because this flux is the only agency through which the eddy current loss is coupled to the power supply, that is, the component of the flux which is $\pi/2$ out of phase with I induces an emf in the coil of such phase as to produce the apparent resistance R. The component of flux in phase with I equals LI. Even in the unlikely event of the conductor producing no alteration in flux apart from a phase change, L will differ from L_0 because of the phase shift. In general then, L will differ from L_0. If a surface crack is present in the conductor, or if the specimen–conductor specimen is changed, or a conductor with different σ or μ substituted, both impedance components may change yet again because of further disturbances to the amplitude and phase of the flux threading the coil.

Eddy current testing, in essence, consists simply of measuring changes in the real and imaginary components of a coil impedance (at constant power of current source). However, interpretation of data is not simple, as quantitative modelling of the effect of variations of σ, μ and volume of material in the impedance plane is cumbersome even with the most straightforward geometry. A general feel for the impedance behavior can be obtained as follows. Given that $N_0 = L_0 I$ is the flux threading the coil when located in free space, in the presence of other material with permeability μ and conductivity σ we can allow for the possible phase shift (relative to I) of the modified flux threading the coil by defining a complex inductance L^* and flux N^*, where $N^* = L^* I$. The

corresponding coil impedance is $j\omega L^*$; thus from Eqn. (34) one can make the following equalities:

$$\begin{aligned} Z^* = R + j\omega L = j\omega L^* = j\omega N^*/I \\ = j\omega(N_R + jN_I)/I \end{aligned} \tag{36}$$

It is convenient to normalize Z^* relative to the free-space case, giving

$$\begin{aligned} Z^*/\omega L_0 = R/\omega L_0 + j\omega L/\omega L_0 \\ = (j\omega N^*/I)/(\omega N_0/I) = jN^*/N_0 = j\mu^* \end{aligned} \tag{37}$$

where μ^* is defined as the complex relative permeability. We have

$$R/\omega L = -\mu_I \tag{38}$$

$$\omega L/\omega L_0 = \mu_R \tag{39}$$

The general behavior of μ when the conducting material effectively fills all of space, that is, the material fills the coil core, can be predicted from the above. Families of curves, required to sort out the effects of these variables, can be sketched intuitively (see Fig. 2), as follows, where the liftoff LO is the change in coil–specimen separation (volume of space occupied by the material). In all cases it will be assumed that, as with the boundary curve, μ^* varies smoothly within the boundary limits. (For metals $\mu/\mu_0 = 1$.)

(a) *Variation of m^* with LO at constant $\sigma\omega$ (see Fig. 2a)*. With increasing LO, a maximum value of LO_∞ must be reached beyond which there is no disturbance to the coil field by the metal, that is $\mu^* = \mu_R = \mu/\mu_0$ (in the case of an encircling coil this is reached when LO is the core radius, that is, when there is no material at all). Thus, from each point of different $\sigma\omega$ along the boundary, at which LO = 0, one can draw smooth curves along which LO increases, each corresponding to a different $\sigma\omega$, and which all meet at the point $\mu^* = \mu_R = \mu/\mu_0$. In the absence of more exact information these curves are drawn as a series of straight lines which are arcs of the boundary curve.

(b) *Variation of $\sigma\omega$ at constant LO (see Fig. 2b)*. At $\sigma\omega = \infty$ there is no loss (i.e., $\mu^* = \mu_R$), but as LO increases from zero μ_R must increase from zero, reflecting the presence of electromagnetic field energy outside the material. Therefore from each point along the μ_R axis one can draw curves anticlockwise which follow the shape of the boundary curve, along which $\sigma\omega$ decreases, each corresponding to a different LO, and which all meet the boundary curve at $\mu^* = \mu R = \mu/\mu_0$, when $\sigma\omega$ is zero.

(c) *Ferromagnetic materials*. The ratio μ/μ_0 is a variable ranging from 1 to some maximum value.

(d) *Variation of $\sigma\omega$ at constant μ/μ_0 and LO (see Fig. 2d)*. These curves are simply replica shapes of the boundary curve, except that for $\sigma\omega = 0$ there are

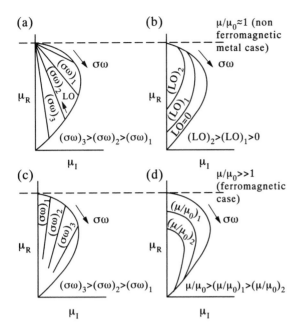

Figure 2
Schematic form of the behavior of the normalized impedance, $\mu^* = \mu_R + j\mu_I$, of a coil in the vicinity of a conductor as a function of coil–metal separation LO (liftoff), conductivity, permeability and frequency, sketched on the assumption that $|\mu^*|$ must vary smoothly and monotonically with each variable, between extremal values

different intercepts of $\mu^* = \mu/\mu_0$ on the μ_R axis varying between 1 (corresponding to saturation magnetization) and $(\mu/\mu_0)_{max}$. As the latter is so much greater than 1, typically 100 or more, the lower intercept on the μ_R axis for $\sigma\omega = \infty$ is negligible in comparison (it varies only from 0 to 1 with increasing LO). An increase in lift-off will lower the upper intercept, thus producing the same apparent effect as a decrease in μ/μ_0.

(*e*) *Variation of μ/μ_0 and LO at constant $\sigma\omega$ (see Fig. 2c).* From each point of different $\sigma\omega$ on the boundary curve, one can draw smooth curves (straight lines in the absence of more exact information) of decreasing μ/μ_0, each one corresponding to a different $\sigma\omega$ and with LO assumed zero; these will converge near the origin, since for $\mu/\mu_0 = 1$, its lowest value, $|\mu^*| \leqslant \mu/\mu_0$ for all $\sigma\omega$. However, increasing LO gradually, at constant μ/μ_0 and $\sigma\omega$, will have the same effect, with all lines converging to the point $\mu^* = 1$, just as in the case of nonferromagnetic metal.

The impedance plane can be divided into a low-frequency (LF) regime and a high-frequency (HF) regime by a change in sign of the curvature of the boundary curve. From inspection of Fig. 2 one can

identify the following salient features for constant, ω, as follows.

(*a*) *Nonferromagnetic metal.* At LFs the direction of changes in μ^* due to changes in liftoff and conductivity are similar. Therefore LF impedance measurements alone with a single coil will not permit unambiguous indications of cracks or material conductivity changes unless LO is precisely controlled. However, even with a nominally zero LO, contact pressure and slight variations of probe orientation can produce liftoff effects, as will dirty or rusty surfaces. For this reason twin coils, rigidly connected for identical liftoff, are frequently used, so that when one coil is over a crack, any difference between the impedance vectors of the coils is due solely to the crack. At HFs the changes in μ^* due to liftoff and conductivity changes are in very different directions, so that material structure changes and cracks can be reliably detected from impedance vector measurements with a single coil. A variety of options exist for processing data at more than one frequency to improve the quality of discrimination between different variables. LO sensitivity is exploited for measurement of the thickness of dielectric coatings, for example, paint on metals, and also for surface roughness measurement.

(*b*) *Ferromagnetic material.* The situation described for nonferromagnetic metal applies if the material has been magnetized to saturation, but otherwise $\mu/\mu_0 > 1$ and relative permeability and liftoff changes cannot be discriminated by impedance vector measurements alone, as the vector angles associated with these changes at any $\sigma\omega$ value are virtually the same. At LFs only the directions of change in μ/μ_0 due to permeability and conductivity are very different, but in cracks local decreases in both of these quantities are involved so that the impedance change due to cracks is not necessarily in the direction of the conductivity change. For example, the skin depth δ depends on the product of μ and σ, therefore for crack depths less than δ, both μ and σ affect the fraction of eddy currents crossing the plane of the crack whose progress is impeded. In the ferromagnetic case, therefore, detailed information on known cracks or mathematical modelling is needed for any sensible predictions on the angles of impedance changes associated with cracks. In any event, it seems clear that there is little to be gained from the use of high frequencies, especially as the sensitivity of μ_R to changes in m caused by cracks, and so on, is greatest at the lowest frequencies, and the use of mains frequency will be the most convenient option.

3.2 Instrumentation and Probe Configurations
Configurations with coils encircling the specimen under test, or with coils in contact or stood-off from the test piece with axis normal to its surface are both used in practice, and the general theoretical analysis previously described applies in all cases. The encirc-

ling variety is used extensively for cooling tube inspection in conventional and nuclear electric power generation plant, and on-line inspection of drawn tubing and rod. The stand-off type is typical of aircraft inspection, especially bolt holes, rivets and turbine blades. It is readily possible to wind coils with a 1 mm cross section. Assuming a length of a few millimeters in this case to confine the field, and given that an impedance change of 1 in 10^4 is readily obtained, then a crack of volume down to about 10^{-3} mm (coil volume/10^4) should be detectable, for example, $\geqslant 1$ mm long, 0.1 mm deep, and 10 μm wide or 1 mm deep and 1 μm wide. For more rapid inspection at lower resolutions, coils of one to two orders of magnitude larger are used, at lower operating frequencies.

3.3 Low-frequency Eddy Currents: Magnetic Methods

Electromagnetic tests on ferromagnetic or ferrimagnetic material are usually performed at mains frequency or quite rarely with dc fields, and almost invariably involve separate transmitter and receiver probes. These tests, which rely on magnetizing the material significantly, are commonly known as "magnetic methods" in the NDT industry. This terminology perhaps disguises the fact that, except in the dc case, eddy current effects are always a significant component in the defect detection in ferromagnetic materials as σ is high and it is the product of μ and σ which governs the loss level. Arguably the terminology could be justified as follows. As **M** in Eqn. (6) is large, thus μ is large and **B/E** is large by combining Eqns. (10, 28); the latter approaches infinity in the case of ω approaching 0 (dc magnetization), that is, **E** approaches zero. Moreover, the coil detection routine used in eddy current tests would still work for an ideal ferrimagnetic with infinite conductivity. The changes in coil impedance attributable to local changes in μ due to a crack or other defects would take place entirely along the μ_R axis, μ_I always remaining zero.

As in Sect. 3.2, the usual test situation in "magnetic flux leakage methods" is that radiation is confined mainly within the order of the skin depth δ, which is typically 1–2 mm at 50 Hz, and propagates normally into the surface with **E** and **B** being orthogonal and parallel to the surface. The distortions in **B** and **D** produced by a surface-breaking crack are as illustrated in Fig. 1, for reasons already described in Sect. 1.3. This technique is used for wire rope inspection, and automated testing of very long lengths of oil and gas pipeline using solenoids mounted on so-called "intelligent pigs" which travel internally at high speeds. Crawler systems also exist for external use (e.g., for subsea inspection of offshore oil platforms), and corrosion monitoring of the floors of large ferritic storage tanks is performed with hand-held lawnmower like systems.

Low-frequency magnetic testing is also used for material sorting and quality evaluation. **M** (and thus μ) in ferromagnetic materials is very sensitive to the variables in material structure that arise during heat treatment—namely the phases present, grain structure, carbide content and precipitation at grain boundaries—and therefore to hardness and mechanical properties. The sensitivity also applies to the hysteresis loss associated with the nonlinearity and nonreversibility of μ at high magnetization levels. Small differences between the cyclic variation of **M** (and thus **B**) in two geometrically identical samples subject to identical alternating magnetic fields **H** are readily observed. Characterization of material microstructure has also been carried out with impressive success by observation of the discontinuous movement of magnetic domains during magnetization, via the observation of transient emfs (Barkhausen effect) or stress waves (magnetoacoustic emission), a rare example of a magnetic technique which is performed at frequencies up to several megahertz (Buttle *et al.* 1986).

3.4 Low-frequency Eddy Currents: AC Potential Difference Method

Low frequencies are used, typically 6 Hz, and the electromagnetic field to be propagated across a crack is launched via two flat electrodes placed on the surface under test. Needle-like electrodes are used to measure the potential difference across any two points of interest on the same surface. The distortion of the field lines (as in Fig. 1) produced by a crack between the electrodes will obviously alter the potential difference. This kind of field geometry has been very successfully modelled (Collins *et al.* 1985, Dover *et al.* 1986). The technique is much in favor as a laboratory tool for providing crack size data for fracture mechanics calculations. Versions of this technique exist called ac field measurement, in which the needle–electrode probe is replaced by a noncontacting magnetic field probe.

4. Miscellaneous Techniques: Mains Frequency Down to DC

Halmshaw (1991) has provided concise introductions to various other techniques with relatively infrequent use, including electrostatic and exoelectron effects.

See also: Nondestructive Testing: General Principles; Nondestructive Testing, Radiographic; Nondestructive Testing, Ultrasonic

Bibliography

American Society of Non-Destructive Testing 1989 Electromagnetic field modelling. *Non-Destructive Testing Handbook (2E)*. American Society of NDT, Columbus, OH, pp. 136–45
Becker R 1986 *Br. J. Non-Destr. Test.* **28**(5) 286

Bonnin O, Chavant C, Goiordino P 1993 Tridimensional numerical modelling of an eddy current nondestructive testing process. In: Hallai and Kulscar (eds.) *NDT 92, Proc. 13th World Conf. Non-Destructive Testing*. Elsevier, Amsterdam, pp. 239–48

Bossavit A, Verite J C 1983 The Trifou code: solving the 3D eddy current problem by using **H** as a state variable. *IEEE Trans. Magn.* **19**(6), 1464–70

Bridge B *Guidance Notes on NDT Laboratory Experiments*. A textbook for the MSc course in Nondestructive Testing of Materials. Department of Physics, Brunel University, Uxbridge, UK (Brunel Library Cat. No. Qto, TA 417.2.B73)

Bridge B 1987 Nondestructive monitoring of the degree of cure of decorative laminates using 3 cm microwaves. *Nondestructive Testing Communications*. Gordon and Breach, London, pp. 47–55

Bridge B, Forde M C, Mclernon D C 1990 Image processing using a parallel processor for impulse radar testing of masonry and concrete bridges. SERC application for a research grant

Buttle D J, Briggs G A D, Jakubovics J P, Little E A, Scruby C D 1986 Magnetoacoustic and Barkhausen emission in ferromagnetic materials. *Phil. Trans. R. Soc. London A* **320**, 363–78

Collins R, Dover W D, Michael D H 1985 In: Sharpe R S (ed.) *Research Techniques in Non-destructive Testing*. Academic Press, London, pp. 211–67

Dean D S, Kerridge L A 1970 Microwave techniques. In: Sharpe R S (ed.) *Research Techniques in NDT*, Vol. 1. Academic Press, London

Dover W D, Collins R, Michael D H 1986 The use of AC-field measurements for crack detection and sizing of defects underwater. *Phil. Trans. R. Soc. London A* **320**, 217–83

Foerster F, Breitfield H 1954 *Z. Metall.* **45**(4), 188

Forde M C, McCavitt N 1993 Impulse radar testing of structures. *Proc. Inst. Civ. Eng. Struct. Buildings* **99**, 96–9

Halmshaw R D 1991 *Non-Destructive Testing*, 2nd edn. Edward Arnold, London, pp. 201–61

Libby H L 1971 *Introduction to Electronic Test Methods*. Wiley, New York

McNab A 1988 *Br. J. Non-Destr. Test.* **30**(4), 249

B. Bridge
[South Bank University, London, UK]

Nondestructive Testing: General Principles

Nondestructive testing (NDT) is the science of diagnosing the presence or confirming the absence of defects in objects such as engineering structures, components or materials, without the introduction of any changes which degrade the intended purpose of the object. Quantitatively a minimum requirement of an NDT operation is that it identifies defects before they are too large to render the object dangerous or otherwise unsuitable for its intended use. In industrial operations NDT is part of the wider area of quality control.

There is no well-defined boundary between NDT as just defined and similar terms such as nondestructive characterization of materials or nondestructive material property measurement, especially when defect is broadly defined in Sect. 1. However, the adopted definition reflects the fact that historically, in industry, the term NDT has been associated with the evaluation of the fitness of an object for a specific purpose rather than with the acquisition of new knowledge about a material, *per se*.

NDT is considered essential for the safety of aircraft, offshore oil installations, railroads, nuclear and conventional power stations, civil engineering structures and a wide range of industrial products. With the increasing emphasis on safety, quality control and product liability throughout industry, which has been accelerated with the maturation of the EC, NDT practice is becoming increasingly widespread beyond the confines of the traditional key users specified in Sect. 3.

The practice of NDT by any particular user is usually broadly driven by external factors: for example, the requirements of government statutes, conditions imposed by insurers, the specifications of products by clients, and public relations in respect of safety-critical areas. Even if the above factors are not involved, NDT may also be internally initiated on economic grounds, for example, if the use of NDT lowers overall costs.

The NDT industry can be loosely divided into three sectors: the users, contract suppliers of NDT services, and equipment manufacturers and their agents. It is in the nature of NDT problems that equipment is invariably specialized and dedicated to a specific niche in NDT. Therefore, the worldwide equipment market, at $£2 \times 10^8$ per annum, is very small compared with, say, the sales of medical instrumentation, and custom-designed systems are quite common. Large companies who are also key users frequently design and build their own prototypes, do their own NDT, train their operators in-house and as a spin off even act as contract suppliers of NDT.

1. Types of Defect

Most generally a defect can be defined as the presence of some structural feature or physical or chemical property value, local or widespread, which prevents or might subsequently prevent a material or component from performing as required. Examples are as follows:

(a) cracks;

(b) porous voids;

(c) inclusions of foreign material;

(d) corrosion;

(e) erosion by abrasive action;

(f) unsuitable hardness levels (too high or too low);

(g) unsuitable elastic moduli values (too high or too low);

(h) unsuitable coating thickness (too thick or too thin);

(i) surface finish too rough;

(j) lack of fusion in welds;

(k) undercutting in welds;

(l) yield stress too low;

(m) delaminations in rolled metal, composites and multiple-layered structures and coating;

(n) poor adhesion and delaminations between a coating and substrate;

(o) poor filler–matrix adhesion in polymer composites;

(p) poor filler dispersion (uneven distribution of particles or fibers in polymer composites); and

(q) missing or displaced parts in a multicomponent structure (for example, incorrect turbine blade spacings in a jet engine).

2. Key User Industries

The industries that make the greatest use of NDT and a representative selection of their main testing problems are as follows.

(a) *Oil and gas supply*: overland, buried and subsea pipelines, seam and circumferential weld defects, and corrosion and erosion wall thinning.

(b) *Oil and gas offshore platforms*: nodal joint inspection on steel jacket-type structures, corrosion and erosion wall thinning of oil and gas risers, concrete repair inspection, flooded tubular member inspection (indicating through-thickness defects) and wire rope inspection.

(c) *Petrochemical and other chemical plant*: large ferrous tanks (for liquid and gas storage) and plant pipework, wall thinning and pitting due to corrosion and erosion, weld defects, and integrity of large glass fiber reinforced vessels for storage of corrosive liquids.

(d) *Electric power generation nuclear and conventional plant*: reactor pressure vessel welds, and cooling tubes in steam turbine generators.

(e) *Aerospace component manufacture*: turbine blades, airframe rivets and boltholes, carbon- and glass-fiber-reinforced composite airframe components, and integrity of adhesive joints.

(f) *In-service aircraft inspection (civil and military)*: as above, all safety-critical joints.

(g) *Civil engineering construction*: concrete inspection, strength of matrix material and integrity of steel reinforcement bars, and wire rope inspection.

(h) *Railways*: cracks in track, and wheels and axles.

(i) *Shipping (naval and merchant)*: hulls and cargo holds (especially on oil tankers), inspection for wall corrosion thinning and weld integrity, and reactor pressure vessels in nuclear submarines.

(j) *Steel production*: on-line inspection during production of rolled plate for delamination defects, extruded rod and tubing, and castings and hot billet inspection in foundries (off-line testing also).

(k) *Paper and board manufacture*: on-line monitoring and control of density, water content and thickness.

(l) *Condition monitoring*: this refers to the inspection of industrial plant rather than the product. It overlaps with (b), (c) and (d) to some extent; however, it refers more broadly to any industrial equipment and production apparatus where regular inspection is needed to check for wear of moving parts in machinery, corrosion or fatigue and so on.

3. Broader Aspects of Nondestructive Testing and Trends of Development

Traditionally NDT has been largely known and defined by its application in the areas listed earlier. Although NDT classified in the broadest sense of the term is much more widespread, and becoming increasingly so. There is a tendency for manufacturing industries other than just those indicated under (j) and (k) to incorporate NDT both in the production-line process and in off-line quality control procedures. Examples are the food-processing industry, livestock inspection (measurement of fat thickness), and extrusion of thermoplastic products. Airport baggage inspection is not known as NDT, but the principles and instrumentation involved are just the same as those applied to industrial production lines. The principles and (in broad terms) the instrumentation used in medical imaging by ultrasonic, radiographic and nuclear magnetic resonance techniques are the same as those employed in industrial inspection. Yet medical diagnostics is rarely described as NDT by its practitioners.

Traditionally, plant condition monitoring has involved periodic checks, but the use of continuous monitoring is increasing, with sensors permanently attached to the surface of a structure. Continuous monitoring is becoming more widespread in all areas of NDT; in the research stage are "smart" materials with sensors built into the body material as an integral part of the manufacturing process. Thus, there are prospects of reducing aircraft downtime through the continuous monitoring of airframe components during flight.

4. Nondestructive Testing: Methods and as a Subset of Radiation Physics

It is evident from the definition of a defect (see Sect. 1) that the measurement of almost any property of matter can in theory be exploited as an NDT tool. It follows that most of the types and techniques of measurement and instrumentation described in this encyclopedia could be used, in principle, as most of them are also nondestructive in their mode of action. However, the type of measurement and design of instrument that turn out to be practical in use are usually severally constrained by factors such as ease of usage and interpretation of data, cost-effectiveness, and measurement speed. Such considerations have led to adoption of some 30–40 common types of NDT technique, differentiated by the type of measurement and instrumentation employed, and conveniently classified under the main headings: (a) ultrasonic and acoustic; (b) radiological; (c) electromagnetic (also frequently referred to as electrical and magnetic); and (d) visual, optical and thermal. The approximate market shares of equipment under these headings are respectively 40, 27, 23 and 10%.

Most NDT techniques involve the "interrogation" of the test object by the transmission of some kind of radiation through it, or by the reflection or scattering of radiation from its surface. These techniques can also be classified according to the type of radiation and particle or quasiparticle involved (see Table 1).

Defects are detectable because they interact with the incident radiation in a way that is different from the effect of a normal structure. For example:

(a) in metals, defects distort the lines of electric and magnetic fields, and disturb the distribution of eddy currents induced by electromagnetic radiation;

(b) defects reflect or scatter ultrasound;

(c) void-like defects reduce the attenuation and scatter of x and gamma rays; and

(d) inclusion-like defects reduce or increase the attenuation and scatter of x and gamma rays.

4.1 Penetration Depth and Defect Detection Resolution

Usually, objects to be tested are optically opaque so that visual and optical inspection only detect surface-breaking defects and cannot measure their depth. Most defects are surface-breaking, but they may be inaccessible to visual and optical tests because of a dirty or rusty surface. Also, if the depth of defects need to be known, one of the other test methods needs to be applied, although it may still be appropriate to do a visual or optical test first since this is frequently easier.

The depth in a material at which defects can be detected to the required specification (i.e., the penetration depth) can be estimated from a knowledge of the radiation attenuation coefficient in the material, which is usually an increasing function of frequency, and the maximum possible signal-to-noise ratio available at the radiation detector, that is, the ratio obtained in the absence of any attenuation. The minimum dimensions of a defect that can be detected and the precision with which it can be located in space (resolution) are of the order of the radiation wavelength. When the wavelength is very short, that is, of the order of an atomic spacing or less, as with x,

Table 1
Classification of NDT techniques

Technique	Radiation	Particle/Quasiparticle
Ultrasonic and acoustic	mechanical waves	phonon
Electromagnetic	low-frequency electromagnetic radiation (0 to microwave frequencies)	photon (low energy)
Radiological	x and gamma rays	photon (high energy)
	beta rays	electron
	neutron beams	neutron
	positron annihilation	positron
Visual, optical and thermal	visible, infrared and ultraviolet light	photon (intermediate energies)

beta or gamma rays, the resolution is limited by quantum statistics and instrumentation performance rather than by wavelength.

4.2 Complementarity of Techniques

For a given frequency and type of radiation the scattering, absorptive and reflective power of different materials varies enormously from negligible to infinite. The same is true of defects because their size and shape, as well as their nature, can vary so widely. For this reason the techniques as classified in Table 1 are often complementary rather than competitive alternatives, for example:

(a) megahertz-frequency ultrasound can penetrate several meters of steel using cheap, simple apparatus, whereas the use of x rays would require a vast, expensive and immobile particle accelerator, and for electromagnetic waves the skin penetration in metals is very small, except at frequencies so low ($\sim$50 Hz or less) that resolution is poor;

(b) microwaves penetrate thick concrete readily, while the ultrasound technique requires very low frequency (thus limiting resolution) and x rays must have very high energies as explained in (a);

(c) megahertz-frequency ultrasound and normal x rays penetrate water readily while radio waves and microwaves do not; and

(d) x-ray gamma-ray and electromagnetic techniques are noncontact methods, which can work on rough inspection surfaces, whereas ultrasound methods usually require close mechanical contact and a smooth surface of entry; furthermore ultrasound, unlike these other radiations, cannot penetrate multiple solid-air interfaces.

5. NDT Probes (Transducers, Sensors)

Here the terms probe, transducer and sensor are used synonymously, although in the NDT industry the first term has always been common parlance in ultrasonic, electromagnetic and some optical techniques, and there has been little enthusiasm for borrowing the latter two terms from the wider realms of measurement and instrumentation science, notwithstanding the fashionability of sensor. Most transmitting probes have the common feature that electric power is converted into some form of interrogating radiation, while most receiver probes convert received radiation into an electrical or optical signal which is then subsequently processed. It is instructive to list the variety of probes in use as follows:

(a) piezoelectric transducers (for ultrasonic and acoustic techniques);

(b) EMATS for noncontact electromagnetic generation of mechanical waves;

(c) pulsed lasers for noncontact ultrasound propagation;

(d) laser interferometers for noncontact detection of ultrasound;

(e) fiber-optic light guides for visual techniques;

(f) continuous-wave lasers for holography, photoelasticity and other optical inspection techniques;

(g) conventional light and infrared sources;

(h) solenoids for electrical and magnetic techniques;

(i) x-ray tubes and beta, gamma and neutron sources; and

(j) photon detectors (light, infrared, ultraviolet x, gamma): film, photodiodes and charge-coupled devices (CCDs)

In some NDT techniques, for example, electromagnetic and ultrasonic/acoustic methods, combined transmitter/receiver probes are possible; in the acoustic technique combined probes are more commonly used than separate ones. In some other techniques (e.g., x and gamma radiology), the radiation transmitters are essentially separate: all techniques are possible using separate probes. Probe arrays are used with increasing frequency instead of or in addition to mechanical scanning mechanisms. For example, phased excitation of an array of ultrasound transmitters can be used for electronic beam steerage. The most common and oldest example of a detector array is photographic film, which can accommodate up to 100 image pixels mm^{-2} at low x-ray energies. A modern CCD array can have pixel dimensions as small as 15 μm.

6. Modelling of the Interaction of Radiation with the Sample under Test

Traditionally, the identification and sizing of real defects have been aided by comparing received probe signals with those obtained using calibration samples containing a range of simulated defects with a spectrum of sizes and shapes. A substantial number of such samples may be needed to cover all required cases, and the storage space requirements and production costs can become prohibitively expensive when large samples are involved, for example, simulations of defects in nodal joint weldments on pressure vessels or offshore oil platforms.

In principle, a nondestructive measurement process can be theoretically modelled. Thus, the radiation field produced by the transmitting probe can be computed, the effect on that field of, for example, a defect of specified size or shape can be modelled, and the corresponding amplitude, profile, phase, and so on, of the signal at the receiver probe can be simulated. By comparison of actual signals with simulated signals from a library of hypothetical defects covering a wide range of expected defect sizes and shapes, real defects can be identified and classified by size and shape. Thus a library of simulations can replace a physical collection of test samples. Within given constraints on production time and costs, and available physical space, it can become cost-effective to generate far more computer simulation cases than experimental ones, thus improving the precision of interpolation between cases, and therefore the overall coverage. The continuously falling costs of the necessary computing power and associated improvements in computer modelling techniques have meant that the above ideas have become a reality in some specialized cases, for example, in the testing of reactor pressure vessel nozzle welds. Here some impetus for computer modelling arose partly because of the great variability of results on defect sizing produced by different operators in the PISC I and II trials, and partly through the bulk and production expense of experimental mock-ups (Farley and Dikstra 1993).

7. Design of the Scanning Mechanism for the Probes: Automated NDT

Most NDT requires the rapid movement of a probe over a sample, either continuously or in discrete steps. Manual methods are still most often used, particularly in field inspection, for example, of geographically remote pipelines to which it may not be cost-effective to transport an automated system. The human operative excels in being able to scan readily over objects of complex contour. The feedback mechanism between the human eye and hand ensures, for instance, that the optimum orientation and position of an ultrasound probe for detection of a given defect is obtained with great precision as the eye "locks in" on the maximum echo, and the hand can maintain these coordinates for sufficient time for the desired signal to remain stable and to be recorded. It is extremely difficult to replicate these best features of the human operative with a mechanical scanner: a robust device which does not permit mechanical vibrations to affect signal stability is usually relatively bulky, purpose-designed to scan a specific type of structure, and not capable of being used on other structures of different shape.

In recognition of the skills of the human operator, hybrid manual–automatic scanning systems exist. Here the probe is manipulated and scanned by a human operator but, like a computer mouse, the probe holder contains a probe coordinate-sensing system so that images can be reconstructed by computer (as described in Sect. 8) from probe and coordinate data fed along a holder–receiver umbilical cable. However, fully automated scanning systems tend to be used in the following situations.

(a) Human limitations such as fatigue and lapses in concentration would occur, for example, when immense lengths of pipeline have to be tested or when 100% coverage of smaller objects is needed in fine detail, for example, safety-critical aerospace components such as carbon fiber reinforced composite sheets.

(b) The environment is hazardous (e.g. radioactive areas of nuclear power plant or subsea oil installations).

(c) The test object is physically inaccessible to a human operative, for example, the interior of a pipeline or a cooling tube in a power generation plant.

8. Image Reconstruction

In manual techniques a human operator interprets the changes in the response of a detector probe as it moves relative to the sample. In automated methods, a common modern technique is for the probe position coordinates (frequently x and y) to be represented by the position of a light spot on a VDU while the light intensity (on a gray scale) or color is a measure of the probe response. Thus a two-dimensional projection of some physical property variation within the sample is obtained.

In some special techniques a three-dimensional image can be built up as a family of two-dimensional projections; examples are computed axial x-ray tomography and Compton scatter tomography. For transparent objects or for surface inspection only, there is also optical holography. In a three-dimensional image the contrast at any given point corresponds to a physical property variation at a unique point in the sample.

Other image presentations are also possible, for example contour mapping of property variations, two-dimensional histogram mapping and isometric projections.

Image resolution, that is, the minimum pixel size, is ultimately limited either by the wavelength of the radiation employed or by quantum effects. Ultrasound technique tends to be wavelength-limited and lengths can vary from several centimeters to one micrometer as the frequency changes from a few kilohertz to a few gigahertz, which is at present the practical upper limit. In the case of x and gamma rays the limitation tends to be quantum in nature because wavelengths are so short. In all cases, instrumentation may impose more stringent limits.

9. Defect Sizing and Fracture Mechanics

Crack-like defects grow in the course of time under the influence of the intense local stress fields that they engender. Unless the growth is halted, the component will eventually fail when the defect traverses a complete wall thickness. If a crack depth is known at a given time, its depth at a subsequent time can be predicted using the theory of fracture mechanics, provided that the initial stress field is also known. Thus precise defect sizing is a vital part of NDT because it allows prediction of the lifetime of the component subsequent to the test and a decision can then be made as to whether there is any immediate danger. A typical testing strategy reflecting these considerations is as follows.

(a) Find the defect (this is a matter of pattern recognition in the image reconstruction);

(b) size it;

(c) predict the component lifetime using fracture mechanics; and

(d) if the lifetime is shorter than required, or if the size exceeds permitted standards (e.g., BSA, EC), either
  (i) Grind away the crack or if this would still leave the structure too weak, that is, too thin, or
  (ii) replace or repair the structure.
    (This is sometimes not a decision to be made lightly—a subsea nodal joint weld repair on an offshore oil platform might cost in excess of £1 × 10^6.)

10. Future Trends

Notwithstanding the relatively small market compared with, for example, medical diagnostic instruments, NDT instrumentation has kept reasonable pace with the latest advances in digital signal processing, and computer hardware and software, and will doubtless continue to do so. Expert-system and neural-network approaches to automated data interpretation have made only minor inroads, but seem set to increase. The impetus will continue to make NDT more quantitative in the characterization of the size and shape of defects (for an example, see Silk *et al.* 1987) for reasons that are obvious from Sect. 9.

See also: Nondestructive Testing, Electromagnetic; Nondestructive Testing, Radiographic; Nondestructive Testing, Ultrasonic

Bibliography

Coffey M J 1988 Mathematical modelling in NDT, Vol 1. *Proc. 4th European Conf. NDT*. Pergamon, Oxford, p. 79
Farley J, Dikstra B J 1993 Performance demonstration of ultrasonic inspection—the value of computer modelling. In: Hallai C, Kulscar P (eds.) *NDT 92. Proc. 13th World Conf. NDT*, Vol. 2. Elsevier, Amsterdam, pp. 845–9
Halmshaw R 1991 *Non-Destructive Testing*, 2nd edn. Edward Arnold, London
Iverson S E 1993 New developments of the ultrasonic P scan techniques and new perspectives in automatic ultrasonic inspection In: Hallai C, Kulscar P (eds.) *NDT 92: Proc. 13th World Conf. NDT*, Vol. 2. Elsevier, Amsterdam, p. 896
Jensen K 1993 UltrasSIM, A new method for simulation of ultrasonics of complex geometry. In: Hallai C, Kulscar P (eds.) *NDT 92: Proc. 13th World Conf. NDT*, Vol. 2. Elsevier, Amsterdam, pp. 901–5
Silk M G, Stoneham A M, Temple A J G 1987 *The Reliability of NDT*. Hilger, Bristol, UK

Further information on nondestructive testing can be obtained from the following organizations.

American Society of Nondestructive Testing (ASNT), (advice on training and certification of NDT operatives) British Institute of Nondestructive Testing (BINDT) 1 Spencer Parade, Northampton, UK (advice on training and certification of NDT operatives) British Standards Institution, 2 Park Street, London, W1A 2BS (information on acceptance standards for various NDT tests) National NDT Centre, UKAEA (UK Atomic Energy Authority), Harwell, Oxfordshire, UK (wallchart of information on NDT techniques and suppliers of NDT services)

B. Bridge
[South Bank University, London, UK]

Nondestructive Testing, Radiographic

Historically the term "radiography" has been used to describe testing by means of x-ray and gamma radiation, but with modern techniques the term is now conveniently extended to cover all cases in which the interrogating radiation consists of atomic, molecular or subatomic particle beams and photon beams which are energetic enough to cause ionization. The term "radiology" is also used by some authors to cover the general case (Halmshaw 1966, 1982, 1991). In radiographic methods the transmitting transducer, which generates the radiation, is always separate from the receiving transducer (radiation detector). The incident beam is attenuated (both absorbed and scattered) by interaction with matter. As an example, photon attenuation increases with atomic number (electron density) and, at sufficiently high photon energies, is linearly proportional to material density. Thus, gas-filled defects such as cracks or voids cause a local attenuation reduction. In contrast, neutron attenuation does not correlate with atomic number, so that neutron beams can penetrate immense thicknesses of steel but are heavily attenuated by certain atoms of low atomic

Table 1
Source types

Source	Photon/particle energy (keV)	μ (cm^{-1})	Useful penetration depth in steel (with film detection)[a] (mm) minimum	maximum
X-ray tubes	30	63		
	150			25
	400	28		75
	1000			125
Y isotopes				
^{192}Ir	296–470 (several lines)		6	100
^{137}Cs	662		12	110
^{60}Co	1.2×10^3	0.47	12	180
Linacs	4–	0.25		300
	24×10^3	0.253		
Betatrons	10–	0.237		
	30×10^3			325
Neutron sources	0.01 eV–100			very thick[b]
β isotopes				
^{90}Sr	540			thin paper
^{204}Tl	770			boards
^{90}Y	770			and metal
				coatings

a for aluminum and other light alloys increase this figure by a factor of five (at the lowest energies), decreasing to two at highest energies   b unless heavy absorbers of neutrons are present, such as H, ^{10}B, Gd, C: for example, penetration in polymeric materials and steel subject to hydrogen attack is much lower

number (see Table 1). The wide range of source and detector types used in radiography are indicated in Tables 1 and 2, respectively. General textbooks on radiographic nondestructive testing (NDT) written at a substantial theoretical level are relatively rare so that some old texts still remain the best available for certain aspects of the theory (McMaster 1959, Halmshaw 1966, Herz 1968). Examples of recent texts are given in the bibliography (ASNT 1985, Halmshaw 1991), which are aimed at a more practical level and contain extensive references to industrial standards (e.g., British Standards Institution and US equivalents), handbooks published by national NDT institutes and equipment manufacturers, and specialized literature for which there is not the space to quote here. Medical diagnostics are not discussed, although all the principles discussed here apply, as it would duplicate coverage in other articles in this Concise Encyclopedia.

Given that single photons and particles interacting with a detector can be counted, it is possible, at best, to achieve radiographic images whose resolution is limited solely by quantum effects, that is, by the standard deviation in incident photon and particle flux which is determined by Gaussian statistics. Since interaction cross sections for a given photon or particle energy and material are usually reliably known and tabulated, theoretical modelling of radiographic images and equipment performance can be computed with great precision by Monte Carlo methods. In all other NDT techniques, lower precision is reached: for example, absolute ultrasound absorption coefficients are usually unpredictable purely from theory (there are invariably unknown variables in theoretical formulae) and experimental values are frequently unreliable. In Sects. 1 and 2, it is shown that a general understanding of the performance available in radiography can be obtained in terms of Gaussian statistics using a simple first-order approximation, without invoking Monte Carlo methods.

Radiography is still the most commonly used NDT technique considered in terms of its world market share of equipment sales, which stands at 37%. It is possible to select a photon or particle of energy sufficient to penetrate almost any engineering structure; hence, radiography is used extensively in "whole body" aircraft inspection, oil and gas pipeline testing and pressure vessel inspection (particularly in nuclear power plants). Film is still the most common detector, increasingly with the help of digital image analysis and data storage based on light spot scanning of films.

For brevity in all subsequent discussions, only

Table 2
Detectors

Type	Quantum efficiency		Spatial resolution, U_d[b]	
	low kV	1 MeV	low kV	1 MeV
Film (without screen)	few %	<0.1%	0.1 mm	0.5 mm
Film (with lead screens)	few %	0.5%	0.1 mm	0.5 mm
Film (with salt screens)	few %[a]	0.15%	0.3 mm	0.5 mm
Fluoroscopic screens	few %[a]	0.15%	0.4 mm	
Image intensifiers with fluoroscopic screens	few %	0.15%	0.4 mm	
Image intensifiers with TV monitors	few %	0.15%	0.4 mm	
Image intensifiers coupled to scintillators	up to 70%	up to 70% (with thicker scintillator)	strongly design dependent	
Photodiode detector arrays	few %	0.15%	few mm	
Photodiodes coupled to scintillators, e.g., CsI	up to 70%	up to 70% (with thicker scintillator)	few mm to 10 mm	few mm to 10 mm

a fluorescent screens increase the detector gain rather than Q b quoted for fine grain film

ionizing photon radiation will be referred to, but almost all of this article is also applicable to other radiographic sources.

1. Through Transmission Technique

The most common method of radiological imaging consists of placing a source of penetrating radiation near one side of a test object and some form of radiation detector near the opposite side (see Fig. 1). A local reduction in object density or atomic number caused by a defect produces a reduction in the absorption of radiation and, correspondingly, an increased detector response, for example, increased darkness if film detection is employed. The images of all defects along a straight line joining the source and detector are superimposed at the same point on the image plane, hence a shadowgraph image is produced. Multiple scattered radiation is non-image forming and fogs the images of defects. The effect of scatter is most serious when using films, fluoroscopic screens or conventional image intensifiers for detection, especially because the scattered photons have a lower energy than the primary beam (source photons), and are therefore selectively absorbed. When photon detector arrays or individual photon counters are employed instead, the energy resolution capabilities of these devices allow the effect of scatter to be reduced.

Traditionally, film has been the most commonly used detector for most industrial testing, including oilfield pipeline inspection, partly because of statutory requirements for a permanent record and partly be-cause film gave the best spatial resolution, all other factors being equal. Dramatic reductions in the costs of computer image storage and improvements in source and detector technology are causing this situation to change rapidly (British Standards Institution 1988).

1.1 Thickness Sensitivity: Performance Equation

With reference to Fig. 1, if N is the detector count in time t over a cross-sectional area A of the image plane, and inverse square law effects are ignored for clarity:

$$N = Qtn_0 AB \exp(-\mu x) \tag{1}$$

where Q is the detector quantum efficiency (ratio of

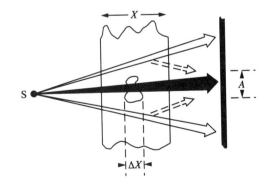

Figure 1
Schematic diagram illustrating through-transmission radiography

incident photons to detected photons, assumed energy independent for simplicity); n_0 is the photon flux (photons per unit per second) incident on the object, assumed monochromatic for simplicity; B is the factor by which the count rate is increased due to scattered radiation (build-up factor); μ is the primary beam attenuation coefficient; and x is the specimen thickness along the beam axis.

For an isotropic and monochromatic source with an activity of n_a photon emissions per second, $n_0 = n_a/4\pi x$(source–specimen distance)^2. If ΔN is the change in count N, due to a thickness change of material Δx caused by a defect:

$$\Delta N = Qtn_0 A\mu \exp(-\mu x)\Delta x \qquad (2)$$

For a high probability of detecting the defect, ΔN must exceed the standard deviation in N, which is $N^{1/2}$, by a factor S, the signal-to-noise ratio, which is typically between three and five, depending on what detection probability is considered acceptable.

Thus,

$$(\Delta x)_{min} = \left(\frac{S}{\mu}\right)\left[\frac{B}{Qtn_0 A \exp(-\mu x)}\right]^{1/2} \qquad (3)$$

where $(\Delta x)_{min}$ is the minimum thickness change that can be detected with a probability $P > \mathrm{erf}(S/2^{1/2})$. Equations (1–3) assume that $A < A_d$, the area of the defect image. If the reverse is the case A must be replaced by A_d^2/A in Eqn. (3). For a given detector (i.e., constant Q and A) and flux incident on the detector (i.e., constant $tn_0 \exp(-\mu x)$):

$$(\Delta x)_{min} = \mathrm{const}(S/\mu)B^{1/2} \qquad (4)$$

Starting from a minimum value of 1 at $x = 0$, B increases indefinitely with x, all other factors remaining the same. Therefore, to maintain a good thickness sensitivity with increasing specimen thickness it is not enough just to increase the measurement time t and the source flux n_0, even supposing that it is practical to do so: the photon energy must also be increased. Increasing the photon energy within the range ordinarily used (up to several MeV) decreases both μ and B monotonically. On balance, the quantity $B^{1/2}/\mu$, for a given x, exhibits a minimum at some finite value of μ. Therefore, under the conditions implicit in Eqn. (4) there exists an optimum photon energy which minimizes $(\Delta x)_{min}$. As x increases, the optimum energy shifts upwards (i.e., the optimum μ shifts downwards) and the minimum value of $(\Delta x)_{min}$ tends to increase.

In summary, the effect of scattered radiation is such that the thicker the object, the higher must be the source photon energy to maintain good thickness sensitivity.

These general principles apply no matter what

radiation detector is used, but B is a function of the radiation detector. B is low for detectors with high-energy resolution capabilities and the highest of all for film detectors. Build-up factors for virtually any situation are computable from atomic scattering probabilities by Monte Carlo techniques. Extensive tables also exist in the published literature.

The useful penetration depths of various radiation sources in steel, based on the earlier considerations, are given in Table 1.

It is also important to note that the basic performance equation (3) is modified by detector limitations, such as long-term electronic instabilities and drift in real-time counting systems, inhomogeneities in the film and overexposure of the film leading to reduced Q (Herz 1966, Bridge 1986c), or the limitations of the human eye (Bridge 1986a).

Typically $(\Delta x)_{min}/x$ values of 0.5–5% can be achieved, depending on the many factors implicit in Eqn. (4), for $S = 5$ (corresponding to $P = 0.999\,999\,94\%$ detection probability). Detection probabilities of at least this magnitude are essential for high-resolution imaging of large areas, since the number of failures required to detect Δx_{min} over a cross section A in an image of N pixels will be $(1 - P)N$. It follows that cracks whose planes run at right angles to the beam cannot be detected, since then $\Delta x \simeq 0$. If the beam is directed down the crack plane, Δx is large, but $(\Delta x)_{min}$ is now larger than the above estimates, that is, the cracks must be quite deep in order to be detected, because A in Eqn. (3) is now very small.

1.2 Spatial Resolution in Through Transmission (Shadowgraph) Images

Since A is the area over which photon counts are averaged to indicate a local change in received photon flux, the corresponding lateral resolution is $A^{1/2}$, that is, this is the smallest resolvable element (pixel) in the image. It will be observed that Eqn. (4) can be written in the form $(\Delta x)_{min}A^{1/2}$ (lateral resolution × depth resolution) = a constant, when A is the only independent variable. This is a familiar result, which is to be expected of any quantum-limited imaging technique. In many detection systems A can be widely varied to provide many image processing options. For example, a film can be scanned with a light spot of variable area and the responses of different elements in a photodiode array can be processed separately or summed. The human eye, which is a photon-limited device, is a notable exception in which A can be varied only over a relatively small range: the eye averages signals over an angular range of 0.5–8.5 m, corresponding to areas of diameter 0.12–2 mm at a normal viewing distance of 25 mm (Bridge 1986a). The smallest attainable value of the lateral resolution in the detector plane (i.e., the minimum pixel size) will be subsequently indicated by U_d, and typical values for different types of detector are indicated in Table 2.

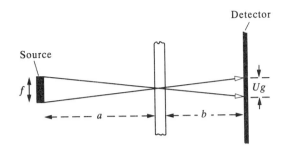

Figure 2
Geometrical factors affecting image resolution in through-transmission radiography

Using the quantities shown in Fig. 2, the geometrical unsharpness is defined as

$$U_g = \frac{fb}{a} \qquad (5)$$

To obtain images with the highest spatial resolution from a system, U_g must be less than U_d. In the case of film, the latter is called the film unsharpness U_f, and is caused by secondary electron emission in the film and/or screens. U_f tends to increase with photon energy and film grain size. It is a minimum of 0.1 mm for fine grain film and photon energy <100 keV, and 0.5–0.6 mm for 1 MeV photons and any film grain.

The reduction of f (at constant photon flux) reduces the value of a and therefore the measurement time required to achieve the maximum (detector-limited) spatial resolution. In this respect developments of minifocal and microfocal x-ray tubes with f as low as 8–10 μm are very important.

Assuming that U_g is arranged to be less than U_d, the spatial resolution in the object plane is

$$U_0 = \frac{aU_d}{(a+b)} \qquad (6)$$

Assuming that U_g is set to be $<U_d$, combining Eqns. (6) and (5) yields

$$U_0 > \frac{fb}{(a+b)} \qquad (7)$$

Even though b/a may have to be large to make $U_g < U_d$, low values of U_0 can be achieved even with detectors of poor spatial resolution (higher U_d), by making f small. Thus, the recent developments in microfocal tube technology have meant that images of resolution good enough for almost all purposes can be obtained even with real-time detectors whose U_d is much larger than U_f. This is one of a number of reasons why real-time radiographic imaging is being increasingly used.

Although real-time detectors generally have lower spatial resolution than film (Table 2), a notable exception is charge coupled devices (CCDs), for which U_d can be as low as 5 μm. CCD optical cameras are available and soft x-ray versions have been envisaged for industrial application, but high costs are a limiting factor.

1.3 Thickness Gauging by Through Transmission Technique

It is sometimes of interest to measure a thickness change over a large area, where lateral resolution is unimportant. Examples include the monitoring of uniform pipe wall corrosion thinning, and production line monitoring of the mass per unit area of board, paper, laminates and coatings. In such cases a single-aperture real-time photon counter is often used. The aperture area A can be made very large, for example, several square centimeters to make $(\Delta x)_{min}$ very small. Typically, a scintillator such as NaI coupled to a photomultiplier will be used with gamma or beta sources for thick and thin structures, respectively (see Sect. 2).

2. Imaging with Scattered Radiation and Backscatter Gauges

By the use of heavily shielded source and detector collimators (see Fig. 3) it is possible to arrange that most of the detector count rate comes from photons singly scattered within a well-defined volume of the test object (the pixel volume). The residual background count rate arises mostly from multiple scattering within the shielding material rather than primary photons. A void coincident with, and larger than, the pixel volume will cause all the single-scatter count rate to disappear, thus producing a high "image contrast." Three-dimensional indexing of the relative positions of the object and the source–detector assembly allows a three-dimensional image to be built up point by point.

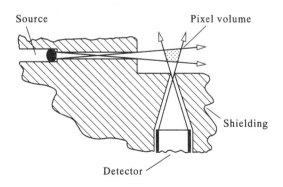

Figure 3
Schematic diagram illustrating three-dimensional imaging with scattered radiation

By three-dimensional it is meant that information from each pixel volume is displayed separately in the image reconstruction.

Systems with finer resolution (1.5 mm^3 pixel volumes) exist, but they are large, cumbersome and expensive, laboratory-based structures. Mobile systems with intermediate resolutions (3–5 mm^3) are under development for field applications, including offshore oilfield inspection (Bridge 1986b, Bridge *et al.* 1991).

2.1 Performance Equation

Assuming, for simplicity, a monochromatic x-ray or gamma source, the detector count C in time t due to scattering within the pixel volume V can be expressed in the form

$$C = kn_a V\Omega t \qquad (8)$$

where n_a is the source activity defined previously, Ω is the solid angle subtended at the pixel volume by the detector, and k is a constant for fixed relative positions of the source, detector and sample. This constant varies only slowly as the sample is indexed through the pixel volume and can be estimated by simple methods (Bridge 1986b) or accurately computed by Monte Carlo techniques (Bridge *et al.* 1991). The count from the pixel volume falls by a factor $1 - v/V$, when a void of volume v is located within V. If S is the change in the pixel count divided by the standard deviation of the change in detector count, and the background count per unit source activity, β, is assumed to be unaffected by changes in v, then it is easy to show (Bridge 1986b, 1991) that

$$\frac{V}{t} = \frac{kn_a v^2 \Omega}{S^2}\left(\frac{2-v}{V} + \frac{2\beta}{kV\Omega}\right) \qquad (9)$$

S determines the void detection probability obtainable with the volume scanning rate V/t.

3. Computed Axial Tomography

Three-dimensional images can also be obtained by a through-transmission technique, by passing a source beam in many directions through each part of an object and recording the total attenuation of each path length by means of a detector array. Figure 4 illustrates this principle. Measurement of A_1, A_2, A_3 and A_4 allows the absorptions a_1, a_2, a_3 and a_4 in the four quarters of the sample to be determined by solving the simultaneous equations

$$A_1 = a_1 + a_2$$
$$A_2 = a_1 + a_3$$
$$A_4 = a_3 + a_4$$
$$A_3 = a_2 + a_4$$

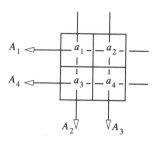

Figure 4
Schematic diagram illustrating three-dimensional imaging by computed axial tomography

Dividing the sample into smaller pixels by taking many other measurements and solving many equations allows two-dimensional absorption maps of high spatial resolution (0.1–1 mm) to be obtained. Three-dimensional images are then built up in successive two-dimensional slices. This system can be used to detect cracks narrower than is ordinarily possible by x-ray methods, although the crack image will be blurred out to the pixel size.

Detectors employed are usually miniature solid-state photoconductors and miniature photomultipliers or scintillators. However, any detector can be used in principle, including film, provided that the recorded data is subsequently digitized for solution of the matrix equations.

See also: Ionizing Radiation Measurements; Nondestructive Testing, Electromagnetic; Nondestructive Testing: General Principles; Nondestructive Testing, Ultrasonic; Transducers, Ionizing Radiation

Bibliography

American Society of Nondestructive Testing 1985 *Nondestructive Testing Handbook*, Vol. 3. ASNT, Columbus, OH
Bridge B 1986a The performance of the human eye in relation to the performance of film, a new appraisal. *Br. J. Non-Destr. Test.* **28**, 277–86
Bridge B 1986b Compton scatter imaging with low strength sources for the inspection of small components or field applications. *Br. J. Non-Destr. Test.* **28**, 216–23
Bridge B 1986c A theoretical model of the performance of film at gamma photon energies. *J. Photogr. Sci.* **34**, 95–105
Bridge B, Harirchian F, Imrie D C, Mehrabi R, Meragi A R 1991 Compton scatter imaging: corrosion, thickness monitoring and void detection. In: MacCuaig N, Holt R (eds.) *Tomography and Scatter Imaging*, Institute of Physics Short Meetings Series, No. 19. Institute of Physics, Bristol, UK
British Standards Institution 1988 Application of real time inspection to weld inspection, BS 6932. BSI, London

Garratt D A, Bracher D A (eds.) 1980 *Real Time Radiological Imaging*. American Society for Testing and Materials, Philadelphia, PA

Halmshaw R (ed.) 1966 *Physics of Industrial Radiology*. Heywood, London

Halmshaw R 1982 *Industrial Radiology: Theory and Practice*. Applied Science, London

Halmshaw R 1991 *Non-Destructive Testing*, 2nd edn. Edward Arnold, London

Herz R H 1968 *The Photographic Effect of Ionising Radiation*. Wiley, New York

McMaster R C 1959 *NDT Handbook*. Ronald, New York

B. Bridge
[South Bank University, London, UK]

Nondestructive Testing, Ultrasonic

The underlying principles of ultrasonic nondestructive testing (NDT) techniques are commonplace in the everyday world. Everyone is probably aware of sounds from beyond the immediate environment—sound can propagate from one medium to another; yodellers' calls echo across a valley—sound can be reflected or scattered at boundaries. Ultrasonic NDT harnesses these phenomena to probe within materials to detect the existence of any discontinuities or flaws that reflect or scatter the sound waves. Furthermore, since the velocity of sound is related to physical properties such as Young's modulus, its measurement using ultrasonic techniques provides a useful tool for the materials scientist.

1. Fundamental Physics

A beam of ultrasound propagating in a medium will be reflected or scattered whenever it encounters a discontinuity in acoustic properties. For smooth discontinuities that are large compared with the beam width, the reflected ultrasound obeys Snell's law, whereas for smaller defects the beam is scattered over a wide range of angles. With large flat reflectors, the amplitude of the reflection of a normally incident plane wave is governed by the ratio of the acoustic impedance of the two media, defined as the ratio of the particle pressure to particle velocity. The pressure reflection coefficient for plane acoustic waves at normal incidence to a plane boundary is given by

$$R = \frac{Z_2 - Z_1}{Z_2 + Z_1} \qquad (1)$$

where the acoustic impedance Z is

$$Z = \rho c$$

where ρ is the density of the medium and c is the velocity of sound.

With small defects, size, shape and acoustic impedance all determine the nature of the scattering, which can only be modelled for a few simple cases.

In fluids, only longitudinal compression waves can propagate, for example, the velocity of sound in water is 1.5×10^3 m s^{-1}, that is, 1.5 mm μs^{-1}. In solids, however, both compression and transverse shear waves can be supported, the velocity of the shear wave being around half that of the compression wave. Both types of wave are routinely used in NDT. At around 4–6 mm μs^{-1} in most solids, the velocity (compression wave) of sound is low enough to allow the direct measurement of the transit time of a short pulse propagating over very small distances. Axial resolution is limited more by the length of pulse that can be generated rather than by any problems with the measurement of small time intervals. In most materials of interest to the NDT engineer, the absorption of ultrasound is acceptably low at frequencies up to a few megahertz, and therefore wavelengths of the order of millimeters are feasible and widely used for routine testing. For specialized applications involving small components, much higher frequencies can be employed, approaching 1 GHz.

Shear waves can be generated by the process of mode conversion. Depending on the angle of incidence, a compression wave crossing a fluid–solid boundary can partially mode convert into a shear wave. At a solid–solid boundary, both reflected and transmitted shear waves can arise from an incident compression wave. Although such behavior is utilized to generate shear waves from more easily produced and coupled compression-wave transducers, the existence of mode conversion greatly complicates the angular dependence of plane-wave reflection coefficients and the form of scattering at defects.

2. Ultrasonic Transducers

The simplest type of transducer used in NDT consists of a single plate of piezoelectric material mounted so that it can vibrate at its half-wavelength ($\lambda/2$) resonant frequency f_r, given by:

$$f_r = \frac{2c_l}{l}$$

where l is the thickness of the plate and c_l is the compressional wave velocity. Wideband transducers are constructed by damping the resonant mode with a backing of a material having similar acoustic impedance to that of the piezoelectric element. To avoid unwanted reflections from within the backing, its ultrasonic attenuation is made as high as possible, typically by using a tungsten powder/epoxy composite. Most modern transducers employ poled ceramic materials such as lead zirconate titanate and lead metaniobate as the piezoelectric elements. The acoustic impedance of such materials is so much greater than that of gases, that virtually all the ultrasound generated within a transducer coupled to air, for instance, will be reflected back into the transducer (see Eqn. (1)). This is true even for the very thin layers of air between a flat, highly polished transducer and a similarly prepared coupling surface on a solid. Either a thin coupling layer of liquid or gel must be used, or the test piece must be immersed in a water bath. One advantage of the latter method is that the transducer may be positioned a convenient "stand-off" distance from the front surface. This aids in the detection of subsurface defects using a single transducer as both emitter and receiver, the problem being that the receiving amplifier takes a time (the "dead time") to recover from saturation by the transducer excitation pulse (typically a few hundred volts). Normally the front surface of the transducer is protected by a thin-wear plate, which can sometimes be designed to act as a $\lambda/4$ matching layer, and/or as a $\lambda/2$ interference filter to aid in reducing the energy coupled at the frequency of the thickness mode resonance.

For specialized NDT applications, there is some move towards employing the far more sophisticated transducers used in medical imaging techniques (electronically switched, linear and phased arrays), but at present the vast majority of NDT is carried out with variants of the simple devices described above.

3. The Radiated Beam

In the most straightforward pulse–echo technique, a pulse of ultrasound is emitted from a transducer and the scattering or reflection is received by the same transducer. To obtain a directed beam, the aperture of the transducer is chosen to be several *center*-frequency wavelengths. The italics serve to emphasize that very short ultrasonic pulses having a wide frequency content can be generated: center frequency/half-amplitude bandwidth ratios of around unity are readily obtained. With such short pulses, classical continuous-wave diffraction theory is not convenient to describe the structure of the radiated beam, a process of harmonic synthesis being required. A more useful approach is to consider directly in the time domain the waves radiated by a piston like source. Taking first the simpler case of propagation in a fluid, a summation of the Huyghens' wavelets radiated from each point on a source a few wavelengths in width shows that it radiates a (locally) plane wave into the region straight ahead of its face, together with a diffracted edge wave which propagates in all directions from its periphery (Stepanishen 1971, Weight 1984). In the case of a transducer emitting a short pulse, these waves can be time-separated, giving rise to complicated field point

pressure waveforms which vary dramatically with position in the beam (Weight 1984).

For continuous-wave emission, the plane and edge waves overlap, the resulting interference determining the beam structure. Take the classical case of a circular transducer undergoing continuous sinusoidal motion. At points along the axis of the transducer, theory predicts equal amplitudes for the plane and edge-wave components, their interference giving a series of equal-amplitude maxima and nulls depending on the path difference between the directly arriving plane wave and the edge wave from the rim; allowance is made for the opposite phase between the plane and inward-directed portion of the edge wave at the transducer face (Weight 1984). At an axial range where the path difference is $\lambda/2$, the two waves reinforce to give an axial maximum. Beyond this, the two waves interfere destructively to give a pressure amplitude that decreases inversely with range. The range where the path difference is $\lambda/2$ marks the end of the "near field," the near-field length being given by

$$x_{\text{nf}} = \frac{R^2}{\lambda} - \frac{\lambda}{4} \approx \frac{R^2}{\lambda} \tag{2}$$

where R is the transducer radius.

Outside the geometric region, only the edge wave exists, the interference between the portions of the wave arriving from the nearer and further parts of the rim giving a cylindrically symmetric pattern of side lobes (Weight 1984).

The concepts of near and far field are also useful when describing the behavior of short-pulse transducers, but the near-field length in terms of wavelength is not then a constant. For circular transducers, a convenient definition is to take the near-field length as that given by Eqn. (2) for the *center* frequency of the pulse spectrum.

The existence of plane and edge waves can affect both axial and lateral resolutions. In the near field, the later-arriving edge-wave contributions lengthen the field-point pressure waveform and hence impair axial resolution, the effect being particularly noticeable for points on or near the axis of a circular transducer. Lateral resolution is governed by the width of the beam, which in the near field approximates to the width of the transducer. Further away, in the far field, the beam increases in width, again as a result of the interference of plane and edge waves (Weight 1984).

Considering now the more complicated case of propagation in a solid, the existence of both compression and shear waves further complicates the field pattern (Kawashima 1984, Weight 1987, Schmerr and Sedov 1989). A simple model with accuracy adequate for many practical applications describes the field radiated from a directly coupled compression-wave transducer in terms of a compression plane wave and a compression edge wave which partially mode-converts

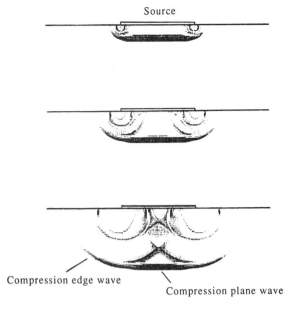

Source

Compression edge wave

Compression plane wave

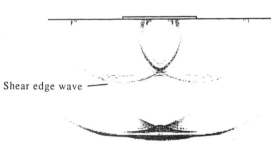

Shear edge wave

Figure 1
Wave profiles at four instants in time after the short-pulse excitation of a circular transducer directly coupled to an aluminum half-space

into a shear edge wave (Weight 1987). This behavior can clearly be seen in Fig. 1 which shows finite difference calculations (Stacey and Weight 1993) of the wave profiles from a circular transducer directly coupled to an aluminum block.

4. Echo Responses

In understanding the form of echo responses, we must certainly consider the diffraction effects outlined earlier. Again, an explanation in terms of plane and edge waves has proved useful (Weight 1984, Stacey and Weight 1993). For small targets, scattering of the incident plane and edge waves gives rise to further multiple pulses when received back at the transducer.

These pulses can be time separated if the target is in the near field. Such behavior is responsible for the poor relationship between the size of such targets and measurements of their echo response amplitude. Further away, the various plane- and edge-wave contributions overlap, and with increasing range in the far field the echo pulse shape becomes less sensitive to field-point position.

Fortunately, with larger flat targets, aligned perpendicular to the transducer axis, echo responses are much simpler. The received reflection of the plane wave is integrated over the area of the transducer to give a contribution to the echo pulse that is much larger than that due to the reflected edge waves, since these are smeared in time on reception over the whole transducer. This behavior allows accurate measurements to be made of material properties that affect the shape of the pulse propagating (the frequency dependence of attenuation, for instance), provided that the samples are carefully prepared with correctly aligned flat faces.

5. Ultrasonic Imaging

Ultrasonic "images" can be obtained using a number of techniques, the most common in NDT being the C scan method, with some application of the B scan technique widely used in medical imaging. In a pulse–echo C scan, the transducer scans over the whole area of the test piece in a raster scan, the position of the transducer being followed by some recording device. The echoes arising from within a particular depth range can be electronically gated and their amplitude detected and used to intensity-modulate the display medium. Such images give a "map" of ultrasonic scattering. Where warranted, high lateral resolution can be obtained using highly focused transducers. The ultimate development of this technique is the "ultrasonic microscope" (Briggs 1992), which can use frequencies up to 1 GHz and obtain resolution approaching that of optical microscopes. Such instruments have application in the integrated-circuit fabrication industry, where they can detect disbonds between substrates and the "pads" for electrical connections. Through-transmission C scan techniques are also used, especially for testing thin plates.

In the B scan method, the transducer is scanned over a line, its position being recorded along one axis of the display while depth into the material is recorded on the other. As in the C scan method, the amplitude of the ultrasonic scattering is shown as an intensity modulation.

Another, more specialized, approach is to form reconstructed images of the ultrasonic velocity or attenuation profile over a plane through the test piece, using techniques and algorithms entirely analogous to those of x-ray computed tomography.

6. Applications

There is hardly an area of industry where ultrasonic NDT of one kind or another does not have a role to play. A number of the more common applications are mentioned below. A point to bear in mind throughout is that much in-service testing is plagued by the difficulty of access—no matter how strong an echo is given by a particular defect, ultrasonic techniques are precluded if the component geometry and surface roughness restrict coupling.

The examination of welds is a major application, though because the geometry is often difficult it is less than straightforward. In testing welded plates and pipes, access to the weld can often only be obtained by using angled transducers positioned to one side, the ultrasound reaching the weld after a number of skips across the plate. Such techniques require great skill on behalf of the NDT inspector, both to position the transducer and to interpret the returned echoes.

The measurement of plate thickness is commonly carried out using ultrasonic methods and a large range of thickness gauges, some handheld, are available. Automatic versions of such systems are widely used to monitor the corrosion of pipelines. In a typical arrangement, a mobile instrumented unit known as an "intelligent pig" is propelled along the inside of the pipe, often by means of gas pressure. On-board computers process and store the data for later interpretation.

With the wider use of bonded structures, especially in the aerospace and automotive industries, there is a growing interest in NDT techniques to detect adhesive and cohesive qualities. The detection of adhesive problems such as glueline disbonds is relatively straightforward using ultrasonic reflection or, often for thin samples, through-transmission techniques. However, evaluation of cohesion is more difficult, and usually requires the use of sophisticated signal processing.

As might be expected, the more safety-critical and expensive the component, the more rigorous and sophisticated is the NDT. A typical example is the very highly stressed disk holding the turbine blades within a jet engine. Here failure is catastrophic and each component undergoes a particularly rigorous series of tests, including a complete ultrasonic scan which can detect defects a small fraction of a millimeter in size. If a defect is detected, further testing ensues, using higher resolution and more quantitative techniques in just the area of the defect. Other similar examples include the testing of pressure vessels and components used in nuclear power reactors.

Ultrasonic techniques are also widely used in on-line testing applications. These fall into two main groups: those to detect defects and those to check material properties, such as the state of polymerization during the manufacture of plastics. A typical example of the former group is the testing of steel plate in a rolling mill. Here, banks of transducers are used, often in through-transmission, to cover the whole width of the plate simultaneously. Coupling is achieved by jets of water continually being sprayed to maintain a column of fluid between the transducers and the rapidly moving plate. Since the mill can only be stopped at great expense, areas containing defects are noted (either by spraying with paint or by logging on a computer) to be cut out later.

7. Signal Processing

Signal processing techniques employed in ultrasonic NDT fall into two main areas: those to improve the detection of defects and those to improve characterization, either of a defect or of overall material properties. In many applications, the smallest defect that can be detected is limited by the presence of unwanted ultrasonic signals, rather than by random electronic noise. Typical examples include coherent scattering such as that from grain boundaries in ultrasonically "noisy" materials and large echoes from surfaces close to the defect. The latter can saturate the receiving amplifier and mask the presence of a nearby defect. Adaptive filtering and detection techniques similar to those used to reduce radar clutter are being applied to reduce the effect of grain scattering (Saniie and Nagle 1992). With some *a priori* knowledge of the grain structure, split spectrum processing (Newhouse *et al.* 1982) can give greatly improved defect signal/grain scattering ratios. A useful improvement in defect detection can sometimes be obtained by spatial averaging techniques that take the point-by-point average of the times-series waveform as the transducer is scanned over a defect.

Techniques to improve target characterization include time-of-flight methods (Silk 1984), where the arrival times of the diffracted waves from the tips of a crack, for instance, are measured as a pair of transducers is scanned over the crack. The crack length can be calculated from the change in arrival times with transducer position. Ultrasonic spectroscopy (Fitting and Adler 1981) has been applied for the measurement of material properties, grain size for instance. Other processing to improve characterization involves synthetic aperture techniques to give improved resolutions (Seydel 1983).

8. Current Trends

The majority of in-service techniques to characterize defects rely on relating defect size to echo amplitude, notwithstanding the limitations outlined above in considering the form of echo responses. There has been some success in developing array and nonuniformly excited transducers to overcome some of these limitations, but these are not yet commercially available. Other new transduction methods include the use of lasers to ablate or thermally shock the material surface and hence to generate an ultrasonic wave (Scruby *et al*. 1982).

Currently, many investigators (see, for instance, (Thompson and Chimenti)) are seeking to make NDT more quantitative, not the least because fracture mechanics shows just how crucial the size of a crack or other defect is to the eventual failure of the component. Growing use is being made of the "design for testing" approach, whereby the difficulties of access for testing purposes are acknowledged and the structure designed and/or instrumented accordingly. As the cost of computers falls, signal processing techniques are increasingly used, allied to greater use of scanning and imaging techniques (McNab and Dunlop 1993, Windsor *et al*. 1993).

See also: Nondestructive Testing, Electromagnetic; Nondestructive Testing: General Principles; Nondestructive Testing, Radiographic

Bibliography

Briggs A 1992 Acoustic microscopy—a summary. *Rep. Prog. Phys.* **7**, 851–909
Fitting D W, Adler L 1981 *Ultrasonic Spectral Analysis for Nondestructive Evaluation*. Plenum, New York
Kawashima K 1984 Quantitative calculation and measurement of longitudinal and transverse ultrasonic wave pulses in solids. *IEEE Trans. Sonics Ultrason.* **SU31**, 83–94
McNab A, Dunlop I 1993 A framework for advanced visualisation and automated evaluation of ultrasonic flaw data. In: Hallai C, Kulcsar P (eds.) *Proc. 13th World Conf. Non-destructive Testing*, Vol. 2. Elsevier, Amsterdam, pp. 974–8
Newhouse V L, Bilgutay N M, Saniie J, Furgason E S 1982 Flaw to grain echo enhancement by split-spectrum-processing. *Ultrasonics* **20**(2), 59–68
Saniie J, Nagle D T 1992 Analysis of order-statistic CFAR threshold estimators for improved ultrasonic flaw detection. *IEEE Trans. Ultrason. Ferroel. Freq. Contr.* **39**, 618–29
Schmerr L W Jr, Sedov A 1989 An elastodynamic model for compressional and shear waves transducers. *Am. J. Acoust. Soc.* **86**(5), 1988–99
Scruby C B, Dewhurst R J, Hutchins D A, Palmer S B 1982 Laser generation of ultrasound in metals. In: Sharpe R S (ed.) *Research Techniques in NDT*. Academic Press, London, pp. 281–327
Seydel J A 1983 Ultrasonic synthetic aperture focusing technology in NDT. In: Sharpe R S (ed.) *Research Techniques in NDT*, Vol. 6. Academic Press, London
Silk M G 1984 The use of diffraction-based time-of-flight measurements to locate and size defects. *Br. J. Non-Destr. Test.* **26**(5), 208–13
Stacey R, Weight J P 1993 Ultrasonic echo rsponses from targets in solid media using finite difference methods. *IEE Proc. Pt A*
Stepanishen P R 1971 Transient radiation from pistons in an infinite planar baffle. *J. Am. Acoust. Soc.* **49**, 1629–38
Thompson D O, Chimenti D E (eds.) *Review of Progress in Quantitative Nondestructive Evaluation*, published annually. Plenum, New York
Weight J P 1984 Ultrasonic beam structures in fluid media. *J. Am. Acoust. Soc.* **76**, 1184–93
Weight J P 1987 A model for the propagation of ultrasonic pulses in a solid medium. *J. Am. Acoust. Soc.* **81**, 815–27
Windsor C G, Anselme F, Capineri L, Mason J P 1993 The classicifaction of weld defects from ultrasonic images: a neural network approach. *Br. J. Non-Destr. Test.* **35**, 15–22

J. P. Weight
[City University, London, UK]

Nuclear Magnetic Resonance

The phenomenon of nuclear magnetic resonance (NMR) can occur when the magnetic component of electromagnetic radiation interacts with matter containing atomic nuclei which possess magnetic moments. Provided that the nuclear energy levels have been separated by ΔE, as occurs in a steady magnetic field, application of an appropriate radio frequency ν can stimulate absorption or emission of quanta $\Delta E = h\nu$, where h is Planck's constant, in spectroscopic transitions. The frequency ν required depends on the species of nucleus and the strength B of the external magnetic field, but is typically in the radio–TV region, 30–600 MHz; thus, frequencies and energies in NMR are much smaller than in other branches of spectroscopy (see *Spectroscopy: Fundamentals and Applications*).

Following its first observation in bulk material in 1946, NMR has been extensively studied in physics and chemistry laboratories since the 1950s. Significant instrumental advances and experimental innovations in the 1970s and 1980s have ensured that so-called high-resolution NMR spectroscopy continues, despite the high cost of equipment, as the most versatile and widely used physical experimental technique for the investigation of molecular structures and processes in liquids and solutions. Whereas applications to gases have been very limited, solid samples have long been studied by NMR and are now, with more specialized instrumentation, susceptible to fairly high-resolution spectroscopy.

1. Nuclear Properties and NMR Spectroscopy

In addition to having the properties of mass and charge possessed by all species of atomic nuclei, nuclei of certain isotopes behave as though they were spinning, although this property is only detected in the presence of an external field. In general, each species of nucleus is characterized by a nuclear spin quantum number or spin I, which may be zero or an integer or half-integer; this specifies the nuclear angular momentum in units of $\hbar = h/2\pi$. The nuclear magnetic moment μ, is proportional to the angular momentum, with the constant of proportionality γ (called the magnetogyric ratio) a characteristic of the species of the nucleus. Nuclear moments are small, typically less than 10^{-3} of the electron magnetic moment. Nuclei for which both mass number and atomic number are even have both I and μ zero and so cannot undergo NMR; that this applies to the common isotopes ^{12}C and ^{16}O is an important simplifying factor in the NMR of organic compounds. Nuclei that can be studied by NMR have nonzero I, either half-integral (mass number odd), such as ^{1}H and ^{13}C with $I = \frac{1}{2}$, or integral (mass number even, atomic number odd), such as ^{2}H with $I = 1$. Nuclei with $I > 1$ also possess an electric quadrupole; interaction with electric field gradients can markedly affect the NMR spectra of such nuclei.

In the presence of a fixed (Zeeman) magnetic field B_0 (typically several tesla, where $1\,T = 10^4$ Gauss), magnetic nuclei ($I > \frac{1}{2}$) in a sample distribute themselves between the $(2I + 1)$ allowed orientations and hence energy levels in accordance with the exponential Boltzmann factor. Transitions between these levels may then be induced by means of an oscillating radio-frequency field B_1, perpendicular to the B_0 field. The resonance condition for NMR, $2\pi\nu = \gamma B_0$, indicates that a given species of nucleus will resonate in a different frequency range according to the magnitude of the steady field B_0 at which the experiment is conducted, for example, ^{1}H nuclei (protons) would resonate at 42.6 MHz in a field of 1 T and at 426 MHz in a field of 10 T. Conversely, in a given B_0 field of, say, 1 T, the corresponding frequencies for other nuclear species, each with characteristic magnetogyric ratio γ, will typically differ by many megahertz.

At any realizable B_0, the equilibrium populations of lower nuclear magnetic energy levels only slightly exceed those of upper levels. Since both upward and downward transitions are induced, only the small net absorption gives rise to the observed NMR effect, an intrinsic adverse influence on sensitivity. Although about 100 nuclides have been studied by NMR, some have very low relative sensitivity (or "receptivity") since this depends on the relative natural abundance of the isotope as well as on its magnetic properties (γ^3). Thus, ^{13}C nuclei, for example, constitute only 1.1% of an unenriched sample and ^{13}C is less than 0.000 2 times as receptive as ^{1}H in the same field. The nuclei most studied by NMR continue to be ^{1}H and ^{13}C, followed by ^{19}F, and ^{31}P. The nuclei ^{2}H, ^{3}H, ^{11}B, ^{15}N, ^{17}O, ^{13}Na, ^{27}Al and ^{29}Si have already been considerably utilized for NMR, and several others have received appreciable attention. Work on the previously little studied nuclei is being stimulated by improved access to powerful multinuclear spectrometers.

2. Equipment and Experimental Measurements

Observation of NMR requires a B_0 magnetic field to separate the magnetic energy levels. This may be provided by a permanent, electroconducting, or superconducting magnet; in addition, small shim coils around the sample (volume about 0.5 cm^3) may be fitted to reduce field variations over the volume of the sample. For many chemical structure applications, especially for complex mixtures, the use of very high B_0 fields (9–10 T or more) from superconducting magnets is desirable to improve resolution between nuclei of the same species with slightly different chemical shifts (different extents of nuclear shielding by fields induced in associated electron distributions). High fields also enhance signal-to-noise ratios but much more moderate fields (no more than 1 T) are often adequate for process control NMR (see Sect. 3).

The oscillating radio frequency B_1 from the transmitter may be applied as a continuous wave (cw) or as a pulse. Early NMR spectrometers were nearly all of the cw type, in which the net absorption of energy from the monochromatic source yields the spectrum (i.e., a record of absorption intensity versus frequency) directly, after appropriate amplification and detection. In the majority of contemporary spectrometers, the B_1 field is supplied in short (1–50 μs) pulses, that is, effectively a range of frequencies is applied simultaneously, leading to the possibility of great enhancement of the signal. For many chemical applications, the transient free induction decay (FID) response signal is Fourier transformed (FT) from the time domain (i.e., as a function of time) to the frequency domain, and presented as a conventional spectrum (i.e., a function of frequency).

As in other branches of spectroscopy, the quantities that can be observed in NMR include the intensity, position and structure (shape and multiplicity) of lines; in NMR, relaxation times can also be measured. The intensity of an NMR line depends on the excess number of nuclei absorbed over those emitted so that, for example, integration of ^{1}H NMR peaks can yield a count of the relative proportions present of chemically different kinds of hydrogen in a solution of a compound or mixture. The position or frequency at which resonance occurs depends primarily on the species of nucleus and on the strength of the B_0 field, so that the species can be identified unequivocally. Since the resonating nucleus senses not only the large laboratory magnetic field but also the contribution to the total magnetic field from within the sample, the exact resonance frequency depends also on the disposition and precise chemical nature (e.g., for a high-resolution ^{1}H NMR spectrum, whether hydrogen is in a –CH$_2$ group or a –CH$_3$ group) of the nucleus.

The breadth of the resonance in a solid may extend over tens of kilohertz unless the experimental technique is able to overcome broadening due to direct magnetic dipolar interactions of one nucleus with its near neighbors. In solution, such effects are averaged out so that, with line widths below 0.5 Hz, the more subtle high-resolution effects of the chemical shift and spin coupling may be observed. Two other quantities that affect the observation of NMR, and which are best determined in time domain experiments, are the spin-lattice relaxation time T_1, and the spin–spin relaxation time T_2. Simple definitions of most of these terms are given in a short review (Jones 1983); Akitt (1983) and Kemp (1986) provide nonspecialist introductions to Fourier transform and multinuclear NMR, while Martin *et al.* (1980) treat experimental aspects, including quantitative measurements, more extensively. Freeman (1987) provides an entertaining but valuable and authoritative glossary of explanations for many of the keywords and terms encountered in NMR.

In high-resolution NMR spectroscopy of solutions, whereby chemical structural information may be extracted mainly from chemical shifts (the magnitudes of which are proportional to the B_0 field) and spin–spin coupling constants (independent of B_0), there has been an inevitable tendency to operate at as high fields as practicable. High fields enhance both resolution (especially valuable in the study of mixtures or, for example, proteins with many closely similar chemical moieties with overlapping chemical shifts) and, by increasing the energy level separations and hence the Boltzmann factor, sensitivity (valuable for spectroscopy with less receptive nuclei). Superconducting solenoid magnets incorporating an alloy or a compound of niobium and titanium or tin can generate steady B_0 fields of up to 15 T, but are expensive and must be maintained in liquid helium at 4 K.

The advent of Fourier transform NMR spectrometers has provided an enormous stimulus to high-resolution NMR spectroscopy by greatly reducing the time per experiment and/or enabling dilute samples or nuclei of low receptivity to be studied (Wayne 1987).

For simplification and interpretation of complicated or overlapping normal or one-dimensional (FID a function of a single time variable) high-resolution NMR spectra, distinguishing between chemical shifts and lines arising from spin–spin (J) coupling can often be achieved by decoupling, that is, irradiating at an additional frequency or range (or broadband) of frequencies; this may be at the cost of losing some information. Many kinds of two-dimensional NMR experiments (designated by acronyms) are well described by Croasmun and Carlson (1987) and Schraml and Bellama (1988). The advantages of two- over one-dimensional NMR in chemical structure elucidation, reviewed by Morris (1986), are such that the availability of pulse programs in commercial spectrometers and the simplification of software is expected to encourage wider use among nonspecialist chemists.

In principle, a solid sample, with its molecules ordered, should contain more NMR shielding information than a solution with molecules in randomly tumbling motion.

Brief illustrations are available of the applications of solid-state high-resolution NMR to catalysts (Klinowski and Thomas 1986) and fossil fuels (Jones 1983).

3. Applications to Industrial Analysis and Control

In a relatively simple wide-line or low-resolution NMR spectrometer, the relative proportions of two components in a sample with very different degrees of motion can readily be evaluated. Rapid batch determination of moisture can be commercially important and is effected by measuring relative areas of a comparatively narrow ^1H NMR absorption line from mobile fluid protons superposed on a broader line from more rigid protons in the solid. Continuous-wave NMR has long been applied for this purpose to quality control in a range of hygroscopic agricultural and industrial products and, analogously, to solid–liquid ratios of fats in the food and confectionery industries. Early applications are outlined by Jones (1966), but many reports are available of quantitative analysis with the Newport 4000 Analyzer (Oxford Analytical Instruments) in, for example, the explosives, polymer, olive oil, chocolate, petroleum and surfactant industries (Hearmon and Rhodes 1986). Sample preparation is minimal, modest temperature variation is possible, and measurement times are of the order of 0.5–1 min. The Brüker Minispec 120 pulsed instrument (operating in a 0.47 T field) is utilized for oil, fat and moisture determination in process control and monitoring in an equally wide range of industries (Weissner and Harz 1984, Jones 1987). Another long-established pulse instrument for analysis of moisture, solid-fat content and so on, the Praxis, can now perform multiple sampling with a robot arm attachment.

One recent example of the use of spin–echo sequences (J echo modulation) in a low-resolution pulse spectrometer is the rapid and nondestructive determination of ethanol from both CH_2 and CH_3 multiplets in aqueous alcohol beverages to an accuracy of 0.5% v/v (Guillon and Tellier 1988); this is suggested as a cheaper alternative to high-resolution NMR for on-line control of industrial fermentation. Quite a different NMR technique in wine analysis arises from the ability of a sensitive spectrometer and technique to detect very small variations (100–160×10^{-6}) in the natural abundance of the ^2H isotope, which arise due to the metabolic process of synthesis. Site-specific natural isotope fractionation (SNIF) NMR with quantitative ^1H at 61.4 MHz enabled Martin and Martin (1987) to establish the precise geographical origin of wines. This ability to detect the results of different biochemical pathways may have other applications in the detection of food and drink adulteration, and in the chiral recognition of enantiomers.

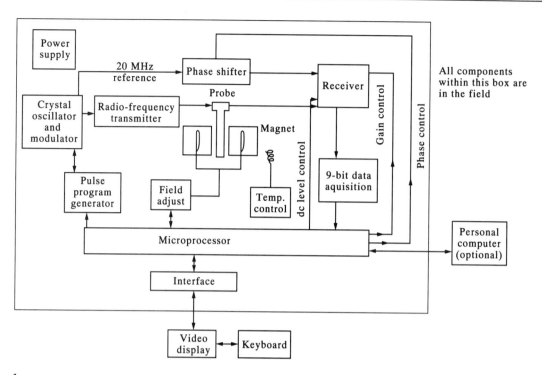

Figure 1
Block diagram of automatic pulse ^{1}H NMR on-line moisture analyzer at an aluminum oxide plant (after Pearson *et al.* 1986 © *Light Metals*. Reproduced with permission)

While measurements of flow and diffusion are dealt with in Sect. 4, a few examples of on-line analysis by NMR will be mentioned here. Some early schemes for process monitoring on pilot plant and industrial scales are summarized by Jones and Child (1976) and Jones (1987). For continuous measurement of water content in reduction-grade alumina in a rotary kiln, Pearson and Parker (1984) modified a Brüker 0.47 T pulse spectrometer in order to effect automatic feedback control of the kiln; the magnet was installed 7.5 m down an air slide in the plant, 30 m away from the control console. A purpose-built on-line analyzer (see Fig. 1), despite location in a dusty humid plant, sampled the material automatically every few minutes over periods of months without operator intervention (Pearson *et al.* 1986). Multiple-pulse transient NMR by spin echoes is also recommended by King (1985) for on-stream measurements of concentrations of flowing liquids and solids using a magnet of only moderate homogeneity. Extensive studies (King and Rollwitz 1984) have been made on the feasibility of applying this approach to the determination by ^{1}H NMR at 10 MHz, of the quality and moisture content of streams of powdered coal moving at 4.5 m s^{-1} through a sensor incorporating a 0.23 T permanent magnet. Measurement in real time of the much more subtle variations, within minutes, of reactant concen-

trations in a chemical process may require the reaction of a much more sensitive and costly high-resolution high-field spectrometer, which would not normally be directly connected on-line in an industrial plant. As an aid to reaction optimization in process engineering, Neudert *et al.* (1986) have linked a semibatch reaction vessel to an 8.7 T spectrometer with a bypass flow tube, the contents of which change every 6 s (see Fig. 2). Spectra of ^{1}H at 360 MHz concurrently recorded with either ^{31}P or ^{13}C enable concentrations of components to be monitored (half-lives > 15 min) much more effectively than by batch sampling.

4. Measurement of Fluid Transport

For the measurement and characterization of flowing systems, NMR offers, in principle, the significant merits of being nondestructive and noninvasive, that is any sensor can surround, but need not be in direct contact with, the fluid. These features were soon recognized as potentially advantageous in flow measurement in such diverse fields as meters for large-scale flow of hostile (e.g., corrosive or contaminated) liquids in industry and the flow of blood in intact veins of human patients in hospitals. Some of these early applications of NMR flow measurements in

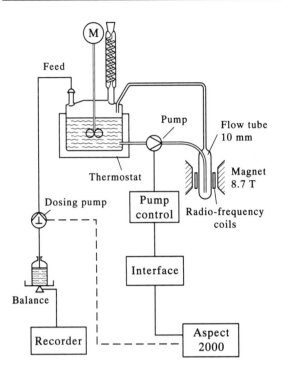

Figure 2
Schematic diagram of chemical reactor coupled to Brüker WM360 spectrometer for on-line analysis by $^{31}P/^{1}H$ or $^{13}C/^{1}H$ high-resolution NMR; pumps are controlled by computer (after Neudert *et al*. 1986. © Wiley, Chichester, UK. Reproduced with permission)

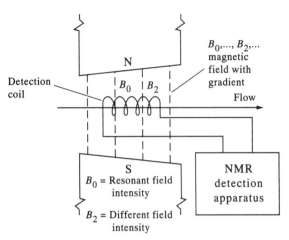

Figure 3
Configuration of simple transient ^{1}H NMR apparatus for measurement of flow and composition of fluids or moving solids (after King 1985. © Instrument Society of America. Reproduced with permission)

industrial process control, chemical kinetics, biology and physiology have been reviewed by Jones and Child (1976). Many of the above examples of using nuclear spins to follow the spatial migration of molecules involve macroscopic flow, which will be considered first (apart from medical applications, dealt with briefly in Sect. 5); subsequently, reference will be made to NMR study of microscopic flow or self-diffusion. In his review, Stepišnik (1985) has presented the relevant parameters in density matrix terms and classified the methods of flow velocity measurement and flow imaging.

NMR methods of flow measurement may be classified approximately as: (a) inflow and outflow (one-coil); (b) time-of-flight or magnetic labelling (two-coil); and (c) field gradient (inhomogeneous field, in contrast to the homogeneous fields in (a) and (b)). In (a) the signal amplitude, traditionally measured in a cw experiment, provides, with appropriate calibration (Jones and Child 1976), a measure of the spin longitudinal magnetization, partial or complete, and therefore of the relative flow velocity; pulsed methods are preferable with the flows in the narrow tubes of biological systems (Hemminga 1984). In (b) fresh nuclei passing through an upstream coil are

tagged by a polarizing pulse of some kind, and the time for this bolus of spins to reach the detector coil, a distance l downstream, can indicate the flow rate fairly directly if the flow is plug type. Relaxation times T_1 and T_2 are involved, and the magnitudes and variation of these in solids renders the method unsatisfactory for robust meters in large pipes carrying heterogeneous fluids and solids (King 1985). For such materials, field gradient methods (c) are favored, where the gradient may be parallel or perpendicular to the flow. In the simplest case the NMR echo frequency (or frequencies, with more than one component present) will change with position along the gradient, but usually repetitive pulse sequences are used (see Fig. 3). In the coal analyzer, moisture and hydrogen content as well as velocity of flow (by phase sensing) are measured by 10 MHz ^{1}H NMR.

NMR has been applied successfully to the measurement of velocity distribution patterns and random microscopic motion of nuclei (self-diffusion) since the 1950s (see Jones and Child 1976, for a summary). Continuous-wave methods apart, the pulse techniques for measuring molecular intradiffusion coefficients include: (a) static-field gradient spin–echo (SGSE); (b) pulsed-field-gradient spin–echo (PGSE); and (c) Fourier transform (FT) PGSE. These methods have been reviewed briefly by Lindman and Stilbs (1987) and, together with illustrations of polymers, gels, surfactants, microemulsions, and so on, by Stilbs (1987), they provide, despite some problems with scalar spin–spin coupling, rapid and fairly accurate means for the measurement of coefficients over a wide range of magnitudes. Applications of SGSE and PGSE methods to measurement of self-diffusion in polymer melts and solutions and their penetrants have been reviewed by von Meerwall (1983). Other

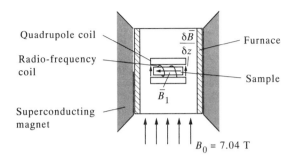

Quadrupole coil

Radio-frequency coil

Superconducting magnet

$\dfrac{\delta \bar{B}}{\delta z}$

Furnace

Sample

$\bar{B}_1$

$B_0 = 7.04$ T

Figure 4
Schematic diagram of magnetic fields for high-temperature self-diffusion measurements by high-field ^{2}H and ^{23}Na NMR (after Herdlicka *et al.* 1988 © *Verlag Z. Naturforsch.* Reproduced with permission)

examples of extension of the SE method with a 7.04 T magnet to self-diffusion measurements under more extreme conditions are provided by Meckl and Zeidler (1988) for alcohols at pressures of 0.1–400 MPa with 300 MHz ^{1}H and for ^{23}Na in molten NaNO$_3$ by 79.4 MH ^{23}Na (see Fig. 4) at up to 673 K (Herdlicka *et al.* 1988). Krüger *et al.* (1984) have designed a system, involving a 5 T polarizing magnet followed by a Brüker CXP spectrometer, aimed especially at the measurement of highly turbulent flow with a velocity range of 0.3–100 m s^{-1}. For rotational, as distinct from translational, diffusion coefficients, NMR also provides a means of measurement through the spin–lattice relaxation times, T_1 (Tyrrell and Harris 1984).

5. Biological and Physiological Applications—Flow Imaging

For flow measurement in biological and physiological systems, the nonintrusive nature of NMR is highly advantageous, despite the complications of non-plug and pulsatile flow of a non-Newtonian fluid such as blood through elastic branched tubes, which, *in vivo*, are both narrow and of variable cross section. Earlier examples are dealt with by Jones and Child (1976), Battocletti *et al.* (1981) and Jones (1987), and later methods, including blood flow in the human brain and water flow in plants, by Hemminga (1984). It is even possible to investigate water balance in plants by automatic noninvasive 20 MHz ^{1}H NMR determination of both water content and flow through capillaries in intact plants (Reinders *et al.* 1988).

In medical applications, particularly, there is increasingly close interaction between NMR spectroscopy and imaging (Morris 1986), whereby NMR signals can be identified from closely defined spatial regions of the specimen. For imaging flowing blood, by time of flight, Gullberg *et al.* (1986) proposed the use of two inversion recovery sequences, such that a

subtraction image presents just the blood flowing in vessels (whether plug, as in arteries, or laminar, as in veinous flow), free of signals from stationary anatomy. For the noninvasive NMR spectroscopic study of metabolism in intact cells, ^{31}P has been most used, whereas ^{13}C and ^{15}N NMR generally require costly enriched samples; high-field pulse techniques for ^{1}H enable the dominant haemoglobin resonances to be largely suppressed or eliminated and so allow small-molecule high-resolution NMR resonances to be observed (Rabenstein 1984).

6. Summary and Prognosis

NMR can sense only certain species of atomic nuclei, some, such as ^{1}H, more easily than others, but the range of nuclei accessible for useful measurements is now wide. Most experiments are on stationary solutions, but NMR can be carried out on solids, liquids and gases and flowing systems; accessible ranges of temperature and pressure (e.g., Van der Putten and Prins 1988) are being extended and, volume-selected spectroscopy is practicable. The NMR technique is nondestructive and can operate nonintrusively on difficult or sensitive samples. It can measure a wide range of parameters, thus sensing molecular order, motion, flow, detailed chemical structure, and so on. Its technology continues to be in a rapid state of development.

See also: Analytical Physical Measurements: Principles and Practice; Process Instrumentation Applications

Bibliography

Akitt J W 1983 *NMR and Chemistry*, 2nd edn. Chapman and Hall, London
Albert K, Dreher E-L, Straub H, Ricker A 1987 Monitoring electrochemical reactions by ^{13}C NMR spectroscopy. *Magn. Reson. Chem.* **25**, 919–22
Battocletti J H, Halbach R E, Salles-Cunha S X, Sances A 1981 The NMR blood flowmeter—theory and history. *Med. Phys.* **8**, 435–43
Bell J D, Brown J C C, Sadler P J 1988 NMR spectroscopy of body fluids. *Chem. Br.* **24**, 1021–4
Croasmun W R, Carlson R M K 1987 *Two-Dimensional NMR Spectroscopy for Chemists and Biochemists.* VCH, Weinheim, Germany
Freeman R 1987 *A Handbook of Nuclear Magnetic Resonance.* Longman, Harlow, UK
Fukushima E, Roeder S B W 1981 *Experimental Pulse NMR, a Nuts and Bolts Approach.* Addison-Wesley, Reading, MA
Gadian D G, Frackowiak R S J, Crockard H A, Proctor E, Allen K, Williams S R, Russell R W R 1987 Acute cerebral ischaemia: concurrent changes in cerebral blood flow, energy metabolites, pH and lactate measured with hydrogen clearance and ^{31}P and ^{1}H nuclear magnetic resonance spectroscopy. I, methodology. *J. Cereb. Blood Flow Metab.* **7**, 199–206

Guillon M, Tellier C 1988 Determination of ethanolic beverages by low-resolution pulsed nuclear magnetic resonance. *Anal. Chem.* **60**, 2182–5

Gullberg G T, Simons M A, Wehrli F W, Roy D N G 1986 Time-of-flight NMR imaging of plug and laminar flow. International workshop on physics and engineering of computerized multidimensional imaging and processing. *Proc. Soc. Photo-Opt. Instrum. Eng.* **671**, 314–19

Hazlett R N, Dorn H C, Glass T E 1984 Analysis of middle distillate fuels by flow liquid chromatography/proton magnetic resonance. In: Petrakis L, Fraissard J P (eds.) *Magnetic Resonance. Introduction, Advanced Topics and Applications to Fossil Energy.* Reidel, Dordrecht, The Netherlands, pp. 709–20

Hearmon R A, Rhodes M P 1986 Hydroxyl value determination by NMR. *Tenside Deterg.* **23**, 245–6

Hemminga M A 1984 Measurement of flow characteristics using nuclear magnetic resonance. In: James T L, Margulis A R (eds.) *Biomedical Magnetic Resonance.* Radiology Research and Education Foundation, San Francisco, CA, pp. 157–84

Herdlicka C, Richter J, Zeidler M D 1988 Spin-echo self-diffusion measurements in molten salts. *Z. Naturforsch.* **43a**, 1075–82

Jones D W 1966 Magnetic resonance and process control. *Chem. Proc. Eng.* **47**, 22–7

Jones D W 1983 NMR examination of fossil fuels. *Trends Anal. Chem.* **2**, 83–8

Jones D W 1987 Nuclear magnetic resonance. In: Singh M G (ed.) *Systems and Control Encyclopedia.* Pergamon, Oxford, pp. 3416–21

Jones D W, Child T F 1976 NMR in flowing systems. *Adv. Magn. Reson.* **8**, 123–48

Kemp W 1986 *NMR in Chemistry: A Multinuclear Introduction.* Macmillan, Basingstoke, UK

King J D 1985 NMR for on-stream measurements. *Adv. Instrum.* **40**, 1387–1405

King J D, Rollwitz W L 1984 Magnetic resonance measurement of calorific value and moisture. *Proc. Conf. EPRI Power Plant Performance Monitoring Workshop.* Electric Power Research Institute, Palo Alta, CA, pp. 3-59–3-71

Klinowski J, Thomas J M 1986 The magic angle and all that: probing the structure of solids using nuclear magnetic resonance. *Endeavour* **10**, 2–8

Krüger G J, Haupt J, Weiss R 1984 A nuclear magnetic resonance method for the investigation of two-phase flow. In: Delhaye J M, Cognet G (eds.) *Symp. Measurement Techniques in Gas–Liquid Two-Phase Flows.* Springer, Berlin, pp. 435–54

Lindman B, Stilbs P 1987 Molecular diffusion in microemulsions. In: Friberg S E, Bothorel P (eds.) *Microemulsions : Structure and Dynamics.* CRC Press, Boca Raton, FL, pp. 119–52

Martin G J, Martin M L 1987 *Modern Methods of Plant Analysis*, Vol. 6, Springer, Heidelberg, Germany

Martin M L, Delpuech J-J, Martin G J 1980 *Practical NMR Spectroscopy.* Heyden, London

Meckl S, Zeidler M D 1988 Self-diffusion measurements of ethanol and propanol. *Mol. Phys.* **63**, 85–95

Morris G A 1986 Modern NMR techniques for structure elucidation. *Magn. Reson. Chem.* **24**, 371–403

Morris P G 1986 *NMR Imaging in Medicine and Biology.* Clarendon, Oxford

Neudert R, Ströfer E, Bremser W 1986 On-line NMR in process engineering. *Magn. Reson. Chem.* **24**, 1089–92

Pearson R M, Parker T L 1984 The use of small nuclear magnetic resonance spectrometers as on-line analyzers for rotary kiln control. *Light Met.* **1**, 81–97

Pearson R M, Ryhti L, Job C 1986 Automatic on-line measurement of moisture content and surface area of aluminium oxide. *Light Met.* **2**, 47–50

Rabenstein D L 1984 ^{1}H NMR methods for the non-invasive study of metabolism and other processes involving small molecules in intact erythrocytes. *J. Biochem. Methods* **9**, 277–306

Reinders J E A, Van As H, Schaafsma T J, de Jager P A, Sheriff D W 1988 Water balance in *Cucumis* plants measured by nuclear magnetic resonance. *J. Exp. Bot.* **206**, 1199–1210

Schraml J, Bellama J M 1988 *Two-Dimensional NMR Spectroscopy.* Wiley, New York

Stepišnik J 1985 Measuring and imaging of flow by NMR. *Prog. Nucl. Magn. Reson. Spectrosc.* **17**, 187–209

Stilbs P 1987 Fourier transform pulsed-gradient spin-echo studies of molecular diffusion. *Progr. Nucl. Magn. Reson. Spectrosc.* **19**, 1–45

Tyrrell H J V, Harris K R 1984 *Diffusion in Liquids.* Butterworth, London, pp. 216–34; 242–8

Van As H, Schaafsma T J 1987 Measurement of flow by the repetitive pulse method. *J. Magn. Reson.* **74**, 526–34

Van der Putten D, Prins K O 1988 ^{13}C Cross polarization and spin relaxation in adamantine at pressures up to 7 kbar. *J. Magn. Reson.* **77**, 550–61

Voges R, Von Wartburg B R, Loosli H R 1986 Tritiated compounds for *in vivo* investigations: CAMP and ^{3}H-NMR specroscopy for synthesis planning and process control. In: Muccino R R (ed.). *Synthesis and Applications of Isotopically Labelled Compounds.* Elsevier, Amsterdam, pp. 371–6

von Meerwall E D 1983 Self-diffusion in polymer systems, measured with field-gradient spin-echo NMR methods. *Adv. Polym. Sci.* **54**, 1–29

Wayne R P 1987 Fourier transformed. *Chem. Br.* **23**, 440–6

Webb G A (ed.) 1988 *Nuclear Magnetic Resonance.* Specialist Periodical Report, Royal Society of Chemistry, London

Weisser H, Harz H-P 1984 NMR studies of foods at sub-freezing temperatures. In: McKenna B M (ed.) *Engineering and Food.* Engineering Sciences in the Food Industry, Vol. 1. Elsevier, London, pp. 445–54

D. W. Jones

[University of Bradford, Bradford, UK]

O

Operational Environments

In general, it must be assumed that an instrument system in use will be exposed to hostile environments; the adverse conditions may be mechanical, climatic or electrical, and may affect the complete system or may affect only part of the system. Normally it would be expected that electronic units and control systems would be situated in a more "friendly" environment in the control room, but this may not be the case, and the sensor unit will almost certainly be installed under conditions which could be very detrimental to its performance. Great care should therefore be exercised in the design and construction of all modules of the system, in particular the sensor.

Effects of environmental conditions on instrumentation can be divided into several categories. The adverse conditions may arise during transport, storage and use. They may also cause damage mechanically or electrically; they may result in the equipment operating outside the environmental limits specified by the manufacturer. They may even cause more fundamental effects, since the actual parameter being measured may be affected.

1. Climatic Conditions

The principal climatic conditions affecting the operational environment are high temperature, low temperature, high relative humidity (usually accompanied by high ambient temperature) and barometric pressure, and rapid changes in ambient temperature (thermal shock).

Most electrical and electronic components in the instrument system will probably be specified by the manufacturers to be suitable for a particular range of temperature. Frequently, two ranges are quoted, one for storage and one for operating conditions. If these temperature ranges are exceeded, the accuracy and performance of the system are likely to be affected. High temperatures may cause electronic components to change, and may also cause mechanical changes (thermal expansion). Low temperature may also cause these kinds of change, and also other mechanical effects; for instance, there may be changes in the elastic properties of materials, which can, for example, cause diaphragm pumps to cease to function.

There are many possible effects of ambient humidity, especially when this is combined with elevated ambient temperature. Again, certain electronic components can be affected and corrosion may be caused, particularly if other gases or vapors are present. Mould growth may also occur in warm humid atmospheres

The effects of thermal shock (rapid changes in temperature) may be complex. The effects on different components of an instrument system will differ; one parameter which will influence the effect is the thermal capacity of the component itself, which will determine how much the temperature of this component will lag behind the ambient change, and this temperature lag will not be uniform throughout the system.

The effects of changes in barometric pressure for an open system, or sample pressure for a closed system are fundamental ones, affecting the actual parameters being measured. An infrared analyzer used for measurement of gas concentrations in a gas mixture normally gives an output in terms of volume units (percentage by volume, or ppm). However, the response of this analyzer for a given gas concentration (in volume units) will change almost linearly with changes in gas pressure, the fundamental measurement for this type of instrument being partial pressure rather than volume. This difficulty can be overcome by incorporation of a pressure sensor in the instrument system and using the output to correct the instrument output for changes in sample pressure. The same remarks apply to variations in ambient temperature, since the measurement given by the analyzer is subject to the gas laws.

2. Contaminated or Corrosive Atmospheres

The industrial environment imposes a variety of aggressive atmospheres on the instrument system. Particulate material can cause malfunction of mechanical components; such material could be siliceous material from a dusty or desert atmosphere or other types of particulates indigenous to the process. Normally adverse effects can be avoided by enclosing the instrumentation in a suitable housing with the appropriate ingress protection (IP) classification. More serious are corrosive atmospheres which may be present in the plant environment. Sulfur dioxide or nitrogen oxides, even in low concentrations, in the presence of water vapor may rapidly cause corrosion of electrical contacts. Similarly, traces of hydrogen sulfide, or related compounds, can cause rapid "blackening" of contacts. In all cases adequate protection against such atmospheric contaminants is essential.

One particular environment in which instrumentation is particularly subject to corrosion is the atmosphere near to coastal regions, which is laden with salt mist. Many effects can be avoided if the electronic units are enclosed in the correct housing. However, in many cases it is not possible to protect the sensors in this way since they need to be exposed to the ambient atmosphere. Adequate protection can, however, be provided by dust filters and membrane filters if these are incorporated in the sensor housing.

3. Mechanical Disturbances

The principal mechanical disturbances are vibration, shock and impact, the first of which is the most common. Electronic units and sensing units may be subjected to vibration during transport, storage, or use, and there are several effects which may be caused. Large components on a circuit board may be brought into resonance, the amplitude of which can be large. Mechanical damage can be caused by the stresses which are set up.

The effects of vibration can be reduced if various precautions are observed in the design of equipment, for example, avoidance of large components connected to circuit boards by long leads, and use of antivibration mounts. These precautions also minimize the effects of other mechanical disturbances, such as impact and shock.

4. Electrical Disturbances

The principal electrical disturbances are voltage and frequency variations, supply interruptions and supply transient over voltages. In general, the tolerances to supply variations are fairly wide, and these variations do not cause significant instrument malfunctions. Malfunctions can arise, however, if there are short-term interruptions or appreciable transients on the mains, but these can be minimized by correct design of the equipment.

5. Magnetic and Electromagnetic Disturbances

The magnetic fields associated with current carrying conductors can cause malfunctions in instruments in the vicinity of the field, but this occurrence is relatively rare and can be avoided by appropriate shielding. More prevalent and more serious are the effects of electromagnetic radiation and electrostatic discharge, the former in particular. Radiofrequency interference can arise in a variety of ways, directly on electronic units and on interconnecting leads. It is necessary to investigate these effects, and to build in the necessary immunity in the design and construction of the equipment, and provide appropriate shielding for the cabling.

6. Substantiation of the Above Effects

The following observations are based on experimental findings of independent laboratories in the UK and Europe. On behalf of instrument users' associations, 93 instruments were tested over a two-year period.

While 33% of the instruments tested failed to meet specification under reference conditions (20 °C, relative humidity 40–60%, no external disturbances), 64% were outside specification under influence conditions (applied singly). Since these findings are based on laboratory tests, the conclusions would be that, under conditions of use, where several disturbances could occur simultaneously, the results could be significantly worse.

A similar analysis of results of tests on behalf of the UK power generation industry showed that 86% of the instruments tested were outside specification under influence conditions. In this series 72 instruments were tested over a period of four years, and the subject instruments were drawn from manufacturers in Europe and North America.

The above evaluation statistics are quoted to give an indication of the importance of giving adequate consideration to environmental effects in the process of design, construction and use, and to ensure that all possible precautions are taken in order to minimize these effects.

See also: Reliability and Maintenance; Safety

Bibliography

International Electrotechnical Commission 1978 Basic environmental testing procedures, IEC 68, Pts 1 and 2. IEC, Geneva

International Electrotechnical Commission 1984 Electromagnetic compatibility for industrial-process measurement and control equipment, Pt 2: electrostatic discharge requirements, IEC 801–4. IEC, Geneva

International Electrotechnical Commission 1984 Electromagnetic compatibility for industrial-process measurement and control equipment, Pt 3: radiated electromagnetic field requirements, IEC 801–4. IEC, Geneva

S. W. J. Hopkins
[Sira Test & Certification, Chislehurst, UK]

Operator–Instrument Interface

This article reviews the human computer interaction aspect of measurement and instrumentation systems. It also covers the development of instrumentation interfaces and the problems inherent in their design, and briefly reviews the methods which may be applied to the development of such user interfaces.

1. Information, Tasks and Instrumentation Interfaces

Instrumentation interfaces are ubiquitous components of command and control systems. Their prime purpose is to provide information to help operators control a system, be that an aircraft, power station or manufacturing process. The human computer interface is a vital part of control systems which often have safety critical properties (Finkelstein 1990). Failures are well known because of their serious and often tragic consequences. Take, for instance, the 1988 Kegworth, UK, air crash. Instrumentation may be blamed because of a mistake made by the aircrew of associating the instrument with the incorrect engine (Gavaghan 1990). The instruments, vibration sensors for the right and left engines, were placed close together at a viewing angle which could allow confusion between the right and left instrument if the pilots did not concentrate. The wrong engine was shut down because a mistake was made in instrument identification.

Failures highlight design problems although, frequently, solutions are not immediately apparent. Design of the operator interface involves understanding how instruments are used in the operator's task. It is necessary to discover what sort of information will be required, and when. Control information can be broadly categorized as monitoring, decision support and planning. Information is necessary to monitor the current state of the system and ensure nothing untoward is happening, to decide what to do next when human intervention is necessary and, finally, to plan what to do in the future. Because the system being controlled is changing, the information must be accurate, appropriate and timely. Deciding what to present to the operator is one key part of the design problem. Another part is how to present the information.

2. Human Factors and the Operator Interface

Designing instrumentation interfaces involves physical features of how information is displayed and logical questions of what sort of information should be displayed, and when. Physical design problems have been well researched in ergonomics, but the knowledge is less complete on logical design.

2.1 Interface Ergonomics

Human factors, or ergonomics, is the applied psychology of machine interface design. Research has produced hundreds of guidelines which advise about the physical features of instrument design from the size of text, to the layout of dials, use of color and so on: for more detail see Bailey (1982), Galer (1987), and Sanders and McCormick (1987). Ergonomic advice covers environmental issues (e.g., lighting, viewing angle, viewing distance), physical details of instrument displays for readability of text, numbers, and design advice on the layout of standard instrument types (e.g., dials, sliding scales, warning signs).

Guidelines cover the layout of instruments, for example, instruments which have to be cross referenced should be placed together and important instruments should be placed in the center of displays. Guidelines are derived from experimental evidence as well as basic human psychology (see Sect. 2.2). However, much of this guidance is concerned with the physical features of the interface. In recent years ergonomics researchers have been progressing from the "knobs and dials" approach to questions about what is shown, and when, which necessitates understanding the cognitive aspects of the instrument interface.

2.2 Human Information Processing

People have certain fundamental limitations which affect instrument design. Our mental processes limit the quantity of information we can process at once, hence, the interface designers have to take human cognitive limitations into account. Consideration of human information processing is complex and beyond the scope of this article, therefore only the key aspects will be highlighted here. Readers may consult Rasmussen (1986), Norman (1988), and Reason (1990) for more detail on human cognition, while Sutcliffe (1988) and Johnson (1992) cover human information processing from a human–computer interaction (HCI) perspective.

The first key limitation is working memory. People can only keep a small amount of information active in working memory at once. Hence, people tend to make errors if asked to remember too many things when making decisions. An example would be trying to remember emergency control operations at the same time as rules for interpreting error signals. Training and learning skills are used to circumvent this problem. Rasmussen (1986) described three levels of problem solving, from knowledge based, which uses heuristics and needs conscious effort, to rule based, when the problem domain is better understood and analyzed as rules, to skill, when problem solutions are encoded as learned procedures.

Most operators have rigorous training so that emergency procedures become automatic. Unfortunately, skill-based operation has its disadvantages. Skill is the human equivalent of a computer program. It is a task procedure which people do automatically when triggered by a particular set of circumstances. Only too often the triggering process is imperfect.

The problem is to interpret the situation and then apply the appropriate behavior. When under pressure people tend to use weak strategies, such as, try the most recent procedure which nearly matches, or try the procedure which is usually used. The recency and frequency of gambling strategies cause people to make mistakes in skilled behavior (Reason 1990). Skill and working memory limitations suggest that designers should attempt to make the triggering context clear. Instrumentation displays need to be designed with an awareness of the operator's task, that is, the activities and procedures which will be carried out, and the information necessary to trigger specific procedures

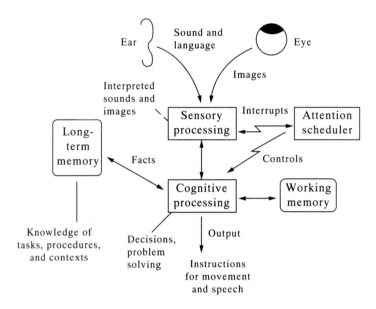

Figure 1
Model of human information processing (after Card *et al.* 1983, Wickens 1984)

(see Fig. 1). The role of task analysis in addressing this problem is covered in Sect. 3.

Another key limitation is attention. If people are forced to switch their attention rapidly they tend to lose the thread of what is going on, mainly because each new stimulus overwrites working memory. Conflict of attention results in having no time to think; the Three Mile Island disaster was partially caused by this problem. The interface had too many alarms which all went off in a cascade, consequently the operators never had time to think about the cause of the problem (US Nuclear Regulatory Commission 1985). As a result, they misinterpreted the state of the system, which had been caused by a faulty valve and a faulty instrument, and made incorrect decisions.

The design implications for warning in instrumentation are clear: (a) try to minimize the number of alarms; (b) beware of cascade effects with system malfunction; (c) allow alarms to be cancelled; and (d) try to link alarms to diagnostic information about the malfunction.

Attention has another implication. People may receive information on different sensory modalities (sound, vision, touch). Perception of information is an active process if the stimulus requires interpretation; in the everyday sense people pay attention or listen carefully. Although research on multimodal communication is still rudimentary, it has been shown that messages on different sensory channels can conflict. Giving different messages on two modalities at the same time is inviting error (Wickens 1984), whereas sending the same message on two sensory modalities usually augments perception. However, how much

detail can be concurrently received on two channels is poorly understood. Working memory is also implicated here. Speech or sound is ephemeral because, once heard, the trace is lost from the working memory. Hence, an important message must be heard the first time, whereas visual messages persist so that operators can reread them. Visual interfaces are therefore better for conveying information content, whereas sound is useful for drawing attention to a message.

2.3 Principles in Interface Design

Human computer interaction researchers have devised design principles based on applied psychology. While no complete consensus exists, some common examples are as follows.

(a) *Consistency*. Interface displays should be consistent in layout. Consistency helps learning because people recognize similar patterns. An example would be to use the same display format and units for instrumentation readings throughout a system.

(b) *Structuring*. Information should be classified, sorted and grouped to help users find and comprehend the information they need. Examples include: (a) classifying hierarchical indicators of system performance—whole system, subsystem, and so on; and (b) sorting readings into priority order.

(c) *Compatibility*. Displays should conform with peoples' expectations, that is, instruments are compatible with the user's prior knowledge, or model, of the system. Compatibility may be seen as consistency over time, therefore the interface uses diagrams to show spatial information and mimics the layout of the plant being controlled.

(d) *Predictability*. Instruments should give information which not only helps people to interpret the current state of the system but also to predict its future state. Examples include: (a) future projections of trends and warnings in air traffic control collision detection; and (b) critical rises in pressure and temperature in power stations.

(e) *Adaptability*. The interface should be designed to suit the changing needs of users. As people learn more about a system they progress from being novices to experts. The interface displays need to be more comprehensive and supportive for novices. This principle creates problems in practice, because interfaces which change to adapt to users run the risk of offending the consistency principle.

From the last example, HCI principles can be seen to create difficulties in interpretation. Principles only give advice about how to design, and need to be combined with methods for analyzing the interface requirements and then designing the user system dialog and displays.

3. Towards a Methodical Approach to Design

Task analysis can give reasonable guidance about what sort of information is necessary for decision support and planning; however, designing instruments for monitoring is more problematic. The designer's dilemma is trying to anticipate the future. When errors or exceptions occur human operators have to take decisions quickly, and the outcome of decisions is often safety critical. Unfortunately, human decision making under stress is not good, so the designer's job is to provide the right quantity of information which is pertinent to the current problem.

The design dilemma revolves around giving too much information, so that the operator "cannot see the wood for the trees," or giving too little information, so that vital data is missing. With ideal foresight, provision of minimal and relevant information may appear to be the answer. In practice, foresight is rarely ideal and the minimal information provided in the glass cockpits of the new generation of airliners has been criticized as running the risk of not giving the pilot sufficient information in an emergency. On the other hand, decision making in emergencies is time critical and people do not have time to search through dozens of instruments.

Instrument interfaces should attempt to give context-sensitive information. Some prediction about the type and quantity of information can be derived from task analysis. HCI task analysis techniques give classifications of the knowledge required by people when carrying out a task—the activity carried out by the operator (e.g., task knowledge structure (TKS), (Johnson 1992)) and analytical techniques for acquiring task models.

Briefly TKS ensures that task knowledge consists of:

(a) plans—high level goals and strategic intentions of what to do;

(b) procedures—lower level goals linked to action sequences which achieve the goal;

(c) actions—the primitive steps of tasks; and

(d) objects—data, information and physical structures which are used in tasks.

Procedures have preconditions and postconditions, which specify the knowledge, or necessary triggering conditions, for a sequence of actions. Task analysis can therefore suggest what type of information should be present in an instrumentation interface to trigger the necessary procedures (see Fig. 2). Furthermore, task analysis techniques can be used in conjunction with standard software engineering methods for display design (Sutcliffe and Wang 1991).

However, HCI methods do not help the designer when planning display details. Detecting and anticipating the operator's context is difficult, therefore current practice is to anticipate as much as possible from a thorough task analysis, then add other information on reasonable assumptions of a "need to know" basis. The instrumentation interface, therefore, aims to combine targeted information, which is predicted from task analysis, with more general information about the system. The latter information should provide sufficient data for the operators to maintain their mental model of the system, so they have sufficient data for decision making and emergency control.

More recent HCI models have suggested that tasks may be composed of a set of contexts, which are a set of recognizable conditions and associated procedures (Woods 1988, Grant 1990). Cognitive task models have also described how the limitations of working memory affect decisions and operation in tasks (Bainbridge 1992). This work has the potential to give designers guidance about the quality and quantity of information in instrumentation interfaces and, possibly, some advice about its layout for effective information seeking. However, as yet, these models do not have sufficient power to predict what sort of information may be required for certain task types.

4. Model-Based Instrumentation Interfaces

One approach to the problem of providing sufficient information is to give the operator a model simulation of the system. This approach views the interface, not

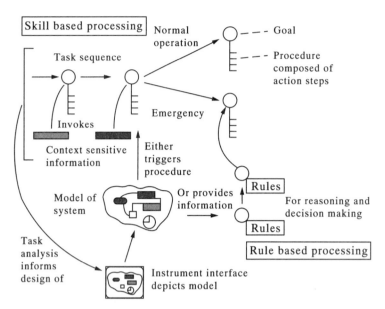

Figure 2
Task analysis, information and interface design

as a set of discrete information displays, but as an integrated virtual world. A semirealistic picture illustrates the system behavior and status of its components. Information is thereby placed in context, so that the operator can see an interconnected set of facts which can be interpreted using a system model. Simulation interfaces were first employed on tutorial systems for process control, for instance, Steamer (Hollan *et al.* 1984). The interface portrayed a map of system components which were instrumented to show changes in boiler pressure, temperature and other physical measures. The operator could interact with the display to change control parameters and then observe the consequent effect on system behavior. The same approach has been applied to interfaces of operational systems.

Advances in graphical displays and multimedia interaction have opened up new possibilities in interface design. Prototype process control interfaces developed in Esprit projects (GRADIENT 1989) have used model-based simulation and different media to convey information more effectively. These displays give continuous monitoring and status information set in the framework of the system; hence, when the untoward happens the operator can locate and interpret faults more easily.

5. Intelligent Interfaces

An alternative to trying to solve the choice of information at design time is to make the interface intelligent enough to make its own choice about what informa-

tion to display for a context. The operator display consists of the minimum information necessary for human intervention. Proponents of this approach claim, quite rightly, that the minimum display reduces human working memory and attention load. Opponents point out that, in theory, this is fine so long as the designer has perfect foresight and can anticipate all the possible operational contexts.

This approach is exemplified by the Airbus Industrie A320 cockpit system with intelligent fly-by-wire controls. Compared with a conventional cluttered cockpit, the instrument displays are simple. Critics claim that in emergency there is insufficient diagnostic information for the pilot to discover what has gone wrong and, more importantly, the monitoring information may not allow the pilot to pick up signs of trouble before emergencies occur. This debate continues between designers who favor trusting the machine and those who favor a lower level of automation with more emphasis on information and giving the operator time to think.

6. Conclusion

Instrumentation interface design has progressed considerably since the early designs. Advances in VDU technology have enabled complex graphical displays to be developed, which integrate several instrument displays into a single representation of the system as a virtual world. The basic ergonomics of instrumentation design are well understood. Task analysis methods can help to predict what information should be required at various stages in a task. Although this can help in designing the interface for normal operation, anticipating emergencies and errors is more difficult. Model-based approaches can provide the background information for decision making even if the display is not tailored exactly to the emergency situation.

The alternative approach of intelligent instrumentation remains to be proven and only experience will tell whether human or machine intelligence is better able to cope with emergency situations. Considerable research is required on human information seeking and problem solving in process control tasks, before instrument interface design can progress beyond guidelines to a more formal engineering approach.

See also: Artificial Intelligence in Measurement and Instrumentation; Digital Instruments

Bibliography

Bailey R W 1982 *Human Performance Engineering: A Guide for System Designers*. Prentice-Hall, Englewood Cliffs, NJ
Bainbridge L 1992 Mental models and industrial process operation. In: Rutherford A, Rogers Y, Bibby P (eds.) *Models of the Mind*. Academic Press, London, pp. 119–43
Card S K, Moran T P, Newell A 1983 *The Psychology of Human–Computer Interaction*. Lawrence Erlbaum Associates, Hillsdale, NJ
Finkelstein L 1990 Measurement and instrumentation as a discipline in the framework of information technology. In: *Knowledge Based Measurement*. VDI Berichte, Dusseldorf, Germany, pp. 275–83
Galer I 1987 *Applied Ergonomics Handbook*. Butterworth, Guildford, UK
Gavaghan H 1990 Human error in the air. *New Sci.* **128**, 1743
GRADIENT 1989 A graphical dialogue environment for process supervision and control. Report Esprit project P 857. Computer Resources International, Birkerod, Denmark
Grant A S 1990 Modelling cognitive aspects of complex control tasks. In: Diaper D, Gilmore D, Cockton G, Shackel B (eds.) *Human–Computer Interaction, INTERACT-90*. North-Holland, Amsterdam, pp. 1017–18
Hollan J D, Hutchins E L, Weitzman L 1984 Steamer: an interactive simulation based training system. *AI Magazine* **5**(2), 15–27
Johnson P 1992 *Human Computer Interaction: Psychology, Task Analysis and Software Engineering*. McGraw-Hill, New York
Norman D A 1988 *The Psychology of Everyday Things*. Basic Books, New York
Rasmussen J 1986 *Information Processing and Human Computer Interaction: An Approach to Cognitive Engineering*. North-Holland, Amsterdam
Reason J T 1990 *Human Error*. Cambridge University Press, Cambridge
Rouse W B 1981 Human–Computer Interaction in the Control of Dynamic systems, *ACM Comput. Surv.* **13**(1), 71–99
Sanders M S, McCormick E J 1987 *Human Factors in Engineering and Design*. McGraw-Hill, New York
Sutcliffe A G 1988 *Human Computer Interface Design*, Macmillan, London
Sutcliffe A G, Wang I 1991 Integrating human–computer interaction with Jackson System development. *Comput. J.* **34**(2), 132–42
US Nuclear Regulatory Commission 1985 Loss of Main and Auxilliary Feedwater at the Davis–Besse Plant on June 9, 1985. Report No. NUREG 1154. National Technical Information Service, Springfield, VA
Wickens C 1984 *Engineering Psychology and Human Performance*. Merrill, Columbus, OH
Woods D D 1988 Cognitive Engineering in Complex and Dynamic Worlds. In: Mancini G, Woods D D, Holnagel E (eds.) *Cognitive Engineering in Dynamic Worlds*. Academic Press, London

A. Sutcliffe
[City University, London, UK]

Optical Instruments

Optical instruments provide qualitative or quantitative measurements of the properties of a flux of light, from which the condition of the source of the flux can be inferred. Throughout history scientific investigation has relied heavily on optics and the modern field of optical instrumentation is expansive (Gopel *et al.* 1992, Williams 1993). The field is extremely active and this is particularly true in the area of optical information processing and computing (Gopel *et al.* 1992).

In this article, some examples of the main types of optical instrumentation are briefly described under three categories: geometric (Gopel *et al.* 1992, Williams 1993); spectrographic (Buican 1990, Tsukakoshi *et al.* 1990); and interferometry, coherent optics and information processing (Gopel *et al.* 1992, Philipp *et al.* 1992, Pedrini *et al.* 1993, Williams 1993). These categories, however, are not totally inclusive of the many different types of optical instruments, but they are representative of the nature of the technology.

1. Geometric Optical Instruments

The earliest instrument exploited the geometric properties of light, starting with alignment markers and sites for triangulation in surveying. The beginning of modern geometric optics started in the seventeenth century with the development of the first astronomical telescope (1608) by Galileo and Hans Lippershey. Instruments in this area, in general, have Fermat's principle as their underlying principle, which states that (Jenkins and White 1957):

"The path taken by a light ray from one point to another through any set of media is such as to render its optical path equal, in the first approximation, to other paths closely adjacent to the actual one."

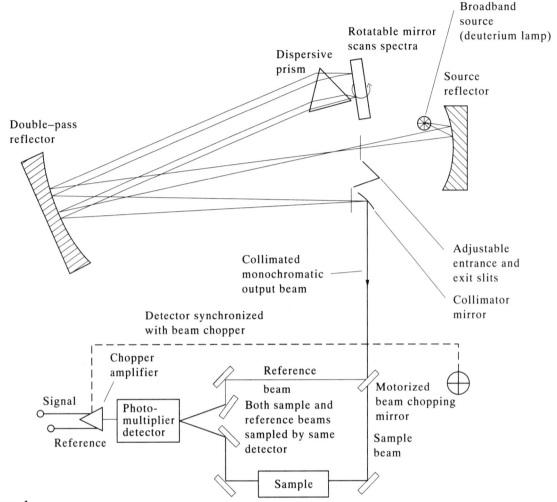

Figure 1
Double-pass prism spectrometer

The discovery of the telescope opened the door to the study of the macrocosm, and had an immediate and profound effect on surveying, navigation and astronomy. The development of the first microscope two years later (1610) by Galileo opened the door to the microcosm, allowing more detailed study of botanical and biological samples. Chromic aberration in optical lenses led Newton to the design of a telescope using reflecting mirrors instead of lenses (1668), the basic design of which is still used today.

There are many modern optical instruments which depend mainly on geometric optics, for example the camera, bifocal binoculars, microscopes, range finders and complex telephoto lenses. An example of the complexity of modern geometric optical instruments is the adaptive lens system used in astronomic telescopes. Here active control of the shape of the mirror under computer control is used to compensate for vibrationally, thermally and mechanically induced deformations.

Other instruments based on geometric principles are optical triangulation, used for noncontact measurements of distances, and optical shaft encoders, which are used to determine the position of motor shafts accurately. However, many developments in optics incorporate geometric optics with other optical principles to produce instruments for spectroscopic and interferometric analysis.

2. Spectrographic Instruments

Optical spectrometers are used to measure the spectral content of a beam of light. Various methods exist for achieving this. Predominant among these are dispersion-based instruments where light is spatially resolved into its different component colors, using

Table 1
Comparison of spectral analysis systems

Type and main use	Advantages	Disadvantages
Prism instruments: monochrometers and spectrophotometer	Wide spectral coverage without spectral overlap problems	Lowest throughput–resolving power product; nonlinear dispersion; material limitations in the infrared
Grating instruments: monochrometers, spectroradiometers, spectrophotometer	Blazed gratings give high efficiency; dispersion linear with wavelength; high resolving power	Given grating optimum for one order and wavelength region; overlap of orders requires modest predisperser system
Fabry–Perot instruments: spectroradiometers	Compact, relatively rugged; high resolving power; wavelength range selectable with predisperser/filter	Very limited free spectral range, very-narrow-bandpass filter or predisperser needed
Fourier transform spectrometers: IR spectroscopy	Highest throughput–resolving power product of all systems; covers broad spectral range without predisperser/filter	Complicated interferometric/servo control of mirror motion; problem with vibration

either a prism or diffraction grating to disperse the light. Other widely used spectrometer instruments, the scanning Fabry–Perot etalon and the Fourier transform infrared (FTIR) spectrometer, are interferometers. A comparison of the advantages and disadvantages of various types of spectrometer (Slater 1980) available is given in Table 1. The historical development of the instruments followed the order in Table 1, which is also the order of increasing technical difficulty.

In prism instruments, variation of the refractive index of the prism glass with light wavelength splits light transmitted through the prism beam into its component parts. Newton provides the earliest reported use of a prism system to disperse light and produce clear spectra (of the sun and of Venus). Figure 1 shows the arrangement of a simple modern spectrophotometer based on a dispersive prism.

The second type of dispersion-based spectrometer is the grating spectrometer. In this spectrometer, when a monochromatic beam of the light under examination is reflected from a diffraction grating it does so into several different orders, the angles between which depend on the rule spacing of the grating and the wavelength of the incident light (Jenkins and White 1957). For a high dispersion and resolution the rules

on the grating should be about the same spacing as the wavelength of the incident light. Examples of modern grating spectrometers are discussed by Robinson (1988).

The scanning Fabry–Perot etalon spectrometer operates by multiple interference between beams of light in an optical cavity (Slater 1980, Vaughan 1989, Williams 1993). This spectrometer has the highest resolution of all the spectroscopic instruments and this gives the interferometer a central role in studies in physics. Its uses include measurements of fine and hyperfine structure in atomic spectra, precision measurements of the velocity of light, and studies of isotope shift and nuclear structure. It has contributed to recent work in remote sensing, optical bistability, plasma physics, atomic physics, light-scattering spectroscopy and astrophysics.

The FTIR spectrometer (Slater 1980) normally consists of a Michelson or other similar two-beam interferometer. In operation, two optical beams originating from the same source are split apart to follow different paths and then brought together on a detector where they interfere. As one of the paths is adjusted away from balance an interference pattern is observed which is characteristic of the spectral content of the source. Fourier transform of the relationship between

271

the detected signal intensity and the path imbalance yields the spectrum of the source (see *Spectroscopy: Fundamentals and Applications*).

3. Interferometric and Coherent Light-Optic Instruments, and Information Processing

The Fabry–Perot and Michelson interferometers discussed earlier are examples of interferometers used to analyze incoherent light with broad spectral width. In such interferometers, the optical paths must be kept in balance or very near balance because of the short coherence length of the source. In other applications, monochromic light with longer coherence length is normally employed and fringe counting of the interference beats between the beams is used for the determination of optical path differences between the beams. An historical example of this is the use of the Michelson interferometer to determine the number of wavelengths of cadmium red line in the standard meter. Other interferometers are the Jamin and Mach–Zehnder interferometers used for determining the refractive index of gases (Jenkins and White 1957).

Currently, the main application of interferometers is in the analysis and manipulation of coherent laser radiation. Coherent light-optic instruments invariably consist of one or more laser sources in an interferometric system, in which a light beam from a laser, after interacting with an object under interest, is brought into interference with a reference light beam from the same laser. This is a large area that ranges from optical-fiber communication and sensor systems, through laser Doppler anemometry, speckle interferometry and holography. This is now one of the most significant areas of modern technology, set to influence most aspects of engineering, communications, computing and development of media systems. (Optical-fiber components and systems are discussed in the article *Transducers, Fiber-Optic.*)

Laser Doppler anemometry is used to measure the velocity of objects (Humphrey *et al.* 1975). Its main application is in measuring the velocity profile of liquid in pipes. The laser provides two beams from the optical unit which intersect at their focal region and provide an interference pattern. Particles present in the flow scatter light from the incident beams. This scattered light is Doppler shifted in frequency by the component of particle velocity in the direction of the light. Scattered light from the particles is then collected and focused onto an electronic detector, where scattered light from the two beams mixes and an electrical signal is obtained with a frequency of

$$\nu_d = \frac{2U\nu_0 \sin\phi}{c}$$

where U is the velocity of the particle, ν_0 is the frequency of the laser light, c is the speed of light, and ϕ is the half-angle between the two incident beams.

In speckle interferometric instruments, when an optically rough surface is illuminated with laser light the scattered radiation interferes, forming speckle patterns in all directions and at all distances from the surface. The pattern is related to the object and is displaced if the object is moved. A modern example of speckle interferometry is double-pulsed electronic speckle interferometry, used to investigate transient vibration in engineering parts. With this technique transient vibrations of 0.01–1 ms can be studied.

In holographic photography light from an expanded laser beam is mixed with reference light from the same laser, reflected from an object of interest. Combination of these beams forms a three-dimensional interference pattern containing both spatial amplitude and phase information on a thick photographic film. A three-dimensional view of the object is obtained by illuminating the developed photographic film with a laser beam from the same direction as the beam used for exposure.

Holographic techniques have many applications in optical information processing. A recent example is the implementation of the single-layer perception, the building block of neural networks, in a fully parallel and analog fashion. Further applications of holography can be found in a number of texts (e.g., Philipp *et al.* 1992, Pedrini *et al.* 1993).

See also: Analytical Physical Measurements: Principles and Practice; Microwave Measurements; Optical Measurements

Bibliography

Buican T N 1990 Real-time Fourier transform spectrometry for fluorescence imaging and flow cytometry. *Proc. Soc. Photo-Opt. Instrum. Eng.* **1205**, 126–33

Gopel W, Hesse J, Zemel J N (eds.) 1992 Sensors: a comprehensive survey. In: Wagner E, Danliker R, Spenner K (eds.) *Optical Sensors*, Vol. 6. VCH, Weinheim, Germany

Humphrey J A C, Melling A, Whitelaw H 1975 Laser-Doppler anemometry. *Proc. Conf. Engineering Uses of Coherent Optics.* Cambridge University Press, Cambridge

Jenkins F A, White H E 1957 *Fundamentals of Optics*, 3rd edn. McGraw-Hill, New York

Pedrini G, Pfister B, Tiziani H 1993 Double-pulse electronic speckle interferometry. *J. Mod. Opt.* **40**(1), 89–96

Philipp H, Neger T, Jager H, Woisetshlager J 1992 Optical tomography of phase objects by holographic interferometry. *Measurement* **10**(4)

Robinson L B (ed.) 1988 Instrumentation for ground-based optical astronomy, present and future. *9th Santa Cruz Summer Workshop Astronomy and Astrophysics.* Springer, Berlin

Slater P N 1980 *Remote Sensing—Optics and Optical Systems.* Addison-Wesley, Reading, MA

Tsukakoshi M, Nishida S, Yamada Y, Misu A, Kasuya T 1990 Laser microfluorometer with intensified photo-diode array detector and scan stage. *Meas. Sci. Technol.* **1**(12), 1311–13

Vaughan J M 1989 *The Fabry–Perot Interferometer: History, Theory, Practice and Applications*. Hilger, Bristol

Williams D C (ed.) 1993 *Optical Methods in Engineering Metrology—Engineering Aspects of Lasers*. Chapman and Hall, London

W. Boyle
[City University, London, UK]

Optical Measurements

In recent years there has been a dramatic increase in the use of optical and fiber-optic techniques in numerous scientific and technological applications for measurement, and the instrumentation on which it depends. The recent expansion of the field of fiber-optic sensors has been discussed elsewhere (see *Transducers, Fiber-Optic*) and this has been complemented by development of the use of optical techniques which do not necessarily rely on the employment of an optical fiber to connect the interrogating light to the sample to be investigated. Thus, while there is little overlap with intrinsic fiber-optic techniques which rely on interactions in an optical fiber, many extrinsic fiber-optic techniques have their origin in bulk optical and open airpath optical meansurements which predate the invention of the optical fiber. Such techniques have been refined and expanded and, together with others which have subsequently been devised, they owe at least part of their popularity to the very rapid expansion in optoelectronics since 1970, as a complement to the rapid development of optical communications and optical media in the domestic and office environment (the compact disk and associated computer data storage). Even classical optical measurement instruments have been influenced by new and convenient sources and detectors, which have been made available by the demands of what is now one of the largest industries worldwide.

However, this article will concentrate on those developments in the field of optical measurements which have both exploited the recent developments in technology and have shown, or are showing, significant potential for influencing the world of instrumentation in the twenty-first century. Thus, topics such as integrated optics in measurement, remote sensing, optical information processing and coherent optical systems, as applied to measurement and instrumentation, optical storage and data manipulation, will be covered, in addition to thermal imaging, holography and laser gyroscopes. This is by no means an exhaustive list, but it is one that is representative of emerging and expanding technologies. The references given will enable the reader to investigate the subject in more detail, through further reading.

The scope of optical measurements is often not clearly defined in terms of the wavelengths (λ) used. However, for most practical purposes, the wavelength regions involved include the visible (400–700 nm), the ultraviolet (200–400 nm) and the near infrared (700–2500 nm and beyond), due to the availability of sources and detectors to cover these regions conveniently (Wilson and Hawkes 1983). Beyond those wavelengths, in the vacuum ultraviolet region, absorption by air is a serious problem and such applications are often limited to those where the shorter wavelength really does offer a significant advantage for overcoming the inconvenience of the necessary optical arrangement (e.g., photolithography using excimer lasers at $\lambda \leqslant 190$ nm). The field is now dominated by optoelectronics, relying heavily on the twin disciplines of optics and solid-state physics and thus, lasers, which are used in many aspects of industrial processing, engineering, metrology, scientific research, communications, holography, medical measurements and treatment, together with military and defense uses. The diversity of the laser as a tool is such as to make it a separate study, and reference is only made to the application of lasers, and not a detailed evaluation of their comparative uses in particular technologies.

1. History and Background

As has been discussed, the history of optical measurements is long and distinguished. For many centuries optical techniques have been used for navigation and inspection, quality control and assessment. Thus, prior to the era of formal optical instruments, the color and appearance of an object was a (nonquantitative) measure of its quality and acceptability, and there is early evidence for the evaluation of wine and spirits by examination of their color and clarity with sunlight. The expansion of the field in the seventeenth century with the invention of optical instruments for astronomy and microscopy opened up the use of optical measurements for scientific research, and much of the historical development of science and the understanding of matter is due to optical methods of spectroscopy, based upon the differential in absorption over the wavelength band of a source, with a quantitative assessment yielding important data. The development of the laser in 1960, based upon the discussion of stimulated emission by Einstein, earlier in the century, led to a new field of optoelectronic measurement, using the power and capabilities of lasers. Complemented by other developments in optoelectronics, there has been much expansion of the use of optical metrology in the years subsequent to the development of the laser.

2. *The Influence of the Laser*

Lasers are as diverse in size and capability as any other scientific instrument, but there are several properties, familiar in various degrees to many lasers, which make them useful for modern optical instrumentation for metrology. These are directionality, narrow linewidth, coherence characteristics, brightness and focusing capabilities and, for some lasers, tunability.

Directionality is a feature where most lasers (with the notable exception of the semiconductor laser) emit radiation in a highly directional, collimated beam with a low angle of divergence. This is important because it means that the energy of the laser can be collected easily and directed onto a small area of the subject. This contrasts with the use of conventional sources, where such efficient light collection is almost impossible. The extent of beam divergence is set by diffraction limits.

The laser is highly monochromatic. This narrow linewidth feature is most readily used in optical measurements in the laboratory, for example, spectroscopy for pollution monitoring, where the narrow linewidth can enhance the degree of interaction with a specific molecule or species. Even for optical inspection, the narrow linewidth and, thus, specific color can enable the easy identification of one beam from another, with consequences on the collection of data from specific objects.

The coherence of the laser, both spatially and temporally, relates to its use in interferometry, a valuable technique for the identification of movements and changes on a nanometer basis, important in the rapidly developing area of nanotechnology. This field is one where the capabilities of machining to nanometer precision for such applications as video recorder head fabrication are complemented by the ease of assessment of the quality of the finish of the surface, by such laser means.

The brightness of the laser is a primary characteristic, defined as power emitted per unit area per solid angle. This, coupled with its directionality, makes it a valuable device for distant probing of objects, and for measurements relying on time of flight, using pulsed laser sources. The focusing properties of the laser enable it to be used for microprobing, drilling, welding and heat treatment in surface inspection and preparation. The focusing aspects of the laser relate to its modal characteristics, and their tailoring to achieve a particularly narrow beam have led to two specific advances in modern metrology: (a) the laser scanner (most familiar from point-of-sale use in supermarkets, but equally valuable in other aspects of metrology), and (b) the optical disk. Information is imprinted on the disk in digital form by forming small pits on its surface, using a "writing" laser, and these pits are subsequently "read" by a lower power laser (usually a semiconductor type) to provide the data stream, which may, of course, be music from a compact disk (CD), pictures from a video disk, or other data to interface to a computer.

Tunability is a feature of some solid-state or dye lasers, particularly valuable for applications in scientific metrology and photochemistry. Thus, atomic and molecular absorptions can be tuned across a wide wavelength band, to encourage specific and selective absorption, to measure such features as absorption or ionization cross sections, and other features on an atomic scale (Wilson and Hawkes 1983).

3. *Developments in Optical Metrology*

3.1 *Distance Measurement*

Distance measurement, on a scale ranging from subnanometer to multikilometer, can be achieved using several different optical metrology methods. Pulse echo techniques measure distance by timing the transit time for a very short pulse reflected from a distant target. A spectacular illustration of the use of this method has been in the monitoring of the earth–moon distance, using a laser beam reflected from a device left on the moon by the Apollo astronauts. The earthbound version of the technique is termed lidar (by analogy with radar), and may be used in conjunction with the monitoring of atmospheric pollutants, where both the time-of-flight and the spectral nature of the laser beam are used to determine the concentration and relative position of the pollutant. Interferometric methods, as illustrated by Fig. 1, use varia-

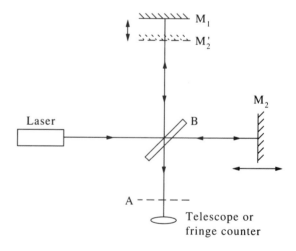

Figure 1
The Michelson interferometer: M_1 is a fixed mirror, M_2 is movable and M_2' is its image in the beamsplitter B. Interference fringes can be seen through the telescope or alternatively the number of fringes crossing an aperture placed in position A can be detected using a photodetector and counted electronically

tions of fringe counting for applications including machine tool control and length standard calibration, and for seismic and geodetic purposes.

3.2 Integrated Optics in Measurement

In the late 1960s, a form of optical circuitry, which held the possibility of guiding laser beams along miniature transmission lines, in a form analogous to miniaturized electronic circuitry, was devised, primarily to complement the associated developments in optical communications. However, the applications of these devices are possibly greater, in their potential to manipulate light for measurement purposes (Andonovic and Uttamchandani 1989). Common to all the materials technologies which may be used are basic planar fabrication methods to develop both waveguides and planar components on a single optical chip. Devices such as electrooptic modulators, optical switches and interferometers can be fabricated in this way and the extension of the open air or fiber Mach–Zehnder interferometer to integrated optic use has been possible due to work in this field (Auracher and Keil 1980). Other devices can both be fabricated and modelled, using advanced finite-element methods, for example, integrated optical switches. Numerous potential uses of the technology have been proposed for measurement: as diverse as consumer electronics, microwave signal processing, inertial navigation and sensors, in addition to communications. The opportunities with these devices for cheap and readily fabricated sensor elements offers a potential new dimension to optical metrology, and its even more widespread adoption in industry.

3.3 Remote Sensing

In addition to applications in distance measurement, as discussed, optical techniques have a wider application in remote sensing. The field can be divided into those with cooperative targets (which provide a deliberately enhanced return) and noncooperative targets where the subject to be interrogated does not provide such an enhancement of the reflected light.

Further, there are two major categories of such sensor, active and passive sensors. Active sensors are those which are categorized as using deliberate target illumination for the measurement of characteristics such as distance from the sensor, reflectivity, radial velocity or internal motion such as rotation or vibration. Passive sensors are systems where the object under scrutiny is not directly interrogated by a deliberate optical analysis, but instead, either emission (such as blackbody radiation), fluorescence or reflected sunlight, or other ambient illumination is used.

Rangefinders using pulsed lasers may be employed in various configurations or they may be used, exploiting their coherence characteristics, in a heterodyne system. Techniques developed for radar applications are often applicable here. Examples may use a pulsed carbon dioxide laser at a wavelength of $10.6\,\mu m$, exploiting the many transitions of the gaseous molecule which can be made to lase. Alternatively continuous-wave (cw) coherent techniques may be employed using acoustooptic modulators to provide an identifiable frequency for detection purposes. Passive sensors include those relying on thermal imaging (as discussed later) for feature detection and evaluation, for example, length or other dimensions. The major uses of such rangefinding techniques are for military metrology, defense systems and in aerospace applications. Such sensors could be envisaged to have wider commercial applications, but cost considerations may preclude the immediate use of many of these.

3.4 Optical Image Processing

Such techniques have been available for some time, but have been slow to be employed in many real applications (Uttamchandani and Andonovic 1992). The processing of images is a wide subject, but may be summarized as the study of two-dimensional data, with particular applications in robotics and machine vision, where such metrology is essential to the operation of sophisticated robots. As such, the broad aspects of the subject are dealt with separately, elsewhere in this encyclopedia, but in summary the field owes as much to advanced computer techniques as it does to optical methods. Indeed, the optical techniques are relatively simple, being based on television and optoelectronics which are widely available, although the use of all-optical methods based on the employment of spatial light modulators and high-performance, two-dimensional devices has been slowed by the rate of development of such devices. However, the use of neural networks and the potential for fully optical computing will have an impact on this field, especially in military applications of metrology, such as high-resolution radar target recognition, texture classification in synthetic aperture radar and image restoration. If the material and device limitations of today's devices are overcome, an all-optical, totally parallel processor should be able to perform recognition tasks in a fast and efficient way.

3.5 Optical Storage

Optical recording methods have been used for a number of years, from the photographic plate to the laser disk (Bell 1983). The optical disk, especially the CD, which is widely used, represents a clear example of optical metrology in the recording and playback, via optical means, of a data stream. This requires nanometer precision in the operation of the device and the recording of the information, but its widespread use in optical metrology is more likely to be in the area of recording data from other systems described. The sophisticated technology could have applications in other areas of dimensional metrology, at relatively low cost, through the exploitation of known methods.

3.6 Thermal Imaging

Thermal imaging systems are widely used for optical measurements of the most coarse type, that is, the presence or absence of a subject. With the employment of machine vision techniques, dimensional information may be obtained, but passive infrared methods are widely used for military and security applications to search a scene for camouflaged or straying objects or personnel. The coupling of advanced signal processing has enabled the production of a useful infrared search and track system (IRST), which is designed to identify and lock on to an object against a noisy or cluttered background. The main performance parameters of such a system are the noise equivalent temperature difference, which is a measure of the ability of the system to provide a useful image of low temperature difference between the object and the surrounding scene ("the contrast"), and the minimum resolvable temperature difference, which is essentially the signal-to-noise limited thermal sensitivity of the image as a function of the spatial frequency. Improvements are linked to the development of higher performance detector arrays and associated signal processing to stretch the use of such systems in more hostile environments.

3.7 Holography

Holography is a technique for recording an optical wavefront; its importance developed with the invention of the laser as a source of the coherent and directional radiation, which made the creation of a practical hologram a reality. Since then, holography has been a tool for use by the engineer to derive quantitative data. The hologram is produced by illuminating the object, and recording the collected reflected wavefront, together with the beam which causes the illumination. The hologram may be used to produce a three-dimensional image of the object from which dimensions may be extracted. The wavefronts may be reproduced for subsequent analysis in interferometry, using the optical recording medium: a photographic film in conventional use. Television cameras may also be used as the recording medium, in spite of the fact that they fall short of the resolution required for conventional holography. They are well suited, however, to the closely related subject of speckle interferometry, from which much useful information on the subject may be derived. In particular, such a technique is valuable in the determination of information on surface shapes and displacements, static strains, and dynamic vibrations.

A limitation of the use of holography is the need for a careful and stable optical alignment, and thus the optical arrangement requires a high degree of precision in adjustment. Thus, the use of an optical table is preferred, which may interfere with the operation of the techniques discussed outside the laboratory. The use of optical fibers can help to overcome some of these problems. The field has developed well, to the

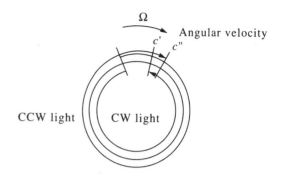

Figure 2
Optical loop with guided light: CCW, counter clockwise; CW, clockwise

extent to which it is in widespread use for the evaluation of engineering components in aerospace and automotive applications. The advance of solid-state systems and the use of optical fibers are giving renewed impetus to the field.

3.8 Laser Gyroscopes

The term gyro(scope) refers to a class of sensor for angular measurements, for applications in aerospace, in particular, and navigation, in general, to determine the total angle turned through, or the angle to which the object points with respect to a fixed reference. Normally three gyros are used in navigation, one for each axis. Optical gyros have been applied with the advent of laser technology to produce lightweight, long-life and reliable devices. The measurement principle is based on the Sagnac effect, where a phase difference is generated in a rotating ring optical structure, as shown in Fig. 2, as a result of the fact that light takes longer to go round in one direction than another, and thus to complete one full revolution. This is a measure of the rotation rate of the gyro, and thus of the vehicle to which it is fitted.

Three principal types of gyro have been developed, the familiar ring-laser gyro, the interferometer fiber gyro and the optical resonator gyro. The fiber gyro is beyond the scope of this article (see *Transducers, Fiber-Optic*). The ring-laser gyro was the first to be developed and is found on commercial aircraft, missiles, and so on. The interferometer gyro is under development, and looks promising for the future. The optical resonator gyro is at an early stage of development at the moment. A typical ring-laser gyro is shown in Fig. 3. It consists of a series of helium–neon lasers in an integral monoblock construction of perimeter 30–40 cm. The mirrors are specially optically contacted to the block to provide a gastight seal. The details of the design are critical to performance and the avoidance of errors in the measurement, and thus features such as the aperture size, mode propagation, gas pressure, purity, electrical characteristics of the

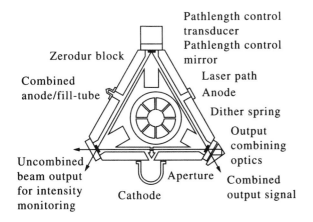

Figure 3
Typical layout of a ring-laser gyro

discharge, and so on, have to be carefully determined and maintained in use. The monoblock is made from low expansion glass, typically Zerodur (Schott Ltd), with internal features drilled out of the block. In these ways, errors are minimized and thus the gyro represents one of the recent success stories of the use of optical metrology, reflecting the degree of effort expended in its development and the sophistication of the resultant product, now supplanting the nonoptical device.

4. Conclusion

Modern optics, applied to optical measurements, is a multidisciplinary subject which has been successful in many fields and which offers considerable promise in others. It is diverse, has gained the maximum from developments in other related technologies, that is, communications, and offers considerable potential for the future, with low cost and compact size a possibility for many of the systems discussed.

See also: Analytical Physical Measurements: Principles and Practice; Microwave Measurements; Optical Instruments

Bibliography

Andonovic I, Uttamchandani D 1989 *Principles of Modern Optical Systems*, Vol. 1. Artec House, Norwood, MA
Auracher F, Keil R 1980 Design considerations and performance of Mach–Zehnder waveguide modulators. *Wave Electron.* **4**, 129–40
Bell A E 1983 Critical issues in high density magnetic and optical storage data. *Laser Focus* Pt 2. 125–36
Uttamchandani D, Andonovic I 1992 *Principles of Modern Optical Systems*, Vol. 2. Artec House, Norwood, MA
Wilson J, Hawkes J F B 1983 *Optoelectronics: An Introduction*. Prentice-Hall, London

K. T. V. Grattan
[City University, London, UK]

P

Position Inertial Systems

Since the 1940s, inertial navigation systems have progressed from crude electromechanical devices that guided the early V-2 rockets, to the present solid-state devices that are in many modern vehicles. The impetus for this significant progress came during the ballistic missile programs of the 1950s, in which the need for high accuracy at ranges of thousands of kilometers using self-contained navigation systems was apparent. By "self-contained" it is meant that no signals from outside the vehicle are required to perform navigation. One of the early leaders in inertial navigation was the MIT Instrumentation Laboratory (now the Draper Laboratory), which was asked by the US Air Force to develop inertial systems for the Thor and Titan missiles, and by the US Navy to develop an inertial system for the Polaris missile (Draper 1981). The notable success of those early missile programs led to further application in aircraft, ships, missiles and spacecraft, such that inertial systems are now almost standard equipment in military and civilian navigation applications.

This article deals with the concepts of using inertial navigation systems in vehicles. Present day components, systems and synthesis techniques are described.

1. Components of Inertial Navigation Systems

Inertial systems require components for sensing angular rates and specific forces, as well as a device for computing the indicated position (usually velocities and attitude are also desired outputs). Conceptually, the rate sensing devices (gyros) can be used to maintain a coordinate system in which the indicated specific forces measured by accelerometers can be integrated (using Newton's law) to produce position and velocity after a correction for gravity.

Some gyros currently used in inertial systems are, in approximate order of demonstrated increasing accuracy: the fiber-optic gyro, the ring-laser gyro, the dry-tuned rotor gyro, the magnetic gyro, the electrostatic gyro and the floated rate-integrating gyro. The last four are mechanical gyros that create angular momentum via a spinning mass. Two of these suspend the spinning mass in gimbals, while the electrostatic gyro uses an electrostatic suspension and the magnetic gyro uses a magnetic suspension, so that there is no physical contact with the spinning masses. All of the mechanical gyros have been applied in various aerospace vehicles, such as aircraft and missiles, as well as naval vessels.

The spinning mass gyroscope was first produced as a single-degree-of-freedom rate gyro used as an instrument for aircraft piloting. The basic configuration of a rate gyro was a ball bearing rotor housed in a gimbal, whose gyroscopic precession in response to angular rate was restrained by a mechanical spring. Improved performance was achieved in the rate-integrating gyroscope which, instead of using a spring, a restraining torque on the gyro gimbal is produced by a damping reaction with a servoloop to maintain the gimbal at null. In addition, the gas bearing was introduced for improved performance and lower noise. A recent application is in the Hubble space telescope. This type of gyro and other spinning mass gyros are discussed in Barbour *et al.* (1992).

The ring-laser gyro operating principle is based on the relativistic properties of light. The basic operating element in a ring-laser gyro is a closed optical cavity containing two beams of single-frequency light. The beams travel continuously in opposite directions around the closed cavity. Each beam must travel an integral number of wavelengths around the path at any instant of time, since they are continuous and close on themselves. Relative to an observer fixed to the optical cavity, a rotation of the cavity causes the light to take longer to traverse the cavity in the direction of rotation than in the opposite direction. Because the speed of light is constant, a frequency difference between the two beams exists and becomes a measure of the rotation rate. Laser gyros have been applied in aircraft, missiles and ships.

Systems are also based on fiber-optic gyros which operate on the same principles as ring-laser gyros, except that the optical path is contained in a glass fiber rather than in a vacuum. The fiber is wound on a bobbin in a manner such that the light beam travels around a circular path. Also, a major portion of the signal processing is done optically in special glass chips containing modulators, phase shifters and similar devices, rather than being done electronically. The technology involved in the fiber-optic gyro is inherently low in labor-intensive operations, and amenable to batch processing techniques, making the fiber-optic gyro a potentially low-cost method of making inertial rotation measurements. The techniques used to make these gyros borrow heavily from the techniques developed by the semiconductor industry and are generally described as being solid-state instruments. There are two fundamentally different implementations of fiber-optic gyros: resonant (RFOG) or interferometric (IFOG). Resonant gyros utilize a short closed-loop fiber as a resonant wave guide and, thus, require a very narrow line laser diode. In contrast, the IFOG uses a longer, multiturn fiber coil with a nonresonant, very broad optical structure with a broadband diode as the light source. To date, fiber-optics gyros have been applied in low-accuracy applications such as in tactical missile navigation.

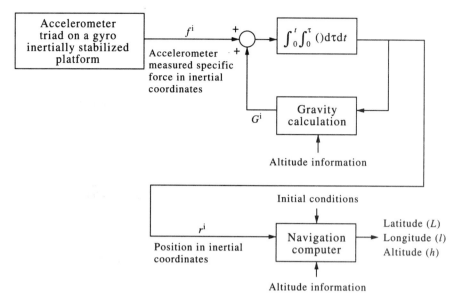

Figure 1
Gimbal system space-stable inertial navigation mechanization

Most accelerometers used in inertial navigation systems involve either a proofmass supported by a pendulum with an electromagnetic restoring loop to null, or a proofmass supported by a beam or string or tuning fork, where specific force is indicated by a change in oscillator system frequency. The use of the word "accelerometer" is, in a sense, a misnomer. Because of the equivalence of gravitation and inertial acceleration, an accelerometer will measure a combination of inertial acceleration and gravitational attraction in the presence of a gravitational field. For this reason, an accelerometer may alternately be called a specific force meter (Wrigley 1969). In the presence of a gravitation field, the instrument outputs must be corrected for gravity (see Sect. 2). Thus, the process of inertial navigation requires the *a priori* knowledge of the gravity field.

Detailed descriptions of gyros and accelerometers, with error models for each, are given by Savage (1978, 1984a) and Barbour *et al.* (1992). The current accuracy of inertial navigation systems based on the previously mentioned gyros and accelerometers appear to meet all military and civilian requirements. It appears that future developments in inertial instruments will be driven by technologies that enable lower power, weight and cost and improved reliability.

2. Gimbal and Strapdown Navigation Mechanizations

Gimbal and strapdown navigation mechanizations are the two most popular implementations of inertial navigation system mechanization. In an inertial platform gimbal mechanization, the gyroscopes mounted on a stable element in a gimbal system measure angular rates and the gimbal drive system can use the angular rate information in a feedback manner to rotate the gimbals and null the angular motion sensed by the gyroscopes. In this manner, the gyroscopes and accelerometers mounted on the stable element are inertially stabilized from the vehicle motion and the stable member physically represents an inertial reference frame. By double integration of the specific force indications from the accelerometers, with a correction for gravity, position determination is possible. Figure 1 illustrates this approach.

Gimbal systems provide a good dynamic environment for inertial instruments, particularly in vehicles that are to perform high maneuver levels, since the gimbals isolate the gyros from the rotational environment. In strapdown inertial navigation systems, the sensors are mounted directly on the vehicle without using isolating gimbals. Inertial sensor outputs now represent specific force and angular rate with respect to inertial space with vehicle body axes as coordinates. Figure 2 illustrates this mechanization and modern strapdown algorithms are discussed in Savage (1984b). The inertial sensors for strapdown navigation applications require a large dynamic measurement range. They are subject to the entire vehicle dynamic environment and, in general, will not perform as well as if they were isolated from it.

Strapdown navigation systems are of interest for all but the highest accuracy demanding missions, since elimination of the gimbals could result in easier maintenance, lower cost and improved reliability. The

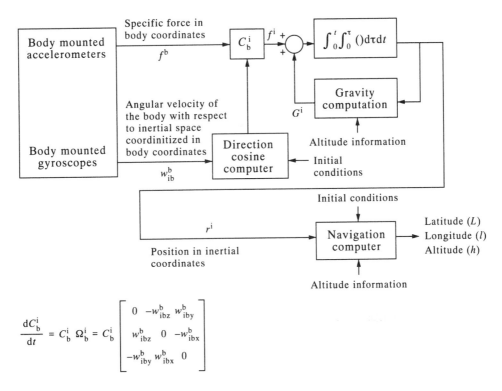

$$\frac{dC^i_b}{dt} = C^i_b \ \Omega^i_b = C^i_b \begin{bmatrix} 0 & -w^b_{ibz} & w^b_{iby} \\ w^b_{ibz} & 0 & -w^b_{ibx} \\ -w^b_{iby} & w^b_{ibx} & 0 \end{bmatrix}$$

Figure 2
Strapdown inertial navigation system computing in inertial coordinates

laser gyro/electronics assembly currently in use in many commercial aircraft has been demonstrating a mean time between failure of tends of thousands of hours. If external information (such as position information derived from a satellite navigation system) is allowed to aid the inertial system, performance differences between gimbal and strapdown systems are even less.

3. Navigation Mechanizations

As was illustrated in Fig. 2, the mechanization concept for a strapdown navigation system computing in an inertial reference frame requires the use of the gyro outputs in the computation of a transformation matrix between body and inertial coordinates. In the lower left corner of Fig. 2 the computations for the direction cosine matrix are illustrated.

Figure 3 illustrates the case when the geographical north pointing local level navigation frame is the instrumented stable member frame for a gimbal system. Figure 4 shows the case where the local-level navigation frame is the computational reference frame for a strapdown navigation system. The local-level frame is the most common reference frame for gimbal systems; for strapdown navigation systems, since the

gyros measure body angular rate with respect to inertial space, the inertial frame seems more appropriate. However, since navigation information in geographic coordinates is almost always a desired output, Fig. 4 is a typical mechanization found in inertial navigation systems.

The local-vertical frame is usually some form of a "wander azimuth" reference frame in which no attempt is made in a gimbal system to point one of the stable member axes north, as illustrated in Fig. 5. This mechanization eliminates difficulties in flying over the earth's poles. Two of the axes are maintained level but free to rotate in azimuth about the third axis, which is maintained parallel to the local vertical.

4. Navigation Errors

Calibration limitations on strapdown navigation systems arise because the inertial sensors, which are rigidly attached to the vehicle, cannot be arbitrarily oriented at different angles to known inputs from the earth's gravity and angular velocity vectors in a stationary vehicle. Conversely, a gimbal navigation system is mechanized so that the inertial sensor outputs measured at different angular orientations with respect to the earth can be compared with the

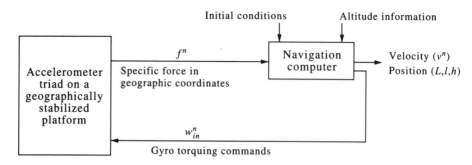

Figure 3
Gimbal system geographic local vertical mechanization

known input components of gravity or earth rate and the sensor errors calibrated in a series of test positions that expose the error sources. Since only a few of the strapdown navigation system sensor errors (compared with a gimbal system) can be calibrated, the system designer must depend on the stability of the instrument's performance between removals of the system from the vehicle for calibration. The errors that cannot be calibrated in a strapdown system propagate into navigation errors when the system begins to navigate.

Sensor inaccuracies that arise from motion of the vehicle are even more difficult to calibrate in a strapdown navigation system. The dynamic response characteristics of the inertial sensor under consideration for use in the navigator must be well understood and the limitations on navigation performance evaluated early in the system design. For example, in single-degree-of-freedom gyros, a gyro drift, called an isoinertia-induced drift, will be present whenever the gyro concurrently experiences accelerations such as vibration about its spin and input axes. To eliminate

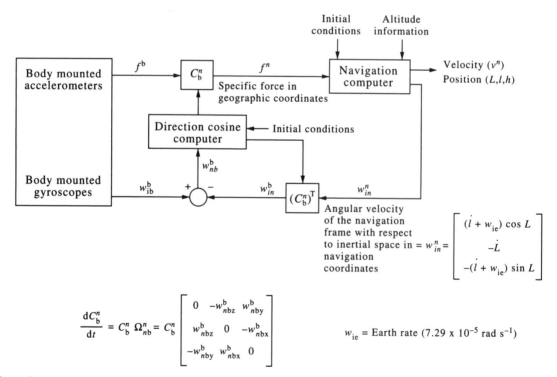

Figure 4
Strapdown navigation system computing in geographic local vertical coordinates

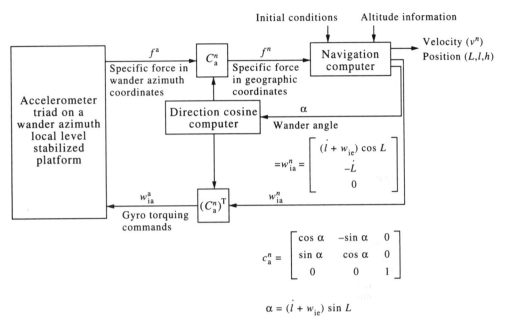

Figure 5
One form of wander-azimuth mechanization

such an error source requires complicated compensation, therefore the system designer must understand the interplay between vehicle environment, sensor design and resulting navigation system performance when selecting a particular type of gyroscope for a strapdown navigation system application.

A source of strapdown navigation system sensor error for some mechanical gyros is gyro torquing scale factor error, particularly asymmetrical scale factor errors in a vibratory environment. In a strapdown navigation system, the gyros must also be torqued for the attitude rate of the vehicle; consequently, scale-factor-induced navigation errors can be quite large. Misalignments of gyro input axes in strapdown navigation systems are also important, for similar reasons.

Another major sensor error source in strapdown single-degree-of-freedom gyros is the output axis rotation error. In this case, the strapdown gyro has a drift error proportional to the angular acceleration along its output axis.

Computation-induced errors in strapdown navigation systems refer to those navigation errors introduced by the incorrect transformation of the body-measured specific forces to the computation frame. The algorithm used to generate the transformation matrix, the speed at which the transformation occurs, the computer wordlength used, and the inertial sensor data quantization level all play a part in contributing to computational errors. Since this type of error also depends on the vehicle's dynamic environment, most designers evaluate it by computer simulation (see Savage 1984b).

Sensor-induced navigation error propagation effects are also different in a strapdown navigation system. For example, in a gimbal navigation system during initial self-alignment, the accelerometer outputs are used to level the stable member, and the mislevelling of the stable member is due, primarily, to accelerometer bias. When the system starts to navigate, no significant position or velocity errors occur due to this error source, since the platform tilt error contribution is exactly cancelled by the accelerometer bias error contribution.

All inertial navigation systems also need to be initialized with correct position, velocity and attitude information. Some initialization quantities can be measured directly by the inertial sensors (e.g., latitude can be determined via measurement of the angle between the earth's rotation vector and the gravity vector); other nonphysical quantities (such as longitude) must be specified at the start of the mission. Any errors in the initialization process will affect performance during the mission.

Generally, designers will use the rule of thumb that for gimbal systems $0.01°\,h^{-1}$ of gyro drift will produce a position error growth rate of $1.85\,km\,h^{-1}$ and a velocity error of $\sim 1\,m\,s^{-1}$ for a terrestrial navigation system; for a strapdown system, the errors would be 20–25% worse. However, as mentioned previously, the difference in errors can be quite small, depending upon the vehicle environment and the use of other aiding sensors (Doppler velocity, position fixing radar, satellite navigation receivers, and so on). Consequently, the process of preliminary system selection

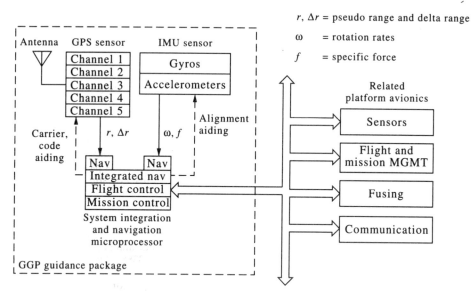

$r, \Delta r$ = pseudo range and delta range

ω = rotation rates

f = specific force

Figure 6
Aided inertial GPS mechanization

involves not only the choice of system components and mechanization, but also the choice of aiding devices and frequency of updates. These tradeoffs are usually conducted with a general purpose covariance analysis program.

5. Performance Tradeoff Analyses

The errors δx in any inertial navigation system mechanization can usually be cast in the state-variable form:

$$\frac{d(\delta x)}{dt} = F\delta x + Gq$$

where the coefficient matrices F and G are functions of the navigation system mechanization and vehicle dynamics (Schmidt 1978). This allows the well-known covariance matrix propagation equations to be used in predicting inertial system performance. The F and G matrices are functions of the vehicle trajectory that will be followed; thus, the performance of the navigation system must usually be evaluated over many representative scenarios. From those performance analyses which depend on accuracies assumed for the gyros and accelerometers, specifications can be placed on the inertial sensors.

If other sensors, in addition to the gyros and accelerometers, are part of the navigation system, the Kalman filter update equations can be applied to study requirements on sensor aiding accuracy, frequency of updates, and so on, to predict accuracies in the estimates of the errors $\delta \hat{x}$ that could be achieved in the actual implementation. Figure 6 illustrates a block

diagram of a typical implementation using the global positioning satellite (GPS) system as part of a navigation system called the GGP (GPS guidance package) (Stotts *et al.* 1989). The integrated navigation represents a system which mutually aids the GPS sensor and the inertial measurement sensors (IMUs) through a Kalman filter. The filter estimates receiver errors as well as inertial navigation errors. However, realistic modelling of all error sources usually results in a state vector dimension that prohibits full implementation in the vehicle computer. Tradeoffs to determine the appropriate suboptimal filter to implement are then required (Schmidt 1976a, Setterlund 1988). From these performance tradeoff analyses, a set of system performance specifications can be developed prior to implementation and integration.

6. System Integration

The integration of the inertial system and external aiding devices is usually performed through a central computer with a suboptimal filter implementation as described in Sect. 5. Many issues must now be addressed, such as how to integrate the state and covariance matrices in the on-board computer. Various techniques continue to be developed (Schmidt 1976b, 1989) that assist in the filter implementation in finite memory, limited wordlength computers.

Modern inertial navigation systems are increasingly meeting the demands for better mission performance, and the range of applications of aided and unaided inertial navigation systems continues to expand. They are being used in aerospace vehicles, guided weapons, ships and land vehicles. The systems are meeting cost, weight and volume constraints as well as providing the required accuracy. Further miniaturization and cost reductions should allow a further expanse of applications in the near future.

See also: Angular Rate Sensing

Bibliography

Barbour N M, Elwell J M, Setterlund R H 1992 Inertial instruments: where to now? *Proc. 1992 AIAA Guidance and Control Conference*, Hilton Head Island, SC. AIAA, Washington, DC, pp. 566–74

Draper C S 1981 Origins of inertial navigation. *J. Guidance Control* **5**, 449–63

Savage P G 1978 Strapdown sensors. In: Schmidt G T (ed.) *Strapdown Inertial Systems*, NATO AGARD Lecture Series, No. 95. NATO Advisory Group for Aerospace Research and Development, Neuilly-sur-Seine, France, pp. 2-1–2-46

Savage P G 1984a Advances in strapdown sensors. In: Schmidt G T (ed.) *Advances in Strapdown Inertial Systems*, NATO AGARD Lecture Series, No. 133. NATO Advisory Group for Aerospace Research and Development, Neuilly-sur-Seine, France, pp. 2-1–2-24

Savage P G 1984b Strapdown systems algorithms. In: Schmidt G T (ed.) *Advances in Strapdown Inertial Systems*, NATO AGARD Lecture Series, No. 133. NATO Advisory Group for Aerospace Research and Development, Neuilly-sur-Seine, France, pp. 3-1–3-30

Schmidt G T 1976a Linear and nonlinear filtering techniques. In: Leondes C (ed.) *Control and Dynamic Systems*, Vol. 12. Academic Press, New York, pp. 63–98

Schmidt G T (ed.) 1976b *Practical Aspects of Kalman Filter Implementation*, NATO AGARD Lecture Series, No. 82. NATO Advisory Group for Aerospace Research and Development, Neuilly-sur-Seine, France

Schmidt G T 1978 Strapdown inertial systems. In: Schmidt G T (ed.) *Strapdown Inertial Systems*, AGARD Lecture Series, No. 95. NATO Advisory Group for Aerospace Research and Development, Neuilly-sur-Seine, France, pp. 1-1–1-10

Schmidt G T (ed.) 1989 *Kalman Filter Integration of Modern Guidance and Navigation Systems*, NATO AGARD Lecture Series, No. 166. NATO Advisory Group for Aerospace Research and Development, Neuilly-sur-Seine, France

Setterlund R H 1988 New insights into minimum-variance reduced-order filters. *J. Guidance Control* **6**, 495–9

Stotts L, Aein J, Doherty N 1989 Miniature GPS-based guidance package. *Proc. NATO–AGARD Guidance and Control Panel Symp. Advances in Techniques and Technologies for Air Vehicle Navigation and Guidance*, Lison, Portugal. NATO Advisory Group for Aerospace Research and Development, Neuilly-sur-Seine, France

Wrigley W, Hollister W, Denhard W 1969 *Gyroscopic Theory, Design, and Instrumentation*. MIT Press, Cambridge, MA

G. T. Schmidt
[Charles Stark Draper Laboratory,
Cambridge, Massachusetts, USA]

Pressure Measurement

There is a very wide range of practical applications of pressure measurement. Among the most important are process control, meteorology and aviation altimetry. Gas and liquid flow rates are often measured by measuring a pressure difference developed across a series impedance such as an orifice. Leak rates are often determined by measuring the rate at which the pressure in a vessel changes.

Pressure is defined as force per unit area. In SI units it is expressed in newtons per square meter ($N\,m^{-2}$), otherwise referred to as pascals (Pa). Many traditional or practical units are still in use, among which are the following: $1\,\text{bar} = 1 \times 10^5\,\text{Pa}$, $1\,\text{mbar} = 100\,\text{Pa}$, $1\,\text{kbar} = 10^8\,\text{Pa}$, 1 standard atmosphere (atm) = $101\,325\,\text{Pa}$, 1 pound force per square inch (lbf in.^2 or psi) = $6894.76\,\text{Pa}$, $1\,\text{torr} = 133.322\,\text{Pa}$ (equal to the conventional mm Hg to within 1 part in 7×10^6), and $1\,\text{in. Hg} = 3386.39\,\text{Pa}$.

It should be pointed out that the last three conversion figures apply for conditions of standard gravity ($9.806\,65\,\text{m s}^{-2}$), while the last two depend on the use of purified mercury having an assigned density value of $13\,595.1\,\text{kg m}^{-3}$.

The range of pressures of practical importance is very large, in the range 10^{-8}–$10^9\,\text{Pa}$. Atmospheric pressure at sea level is approximately $10^5\,\text{Pa}$, and the pressure at the center of the earth has been estimated to be approximately $3.5 \times 10^{11}\,\text{Pa}$.

The measurement of pressure is always relative to a reference pressure and there are three common modes.

(a) The absolute mode is relative to a theoretically perfect vacuum. Absolute pressures below about $10^3\,\text{Pa}$ fall into the vacuum regime and are dealt with in the article *Pressure Measurement: Vacuum*.

(b) The gauge mode is relative to the ambient pressure.

(c) The differential mode is relative to any defined "line" pressure.

Furthermore, pressures may be either static (steady) or dynamic (time dependent).

1. Pressure Measurement Techniques

This section begins with a review of U tube manometers, barometers and pressure balances (piston gauges). U tube manometers and pressure balances are often used as reference standards for the calibration of indicating gauges and transducers used in the field.

1.1 U Tube Manometers

A U tube manometer is shown schematically in Fig. 1. The U tube is frequently constructed of glass and the

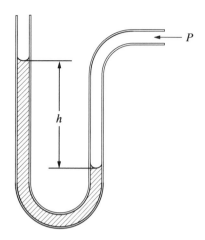

Figure 1
U tube manometer

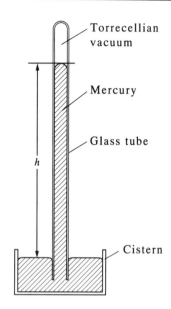

Figure 2
Basic mercury barometer

liquid is most commonly water, oil or mercury. The application of pressure to the top of one limb causes a displacement of the liquid. To a first approximation, the applied pressure may be calculated using the equation:

$$P = \rho g h$$

where ρ is the density of the manometric fluid, g is the local value of gravitational acceleration, and h is the difference between the heights of the columns.

Mercury, being much more dense than oil or water, has the smallest displacement for the application of a given pressure and, consequently, is more suitable for use with relatively high pressures. For the lower pressures, oil or water give greater sensitivity; oil is generally preferred to water because it is less prone to loss by evaporation. It must be remembered that for all liquids ρ is temperature dependent.

Many methods are available for measuring the displacement of liquid surfaces. The simplest is visual observation against a scale, aided if necessary by a Vernier or an optical microscope. More sophisticated methods include mechanical contacts attached to micrometers, laser interferometry, electrical capacitance probes, inductive sensing of the positions of magnetic floats on the surfaces, and the measurement of the transit times of ultrasonic pulses reflected by the undersides of the liquid surfaces.

(a) Meteorological mercury barometers. Meteorological mercury barometers (see Fig. 2) are a special type of U tube manometer, in which one column, the cistern, is concentric with the other and has a much larger cross-sectional area. The upper column is sealed and contains a Torricellian vacuum. As the atmospheric pressure increases, the mercury in the central column rises, while that in the cistern falls, though to a much lesser extent because of its relatively large

cross-sectional area. In the Kew pattern of barometer the cistern volume is fixed and the scale adjacent to the upper column is intentionally foreshortened to compensate for the change in the lower level. In contrast, the cistern of the Fortin design (see Fig. 3) incorporates a flexible bag whose volume can be adjusted so that the lower mercury surface can be set at a definite level indicated by means of a fiducial pointer.

By international agreement the scales of mercury meteorological barometers are inscribed so that they indicate correctly under standard conditions, that is, at standard gravity ($9.806\,65\ \mathrm{m\,s^{-2}}$) and at a temperature of $0\ ^\circ\mathrm{C}$. For all other conditions the indication must be corrected using tables or formulae recommended by the World Meteorological Organization.

When comparing barometric pressures, as, for example, in weather forecasting, it must be remembered that pressure decreases with increasing altitude. Near sea level the pressure reduction is approximately 11.6 Pa per meter increase in altitude.

(b) Aneroid meteorological barometers. In recent years mercury barometers have become less popular because of the need to apply temperature and gravity corrections and because they are not very robust. Their place has been largely taken by aneroid barometers which embody an evacuated capsule or nest of capsules. These flex with changes in ambient pressure and the movement is either amplified by a system of levers attached to a pointer or sensed by an electrical contact attached to a micrometer movement. A good quality aneroid barometer incorporates compensation

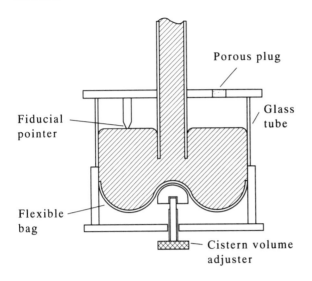

Figure 3
Cistern of Fortin barometer

for changes in ambient temperature. A recent development is the replacement of the capsule by a silicon wafer with an evacuated volume on one side; flexing of the wafer due to changes in ambient pressure is detected by diffused strain gauges in a bridge configuration.

1.2 Pressure Balances

The pressure balance (piston gauge) is widely used for the accurate measurement of pressure in both gases and liquids (Dadson *et al.* 1982, Peggs 1983, Lewis and Peggs 1992). Gas-operated pressure balances are generally used in the pressure range 2 kPa–50 MPa. The usual liquid pressure medium is oil, either mineral oil or a synthetic oil such as a sebacate. The upper pressure limit for oil operation is usually around 800 MPa, by which pressure most oils have become waxy.

The principle is illustrated in Fig. 4. An accurately made piston–cylinder assembly is mounted on a vertical column at the bottom of which is a pressure port. The upward force due to the pressure acting on the base of the piston is balanced by the downward gravitational force acting on the piston and on a stack of circular weights supported by it. The gap between the piston and cylinder is of the order of a few micrometers and the piston and weights are rotated either by hand or by motor to ensure that the piston is centered in the cylinder.

When the forces are in equilibrium the piston is said to "float," and the following relationship exists:

$$PA = g \Sigma m$$

where P is the applied pressure relative to that in the space surrounding the weights, A is the effective cross-sectional area of the piston–cylinder combination (generally the mean cross-sectional area of the piston and the cylinder), g is the value of the local gravitational acceleration, and Σm is the total supported mass.

In practice, many other factors should be taken into account, including:

(a) the effect of air buoyancy on the weights;

(b) the change in effective area of the piston–cylinder assembly due to thermal expansion; and

(c) the change in the effective area due to distortion resulting from the internal pressure.

When the pressure fluid is a liquid two other factors should also be taken into account:

(a) the buoyancy force which the liquid exerts on the piston, and

(b) the downward force on the piston due to the surface tension of the liquid that has leaked through the gap between the piston and cylinder.

Pressure balances are calibrated by a procedure called "cross floating," in which hydrostatic equilibrium is obtained between the test instrument and a reference standard pressure balance of known effective area. When balanced in equilibrium, and provided that both piston bases are at the same height, the pressures under both pistons are the same, hence:

$$A_t = \frac{A_s \Sigma m_t}{\Sigma m_s}$$

Distortion effects make both A_t and A_s pressure dependent, usually in a linear fashion. The results of a calibration therefore usually give two quantities: the effective area A_0 of the test instrument extrapolated to

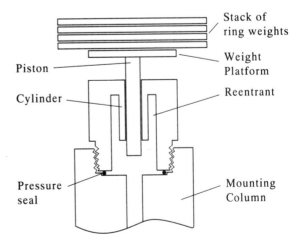

Figure 4
Pressure balance

zero applied pressure; and the distortion coefficient λ. These are related by the equation:

$$A_t = A_0(1 + \lambda P)$$

The distortion may be minimized by cutting a re-entrant in the bottom face of the cylinder (see Fig. 4) or by applying a controlled pressure separately to the outside wall of the cylinder (controlled clearance).

1.3 Ultrahigh Pressures

The term ultrahigh is generally used to denote pressures above 1 GPa. Since most oils solidify at these pressures the choice of pressure medium is rather restricted; light paraffins or light alcohols are normally used.

One common form of gauge for measuring pressures up to about 2.5 GPa is the piezoresistive gauge, which relies on the fact that electrical resistance changes with pressure. The resistance alloy Manganin is frequently used because it has a very low temperature coefficient of resistance. Although III–IV semiconductor compounds have greater pressure sensitivities, this is counteracted by their greater sensitivity to changes in ambient temperature.

Another method used in this range relies on the fact that the melting point of mercury is pressure dependent; the transition from the liquid to the solid phase may be detected by a change in electrical resistance.

For pressures above 2.5 GPa both pressure production and pressure measurement become more difficult. Provided that an experiment can be carried out in a very small volume, it is now common to use a cell consisting of two diamond anvils separated by a metal washer. Very high pressures can be generated within the cell when the anvils are squeezed together in a vice.

The internal pressure is normally measured by one of two methods. One is to use x-ray diffraction to measure the lattice spacing, and hence the compression, of small crystals of alkali halides such as sodium chloride. Another increasingly popular method is ruby fluorescence. A small crystal of ruby is illuminated with blue or green light from a laser. It produces red light by fluorescence; the wavelength of the peak of the fluorescence emission shifts as the pressure is changed.

In order to maintain consistency in the measurement of pressures above 2.5 GPa an International Practical Pressure Scale (IPPS) has been established. Values have been assigned to the pressures at which certain electrically detectable phase transitions occur in metals. The lowest such fixed point is 2.55 GPa, the pressure of the bismuth I–II transition at 25 °C.

1.4 Dial Gauges

By far the most common dial pressure gauge is the Bourdon tube gauge. A metal, or in some cases quartz, tube of noncircular (usually oval) cross section

is bent in the form of an arc, spiral or helix, and sealed at one end. The other end is rigidly fixed. On applying pressure to the inside of the tube it progressively unwinds, producing a movement of the unclamped end. Levers or gears are used to produce amplified movement of a pointer over a graduated scale.

1.5 Pressure Transducers

The vast majority of pressure transducers operate by the measurement of the elastic deformation of a flexible member, which has the pressure medium on one side and a reference pressure on the other (Doebelin 1975, Jones 1982, Neubert 1982). Sensors of this type have two distinct parts, the flexible sensing element (force summing unit) and the means of measuring its movement (transduction element). Pressure sensors with elastic sensing elements may be additionally classified into two groups: (a) the simple type, in which the movement of the flexible element is directly measured; and (b) the force-balance type, in which a measured electromagnetic force is applied to restore the force-summing element to its null position. Force-balance sensors are generally more accurate than simple sensors but since they are more complicated and expensive their main use is as calibration standards.

The simplest geometry for the elastic sensing element is a circular membrane or diaphragm which is usually mounted in a state of tension. The diaphragm is often corrugated to improve the linearity of movement. Higher sensitivity is achieved by employing a capsule, a stack of capsules or a bellows. Such improved sensitivity is generally achieved only at the expense of a reduced speed of response. The second most common form of elastic sensing element is the Bourdon tube, described in Sect. 1.4.

There are many different forms of transduction element to measure the sensor movement. One common method is to bond strain gauges on to the sensing element. They are frequently operated in groups of four, connected in a Wheatstone bridge network. They are positioned and oriented in symmetrical pairs, so that on the application of pressure the change of strain in one pair differs from that in the other. Temperature variations, on the other hand, would affect both pairs equally, producing a minimal effect on the bridge output.

Many pressure transducers have capacitive sensing of the movement of the sensing element. An ac bridge is used to measure the capacitance between the sensing element and a fixed electrode. Another method of measuring the movement of the sensing element is the linear variable differential transformer (LVDT). Optical-fiber or pneumatic transduction elements are sometimes used in hazardous environments.

1.6 Dynamic Pressure Measurement

The response of a pressure transducer to fluctuating or transient pressures depends on its mechanical inertia and on the time constants of the associated electronic circuits. When accurate measurements are required it is sometimes necessary to calibrate a transducer under dynamic conditions.

It is not easy to generate sinusoidally fluctuating pressures, therefore a Fourier analysis is often made of the transient output resulting from a step pressure change. A rapid fall in pressure may be obtained by quickly venting a pressurized vessel to the atmosphere. A rapid rise in pressure may be obtained by mounting the transducer in a shock tube, into which pressure is suddenly introduced by piercing a thin diaphragm.

The most common form of transducer for measuring transient pressures employs the piezoelectric effect. Voltage is developed across the faces of a crystal such as quartz when it is subjected to stress. The transient voltage is measured with a charge amplifier.

See also: Flow Measurement: Principles and Techniques; Pressure Measurement: Vacuum

Bibliography

Dadson R S, Lewis S L, Peggs G N 1982 *The Pressure Balance: Theory and Practice.* Her Majesty's Stationery Office, London
Doebelin E O 1975 *Measurement Systems: Application and Design.* McGraw-Hill, New York
Jones E B 1982 *Instrument Technology*, 4th edn., Vol. 1. Butterworth, London
Lewis S L, Peggs G N 1992 *The Pressure Balance: A Practical Guide to its Use*, 2nd edn. Her Majesty's Stationery Office, London
Neubert H K P 1982 *Instrument Transducers*, 4th edn. Clarendon, Oxford
Peggs G N 1983 (ed.) *High Pressure Measurement Techniques.* Applied Science, London

P. R. Stuart
[National Physical Laboratory, Teddington, UK]

Pressure Measurement: Vacuum

There is no sharp dividing line between pressure and vacuum; strictly speaking, any pressure below ambient can be regarded as a partial vacuum.

Before considering vacuum measurement it is useful to understand the concept of mean free path. This is the average distance travelled by a gas molecule before it collides with another molecule. For nitrogen at atmospheric pressure and 20 °C the mean free path is approximately 10^{-4} mm. At 1 kPa it is about 10^{-2} mm, and at 1 Pa about 10 mm. When the mean free path is much shorter than the typical internal dimensions of the containment vessel any gas flow is said to be viscous. When the mean free path is much longer any flow is said to be molecular. Between the

two regimes there is a transition region. Most of the gauges and calibration techniques described in this article are for use in the molecular flow region, although some gauges such as the capacitance manometer can also be used in the transition and viscous flow regions.

1. Vacuum Gauges

The capacitance manometer is a sensitive version of the diaphragm pressure transducer with capacitance sensing of the diaphragm movement (see *Pressure Measurement*). Unlike many other vacuum gauges its sensitivity is independent of the chemical composition of the gas.

The thermal conductivity gauge is generally used in the pressure region $10-10^{-4}$ Pa. The degree of cooling of an electrically heated thin wire depends on the pressure of the surrounding gas. The temperature of the wire can be estimated either by measuring its resistance (Pirani gauge) or by using an attached thermocouple (thermocouple gauge). The response of a thermal conductivity gauge to pressure is nonlinear and it should be calibrated using the same gas as that in which it will be used.

Ionization gauges are mainly intended for use at pressures below about 10^{-4} Pa. In the hot-cathode, triode version (see Fig. 1), electrons from a heated filament are accelerated towards a positive grid. Most of them pass through the grid into a retarding field beyond, which is produced by a negative electrode. This field repels the electrons back towards the grid. If, while in the space between the grid and the negative electrode, an electron collides with a gas molecule, ionization can occur and the positive ion so

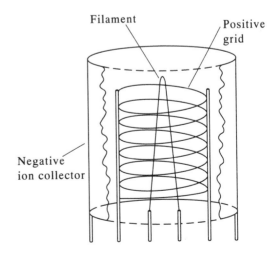

Figure 1
Triode ionization gauge

(a)
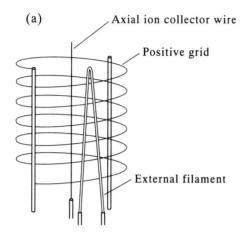

Axial ion collector wire

Positive grid

External filament

(b)
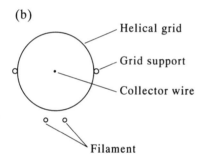

Helical grid

Grid support

Collector wire

Filament

Figure 2
(a) Bayard Alpert ionization gauge, (b) cross-sectional view

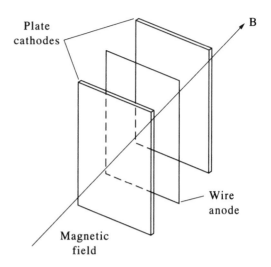

Plate cathodes

B

Wire anode

Magnetic field

Figure 3
Penning cold cathode ionization gauge

created is attracted towards the negative electrode. The ion current to the negative (collector) electrode is proportional to the number of collisions between electrons and molecules, and hence to the pressure.

The lowest pressure at which hot-filament ionization gauges may be used is set by a photoelectric effect. Soft x rays generated by electrons striking the grid wire irradiate the collector electrode and cause photo-electrons to flow from the collector to the grid. In addition, stray ultraviolet radiation from the hot filament causes the emission of photoelectrons from the surface of the ion collector and these also produce an enhanced reading of the collector current. In the Bayard Alpert version of the gauge (see Fig. 2) these effects are minimized by changing the geometry so that the collector is replaced by a fine wire having a very small surface area.

In the Penning ionization gauge (see Fig. 3) a cold-cathode discharge is struck between a pair of cathode plates and a wire anode. A strong magnetic field from a pair of permanent magnets makes the electrons and ions follow spiral paths, thereby increasing their pathlengths so that the discharge can be sustained at lower pressures than otherwise.

A hot filament ion source may be combined with an ion mass analyzer to produce a mass spectrometer. The mass analyzer is usually either of the magnetic sector type or of the quadrupole type, which employs two pairs of rod electrodes connected to a voltage source of variable radio frequency. Mass spectrometers are used for the analysis of gases, including the analysis of residual gases in vacuum systems. When tuned for helium ions they can be used for the detection of leaks in vacuum systems; a fine jet of helium is played around the outside of suspect areas until a response is obtained.

A relatively new type of vacuum gauge is the spinning rotor or molecular drag gauge (see Fig. 4). A stainless steel ball is supported on the axis of a tube by magnetic levitation. A rotating magnetic field is then produced by two pairs of orthogonal coils and this makes the ball spin. After switching the rotating field off, the angular velocity of the ball is measured with the aid of another pair of coils which detects the residual magnetism of the ball. The rate of decrease of angular velocity can be related theoretically to the gas pressure.

2. Vacuum Calibration Systems

Vacuum calibration systems are essentially generators of calculable low pressures. There are two main types: series expansion and orifice flow.

A series expansion apparatus consists of a set of large and small volumes of known volume ratios (see Fig. 5). Initially all the volumes are evacuated via a manifold to a low pressure. The first small volume v_1 is then filled with gas at a relatively high, precalcu-

(a)

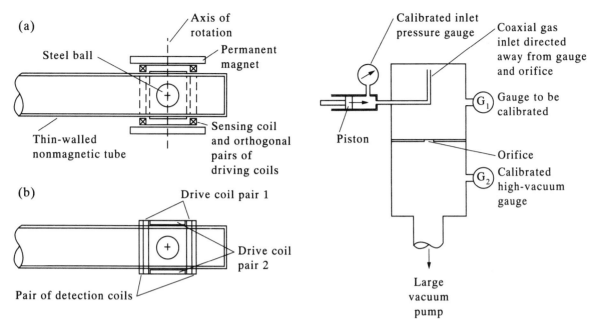

Figure 4
(a) Spinning rotor gauge, (b) top view, permanent
magnets removed

Figure 6
Orifice flow vacuum calibration system

lated, measured pressure. This gas is then allowed to
expand into the combined volumes v_1, V_1 and v_2. The
pressure reduction is calculated using Boyle's Law. v_2
is then isolated and the gas within it is allowed to
expand into the second large volume V_2 and the third
small volume v_3. The process is repeated for as many

stages as are required to obtain the required calibration pressure. The method is restricted to gases such
as nitrogen and the inert gases, which do not significantly adsorb onto the interior surfaces of the
volumes.

In the orifice flow system (see Fig. 6) gas passes

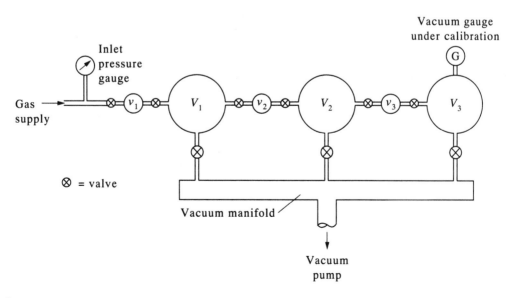

Figure 5
Three-stage series expansion apparatus

from a mass flowmeter, for example, of the positive displacement type, through a low-conductance tube into a large vessel V_A to which the gauge is connected. The gas then diffuses through an orifice of known dimensions, and hence of known gas conductance, into a second volume V_B, which is connected to a large vacuum pump. Knowing the gas mass flow rate, the orifice conductance and the relatively low pressure in V_B, it is possible to calculate the pressure in V_A.

See also: Pressure Measurement

Bibliography

O'Hanlon J F 1980 *A User's Guide to Vacuum Technology*. Wiley, New York

P. R. Stuart
[National Physical Laboratory, Teddington, UK]

Process Instrumentation Applications

Process instrumentation is a means whereby continuous or regular operations may accomplish a result, and is taken to be concerned with measurement or the acquisition of quantitative information necessary for decision making in manufacturing processes, but there are many other applications.

1. Human Needs

Any organism or organization, from the microbe to a nation, needs to have the ability to remain viable in a competitive environment. An industry's survival in business is dependent on its ability to meet human needs, exemplified in Table 1, in an approximate order of immediate priority.

2. Assumptions

A successful and responsible approach to the industrial manufacture and processing of materials will mainly rely on sound engineering design, painstaking construction and commissioning, careful and skilled operation and good maintenance of the process plant, rather than on the associated instrumentation. Thus, nuclear power plants depend on basic design strategies to fail safely. Despite large capital expenditure, absolute perfection of a process is often neither technically feasible nor commercially viable. Therefore instrument systems will usually be employed to provide additional safety, greater efficiency and, increasingly, more environmental protection, particularly the reduction of effluents.

Table 1
Human needs

Clean air
Potable water
Good food
Protection (clothes, shelter, security)
Raw materials
Energy
Tools
Health
Training
Employment
Transport
Recreation

An instrument design checklist, filled out jointly between process management and the measurement specialist, is a useful document for agreeing aims and finding a common understanding of an application. By far the greatest number of industrial measurement systems are for the assessment of physical parameters such as temperature, pressure or flow, rather than of chemical concentration or composition, even in the chemical industry.

However, as the processes they serve, chemical measurement systems are often complex, frequently including the simpler forms of physical measurement, such as pressure or flow. This article may therefore draw most interest from the consideration of chemical stream monitoring.

The traceability of chemical measurements back to national standards such as mass, length, time and the mole is not as easy, and therefore not as rigorous, as is the case for physical measurements. Measurement systems will continue to be manual, semiautomatic or fully automatic, according to technoeconomic needs.

3. Measurement System

Measurement instruments are conveniently considered as having three functionally separate components which, only when acting together, form a complete and viable system:

(a) the sample system;

(b) the transducer; and

(c) the signal processing system.

3.1 Sampling

The sampling system is needed to obtain any or all of the following functions:

(a) a representation of the material to be measured or the measurand;

(b) physical or chemical preparation to change the sample into an acceptable form for the transducer;

(c) transport of the sample to and from the transducer;

(d) calibration or standardization of the whole measurement system; and

(e) maintenance of operational safety.

The sampling system is of particular importance with respect to chemical measurements. By far the greatest number of such instrument system failures are due to sampling difficulties. Special problems in fluid samples are typically, unwanted solid, liquid or gaseous phases or debris, changes of phase due to variation of temperature or pressure, and chemical action or corrosion. It can be particularly difficult to obtain representation of heterogeneous solid materials because of obvious mixing problems, and it may be necessary to fall back on national or international sampling standards. In some difficult cases these give exact instructions on how to obtain a consistent but possibly biased sample, say, from a bulk container.

Compared with manual methods, automated measurement systems usually give less statistical uncertainty or less total error, even if automated results contain nonrandom bias.

Applying automated on-line sampling could improve and perhaps reduce a terrifying global amount of sophisticated and careful manual laboratory work carried out on material, derived by means of poor organization or technique.

3.2 Transduction

The transducer has two roles:

(a) to sense the physical or chemical phenomenon which is varying in the measurand, and

(b) to change the energy associated with measurand variations into, what is often another, more convenient form for information transmittal, such as electrical energy.

Table 2 shows examples of energy inputs to the transducer and therefore gives an indication of the huge range of possible measurement applications.

3.3 Signal Processing

The signal processing system has many possible roles, some of which are given in Table 3.

Table 2
Forms of energy input to the transducer

Chemical
Electrical
Electromagnetic
Gravitational
Magnetic
Mechanical
Nuclear
Thermal

Table 3
Possible functions of signal processing systems

Alarm analysis and control
Data coalescence from other transducers
Data storage
Data transmission
Fault detection
Fault diagnosis
Feedback to the process or the transducer
Information display
Noise reduction
Statistical analysis
System validation

Since it is essential that perfectly working instruments are believed, system validation is important. But such have been the problems often associated with chemical sampling systems, that Shaw (1983) claimed that less than 20% of process operator calls for maintenance checks revealed any faults.

This stimulated not only validation devices for specific instruments but also the consideration of a more generally applicable built-in microelectronic system (Henry and Clarke 1991). This model system learns and assesses internal instrument or signal noise. Continually updated process measurements are compared with those learned from normal conditions to indicate the system status.

3.4 Instrumentation Design

Since all measurement applications are multidimensional in their requirements for accuracy, cost, robustness and so on, and all sampling, transducer (see Table 4), and signal processing systems have their

Table 4
Some characteristics of transducers

Accuracy
Capital cost
Corrosion resistance
Credibility
Dynamic range
Hysteresis
Life
Maintenance requirements
Noninvasive
Precision
Reliability
Response time
Running cost
Safety
Sampling requirements
Sensitivity
Signal processing needs
Signal-to-noise ratio
Size
Weight

own wide range of characteristics, it is essential for the successful instrument designer both to think widely and to be creative.

Once a transducer has been chosen, some limitations necessarily follow with respect to suitable selection of both the sampling and the signal processing systems. Even so, the final instrument is almost always a unique compromise between many desirable characteristics.

4. The Evolutionary Model

As process instrumentation is being broadly used to meet human needs, it is worthwhile considering what instrumentation systems nature has provided for the human race.

Since all our sensing systems are cellular in character, there is enormous replication, which leads to enhanced reliability, even when damage has occurred. Each sense has a relatively small number of types of cell (about 30 in the nose), whose outputs are parallel processed for quality and quantity by the brain. The cognitive fusion of information from different senses such as sight and hearing can be synergistic. As one goes higher up the evolutionary scale, the addition of knowledge, mainly learned from parents, plays a further, increasingly important role in survival.

Many future measurement systems will be based on what evolution has proved to be effective. For example, the huge disability of human deafness should, with automated data reduction and pattern recognition techniques, give stimulus to further process instrumentation and its applications.

5. Air Monitoring: A Common Application

Human needs have spawned a wide range of industries and an enormous number and diversity of instrument systems. Therefore, it is intended to discuss only one of our requirements in fairly superficial detail.

The most immediate human need for clean air leads to a range of important industrial applications. The principle of "best practicable means" requires that a combination of techniques is always considered.

5.1 Vent Monitors

Combustion processes give rise to the most common legitimate plant exit. Domestic chimneys or automotive exhausts considerably extend the potential market for instruments. Such mass demand can often help manufacturers meet the challenge for inexpensive monitors; in this case, for oxygen and carbon monoxide, to aid efficiency, and for smoke, dust and oxides of sulfur or nitrogen, to combat pollution of the environment.

In situ probes and extractive sampling systems are common for oxygen measurements close to the com-

bustion zone, but inadequate mixing of the measurand and leakage of air (containing 21% oxygen) into such vents has led to more optical, across stack concentration averaging methods for polar molecules, such as carbon monoxide, to control the fuel–air ratio.

5.2 Fixed-Point Monitors

Since gaseous effluents are dissipated in the atmosphere, it might generally be argued that if industrial hygiene close to a plant is well maintained, then exposure of the environment outside the factory boundary is also likely to be satisfactory. Apart from cases of extreme immediate or acute hazard, where it is necessary to enforce a rigorous containment policy, it is increasingly common to place fixed-point gas sensors or sampling tubes from a central analytical instrument, such as a mass spectrometer, across a plant site according to:

(a) the probability of a leak at a particular place, and

(b) the probability of human presence at a place or (for flammable gas sensors) the position of possible sources of ignition.

In an attempt to improve the coverage available from fixed-point monitors, recent flammable hydrocarbon gas detectors have used open-path infrared beams. Other, more sophisticated steerable, open-path laser systems can even detect the presence of some toxic gases using the minimal back reflection from atmospheric dust particles as a natural mirror. However, open-path instrumentation is not sensitive enough to detect every toxic vapor at its occupational exposure level (OEL). (This is the concentration considered to be safe if a worker were exposed to it continuously over a working life.)

The signal processing and presentation of data from a variety of fixed-point monitors is becoming an increasingly sophisticated local and central aid to crisis management. Thus, out on a plant, local sensors would activate an audible alarm and each have a flashing light to tell operators which instrument or sampling point was in an alarm condition. With knowledge of local topology, wind direction and velocity, and so on, modern signal processing systems can assess and directly display vapor cloud size, concentration and movement.

5.3 Leak Seeking

The chance of a major vapor escape going undetected for a prolonged time is not great, but this is not the case for smaller leakages. Fixed-point monitors are unlikely to pinpoint the precise location of an emission, therefore they need to be backed up by portable, sensitive and rapidly responsive leak seeking sensors which can be "nosed" down to the source of a vapor plume.

By means of considerable effort and rigorous management, regular leak seeking can be effective on its own in finding vapor weeping from likely places, for example, process valve or pump glands.

5.4 Portable and Personal Monitors

In addition to fixed point instrumentation, process workers are often directly protected on the plant by portable vapor monitors. Unfortunately, the present state of the art, often based on electrochemical transduction, is not always able to provide an acceptably small wearable monitor for every potentially toxic chemical (although the nose itself is by no means always useless as a warning device, especially when used with intelligence).

Most portable instrumentation and many fixed-point systems, are required to be intrinsically safe, that is, not capable, even if faulty, of igniting a flammable atmosphere.

5.5 The Biological Approch

The ideal future personal air monitor would sound an alarm for any toxic vapor and not just for a narrow range of specific chemicals. Such a device might rely on monitoring the metabolism of microorganisms.

Healthy communities of plants and animals, including microorganisms, are indicative of good living conditions and should be considered as an addition to process instrumentation for industrial hygiene or environmental monitoring.

6. Conclusions

Process instrumentation systems are always linked to economic need, whether they are to meet legal requirements for the environment, hygiene or safety, or for the more direct control of process conversion efficiency, the maintenance of product quality, or to ensure continuous operation. These needs of business, rather than the push of technological research, provide the usual main stimulus for specific applications of process instrumentation. Major advances in instrumentation technology often arise from more general developments, such as new materials, electronics, fiber optics or satellite television, which have been devised for other purposes. The multidisciplinary requirements of instrumentation applications and the huge array of possible solutions therefore provide a major challenge for original thought.

See also: Flow Measurement: Principles and Techniques; Level Measurement; Pressure Measurement; Temperature Measurement

Bibliography

British Standards Institution 1991 Guide to selection and application of flow meters for the measurement of fluid flow in closed conduits. BS7405. BSI, Milton Keynes, UK

Considine D M 1971 *Encyclopedia of Instrumentation and Control.* McGraw-Hill, New York

Cornish D C, Jepson G, Smurthwaite M J 1981 *Sampling Systems for Process Analysers.* Butterworth, London

Christian G D, O'Reilly J E 1986 *Instrumental Analysis.* Allan and Bacon, Boston, MA

Finkelstein L, Finkelstein A C W 1987 Instruments and instrument systems: concepts and principles. In: Singh M G (ed.) *Systems and Control Encyclopedia.* Pergamon, Oxford, pp. 2527–36

Health and Safety Executive 1992 Occupational exposure limits. Guidance note, EH40/92. Her Majesty's Stationery Office, London

Henry M P and Clarke D W 1991 A standard for self validating sensors. OUEL 1884/91. Department of Engineering Science, University of Oxford, Oxford

Huskins D J 1981 *General Handbook of On-Line Process Analysers.* Ellis Horwood, Chichester, UK

Huskins D J 1982 *Quality Measurement Instruments in On-Line Process Analysis.* Ellis Horwood, Chichester, UK

Jones B E (ed.) 1982, 1983, 1985 *Instrument Science and Technology,* Vols. 1, 2, 3. Hilger, Bristol, UK

Noltingk B E 1988 *Instrumentation Reference Book.* Butterworth, London

Payne P A (ed.) 1989 *Instrumentation and Analytical Science.* Peregrinus, London

Shaw R 1983 Future instrumentation: its effect on production and research. In: Jones B E (ed.) *Instrument Science and Technology,* Vol. 2. Hilger, Bristol, UK, pp. 132–9

J. R. P. Clarke
[UMIST, Manchester, UK]

R

Reliability and Maintenance

Instrumentation systems are now employed in situations where a malfunction or failure can endanger human life, incur huge financial losses and lead to significant damage to the environment. Reliability is therefore an important feature of the system design. The designer, however, is not in a position to fully test the complete system before it is put into operation; the cost and time involved and the consequence of a failure make this impossible. Reliability of the whole system is predicted using statistically based techniques using data obtained from observations or tests of manageable subsystems or components. These techniques have been extended to allow the analysis of the maintenance and logistic support requirements.

1. Variables and Statistical Relationships

The variables used in reliability analysis are as follows.

(a) Failure density function $f(t)$—a probability density function such that the probability of failure between time t and $t + \Delta t$ is $f(t)\Delta t$.

(b) Cumulative failure function $F(t)$—the probability that failure will occur before (or at) time t.

$$F(t) = \int_0^t f(\tau)\,d\tau \qquad (1)$$

(c) Reliability $R(t)$—the probability that a system will operate for a period t without failure.

$$R(t) = 1 - F(t) = 1 - \int_0^t f(\tau)\,d\tau \qquad (2)$$

$$f(t) = -\frac{dR(t)}{dt} \qquad (3)$$

$R(t)$ is a monotonically decreasing function of time with $R(0) \leqslant 1$ and $R(\infty) = 0$. Usually $R(0) = 1$ but it is possible that equipment will be failed at time zero if a faulty component is used, for example.

(d) Failure rate or hazard rate $H(t)$—the probability that the system will fail between t and $t + \Delta t$ if it has not failed at time t is $H(t)\Delta t$.

$$H(t) = \frac{f(t)}{R(t)} \qquad (4)$$

$$R(t) = \exp\left(-\int_0^t H(\tau)\,d\tau\right) \qquad (5)$$

(e) Mean time to failure MTTF—the average time that a system will operate before failure.

$$\text{MTTF} = \int_0^\infty tf(t)\,dt = \int_0^\infty R(t)\,dt \qquad (6)$$

(f) Mean time to repair MTTR—the average time taken to repair a system after failure.

(g) Mean time between failures MTBF. For a system that works on a continuous operation–failure–repair cycle there is an operating period and a repair period between each failure:

$$\text{MTBF} = \text{MTTF} + \text{MTTR} \qquad (7)$$

(h) Asymptotic availability A — the proportion of time that a system is available to operate:

$$A = \frac{\text{MTTF}}{\text{MTTF} + \text{MTTR}} = \frac{\text{MTTF}}{\text{MTBF}} \qquad (8)$$

(i) Point availability $A(t)$—the probability of the system being available at time t.

(j) Interval availability $A(t_1, t_2)$—the probability of the system being available over the period t_1 to t_2.

(k) Demand availability Ad—the probability of the system being available on demand.

Reliability and availability figures are often close to 1.0 which makes arithmetic manipulation difficult. It is more convenient to deal with the compliments:

(l) Unreliability $U(t)$

$$U(t) = 1 - R(t) = F(t) \qquad (9)$$

(m) Unavailability UA

$$UA = 1 - A = \frac{\text{MTTR}}{\text{MTTF} + \text{MTTR}} \qquad (10)$$

Similarly, for other availability measures.

It must be recognized that all reliability variables will be dependent on operating and environmental conditions.

2. Data

Reliability data is normally obtained from observations or tests on components or small equipments.

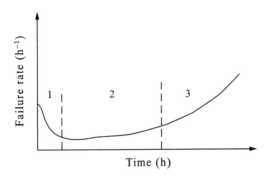

Figure 1
The bathtub curve, the observed form of failure rate for many types of system

Most published data quotes failure rate, usually in failures per 10^6 hours.

Failure rate varies with time, typically exhibiting the well-known bathtub curve, as illustrated in Fig. 1.

Three phases are defined during the life of a component although, in practice, the boundaries are often not well defined.

In phase 1, the wear in phase, the failure rate is initially high but it reduces over a short time. Failures in this phase are due to flaws and weaknesses in materials or faults in manufacture.

In phase 2, the useful life phase, the failure rate is low and approximately constant, representing random failures. The length of this phase will vary with the type of equipment.

In phase 3, the wear out phase, the failure rate increases due to aging, wear, erosion, corrosion and related processes. The rate at which failure rate increases varies with type of equipment. For much equipment, especially electronics, no wear out is observed within the working life and there is some doubt as to whether it exists.

Obtaining failure rate data requires testing many components over long periods of time. For components with a very low failure rate, accelerated testing is usually employed.

Published lists usually contain basic failure rates and a series of factors for modifying the basic rate. These factors reflect the quality of the item, the environmental conditions and service stress.

In order to develop models for reliability assessment it is useful to express reliability functions analytically. The simplest form represents phase 2 of the bathtub curve where the failure rate is approximately constant when the exponential reliability function is used:

$$H = \lambda \tag{11}$$

$$R = e^{-\lambda t} \tag{12}$$

$$f = \lambda e^{-\lambda t} \tag{13}$$

$$MTTF = 1/\lambda \tag{14}$$

An equivalent function for repair can be developed, based on a constant repair rate

$$MTTR = 1/\mu \tag{15}$$

which, by substitution into Eqns. (8) and (10), gives

$$A = \frac{\mu}{\lambda + \mu} \tag{16}$$

$$UA = \frac{\lambda}{\lambda + \mu} \tag{17}$$

Other distributions which reflect variations in failure rate, notably wearout behavior, are also used. Of these the lognormal, gamma and Weibull distributions are the most common.

3. System Structures

Large systems may be broken down into many subsystems which are connected in a number of structured forms. If the subsystem reliability is known the system reliability is calculated using equations relevant to the structure. Care should be taken to differentiate between the physical structure of complex systems and the reliability structure.

3.1 Series Systems

In series systems the system will function only if all components function. Most instrumentation systems are of this type unless measures are taken to increase reliability. A simple example is shown in Fig. 2.

For series systems with n components:

$$R_s = R_1 \times R_2 \times \cdots R_n \tag{18}$$

$$U_s \simeq U_1 + U_2 + \cdots U_n \tag{19}$$

The approximate form, Eqn. (19), should not be used if U_s exceeds about 0.3. If the reliability functions are all exponential, then

$$R_s = \exp - (\lambda_1 + \lambda_2 + \cdots \lambda_n)t \tag{20}$$

$$MTTF = \frac{1}{\lambda_1 + \lambda_2 + \cdots \lambda_n} \tag{21}$$

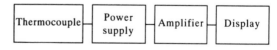

Figure 2
A typical series system; all components must be working for the system to function correctly

3.2 Parallel Systems

Where items are connected in a parallel structure the system will work if any one of the items is functioning. A feature of parallel systems is that all subsystems work together until a failure occurs when the remainder continue to work. Figure 3 represents a typical parallel system.

The unreliability of a system with n parallel elements is:

$$U_p = U_1 \times U_2 \times \cdots U_n \quad (22)$$

This relationship implies that failures of the components are independent and that failure of one component does not affect the unreliability of others.

Parallel items may be similar or dissimilar. If components are similar and they have exponential failure distributions then:

$$\text{MTTF}_p = \frac{1}{\lambda} \left(1 + \frac{1}{2} + \frac{1}{3} + \cdots \frac{1}{n} \right) \quad (23)$$

Eqns. (22) and (23) indicate that parallel systems are effective in increasing reliability at a given time, but are not effective in increasing MTTF.

3.3 Standby Systems

In a standby system a number of items are available to perform some function; at any time one is operating and the others are held in reserve to be brought into operation in the event of failure. The benefit of standby systems over parallel systems is that generally the items on standby have a lower failure rate than the operating components and consume less power.

The disadvantage of standby systems is that a failure sensing and switchover function is required. This may

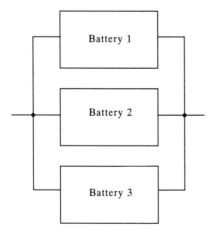

Figure 3
A typical parallel system; power can be supplied if any one of the three batteries is working

be manual or automatic. Standby systems are therefore more complex and expensive than parallel systems. There may also be a loss of function during switchover.

A simple analysis requires the following assumptions:

(a) items do not fail or standby; and

(b) failure sensing and switchover is instantaneous and 100% reliable.

In case (b), for n items in standby mode:

$$\text{MTTF}_{SB} = \text{MTTF}_1 + \text{MTTF}_2 + \cdots \text{MTTF}_n \quad (24)$$

and if all items are similar with exponential failure functions

$$R_{SB} = e^{-\lambda t} \left(1 + \lambda t + \frac{\lambda^2 t^2}{2!} + \cdots \frac{(\lambda t)^{n-1}}{(n-1)!} \right) \quad (25)$$

If the assumptions are not valid the analysis can be made using state transition models (see Sect. 5).

3.4 K out of N Systems

This is a development of the parallel system where there are N items and the system can operate with any K. This is often a cheaper solution. Subject to the same independent failure assumptions as for parallel systems, the unreliability of a K out of N system may be expressed as:

$$U_{K/N} = \sum_{r=N-K+1}^{N} \frac{N!}{r!(N-r)!} U^r R^{N-r} \quad (26)$$

which, if U is small, approximates to

$$U_{K/N} = \frac{N!}{(N-K+1)!\,(K-1)!} U^{N-K+1} \quad (27)$$

A particular example of a K out of N structure is a two out of three system used in many measurement and computing applications. Here, a simple parallel system is inadequate because in the event of a disagreement it may not be clear which item is in error. With a two out of three structure a disagreement is assumed to be an error and is overruled by a voting system.

3.5 K out of N Standby System

This is a combination of the K out of N and standby concepts, common where a plant has many instruments of the same type installed and a stock of spares is held.

Systems of any complexity will be structured using combinations of the basic structure outlined above.

4. Repair and Maintenance

The mean time to system failure of repaired systems can be many times greater than for the elements. In a parallel system the repair of a failed element will usually be completed before the second fails. If two similar components are connected in parallel the mean time to system failure can be calculated:

$$\text{MTSF} = \frac{\text{MTTF}(\text{MTTF} + \text{MTTR})}{2\text{MTTR}} \tag{28}$$

Maintenance can be used to reduce the chance of failure and, hence, increase MTTF. For a system maintained at intervals T, where maintenance restores reliability to $R(0)$ we have:

$$\text{MTTF} = \frac{\displaystyle\int_0^T R(t)\,dt}{1 - R(T)} \tag{29}$$

Note that this is the time that the system is expected to operate before failure; it does not include maintenance time.

If regular maintenance takes a time T_s and repair after failure T_e ($= \text{MTTR}$), the availability of the maintained system is:

$$A_m = \frac{\displaystyle\int_0^T R(t)\,dt}{(1 + T_s/T)\displaystyle\int_0^T R(t)\,dt + T_e(1 - R(T))} \tag{30}$$

Other maintenance strategies may be based on intervals of a variable other than time, number of operations, for example.

5. State Transition Models

State transition models, or Markov models, are a powerful technique for the analysis of complex systems: systems that may exist in a number of states between fully operational and totally failed. A series of system states, X_1–X_M, is defined and transitional probabilities p_{ij} between states over a period δt are calculated in terms of the component failure and repair rates. This leads to the state transition diagram, illustrated in Fig. 4, and the equivalent state transition matrix, Eqn. (31).

$$\mathbf{P} = \begin{bmatrix} p_{11} & p_{12} & 0 & 0 \\ p_{21} & p_{22} & p_{23} & p_{24} \\ p_{31} & 0 & p_{33} & 0 \\ 0 & p_{42} & 0 & p_{44} \end{bmatrix} \tag{31}$$

It is usual to select the interval δt small enough so that

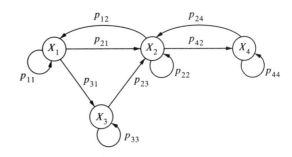

Figure 4
A typical state transition diagram: p_{ij} is the probability of the system making the transition from X_j to X_i in an interval δt

terms in δt^2 can be neglected. This has the effect of making many of the transition probabilities zero and they are not shown in the diagram. The sum of all probabilities leaving each state must be 1.0, or equivalently the sum of each column of $\mathbf{P}$ is 1.0. A state probability vector

$$\mathbf{x}_n = (x_{1,n}\ x_{2,n} \ldots x_{m,n}) \tag{32}$$

is defined such that $x_{i,n}$ is the probability of the system being in state X_i at time $n \cdot \delta t$. The sum of the elements of this vector is 1.0 at all times. State probabilities at any time can be calculated using

$$\mathbf{x}_{n+1} = \mathbf{P}\mathbf{x}_n \tag{33}$$

and ultimately

$$[\mathbf{P} - \mathbf{I}]\mathbf{x}_\infty = \mathbf{0} \tag{34}$$

In Eqn. (33) $\mathbf{x}_0$ is usually known, thus probabilities at subsequent times can be calculated by applying the equation recursively. Eqn. (34) can be solved using the condition that the sum of the elements of $\mathbf{x}_\infty$ is 1.0.

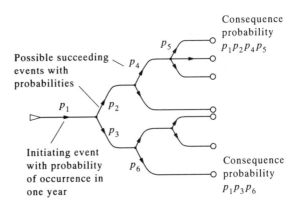

Figure 5
An event tree; the trees can rapidly become very large with much repetition of the succeeding events

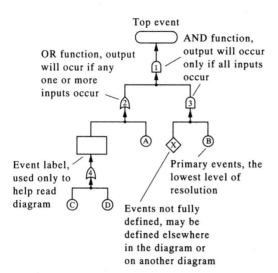

Top event

AND function, output will occur only if all inputs occur

OR function, output will ocur if any one or more inputs occur

Event label, used only to help read diagram

Primary events, the lowest level of resolution

Events not fully defined, may be defined elsewhere in the diagram or on another diagram

Figure 6
A fault tree diagram showing basic elements; large diagrams may contain a thousand elements. Computer programs are available for their analysis

A development of the analysis can be used to calculate the mean time to failure.

6. Event Trees and Fault Trees

Event trees are a bottom-up approach for analyzing a series of events following an initiating event; a simple example is shown in Fig. 5. Probabilities may be assigned to each path and cascaded as shown to calculate the probability of the consequence at the right hand side.

Fault trees are a top-down approach which use logical AND and OR functions to define the causes of a TOP EVENT in terms of PRIMARY EVENTS. Figure 6 illustrates this principle.

7. Human Factors

The possibility of human error is important in many applications; the error may occur in the design, manufacture, operational or maintenance stages. This topic is an area of study in itself; it must consider stress, boredom, and the fact that people tend to see what they expect to see and do what they are used to doing. The possibility of intentional action (sabotage) cannot be discounted.

8. Software Reliability

With the increasing use of computers in monitoring and control applications the problem of software reliability is assuming increasing importance. Software

faults will appear when the system is subject to a particular history of input conditions, often after many hours of correct operation. The number of possible input histories makes exhaustive testing impossible.

See also: Life Cycle; Operational Environments; Safety

F. J. Charlwood
[City University, London, UK]

Resistance, Capacitance and Inductance Measurement

The measurement of the numerical values relating the properties of an elementary electrical component is intimately connected to the concept of impedance, defining the relationship between voltage and current under the assumption that both are sinusoidal and of the same identical frequency f. With this assumption the methods of symbolic (or complex) calculus can be applied, where a differentiation is represented by a multiplication by the imaginary value of the angular frequency $\omega = 2\pi f$, and an integration by a division, thus:

$$\frac{d}{dt} \to j\omega, \qquad \int dt \to 1/j\omega$$

where $j = \sqrt{-1}$, expressing a 90° phase shift.

In this case, the relation between phasors U of voltage and I of current can be expressed by

$$U = Z.I, \qquad Z = \operatorname{Re} Z + j \operatorname{Im} Z = Z e^{j\phi}$$

$$Z = \sqrt{(\operatorname{Re} Z)^2 + (\operatorname{Im} Z)^2} = \frac{U}{I}$$

$$\phi = \frac{\arctan \operatorname{Im}(Z)}{\operatorname{Re}(Z)}$$

where $\operatorname{Re}(Z)$ and $\operatorname{Im}(Z)$ denote, respectively, the real and the imaginary parts of the impedance Z, U, and I denotes the modulus of U and I. According to the usual conventions, an inductive impedance exhibits a positive numerical value of ϕ, a capacitive impedance a negative one. Sometimes, a further quantity, the loss angle δ will be used, where

$$\delta = \frac{\pi}{2} - |\phi|$$

The quality factor Q equals $1/\tan \delta$.

An impedance (or component) is linear if a sinusoidal input of frequency f yields a sinusoidal reaction of the same frequency f and, furthermore, doubling of the forcing signal doubles the response without introducing frequency components differing from the original frequency (superposition is allowed).

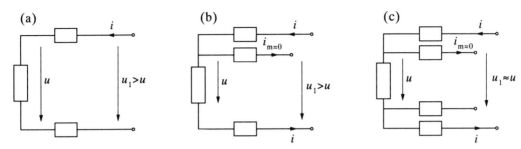

Figure 1
(1) Two-, (b) three- and (c) four-wire connection of a device under test

In the case of a nonlinear impedance, the relation between instantaneous voltage $u(t)$ and current $i(t)$ is to be represented by a nonlinear function $u(i,t)$, where t denotes time. The idea of the linear impedance can generally be used only for small excursions Δu of u, and thus $i(\Delta i)$, around a working point U_0, I_0. This implies the concept of a differential impedance, $Z_d(U_0, I_0, \omega) = \Delta U / \Delta I$ for this working point and an angular frequency ω.

In the following, either linear components (and thus linear impedances) or linearized behavior for a given working point will be assumed. For all elementary components, a simplified model valid for a given working point (case of nonlinear components) and a given frequency (or frequency interval) will be derived, and the corresponding measurement methods outlined.

For the connection of the device under test (DUT) to the remainder of the measurement systems, a two-, three- or four-wire connection can be considered, as shown in Fig. 1. A four-wire connection will thus, under the assumption of sufficiently nonloading voltage measurement, completely eliminate the influence of connecting impedances Z_e.

1. Measurement of Resistors

1.1 Model

Under ideal considerations, an ohmic resistor simply performs the conversion between voltage and current according to

$$u(t) = R \cdot i(t), \qquad U = R \cdot I$$

From the construction point of view, three different types of resistors have to be considered.

(*a*) *Wire-wound resistors.* The component is realized by winding a calculable number n turns of a resistive metallic wire (specific resistance ρ), of given cross section A, onto a (ceramic) support of diameter d.

$$R \approx \frac{\rho n \pi d}{A}$$

This design is used for high-precision and/or high-power (>1 W) resistors.

(*b*) *Film resistors.* A thin film (typically a few micrometers thick) is fused onto a ceramic tube, terminated by terminal connectors. As needed, the value is adjusted by grinding helicoidal grooves into the film.

(*c*) *Mass resistors.* The resistive element is composed of a rod (e.g., carbon composite) of length and cross section corresponding to the desired ohmic value.

For the three kinds of resistors, a similar lumped parameter model can be given as shown in Fig. 2, which generally can be simplified to one of the two models given in Fig. 3. Resistors of low ohmic value tend to have an inductive stray component L_s, while resistors of higher ohmic value are capacitive (stray capacitance C_s). The frequency domains where the components can be considered as being sufficiently ideal must be extracted from the data sheets of the manufacturer.

It should be noted, however, that for power resistors the inductive part can play a noticeable role at critical frequencies f_c as low as some 100 Hz, while for higher ohmic values, critical frequencies in the range of 1 MHz to several tens of megahertz are to be expected. Care must therefore be taken to use these elements at frequencies well below f_c.

1.2 Measurement Methods

(*a*) *Voltage and current measurement.* This is the most direct method (see Fig. 4) but it is seldom used as it

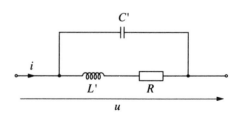

Figure 2
Lumped parameter model of a resistor

(a)   (b)

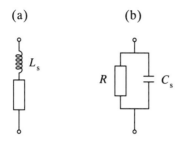

Figure 3
Simplified models of a resistor: (a) wire wound, (b) mass resistor

(a)   (b)

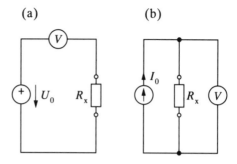

Figure 5
Direct reading resistance measurement using: (a) voltage source, (b) current source

requires computation of the quotient U by I and, in some cases, a correction according to the internal resistance R_{iu} of the voltmeter or R_{ii} of the ampere meter, yielding

Case 1 $R_x = U/(I - U/R_{iu}) \simeq U/I$ for $R_x \ll R_{iu}$
Case 2 $R_x = U/I - R_{ii} \simeq U/I$ for $R_x \gg R_{ii}$

(b) *Direct reading ohmmeters*. These are outlined in Fig. 5. Either a voltage source U_0 or a current source I_0 (simple electronic realization) are used.
   For a voltage source (see Fig. 5a), U_0 is selected such that for $R_x = 0$, full-scale indication is obtained. R_x can be read directly on the dial according to:

$$R_x = R_{iu}(-1 + U_0/U)$$

resulting in a strongly nonlinear scale.
   Correspondingly, the scale is obtained for a current source (the method used in most digital voltmeters) according to:

$$U = I_0 R_x (1 + R_x/R_{iu})^{-1}$$
$$R_x \simeq U/I_0, \quad \text{for } R_x \ll R_{iu}$$

(c) *Bridge methods*. These are commonly used for precision measurements (null method) and for small deviations of a measurand R_x from a rest value R_{x0}. The configuration of this Wheatstone bridge is illustrated in Fig. 6, where R_x is the unknown resistor, R_n

is a known resistor of same order of magnitude, and R_1 and R_2 are parts of an accurately calibrated variable resistor R.
   As a general principle, this bridge performs the comparison of the (unknown) relative voltage drops over R_n and R_x with the known relative voltage drops over R_1 and R_2.
   The voltage ΔU across the (ideal) voltmeter can be calculated as

$$\Delta U = U_0 \frac{R_x R_2 - R_1 R_n}{(R_x + R_n)(R_1 + R_2)}$$

For $\Delta U = 0$ (null method), this relation yields

$$R_x = \frac{R_n R_1}{R_2}$$

Thus, besides R_n, only the relation R_1/R_2 must be known.
   In the active configurations, ΔU is first nulled for a rest value R_{x0} of R_x. Small variations ΔR of R_x then yield, under the simplifying assumption $R_n = R_1 = R_2 = R_{x0} = R$

$$\Delta U = \frac{1}{4} U_0 \frac{\Delta R}{R}$$

   This method is mostly used with resistive sensors for nonelectric quantities, for example, strain gauges,

(a)   (b)

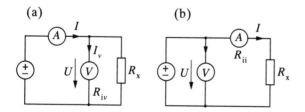

Figure 4
Direct measurement of resistance: (a) low ohmic value, (b) high ohmic value

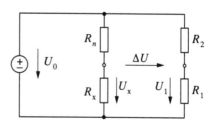

Figure 6
Wheatstone bridge

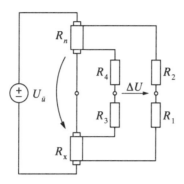

Figure 7
Thompson bridge

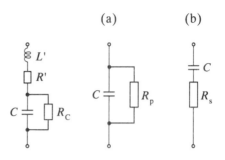

Figure 8
Model of a capacitor: (a) simplified parallel, and (b) series models

where $\Delta R/R_{x0} \ll 1$. According to the number n of judiciously inserted identical transducers, this application yields

$$\Delta U = \frac{n}{4} U_0 \frac{\Delta R}{R}$$

In case a four-wire connection is needed (ohmic resistance of connecting leads and connections to the DUT comparable in magnitude to R_x), a Kelvin bridge (also called Thompson bridge) is required (see Fig. 7).

With the supplementary condition that $R_1/R_2 = R_3/R_4$, realized by a mechanical interconnection between the two variable resistors, the null condition of this bridge yields, as in the previous case:

$$R_x = \frac{R_n R_1}{R_2}$$

with the resistance of the connecting elements being eliminated.

2. Measurement of Capacitors

2.1 Model

The capacity of an ideal capacitor with plate surface A, plate distance d and dielectric constant ε_r of the insulator between the plates can be calculated according to

$$C = \frac{\varepsilon_0 \varepsilon_r A}{d}$$

where ε_0 = permittivity of vacuum = $8.854\,\mathrm{pF\,m^{-1}}$.

Taking into consideration the noninfinite resistivity ρ of the dielectric material, this configuration also corresponds to a resistor R of value

$$R = \frac{\rho d}{A}$$

in parallel with C. The time constant τ_i of this model is

$$\tau_i = RC = \varepsilon_0 \varepsilon_r \rho$$

and is of the order of magnitude of seconds to minutes.

This model will exhibit a resonant frequency f_c, above which the impedance of the element will be inductive instead of capacitive. The order of magnitude of f_c depends greatly on the design of the capacitor and ranges between 1 kHz (electrolytic capacitors of high capacitance) and several hundred megahertz (small capacitors in the picofarad range).

Below f_c, two different models can be assumed according to Fig. 8: (a) a series connection R_s and C_s; and (b) a parallel connection R_p and C_p. As both models are a simplification of reality, the values of R_s and C_s, and R_p and C_p will be dependent on the angular frequency ω at which the measurement is performed.

For practical purposes, normally the capacity at this frequency and the loss factor $\tan \delta$ (equal to the difference to 90° of the angle between U and I) are indicated.

$$\tan \delta = (\omega R_p C_p)^{-1} = \omega R_s R_s$$

In most practical cases, $\tan \delta < 10^{-2}$, and is frequency dependent. In this case, $C_s \approx C_p \approx C$.

2.2 Measurement Methods

(a) *Direct methods.* These evaluate the relationship between U and I and require not only the modulus of U and I but also the phase angle φ between them, from which $\delta = \pi/2 - \varphi$ can be derived. In view of the small magnitude of δ, this method is subject to high errors, and is thus impractical.

(b) *Direct reading instruments.* These are designed for quasi-ideal capacitors, at least for measurements with reduced accuracy, similar to Fig. 5a, using an ac voltage source of known angular frequency ω and

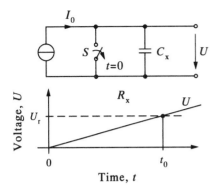

Figure 9
Direct reading capacitance measurement

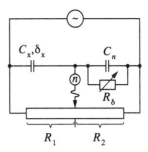

Figure 10
Wien bridge

known rms value U_0. The reading of U will be related to C_x according to

$$C_x = \frac{1}{\omega R_{iu}} \sqrt{\frac{1}{(U_0/U_x)^2 - 1}}$$

Thus, a strongly nonlinear scale will be obtained

A linear scale can be obtained using a circuit as shown in Fig. 9, where C_x is charged with a constant current I_0, until a predetermined voltage U_r is attained at a time instant t_0.

(c) *Bridge methods.* These are mostly used either in a special configuration or in a nonspecific configuration, as discussed in Sect. 5.

For the most commonly used Wien bridge, as shown in Fig. 10, the null method gives:

$$C_x = \frac{C_n R_2}{R_1}$$

$$\tan \delta = (\omega C_n R_\delta)^{-1}$$

The bridge nulling is performed by alternatively adjusting R_1/R_2 and R_δ until a sharp null is obtained.

Other bridge configurations are in use for specific measurement problems (e.g., the Schering bridge for high-voltage capacitors).

3. Measurement of Inductors

3.1 Model

Two very different types of inductors have to be considered:

(a) air inductors, where the magnetic path is closed mainly in the air or in other nonmagnetic materials; and

(b) iron core inductors, where the magnetic path is prescribed mainly by a material, with high magnetic permeability μ_r.

An approximate value for the inductance can be calculated according to

$$L = \frac{n^2}{\oint \frac{dl}{\mu A}} \approx \frac{n^2}{\sum_k \frac{l_k}{\mu_k A_k}}$$

where the closed integration path is selected such that always the whole magnetic flux $\phi = \int B \, dA$ is comprised (for air inductors, a good approximation can be obtained if the flux outside the inductor is neglected), and where n is the number of turns, $\mu = \mu_0 \mu_r$, $\mu_0 = 0.4\,\pi \times 10^{-6}\,\mathrm{H\,m^{-1}}$, and l_k and A_k are contained segments with constant magnetic induction $B \approx \phi/A$.

For air inductors, the primary nonideality is the ohmic resistance of the windings and the stray capacitance between the windings, as indicated in Fig. 11. It must be remembered that, due to skin effects, the ohmic resistance may be dependent on the frequency. Simplified models are indicated in Fig. 11a and b.

The situation is more complicated for iron core inductors, as in this case the nonideal behavior of the ferromagnetic material has to be considered. Main points are hysteresis (magnetization losses, proportional to frequency), eddy current losses and nonlinearity of the magnetization curve. In view of

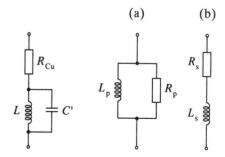

Figure 11
Model of an air inductor: simplified (a) parallel, and (b) series models

this, representative values for iron core inductors must always be measured at values of current (or voltage) and frequency close to the practical application.

3.2 Measurement Methods

(*a*) *Electrodynamic instruments*. For all components described by a differential equation, a knowledge of the modulus of *U* and *I* and of their mutual phase shift is required. In the case of power inductors, these values can be obtained through a simultaneous measurement of *U* and *I* and, for example, the active power *P*, using true root mean square and electrodynamic (wattmeter) instruments.

The values for R_s (or R_p) can be calculated according to

$$R_s = \frac{P}{I^2}$$

$$R_p = \frac{U^2}{P}$$

$$L_s = \frac{1}{\omega} \sqrt{\frac{U^2}{I^2} - R_s^2},$$

$$L_p = \left(\omega \sqrt{\frac{I^2}{U^2} - \frac{1}{R^2}} \right)^{-1}$$

This method is often used even for inductors driven into saturation under large signal conditions. In this case, the physical meaning is not very clear, but the values obtained can be used for intercomparison.

(*b*) *Direct reading inductance meters*. These meters, where L_x is used as a frequency-determining component in an oscillator, are available on the market. Their usability is restricted to inductors with fairly small tan δ.

(*c*) *Bridge methods*. These methods, using either devoted bridge configurations or unspecific methods (see Sect. 5), are widely in use. A possible configuration is illustrated in Fig. 12. It must be stressed here that inductors cannot be used as the normal (reference) component, in view of their definitely nonideal behavior, even under the best design. Instead, a capacitor is used as the comparison impedance. For the circuit shown in Fig. 12, the null condition yields, assuming a series model for L_x:

$$L_s = C_n R_1 R_2$$

$$R_s = \frac{R_1 R_2}{R_\delta}$$

$$\tan \delta = (\omega C_n R_\delta)^{-1}$$

Again, it should be emphasized that for unambiguous interpretation of the results, the measurement voltage

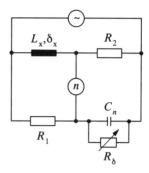

Figure 12
Bridge for inductance measurement

(or current) and measurement frequency must be included.

4. Multifunction Automatable Methods

The methods discussed so far have in common that they are designed for the evaluation of a specific type of DUT (resistor, capacitor, inductor), and that they do not lend themselves to be easily converted to an automated measurement system. In this section more general methods apt for general automated application will be described.

4.1 Electronic Bridge

As mentioned previously, a bridge performs the comparison between the voltage ratios of a known and an unknown impedance, referred to as a well-known divider. This idea can be extended to a structure outlined in Fig. 13, using two voltage sources of the same frequency, but different amplitude and phase. This can be realized efficiently using electronic means, for example, a table of sine coefficients (random access memory) and two digital-to-analog converters. U_1 is fixed and considered as a phase reference, and U_2 is variable in phase and amplitude.

The resolution of this method is solely limited by the resolution of the sine table in phase and amplitude, and by the accuracy of the digital-to-analog

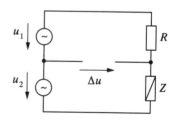

Figure 13
Dual voltage source bridge

converters used. Using state-of-the-art technologies, resolutions better than one part in 10^3 to 10^4 can be obtained.

4.2 Calculating Systems

A general analysis of the properties of a general impedance or one port leads to a set of relationships connecting the moduli of U and I and the phase φ to these moduli and the complex apparent power $S = UI^*$ (where * denotes complex conjugate), as well as its component's active power $P = UI\cos\varphi$ and reactive power $Q = UI\sin\varphi$.

For the measurement process, the values for P and Q are calculated according to

$$P = \frac{1}{mT} \int_0^{mT} u(t)\,i(t)\,\mathrm{d}t$$

$$Q = \frac{1}{mT} \int_0^{T} u^{90}(t)\,i(t)\,\mathrm{d}t$$

assuming sinusoidal waveshapes (where m is an integer). $u^{90}(t)$ denotes a phase shift of $-90°$. For practical purposes, $u(t)$ and $u^{90}(t)$ can be replaced by a signum function or, even simpler, by a function equalling 1 for $u(t) > 0$ and 0 for $u(t) < 0$, thus reducing multiplication to a simple switching operation, assuming a constant peak voltage $\hat{u}$ across the DUT. According to convention, $Q > 0$ denotes inductive impedance, $Q = 0$ resistive impedance, and $Q < 0$ capacitive impedance. The components of a simple model can then be derived, remembering that a resistor dissipates active power $P = RI^2 = U^2/R$, a pure inductive or capacitive element of reactance $X(\omega L$ or $(\omega C^{-1}))$ and reactive power $Q = XI^2 = U^2/X$, where U and I denote the quantities across and through the element, respectively.

4.3 Frequency Dependency of Impedances

For the measurement of the frequency dependency of impedances, the well-known methods of transfer function measurement using either multifrequency signals (such as pseudorandom binary sequences) evaluated by Fourier transform of both $u(t)$ and $i(t)$ or monofrequency signals using, for example, orthogonal correlation, can be used. It should be remembered that in the convention for the phase angle φ, the current is considered as the stimulus, and the voltage u as the response or output signal. Care must be taken in the layout of the measurement setup to judiciously choose the places where voltage and current are to be measured in order to avoid the inclusion of stray impedances.

See also: Bridges; Current Measurement; Voltage Measurement

Bibliography

Cage J M, Oliver B M 1971 *Electronic Measurements and Instrumentation*. McGraw-Hill, Tokyo
Measurement of Electrical Quantities, Acta Imeko Topics. North-Holland, Amsterdam.
Sydenham P 1983 (ed.) *Handbook of Measurement Science*, Vols. 1 and 2. Wiley, New York
van Putten A F P 1988 *Electronic Measurement Systems*. Prentice-Hall, London

J. Weiler
[ETH Zentrum/ETL,
Zurich, Switzerland]

S

Safety

There are many situations when an instrumentation system which does not perform its intended function, due either to poor design or malfunction, can lead to a hazardous situation. Common examples are life support systems, control systems for aircraft, and protection systems for industrial plant. In such cases a method is required to analyze the risk and to design the system so that the risk is reduced to an acceptable level.

1. Causes of Hazard

An instrumentation system can cause a hazard in one of three ways.

(a) Failure to execute an intended action when required, for example, an alarm, a shutdown sequence or a fire extinguishing action.

(b) Taking an unintended action, for example, shutting down an aircraft engine which is working satisfactorily.

(c) By creating a hazard in itself, for example, releasing radiation from a source used for measurement, causing a fire or an electrical hazard.

In each case the hazard may have its source in poor design, faulty manufacture, bad operating procedures, or a failure in service. The first two hazard modes are, in theory, detectable by testing but, as the actions may take place only after a particular sequence of operating conditions, comprehensive testing would take an impossibly long time.

2. Acceptable Risk Levels

Where a hazard results in a financial loss it is possible, in principle, to balance the loss against the cost of providing a safer instrumentation system. In practice the data is rarely good enough to make this calculation precisely. Factors such as inconvenience and public image complicate the decision.

Where a hazard may result in injury or loss of human life, the situation is much less straightforward. At an individual level the risk a person is prepared to accept is higher if they enter the risk voluntarily (e.g., skiing), compared with when it is imposed upon them (e.g., at work). The attitude of society, in general, is not consistent; we accept a level of risk in motor vehicle transport that would not be tolerated in industry. Accepted levels of risk vary widely from country to country. There is pressure to reduce risk levels over time; this occurs gradually by improving regulations and standards, and often more radical changes after a major incident.

An important consideration when assessing risk levels is the difference between perceived risk and the risk observed or calculated. Perceived risk is influenced by the media reporting incidents, and a person's dread of certain events, radioactive release, for example. A designer must often strive to ensure that a system is perceived to be safe.

3. Design Strategies

Many instrumentation systems are designed without any formal safety analysis being carried out. Such systems are designed using components manufactured in accordance with appropriate National and International Standards and installed similarly. The safety analysis is indirect, in that the standards are based on what experience has shown to be safe practice over many years, and are updated regularly as new experience is gained and new technology is developed.

In more critical applications, a formal statistically based analysis may be carried out. Here the instrumentation system is designed to achieve a failure rate which will lead to probabilities of failure which are below stated limits. The limits may be set in a number of ways.

(a) By a statutory authority, for example, the US Nuclear Regulatory Commission specifies limits for various types of incident per reactor year.

(b) To be an improvement on the system that the instrumentation system is replacing, be it a manual or an older system.

(c) To meet a limit that is perceived to be acceptable by the user.

When a basic system cannot meet the failure rate criteria demanded by safety considerations, some form of redundant instrumentation may be required. This may take the form of similar systems working in parallel or standby modes, or a supervisory monitoring system which checks the basic system.

Recognizing that failure will sometimes occur, the designer can strive to concentrate failures towards less severe consequences; ultimately this leads to the fail-safe design. Fail-safe designs require, however, that a safe failure mode exists: it may be safe to shut down a chemical plant, but one cannot adopt this strategy for a life support system or an aircraft control system. Failure mode, effect and criticality analysis is a technique often employed, but it is very time consuming.

In cases where a particular hazard exists, special techniques may be employed to ensure safety in the event of a malfunction. Electronic instrumentation for use in the presence of flammable gas, for example, is often protected by Zenner barriers. In normal use the energy levels in the circuits are too low to cause a spark which will ignite the gas. A fault in the circuits may allow higher energy levels if the protection offered by Zenner barriers were not available.

4. Manufacture, Installation and Test

It is clearly important that an instrumentation system is manufactured and installed in accordance with the designer's intention. Quality assurance programs are now the subject of National Standards, and many customers insist on their implementation.

Test procedures before and after installation are necessary to demonstrate that the system is functioning according to the design intention although, as mentioned in Sect. 1, testing cannot be exhaustive. The tests should be specified by the designer. An extended test may be carried out in an attempt to reveal early life failures, before the system is put into service.

5. Operation

The designer must also give consideration to operation of the instrumentation system. Regular test and maintenance procedures, diagnostic and repair procedures in the event of failure, and the provision of spare parts will all influence the reliability and safety of the system.

Special attention must be paid to the interface between an operator and the system. Discrimination of decision making between man and computer must be considered; a computer may be quicker and more precise, but the operator likes to feel in control. Boredom during routine operation, and stress at times of malfunction can both lead to mistakes. The training of operators is very important; a simulator is often required, as training on the real plant may be expensive and dangerous.

See also: Operational Environments; Reliability and Maintenance

F. J. Charlwood
[City University, London, UK]

Signal Processing

Signal processing is of importance in measurement and instrumentation, because it provides the basis for changing the information content of signals. Such changes may be needed for signal conditioning, perhaps to remove unwanted noise or restrict bandwidth, or for deliberately shaping the information or, perhaps, to implement a controller or compensate an instrument. The modern technology is predominantly electronic, although examples of other methods, such as pneumatic or optical signal processing, are found in some special situations. Electronic signal processing technology divides naturally into the traditional analog methods and the more modern digital methods, which are increasingly competitive in terms of accuracy, flexibility, reliability and cost.

1. Analog Signal Processing

Most natural phenomena are continuous, and therefore transduce directly into electrical analog signals, in which information is conveyed through modulation of either a current or a voltage waveform. In practice, directly modulated bipolar voltage signals in the range $-10\,\mathrm{V}$ to $+10\,\mathrm{V}$ are mostly used for analog signal processing. An analog signal of this kind can be processed by means of a passive network of resistors, capacitors and inductors, and an example of a passive third-order Butterworth filter is given in Fig. 1, but the effects of loading and the demands placed on component specifications, particularly of inductors, make both design and implementation difficult, to the extent that this is rarely a preferred approach.

Analog signal processing is therefore mostly based on active networks which involve electronic operational amplifiers, resistors and capacitors. Several different kinds of operational amplifier (see *Amplifiers*), are available (Clayton 1979), of which the integrated circuit, differential input, single-output type is the most popular, as shown in Fig. 2. Although the signal processing can be accomplished by means of a single operational amplifier in conjunction with

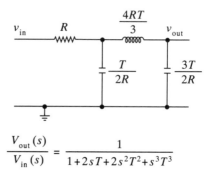

$$\frac{V_{\mathrm{out}}(s)}{V_{\mathrm{in}}(s)} = \frac{1}{1+2sT+2s^2T^2+s^3T^3}$$

Figure 1
Passive third-order Butterworth filter

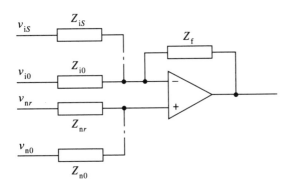

$$V_{\text{out}} = Z_{\text{f}} \left[-\sum_{k=0}^{s} v_{ik} Z_{ik}^{-1} + (Z_{\text{f}}^{-1} + \sum_{k=0}^{s} Z_{ik}^{-1})(\sum_{k=0}^{r} Z_{nk}^{-1})^{-1} \sum_{k=0}^{r} v_{nk} Z_{nk}^{-1} \right]$$

Figure 2
General configuration of operational amplifier

appropriate resistor–capacitor networks, as shown, for example, in Fig. 3, satisfactory designs may be difficult to achieve (Bowran and Stephenson 1979). Generally, it is preferable to use designs based on cascades of operational amplifiers, which are available in integrated circuit form as universal active filters. Among the most popular of these are state-variable filters (Kerwin *et al.* 1967), which are easy to design and are relatively insensitive to component value tolerances. An example of a state-variable third-order Butterworth filter is given in Fig. 4.

2. Digital Signal Processing

Although analog methods are traditional, well established, and often seem to be the natural choice for signal processing, recent advances in digital integrated circuit technology have been such as to make

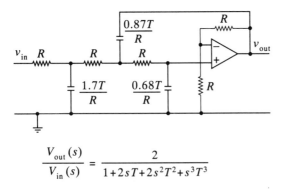

$$\frac{V_{\text{out}}(s)}{V_{\text{in}}(s)} = \frac{2}{1 + 2sT + 2s^2T^2 + s^3T^3}$$

Figure 3
Single-amplifier, third-order Butterworth filter

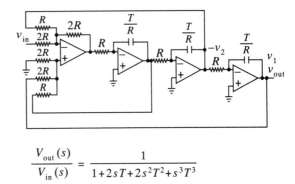

$$\frac{V_{\text{out}}(s)}{V_{\text{in}}(s)} = \frac{1}{1 + 2sT + 2s^2T^2 + s^3T^3}$$

Figure 4
State-variable third-order Butterworth filter

digital methods a better choice for signal processing in many cases. For measurement and instrumentation, 16-bit digital signals, which give a resolution of 0.0015% of full scale, are more than sufficient to represent the majority of physical processes, and such signals can now be processed very effectively by either general-purpose or special-purpose devices.

In a digital signal processing arrangement, such as that shown in Fig. 5, a digital signal processor is normally preceded by an analog-to-digital converter and succeeded by a digital-to-analog converter. In some cases, however, either of these converters may be omitted; for example, there may be no analog-to-digital converter in a signal generator, and no digital-to-analog converter in a display. In the normal arrangement, any limitations are usually imposed by the analog-to-digital converter, which must sample the analog signal at a frequency at least twice that of the highest frequency present to avoid aliassing errors, and must resolve the digital signal to a sufficiently large number of bits to avoid quantization errors (Rabiner and Gold 1975).

Digital methods are characterized by their extreme flexibility, and in signal processing they can achieve a great deal more than analog methods. The most obvious application is digital filtering, but other applications include convolution, correlation, Fourier transforms, Hilbert transforms, signal generation, signal analysis, image analysis, pattern recognition and controller implementation. Efficient algorithms are continually being developed and refined for these, and

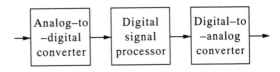

Figure 5
Digital signal processing

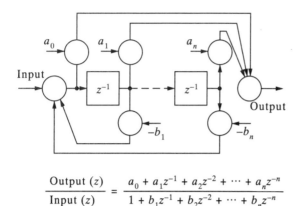

$$\frac{Output\ (z)}{Input\ (z)} = \frac{a_0 + a_1 z^{-1} + a_2 z^{-2} + \cdots + a_n z^{-n}}{1 + b_1 z^{-1} + b_2 z^{-2} + \cdots + b_n z^{-n}}$$

Figure 6
Direct implementation of digital process

many other applications, of which the best known is probably the fast Fourier transform (Oppenheim and Schafer 1975).

Despite their wide applicability, digital signal processing methods are based on only a few kinds of digital operation, particularly those for the storage, addition and multiplication of digital numbers. Dynamical behavior arises through the storage operation, which allows a delay of virtually any required duration to be implemented. The sampling period delay is conventionally denoted by the operator z^{-1}, so that a typical digital process is denoted by

$$\frac{a_0 + a_1 z^{-1} + a_2 z^{-2} + \cdots + a_n z^{-n}}{1 + b_1 z^{-1} + b_2 z^{-2} + \cdots + b_n z^{-n}}$$

which may be implemented in many different forms, for example, by the direct structure shown in Fig. 6. This structure represents a nonrecursive, finite impulse response or transversal filter if all the feedback coefficients $b_1, b_2, \ldots, b_n$ are zero, otherwise it represents a recursive, or infinite impulse response filter (Antoniou 1979).

For the implementation of structures such as these for digital signal processing, conventional computers, microprocessors or microcontrollers, with general-purpose architectures and instruction sets, can be used. Basically, the design is incorporated into the software of a computer program, the core of which is a procedure by which the execution of a sequence of instructions accomplishes the same digital signal processing as the structure to which it is equivalent. The procedure need not be complicated, as seen, for example, in the Pascal (Jensen and Wirth 1974) procedure in Table 1, which contains instructions equivalent to the structure in Fig. 6. Associated with such a procedure are others, at least for obtaining the input and disposing of the output. An increasingly attractive alternative is to use an integrated circuit digital signal processing device with both architecture

Table 1
Procedure for digital signal processing

```
BEGIN
  Output: = 0
  Store[0]: = Input
    FOR r: = n DOWN TO 1 DO
BEGIN
  Store[0]: = Store[0] − b[r] ∗ Store[r]
  Output: = Output + a[r] ∗ Store [r]
  Store[r]: = Store[r − 1]
END
  Output: = Output + a[0] ∗ Store [0]
END
```

and instruction set optimized for signal processing (Chassaing and Horning 1990). Both fixed-point and floating-point architectures are available, the former with 16-bit word length and arrangements to minimize numerical problems and maintain precision: single instructions can implement operators such as z^{-1}.

See also: Networks of Instruments; Signal Theory in Measurement and Instrumentation; Signal Transmission

Bibliography

Antoniou A 1979 *Digital Filters: Analysis and Design.* McGraw-Hill, New York
Bowran P, Stephenson F W 1979 *Active Filters for Communications and Instrumentation.* McGraw-Hill, Maidenhead, UK
Chassaing R, Horning D W 1990 *Digital Signal Processing with the TMS 320C5.* Wiley, Chichester, UK
Clayton G B 1979 *Operational Amplifiers.* Butterworth, London
Jensen K, Wirth N 1974 *PASCAL User Manual and Report.* Springer, New York
Kerwin W J, Haelsman L P, Newcomb R W 1967 State-variable synthesis for insensitive integrated circuit transfer functions. *IEEE J. Solid-State Circuits* **2**, 87–92
Oppenheim A V, Schafer R W 1975 *Digital Signal Processing.* Prentice-Hall, Englewood Cliffs, NJ
Rabiner L R, Gold B 1975 *Theory and Application of Digital Signal Processing.* Prentice-Hall, Englewood Cliffs, NJ

H. A. Barker
[University of Swansea,
Swansea, UK]

Signal Theory in Measurement and Instrumentation

In measurement and instrumentation, due to the two main problems to be solved, signal identification and system identification, a number of different signals appear: input and output signals of both measuring equipment and systems to be tested, noise and other

Table 1
Signal classification

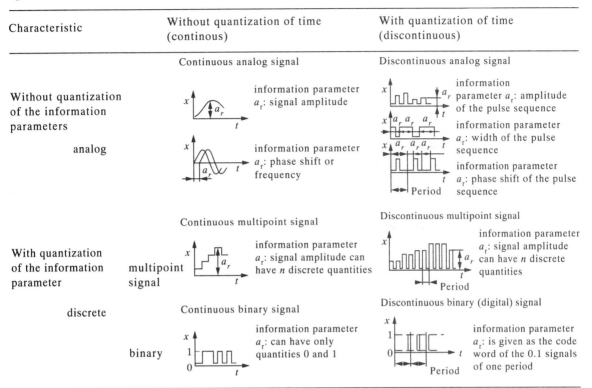

Characteristic		Without quantization of time (continous)	With quantization of time (discontinuous)
		Continuous analog signal	Discontinuous analog signal
Without quantization of the information parameters		x ⟋⌒ a_r information parameter a_r: signal amplitude	x ⊓⊔⊓ a_r information parameter a_r: amplitude of the pulse sequence
	analog	x ⌒⌒ a_r information parameter a_r: phase shift or frequency	x $a_r\, a_r\, a_r$ information parameter a_r: width of the pulse sequence
			x ⊓⊓⊓ information parameter a_r: phase shift of the pulse sequence — Period
		Continuous multipoint signal	Discontinuous multipoint signal
With quantization of the information parameter	multipoint signal	x ⊏⊏ a_r information parameter a_r: signal amplitude can have n discrete quantities	x ⊔⊓⊓ a_r information parameter a_r: signal amplitude can have n discrete quantities — Period
discrete		Continuous binary signal	Discontinuous binary (digital) signal
	binary	x 1 ⊓⊓⊓ 0 information parameter a_r: can have only quantities 0 and 1	x 1 ⊓⊓ 0 information parameter a_r: is given as the code word of the 0.1 signals of one period — Period

Source: Woschni 1988

disturbance signals, which are not only analog but may also be discrete or digital.

Signals are carriers of information, therefore it is necessary that certain information parameters of the signal a_r, mapping this information, are able to change. Table 1 gives a survey of signal designations with respect to several information parameters a_r.

Signals may be represented in the time and frequency (spectral) domains. For the formation of characteristic values, time or statistical means may be used. Furthermore, signal representations may be applied by making use of geometrical relationships: these are not discussed in detail here.

1. Frequency Representations of Signals

A very important fundamental signal, which is also of major importance as a test signal, is the harmonic oscillation

$$x = \hat{X}\sin(\omega t + \phi) \qquad (1)$$

where $\hat{X}$ is the amplitude, $\omega = 2\pi f$ is the angular frequency, $t = 1/f$ is the oscillation period, and ϕ is the phase angle. Representation in the complex plane, as shown in Fig. 1, yields, according to the so-called symbolic method, complex and oriented indicators $\hat{X}^{\angle}$ and, consequently, the following relationship:

$$x = \hat{X}\exp[j(\omega t + \phi)]$$
$$= \hat{X}\cos(\omega t + \phi) + j\hat{X}\sin(\omega t + \phi) = \hat{X}^{\angle}\exp(j\omega t)$$

which is equivalent to Eqn. (1), and which has the oriented complex amplitude $\hat{X}^{\angle} = \hat{X}\exp(j\phi)$.

The advantage of the symbolic method consists of the possibility of having a simple and easily understandable addition of several partial oscillations having the same frequency (Woschni 1988).

According to Fourier, it is possible to represent signals having a cyclic time behavior $x(t)$ by a series of sinusoidal oscillations with frequencies which are multiples of the fundamental frequency ω_0 given by $\omega_0 = 2\pi f_0 = 2\pi/T$, shown as an example in Fig. 2,

(a)

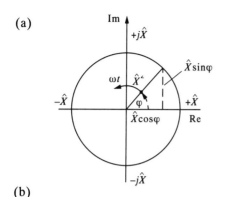

(b)

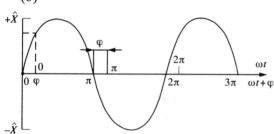

Figure 1
Harmonic oscillation: (a) indicator representation;
(b) time representation

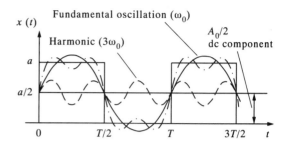

Figure 2
Rectangular periodic signal

which depicts two sinusoids (of many) that form the cyclic rectangular oscillation:

$$x(t) = \frac{1}{2} A_0 + \sum_{n=1}^{\infty}$$

$$[A_n \cos (n \omega_0 t) + B_n \sin (n \omega_0 t)]$$

with the amplitude spectrum

$$C_n = (A_n^2 + B_n^2)^{1/2}$$

The Fourier coefficients may be calculated by the relations

$$A_n = \frac{2}{T} \int_{-T/2}^{+T/2} x(t) \cos (n \omega_0 t) dt \qquad (2a)$$

$$B_n = \frac{2}{T} \int_{-T/2}^{+T/2} x(t) \sin (n \omega_0 t) dt \qquad (2b)$$

From Eqns. (2a, b) it can be seen that there are only cosine terms for even time functions $x(t) = x(-t)$, and only sine terms for odd functions $x(t) = -x(-t)$.

Transformation with the aid of Euler's theorem, leads to the complex Fourier series

$$x(t) = \frac{1}{T} \sum_{n=-\infty}^{+\infty} \hat{X}(jn\,\omega_0) \exp (jn\,\omega_0 t) \qquad (3a)$$

with the complex coefficient

$$\hat{X}(jn\,\omega_0) = \int_{-T/2}^{+T/2} x(t) \exp (jn\,\omega_0 t) dt \qquad (3b)$$

and the amplitude spectrum C_{nk} given by:

$$| C_{nk} | = \frac{1}{T} | \hat{X}(jn\,\omega_0) | \qquad (3c)$$

In addition to the system of orthogonal functions based on sine and cosine functions, other orthogonal systems have been introduced, in particular, the Walsh functions (Harmuth 1970).

In widening the theory to nonperiodic signals, instead of the discrete Fourier spectrum as described by Eqns. (3a, b, c) a continuous spectrum appears. It is obtained from the Fourier series equations (3a, b, c), by passing to the limit

$$T \to \infty, \quad \omega_0 = 2\pi/T \to d\omega, \quad 1/T \to d\omega/2\pi, \quad n\omega_0 \to \omega$$

Therefore, the Fourier transforms are obtained:

$$\hat{X}(j\omega) = \int_{-\infty}^{+\infty} x(t) \exp (-j\omega t) dt = F\{x(t)\} \qquad (4a)$$

$$x(t) = \frac{1}{2\pi} \int_{-\infty}^{+\infty}$$

$$\hat{X}(j\omega) \exp (j\omega t) d\omega = F^{-1} \{\hat{X}(j\omega)\} \qquad (4b)$$

Physically interpreted, $\hat{X}(j\omega)$ represents the complex amplitude, related to $d\omega$, and is therefore termed the spectral amplitude density, having the dimension of amplitude per frequency interval, that is, $V\,s$ or $V\,Hz^{-1}$.

An identical calculation can be made for the Walsh functions. This leads to the sequential amplitude density (Harmuth 1970).

The Fourier transform has a number of important properties and theorems, as explained in Harmuth (1970), Birgham (1974) and Woschni (1988). The algorithm of the fast Fourier transform (FFT) reduces the number of multiplications from N^2 to $(N/2)\,1b(N/2)$, where N equals the number of Fourier coefficients (Cooley and Tukey 1965, Birgham 1974) and $1b$ equals the content of information in a bit of one message. For $N = 1024$, for example, the calculation time may thus be reduced by a factor $1/227$ and often, as with modern measuring devices, on-line processing is possible.

Another characteristic function used for non-deterministic signals is the spectral power density $S_{xx}(\omega)$. It is defined as the part of the power ΔP, which falls into a differentially small frequency range $\Delta\omega$. In contrast to the spectral amplitude density, the spectral power density is a real-valued function of the frequency ω and does not contain phase information. The latter is lost in the calculation of the average value, which is necessary for the formation of the power. This can also be seen from the relationship existing on the basis of Parseval's equation, by averaging over a time domain T (Zadeh and Desoer 1963, Woschni 1988)

$$S_{xx}(\omega) = \frac{1}{2\pi} \lim_{T\to\infty} \frac{|\hat{X}(j\omega)|^2}{2T}$$

The power P of the entire signal existing in the whole frequency domain can be calculated on the basis of Parseval's equation (Zadeh and Desoer 1963, Woschni 1988)

$$P = \overline{x^2(t)} = \lim_{T\to\infty} \frac{1}{2\pi} \int_{-T}^{+T} x^2(t)\,dt$$

$$= \int_{-\infty}^{+\infty} \lim_{T\to\infty} \frac{1}{2\pi} \frac{|\hat{X}(j\omega)|^2}{2T}\,d\omega$$

$$= \int_{-\infty}^{+\infty} S_{xx}\,d\omega \tag{5}$$

From Eqn. (5) it follows that the power density $S_{xx}(\omega)$ must decrease rapidly from a certain critical frequency and must vanish at higher frequencies because of the requirement of boundedness of the power P. Depending on the critical frequency ω_c, a distinction may be made between narrowband and wideband signals.

2. Time Representations of Signals

A signal with the time function $x(t)$ may be characterized in the time domain by time averages, also called moments of the nth order. For periodic signals

$$\overline{x^n(t)} = \frac{1}{2T} \int_{-T}^{+T} x^n(t)\,dt$$

and for nonperiodic signals

$$\overline{x^n(t)} = \lim_{T\to\infty} \frac{1}{2T} \int_{-T}^{+T} x^n(t)\,dt$$

The linear (arithmetic) average is equal to the Fourier coefficient $A_0/2$, and the mean square value ($n = 2$) is equal to the square of the effective value X_{eff}.

A generalized mean square value, the correlation function $\psi(\tau)$, is gained if the function $x(t)$ is multiplied by the function with a time delay τ, that is, the autocorrelation function is defined by

$$\psi_{xx}(\tau) = \overline{x(t)\,.\,x(t \pm \tau)} =$$

$$\lim_{T\to\infty} \frac{1}{2T} \int_{-T}^{+T} x(t)\,x(t \pm \tau)\,dt \tag{6a}$$

and the crosscorrelation function

$$\psi_{xy}(\tau) = \overline{x(t)\,.\,y(t + \tau)} =$$

$$\lim_{T\to\infty} \frac{1}{2T} \int_{-T}^{+T} x(t)\,y(t + \tau)\,dt \tag{6b}$$

Typical properties of the autocorrelation are as follows:

(a) in averaging, the phase information is lost, as in the case of spectral power density;

(b) the value for $\tau = 0$ represents the mean square value and is the maximum value of the autocorrelation function, that is,

$$\psi_{xx}(0) = \overline{x^2(t)} = X_{eff}^2$$

(c) the other limiting value for $\tau \to \infty$ is the square of the linear mean value,

$$\lim_{\tau\to\infty} \psi_{xx}(\tau) = [\overline{x(t)}]^2$$

(d) since it is of no significance whether the function $x(t)$ in Eqn. (6a) is displaced towards positive or negative times, the autocorrelation function is an even function, that is,

$$\psi_{xx}(\tau) = \psi_{xx}(-\tau) = \overline{x(t)x(t + \tau)} = \overline{x(t)x(t - \tau)}$$

315

The crosscorrelation function has the following features:

(a) it is a noneven function;

$$\psi_{xy}(\tau) = \psi_{yx}(-\tau)$$

(b) it contains relative phase information concerning the two events $x(t), y(t)$;

(c) The limiting values are

$$\psi_{xy}(0) = \psi_{yx}(0) = \overline{x(t)\,y(t)},$$

$$\lim_{\tau \to \infty} \psi_{xy}(\tau) = \overline{x(t)} \cdot \overline{y(t)}$$

The crosscorrelation function plays an important role, because with white noise as input the weighting (pulse answer) function $g(t)$ corresponds to the crosscorrelation function $\psi_{xy}(\tau)$. The solution to this measuring problem during normal operation of systems is one of the most important suppositions for adaptive systems (Davies 1970).

The correlation function is measured either using analog or, more commonly, digital methods. In this case, the functions $x(t)$, and $y(t)$ are split up into various values at different times t, the sampling theorem having to be observed. With the aid of the corresponding relationships, it is possible to calculate the correlation function point by point from the sampled values.

Further methods are based on the application of the fast Fourier transform corresponding to the relations given in Sect. 3, that is, calculation of the correlation function from the spectral power density.

3. Transforms Between Time and Frequency Representations

The autocorrelation function, like the spectral power density, contains no phase information; this is lost in both cases because of the averaging operation. In relation to the time behavior of the signal $x(t)$ and the corresponding spectral amplitude density $\hat{X}(j\omega)$, there exists a relationship between the autocorrelation function and the power density via the Fourier transform, known as the Wiener–Chinchine theorem (Davies 1970, Woschni 1988).

$$S_{xx}(\omega) = \frac{1}{2\pi} \int_{-\infty}^{+\infty} \psi_{xx}(\tau) \exp(-j\omega t)\, d\tau$$

$$= \frac{1}{2\pi} F\{\psi_{xx}(\tau)\} \tag{7a}$$

$$\psi_{xx}(\tau) = \int_{-\infty}^{+\infty} S_{xx}(\omega) \exp(j\omega t)\, d\omega$$

$$= 2\pi F^{-1}\{S_{xx}(\omega)\} \tag{7b}$$

A summary of the relationships established so far between both time and frequency representations of signals is given in Table 2. Because phase information is lost due to averaging, conversions are possible only in the direction indicated by an arrow.

4. Probability Representations of Signals

Randomly fluctuating events $\xi(t)$ are described by probability functions, the probability distribution and the probability density.

The probability distribution $W(x)$ indicates the probability p that the signal $\xi(t)$ remains smaller than a barrier value x, that is,

$$W(x) = p[\xi(t) < x]$$

with the limiting values

$$\lim_{x \to -\infty} W(x) = 0 \quad \lim_{x \to +\infty} W(x) = 1$$

The probability density $w(x)$ is the probability related to Δx because the values of the event $\xi(t)$ are within a narrow region near the value x, that is,

$$w(x) = \frac{1}{\Delta x} \quad p\,[x \leqslant \xi(t) < x + \Delta x], \quad \Delta x \to dx$$

with the relations

$$\int_{-\infty}^{x} w(u)\,du = W(x)$$

and

$$\frac{dW(x)}{dx} = w(x)$$

For multidimensional distributions $x_1, x_2, \ldots, x_n$, it is useful to introduce compound probability distributions $W(x_1, x_2, \ldots, x_n)$ and compound probability distribution densities $w(x_1, x_2, \ldots, x_n)$:

$$W(x_1, x_2, \ldots, x_n) =$$

$$p[\xi_1(t) < x_1, \xi_2(t) < x_2, \ldots, \xi_n(t) < x_n]$$

$$w(x_1, x_2, \ldots, x_n) = \frac{\partial^n}{\partial x_1\, \partial x_2, \ldots, \partial x_n}$$

$$\times W(x_1, x_2, \ldots, x_n)$$

Furthermore, conditional probability distributions $W(x_1/x_2)$ and conditional probability distribution densities $w(x_1/x_2)$ are defined. They indicate the probability that the value x_1 occurs on condition that the value

Table 2
Relations between time and frequency representations of signals

	Time representation		Frequency representation	
	function of time		amplitude density	complex function of a real variable
Real function of a real variable	$x(t) = \dfrac{1}{2\pi} \displaystyle\int_{-\infty}^{+\infty} \hat{X}(j\omega)e^{j\omega t}d\omega$	$\Longleftarrow F \Longrightarrow$	$\hat{X}(j\omega) = \displaystyle\int_{-\infty}^{+\infty} x(t)e^{-j\omega t}dt$	
Mean value operation; phase information is lost	$\psi_{xx}(\tau) = \lim_{T\to\infty} \dfrac{1}{2T} \displaystyle\int_{-T}^{+T} x(t)\,x(t+\tau)dt$	$\longrightarrow$	$S_{xx}(\omega) = \dfrac{1}{2\pi} \lim_{T\to\infty} \dfrac{|\hat{X}(j\omega)|^2}{2T}$	only unilateral conversion possible
	autocorrelation function		spectral power density	
Real function of a real variable	$\psi_{xx}(\tau) = \displaystyle\int_{-\infty}^{+\infty} S_{xx}(\omega)e^{j\omega t}d\omega$	$\Longleftarrow F \Longrightarrow$	$S_{xx}(\omega) = \dfrac{1}{2\pi} \displaystyle\int_{-\infty}^{+\infty} \psi_{xx}(\tau)\,e^{-j\omega t}dt$	real function of a real variable

Source: Woschni 1988

x_2 already exists. The following relationships hold for the compound probability density

$$w(x, y) = w(x/y)\,w(y) = w(y/x)\,w(x)$$

Of utmost importance, in practice, is the Gaussian distribution density

$$w(x) = \frac{1}{\sqrt{2\pi}\sigma} \exp\left[\frac{-(x-a)^2}{2\sigma^2}\right] \qquad (8)$$

where $a = \overline{x(t)}$ is the linear mean value and σ is the standard deviation, related to the square mean value, $\overline{x^2(t)}$, by

$$\sigma = \sqrt{[\overline{x^2(t)} - a^2]}$$

If the ergodic theorem is fulfilled, as normally would be in practice, statistical mean values (expectation values) $E\{x^n\} = \bar{x}^n$ are equal to the time mean values $\overline{x^n(t)}$, that is,

$$E\{x^n\} = \widetilde{x^n} = \int_{-\infty}^{+\infty} x^n\, w(x)\,dx = \overline{x^n(t)}$$
$$= \lim_{T\to\infty} \frac{1}{2T} \int_{-t}^{+T} x^n(t)\,dt$$

For the important Gaussian distribution density, from Eqn. (8), one obtains

$$M_1 = \bar{x} = \overline{x(t)}$$
$$= \int_{-\infty}^{+\infty} \frac{x}{\sqrt{2\pi}\sigma} \exp\left[\frac{-(x-a)^2}{2\sigma^2}\right]dx = a$$
$$M_2 = \bar{x} = \overline{x^2(t)}$$
$$= \int_{-\infty}^{+\infty} \frac{x^2}{\sqrt{2\pi}\sigma} \exp\left[\frac{-(x-a)^2}{2\sigma^2}\right]dx = a^2 + \sigma^2$$

As an example with great practical importance, it may be found, by measurement, that the length of workpieces having an average value of $a = 10$ cm satisfies a Gaussian distribution and has a standard deviation of $\sigma = 3$ mm. What matters then, might be the number of workpieces that lie within a tolerance range of 10 cm ± 4 mm. The probability for the fluctuation process to lie within the range $a - x \leqslant \xi(t) < a + x$ is given by:

$$p[-x \leqslant \xi(t) < x]$$
$$= \frac{1}{\sqrt{2\pi}\sigma} \int_{-x}^{+x} \exp\left(\frac{-\xi}{2\sigma^2}\right)d\xi$$

Using the probability integral tabulated by Jahnke (1960), the result that 82% of the pieces are within the tolerance range is obtained.

See also: Information Theory in Measurement and Instrumentation; Signal Processing

Bibliography

Birgham E O 1974 *The Fast Fourier Transform*. Prentice-Hall, Englewood Cliffs, NJ
Blumenthal L 1961 *A Modern View of Geometry*. Freeman, San Francisco, CA
Cooley J W, Tukey J W 1965 An algorithm for the machine calculation of complex Fourier series. *Math. Comput.* **19**, 279–301
Davies W D T 1970 *System Identification for Self-Adaptive Control*. Wiley, Chichester, UK
Harmuth H F 1970 *Transmission of Information by Orthogonal Functions*. Springer, Berlin
Jahnke E F 1960 *Tafeln Höherer Funktionen*. Teubner, Leipzig, Germany
Peterson W W 1962 *Error Correcting Codes*. MIT Press, Cambridge MA
Woschni E-G 1988 *Informationstechnik*, 3rd edn. VEB Verlag, Berlin
Zadeh L A, Desoer 1963 *Linear System Theory*. McGraw-Hill, New York

E.-G. Woschni
[Technische Universität Chemnitz-Zwickau, Chemnitz-Zwickau, Germany]

Signal Transmission

In measurement and instrumentation there is a fundamental requirement for signal transmission to convey information accurately from one point to another. The modern technology is predominantly electrical, although pneumatic transmission still persists and optical transmission takes an increasing market share. The technology divides naturally into the traditional analog methods and the increasingly competitive modern digital methods.

1. Analog Signal Transmission

Most analog signal transmission is by live-zero baseband signalling using either pneumatic pressure signals or electrical current signals which are direct representations of the information conveyed. The standard ranges are 3–15 psi (~20–100 kPa) for pneumatic signals and 4–20 mA for electrical signals. In pneumatic signal transmission, a standard diameter pipe of length L m will satisfactorily transmit signals with component frequencies of less than $4/L$ Hz, a bandwidth that is extremely restricted compared with the 10 kHz achievable in electrical signal transmission using a twisted-pair cable with an electrostatic screen. If this bandwidth is inadequate, then electrical voltage signals in coaxial cable may be used to transmit baseband signals with component frequencies up to 10 MHz. Modulation signalling, in which the information to be conveyed modulates a sinusoidal carrier signal, may be used as an alternative to baseband signalling, the most common arrangements being amplitude modulation, frequency modulation and phase modulation.

2. Digital Signal Transmission

In digital signal transmission, the information consists only of binary digits, or bits, but the signals used to convey this information do not necessarily take only the corresponding binary values. The signalling methods are actually very similar to those used in analog signal transmission, and include both baseband and modulation signalling.

In baseband signalling, the signal conveys the binary information through some simple form of coding, examples of which are shown in Fig. 1. The non return to zero (NRZ) method, which is the simplest and most common, uses full-width plus and minus pulses to represent the bit values 1 and 0, respectively. The alternate mark inversion (AMI) method uses a half-width pulse to represent the bit value 1, the pulses

being alternately plus and minus. The conditioned diphase (Manchester coding) method uses pairs of half-width pulses of different polarity to represent either of the bit values 1 or 0, with the polarity changing at every bit value 1.

In modulation signalling, the signal conveys the binary information through some form of modulation of a carrier signal, examples of which are shown in Fig. 1. The amplitude modulation (AM) method uses the presence or absence of the carrier to represent the bit values 1 and 0, respectively. The frequency modulation, or frequency shift keying (FSK), method uses low and high frequencies to represent the bit values 1 and 0, respectively. The phase modulation, or phase shift keying (PSK), method uses the reference phase of the carrier and its inverse to represent the bit values 1 and 0, respectively. Standardized equipment for modulation and demodulation (modem) is available for digital signal transmission by these methods.

A single communication channel may be used to transmit n different signals by means of multiplexing. In frequency-division multiplexing (FDM), the channel bandwidth is subdivided into n contiguous bands, each of which encompasses the principal components resulting from the modulation of a carrier of appropriate frequency, by a digital signal with specified maximum bit rate. In time-division multiplexing (TDM), each signal is transmitted in one of n contiguous periods of time.

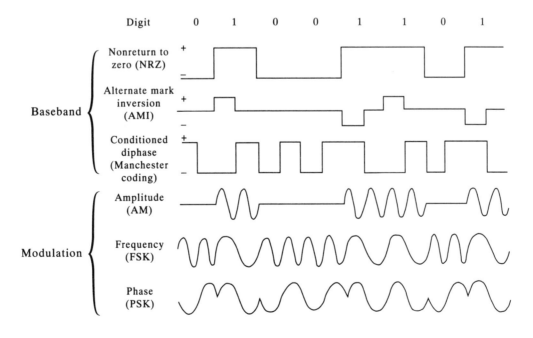

Figure 1
Baseband and modulation methods for digital signal transmission

3. Coded Transmission

In most cases, the bits in a digital signal are meaningful only when they are grouped together to form a word, the meaning of which is conveyed by a code. Examples include the Gray code, where the four bits of a nybble are used to represent the integers 0 to 15, and the American standard code for information interchange (ASCII), which uses seven of the eight bits of a byte to represent 128 different characters.

When ASCII coded characters are transmitted as bytes, the eighth bit may be set to give even or odd parity, that is, an even or an odd number of bit values 1; a parity check on each received byte then allows errors of one bit per byte to be detected. This example of error detection may be extended by adding a byte to each block of data transmitted to give even or odd parity for each of the eight bit positions; parity checks on the received data then allow errors of up to three bits per block to be detected, of which at least one may be corrected. The addition of further bits gives enhanced capability for error detection and correction; for example, the addition of a Hamming code of $\log_2 n + 1$ bits to an n-bit word allows errors of up to two bits per word to be detected, of which at least one may be corrected (Peterson 1961). The most usual arrangement for blocks is the addition of words based on one of the many cyclic redundancy check (CRC) codes, for example, the Bose–Chaudhuri codes (Bose and Chaudhuri 1960).

Words may be transmitted using either parallel or serial transmission. In parallel transmission the bits are transmitted simultaneously on separate lines, collectively known as a bus, while in serial transmission the bits are transmitted sequentially on a single line. In either case, additional buses or lines may be used for controlling the transmission, routing the data or bidirectional communication.

Parallel transmission is appropriate for high data rates over short distances. It is used for the internal interconnection of computer-based systems, where standard buses such as the S100 (MITS), Multibus (Intel) or Unibus (DEC) are invariably used. It is also used for laboratory instrumentation, with up to 1 Mbyte s^{-1} over 15 m, for which the IEEE 488 and 583 standards have been developed.

In the IEEE 488, or Hewlett-Packard interface bus (HPIB) standard, all devices possess one or more of the distinct capabilities of acting as a "listener," a "talker" or a "controller." The devices are interconnected by an eight-line data bus, a three-line data byte transfer control bus and a five-line general interface management bus. At any instant, only one device may act as a controller and only one device may act as a talker, but any number of devices may act as listeners. The controller determines which device is acting as a talker and which devices are acting as listeners. Transmission control is exercised through the data byte transfer control bus, which implements the handshake functions that are necessary when the listening devices accept data transfers at different rates.

In the IEEE 583, or computer-automated measurement and control (CAMAC) standard, the basic subsystem is a "crate," which contains a crate controller module and 24 other modules, each of which is a functional device. The modules are interconnected by a databus comprising 24 read lines and 24 write lines, a five-line address bus and a control bus with three control lines, five command lines, four status lines and two synchronization lines. Communication between crates is accomplished by means of a 24-line bus or by means of a serial link.

Serial transmission is used when data rates lower than those associated with parallel transmission are acceptable, in which case it has several advantages over parallel transmission, particularly its requirement for only a few lines rather than for special cables with large numbers of wires, and its ability to work over greater distances. Serial transmission between two devices may be simplex, with transmission in one direction only, or half duplex, with transmission in either direction but not simultaneously, or full duplex, with transmission in both directions simultaneously. The most common standard for this method of transmission is the Electronics Industry Association (EIA) RS232C standard. Although this actually specifies a total of 25 lines, the majority of these are concerned with the protocols of transmission over telephone networks and as few as three (transmit data, receive data, ground) are required for full duplex transmission. Occasionally a further two lines (request to send, clear to send) are used for control purposes, and a further two (transmitter timing, receiver timing) for synchronization purposes.

In asynchronous serial transmission, characters are transmitted independently, with variable intervals of time between them. To ensure correct reception, it is necessary to add control bits to the transmitted data, in the form of a start bit value 0 preceding the byte and one or more stop bits value 1 succeeding the byte, as shown in Fig. 2. In synchronous serial transmission, there are no intervals of time between successive characters transmitted in a block and, to ensure

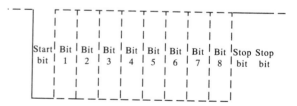

Figure 2
Characters in asynchronous serial transmission

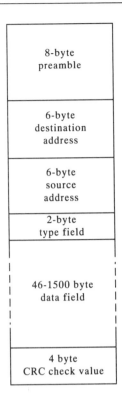

Figure 3
Format of Ethernet packet

correct reception, it is necessary to precede the block with control bytes for character and bit synchronization. Serial transmission with RS232C is limited to $20 \, \text{kbit s}^{-1}$ over 30 m. Higher rates over greater distances are possible with similar standards by using twisted-pair cables, for example, $1 \, \text{Mbit s}^{-1}$ over 1.5 km with RS422, and with coaxial cable, transmission rates of $10 \, \text{Mbit s}^{-1}$ over 3 km may be achieved.

Serial transmission is not restricted to communication between two devices but can also be used in networks with more devices, or nodes. Of particular interest for instrumentation are local area networks (LANs) in which the nodes are separated by less than 10 km. A good example is Ethernet, which uses synchronous serial baseband transmission with conditioned diphase (Manchester coding) at $10 \, \text{Mbit s}^{-1}$ over coaxial links between nodes with a maximum separation of 2.8 km. In Ethernet, data is transmitted in blocks, or packets, with the format shown in Fig. 3. Between 46 and 1500 bytes of data are preceded by an eight-byte preamble for character and bit synchronization, a six-byte source address, a six-byte destination address and a two-byte type field which identifies the format of the data, and succeeded by a four-byte cyclic redundancy check value for the packet.

See also: Information Theory in Measurement and Instrumentation; Networks of Instruments; Signal Processing; Signal Theory in Measurement and Instrumentation

Bibliography

Bose R C, Chaudhuri D K R 1960 On a class of error-correcting binary group codes. *Inf. Control* **3**, 279–90
Grimes R W 1982 Transmission of data. In: Sydenham P H (ed.) *Handbook of Measurement Science.* Wiley, Chichester, UK, pp. 539–89
Lesea A, Zaks R 1977 *Microprocessor Interfacing Techniques.* Sybex, Berkeley, CA
Matthews P R 1982, 1983 Communications in process control. *Meas. Control* **15**, 427–32, 467–71; **16**, 19–22
Miller G M 1978 *Modern Electronic Communications.* Prentice-Hall, Englewood Cliffs, NJ
Peterson W W 1961 *Error-Correcting Codes.* MIT Press, Cambridge, MA
Schwartz M 1977 *Computer Communication Network Design and Analysis.* Prentice-Hall, Englewood Cliffs, NJ

H. A. Barker
[University of Swansea,
Swansea, UK]

Spectroscopy: Fundamentals and Applications

All molecular and atomic species interact with electromagnetic radiation, with the result that the intensity or the power of the incident radiant beam is attenuated. By far the most common form of spectral analysis is based on measurement of the absorption of incident radiation, using the decrease in power (attenuation) of the radiation brought about by the species involved. However, it is also possible to measure reemitted absorbed energy as either radiation (fluorescence or phosphorescence) or heat (photothermal or photoacoustic spectroscopy).

It is convenient to characterize methods of absorption spectroscopy according to the type of electromagnetic radiation involved: x-ray, ultraviolet, visible, infrared, microwave or radiofrequency. The various types of radiation, their spectroscopic techniques and applications are summarized in Table 1.

1. Properties of Electromagnetic Radiation

Electromagnetic radiation (EMR) is transmitted through space at the velocity of light. Of its many forms, the most familiar are light and radiant heat. Since no supporting medium is necessary for the transmission of EMR, it readily passes through a vacuum. Its propagation may be described in terms of

Table 1
Summary of the spectroscopic applications of the various types of electromagnetic radiation

Type of electromagnetic radiation	Wavelength range	Spectroscopic technique	Typical applications
Radio waves	10 m	Radiofrequency absorption	Limited to examination of rotational spectra of gases; used principally in astronomy (Lyndon-Bell and Harris 1969)
	1–100 m	Nuclear magnetic resonance spectroscopy	Mainly for organic liquids and solids: investigation of structure and bonding in hydrogenic systems difficult to measure by other techniques. ^{13}C, ^{14}N and ^{19}F measurements have also become important. This is an off-line technique in routine use in analytical studies. Imaging by nuclear magnetic resonance is an area of growing importance, particularly for medical diagnostic measurements (Griffiths 1978)
Microwaves	1 cm–2 m	Microwave absorption spectroscopy	Used for fluids, mainly gases. Large effective pathlengths are usually required. Has been used to measure the concentration of gases in flowing streams and, for example, the moisture content of organic liquids (Townes and Schawlow 1975, Varna and Hrubesh 1970)
		Electron spin resonance spectroscopy	Extensively used in the study of free radicals and the structure of molecular species, for example, organometallic and other complexes of the transition metals (Ayscough 1967)
Far infrared	25 μm–1 mm	Absorption spectroscopy	Study of the rotational spectra of gas and liquid samples and the vibrational spectra of molecules containing heavy atoms. Principally a laboratory technique requiring Fourier transform spectroscopic analysis (Finch et al. 1970)
Mid-infrared	2.5–25 μm	Absorption spectroscopy	Vibrational structure of solids, liquids and gases and their rotational fine structure. The technique is used routinely for off-line measurements. Gas detection systems are in common use for on-line applications, for example, in monitoring atmospheric pollution and analyzing gases from industrial plant (Kendall 1966)
		Reflectance spectroscopy	Many techniques exist for studying the infrared absorption characteristics of solid samples. Most common are reflectance methods which measure the partially attenuated radiation reflected from the sample. These include specular reflectance, attenuated total reflectance (a multiple reflection technique) and diffuse reflectance. These techniques may be used with powders and with opaque laminar samples, and for the study of surface coatings, usually in off-line analytical systems.
		Emission spectroscopy	Heating solid samples leads to emission of characteristic infrared radiation which may then be observed using an interferometric system. An off-line technique useful for amorphous, opaque samples
Near infrared	0.75–2.5 μm	Absorption spectroscopy, reflectance spectroscopy	An area of limited interest. Absorption and reflectance instrumentation exists for off-line measurements. The technique is applied to hydrogenic systems and compounds of the rare earths. Some semiconductor materials have their absorption band edges in this region of the spectrum

Region	Wavelength	Technique	Description
Visible	400–750 nm	Raman spectroscopy	Used to obtain information on vibrational transitions. A complementary technique to conventional infrared spectroscopy. A laser source is generally used (Freeman 1974)
		Molecular absorption spectroscopy	Study of electronic transitions which may exhibit vibrational fine structure. Most measurements are performed on solutions. One of the oldest classical spectroscopic techniques. A routine analytical method with many applications
		Reflectance spectroscopy	Extension of the absorption technique to opaque solids and surface coatings. A variety of methods: on-line measurement is often possible (Wendlandt and Hecht 1972)
		Luminescence spectrometry	Optically induced emission-fluorescence and/or phosphorescence provides a powerful technique for the detection of many organic molecules. The principal limitation is that not many molecules do exhibit luminescence. Used to determine low levels of organic pollutants and for post-separation detection in chromatographic techniques. A widely applied off-line analytical method (Hercules 1966, Parker 1968)
		Atomic absorption spectroscopy	Electronic spectra of atoms: flame or graphite furnace atom cells are used. One of the most widely used techniques for determining metals and metalloids in trace amounts. A routine off-line analytical technique
		Thermally stimulated emission spectroscopy	Electronic spectra of atoms. Thermal energy is used to create and excite atoms from samples introduced into arc plasma sources. Off-line techniques for trace element determination
		Optically stimulated emission spectroscopy	Atomic fluorescence spectrometry monitors radiatively stimulated resonance emission for analyte atoms created in flame of furnace sources. A specialized technique for off-line trace element determination
Ultraviolet	200–400 nm	Optically stimulated emission spectroscopy	Most techniques listed above for the visible region use instrumentation which allows near ultraviolet to be performed. This extends the range of samples accessible to the instrumentation, particularly for atomic fluorescence and atomic emission
Vacuum ultraviolet	10–200 nm	Optically stimulated emission spectroscopy	In theory, atomic spectroscopy in this region is an extension of techniques used in the near ultraviolet–visible region applied to higher-energy valence electron systems. Practical difficulties (vacuum system, lack of suitable materials for optical components) limit applications in this area. Specialist laboratory technique
X ray	0.01–10 nm	Absorption spectroscopy	It is difficult to obtain continuously tunable radiation in this region: true absorptivity is thus of limited interest (Liebhafsky et al. 1966)
		Emission techniques	X-ray fluorescence is a widely used off-line technique for determining elemental composition of solids in concentration from a few ppm to about 10% (Tertian and Claisse 1982)
γ ray	<0.01 nm	γ-ray emission	This derives from excitation of the nucleus. A specialized technique for laboratory use, it has found application in the quantification of isotopic composition
		Mössbauer spectroscopy	Highly specialized laboratory technique useful for a few elements only, notably iron (Cohen 1980)

wave parameters, such as velocity, frequency, wavelength and amplitude. Details are discussed in a number of texts (e.g., Herzberg 1945).

2. The Electromagnetic Spectrum

Available analytical techniques restrict the region of the electromagnetic spectrum which can be used, to wavelengths from 10^{-10}–10^4 cm. Over this large range of wavelengths, information is obtained by transmission, reflection, fluorescence, refraction or diffraction techniques on electronic transitions within atoms and molecules, as well as on molecular vibrations, rotations and magnetic spin state. Table 2 shows the analytically important areas of the electromagnetic spectrum. Note that a logarithmic scale is used, and that types of radiation as diverse as x rays and radio waves differ from visible light only in their frequency, and hence in their energy.

As the individual atoms of a particular element possess a unique set of quantum levels, that element may be characterized by the way in which its atoms absorb or emit radiation. This forms the basis of an important branch of spectral analysis known as atomic spectroscopy. Molecules have a more complex electronic structure than their constituent atoms, and so have more energy levels. It is possible to classify these levels by considering processes that occur when a transition takes place between two levels. The energy E of the molecule may then be considered to be given by the sum of the energies of the different states. These energies—electronic, vibrational, rotational, nuclear orientation and translational—are discussed in the following sections.

2.1 Electronic Energy

As in atomic systems, the electrons in a molecule occupy quantized energy levels, and transitions between them may be accompanied by the absorption or emission of radiation. The energy gaps between levels are larger for electronic transitions, which are stimulated by high-energy photons. Core electrons, those bound most tightly to the nucleus, can be excited only by x rays ($\nu > 10^{17}$ Hz), while bonding or valence electrons, those less tightly bound to the nucleus, are excited by ultraviolet or visible radiation ($\nu = 10^{15}$–10^{16} Hz). It is not possible to analyze the energy levels since, in general, many-body problems cannot be solved using quantum theory.

2.2 Vibrational Energy

The atoms in a molecule do not occupy fixed positions with respect to one another, but oscillate about mean positions in a way that may be described by quantum mechanics. To a first approximation, each vibrational mode may be treated as a separate harmonic oscillator

which has certain permitted energy levels; the energy of each mode is then

$$E_{vib} = h\nu'(v + \tfrac{1}{2}) \tag{1}$$

where ν' is the frequency of vibration and ν is the vibrational quantum number, which can take the values $0, 1, 2, 3, \ldots$.

The separation between the vibrational energy levels is smaller than between electronic energy levels, and transitions are stimulated by interaction with infrared radiation ($\nu = 10^{12}$–10^{15} Hz).

2.3 Rotational Energy

The rotational energy of molecules is also quantized. For a linear diatomic molecule of moment of inertia I, the energy of each quantized level can be approximately described by the formula

$$E_{rot} = J(J + 1)\left(\frac{\hbar^2}{2I}\right) \tag{2}$$

where $\hbar = h/2\pi$ and J, the rotational quantum number, can take the values $0, 1, 2, 3, \ldots$. This formula applies only to simple diatomic molecules in the gas phase; the greater interaction between molecules in the liquid and solid phases renders the analysis invalid. Rotational energy levels are even more closely spaced than vibrational levels, and a transition may result from interaction with microwave radiation ($\nu = 10^9$–10^{11} Hz).

2.4 Nuclear Orientation Energy

In the presence of a magnetic field, a molecule which contains an atom with nuclear spin has quantized energy levels determined by the orientation of the nucleus in the field. The separation between energy levels is smaller than for rotational energies, and transitions are associated with radiation in the radio-frequency range ($\nu < 10^9$ Hz). The levels are determined by the magnetic field strength and the environment of the nucleus.

2.5 Translational Energy

Some of a molecule's energy is stored as translational motion; however, translational energy may be assumed to have no spectroscopically useful interaction with electromagnetic energy.

3. Instrumentation

The nature of the instrumentation required for a particular spectroscopic technique will vary with the type of radiation employed and the measurement requirements. A detailed description of these components is beyond the scope of this article (see *Analytical Physical Measurements: Principles and Practice*).

Table 2
Analytically important areas of the electromagnetic spectrum

Wavelength (cm)	Type of radiation	Type of transition resulting from interaction	Associated spectroscopic technique
10^{-6}–10^{-9}	X ray	inner electron	x-ray absorption spectroscopy (T)[a] x-ray fluorescence (E)
10^{-5}–10^{-6}	vacuum ultraviolet[b] ultraviolet–visible	outer electron	ultraviolet–visible absorption and reflectance spectroscopy (T) (R) luminescence spectroscopy (E) atomic emission spectroscopy (E)[c]
10^{-1}–10^{-4}	infrared	vibrational	infrared absorption spectroscopy (T) infrared emission spectroscopy (E)[c] infrared reflectance spectroscopy (R)[d] Raman spectroscopy (E)
10^{2}–1	microwave	rotational (electron spin)	rotational absorption spectroscopy (T) electron spin resonance spectroscopy[e]
>10^{2}	radiofrequency	nuclear orientational[f]	nuclear magnetic resonance spectroscopy

a the letters in parentheses indicate that the technique measures the transmitted (T), emitted (E) or reflected (R) radiation   b since ultraviolet light at these wavelengths is absorbed by the atmosphere, an evacuated spectrometer must be used to avoid attenuation   c emission here is usually thermally induced   d reflectance techniques in the infrared include diffuse reflectance, specular reflectance and attenuated total reflectance spectroscopy   e when the molecular species has an unpaired electron, for example, a free radical, the spin of this electron will be parallel or antiparallel to the magnetic field direction   f an external magnetic field must be applied before these transitions can be observed

4. Quantitative Measurements in Spectral Analysis

It is clear, then, that molecules can be classified by identifying the wavelength of the frequency of spectral absorption or emission lines or bands. In order to determine quantitatively the amount of absorbing or emitting material present in a sample, it is necessary to establish a relationship between the reduction in radiant power flux of radiation interacting with the material and the number of molecules of the absorbing species present in the sample. For absorption processes, this relationship is described by the Beer–Lambert law: let P_0 be the radiant power of a beam of monochromatic light of wavelength incident on a solution containing c moles per cubic meter of a species which absorbs radiation of wavelength λ, and let P be the power of the beam after it has traversed b meters of the solution. The Beer–Lambert law gives the relationship between the change in radiant power, the path length and the concentration of absorbing species present as

$$\log\left(\frac{P_0^\lambda}{P^\lambda}\right) = \varepsilon bc = A^\lambda \qquad (3)$$

where ε is a constant known as the decadic molar absorption coefficient (molar absorptivity) at wavelength λ. The quantity $\log(P_0^\lambda/P^\lambda)$ is referred to as the absorbance of the solution, and is given the symbol A.

The Beer-Lambert law can be applied to multi-component systems, provided there is no interaction between the different absorbing species. For a sample containing n independent species, the total absorbance A_T is given by

$$A_T^\lambda = A_1^\lambda + A_2^\lambda + \cdots + A_n^\lambda$$
$$= \varepsilon_1 bc_1 + \varepsilon_2 bc_2 + \cdots + \varepsilon_n bc_n \qquad (4)$$

where $\varepsilon_1, \varepsilon_2, \ldots, \varepsilon_n$ are the molar absorptivities of individual components; and $c_1, c_2, \ldots, c_n$ are their molar concentrations (mol m^{-3}).

In practice, it is difficult to determine the radiant powers P and P_0 directly because of reflection and scattering losses through the material examined. To compensate for this, the power transmitted by the sample is compared with the power transmitted by a reference sample, known as the blank, which should contain all the elements present in the sample except the species being measured. For example, if an aqueous solution of material contained in a quartz cuvette constitutes the sample, and the species of interest is the dissolved material, the blank will consist of an identical cuvette containing pure water.

The terms and symbols used in the description of the Beer–Lambert law are those in general use. Historically, however, a diversity of names and symbols have been employed. Table 3 lists those which have been used for absorption measurements and which may be encountered in the literature.

The Beer–Lambert law is not always obeyed by real systems. Deviations may result either from the absorbing species interacting by direct chemical processes (e.g., dimerization), or by the charge distributions of two molecules interacting with each other. Changes in the refractive index of the sample also alter the

Table 3
Nomenclature for absorption measurements

Parameter	Definition	Alternative names and abbreviations
Radiant power (P_0, P)	Energy of radiation reaching a given area of a detector per second	Radiation intensity (I_0, I)
Absorbance (A)	$\log(P_0/P)$	Optical density (OD,D), extinction (E)
Transmittance (T)	P/P_0	Transmission (T)
Absorptivity (a)	A/bc (with c in g dm^{-3})^a	Extinction coefficient, absorbancy index (k)
Decadic molar absorptivity (ε)	A/bc (with c in mol m^{-3})^a	Molar absorptivity, molar extinction coefficient, molar absorbancy index (a_m)
Pathlength of radiation in cm (b)		(l, d)

a in many older texts c is in units of mol dm^{-3} (mol l^{-1}), b is in units of cm and ε is in units of dm^3 mol^{-1} cm^{-1}

absorption characteristics. The Beer–Lambert law is obeyed most closely by samples which contain only small amounts of the absorbing species and, where possible, quantitative spectral analysis should be performed on materials which contain low rather than high concentrations of the absorbing species.

Quantitative measurements in emission spectral analysis require separate consideration. There are two types of emission:

(a) externally stimulated emission, for example, luminescence or thermal emission from atoms and molecules; and

(b) internally stimulated emission, for example, that observed when a species is excited in arcs, sparks, flames or plasmas.

In external stimulation, a fraction of the radiation which strikes the sample is absorbed, and this absorbed energy is subsequently reemitted radiatively or nonradiatively. When the absorption process can be described by the Beer–Lambert law, the fraction for power transmitted is given by

$$T = \frac{P^\lambda}{P_0^\lambda} = \exp\left(\frac{-\varepsilon bc}{2.303}\right) \tag{5}$$

The corresponding absorbed fraction is thus given by

$$1 - \frac{P^\lambda}{P_0^\lambda} = 1 - \exp\left(\frac{-\varepsilon bc}{2.303}\right) \tag{6}$$

and so

$$P_0^\lambda - P^\lambda = P_0^\lambda\left[1 - \exp\left(\frac{-\varepsilon bc}{2.303}\right)\right] \tag{7}$$

The measured power of radiatively reemitted radiation $P_f^{\lambda_1}$, is given by

$$P_f^{\lambda_1} = k\phi_f(P_0^\lambda - P^\lambda) \tag{8}$$

where k is an instrument constant which depends on the solid angle of the light collection optics and the detection capability of the system, and ϕ_f is the quantum efficiency of luminescence of the system, that is, the ratio of the number of quanta emitted at the same wavelength (or a different wavelength) to the number of quanta absorbed, per unit time.

The term $1 - \exp(-\varepsilon bc/2.303)$ may be expanded as an exponential series:

$$1 - \exp\left(\frac{-\varepsilon bc}{2.303}\right)$$
$$= \varepsilon bc - \frac{(\varepsilon bc)^2}{2!} + \frac{(\varepsilon bc)^3}{3!} - \cdots \tag{9}$$

For weakly absorbing materials the higher order terms in this expansion may be neglected, and the luminescence intensity may be represented by

$$P_f = k\phi_f P_0 \varepsilon bc \tag{10}$$

For optically thin media, therefore, as for dilute solutions, measurements of externally stimulated emission produce a linear growth curve of emitted power versus concentration of absorbing species. When radiation absorbed is reemitted nonradiatively it may appear as heat (kinetic energy), in which case the nonradiative conversion factor β may replace ϕ with the approximation $\beta = 1 - \phi$.

For internally stimulated emission, the energy required to cause radiative emission from the sample is contained within the system in the form of chemical and kinetic energy. The Beer–Lambert law does not apply as there is no direct interaction between electromagnetic radiation and matter; the process taking place is the internally stimulated emission of quantized radiation via collisional excitation with atoms, ions or molecules. The emitting species can be identified from the energy levels involved, while quantitative spectral analysis is possible if the temperature and excitation conditions in the sample are known, and the sample is calibrated by introducing standard materials into the system.

It should be emphasized that the principles of quantitative spectral analysis discussed earlier apply irrespective of the energy (frequency) of the radiation considered. Spectral analysis by absorption and emission techniques is routinely applied in most science and engineering disciplines—from clinical chemistry and metallurgical, agricultural and geological sciences to control and production engineering.

See also: Analytical Physical Measurements: Principles and Practice; Optical Instruments; Optical Measurements; Process Instrumentation Applications

Bibliography

Ayscough P B 1967 *Electron Spin Resonance in Chemistry.* Methuen, London

Bauman R P 1969 *Absorption Spectroscopy.* Wiley, New York

Bovey F A 1969 *NMR Spectroscopy.* Academic Press, New York

Brittain E F H, George W O, Wells C H J 1970 *Introduction to Molecular Spectroscopy.* Academic Press, London

Burgess C, Knowles A (eds.) 1981 *Standards in Absorption Spectrometry (Ultraviolet Spectrometry Group).* Chapman and Hall, London

Cohen R L (ed.) 1980 *Application of Mossbauer Spectroscopy,* Vol. 2, Academic Press, London

Conley R T 1966 *Infrared Spectroscopy.* Allyn and Bacon, Boston, MA

Finch A, Gates P N, Radcliffe K, Dickson F N, Bentley F F 1970 *Chemical Application of Far-Infrared Spectroscopy*. Academic Press, London

Freeman S K 1974 *Application of Laser Raman Spectroscopy*. Wiley, New York

Grasselli J G, Snaveley M K, Bulkin B J 1981 *Chemical Application of Raman Spectroscopy*. Wiley, New York

Griffiths P R 1978 *Transform Techniques in Chemistry*. Heyden, London

Grove E L (ed.) 1978 *Applied Atomic Spectroscopy*. Plenum, New York

Hercules D M (ed.) 1966 *Fluorescence and Phosphorescence Analysis*. Wiley, New York

Herzberg G 1945 *Infrared and Raman Spectra of Polyatomic Molecules*. Van Nostrand Reinhold, New York

Kendall D N 1966 *Applied Infrared Spectroscopy*. Van Nostrand Reinhold, New York

Kirkbright G F, Sargent M 1974 *Atomic Absorption and Fluorescence Spectroscopy*. Academic Press, London

Kolthoff I M, Elving P J (eds.) 1964 *Treatise on Analytical Chemistry*, Vol. 5, Pt 1. Wiley, New York

Liebhafsky H A, Pfeiffer H G, Winslow E H, Zemany P D 1966 *X-Ray Absorption and Emission in Analytical Chemistry*. Wiley, New York

Lyndon-Bell R M, Harris R K, 1969 *Nuclear Magnetic Resonance Spectroscopy*. Nelson, London

Parker C A 1968 *Photoluminescence of Solutions*. Elsevier, New York

Tertian R, Claisse F 1982 *Principles of Quantitative X-Ray Fluorescence Analysis*. Heyden, London

Townes C H, Schawlow A L 1975 *Microwave Spectroscopy*. Dover, New York

Varna R, Hrubesh L W 1970 *Chemical Analysis by Microwave Rotational Spectroscopy*. Wiley, New York

Wendlandt W W, Hecht H G 1972 *Reflectance Spectroscopy*. Wiley, New York

Winefordner J D, Schulman S G, O'Haver T C 1972 *Luminescence Spectrometry in Analytical Chemistry*. Wiley, New York

G. F. Kirkbright[†]
[UMIST, Manchester, UK]

D. E. M. Spillane
[University of North London, London, UK]

T

Temperature Measurement

The concept of temperature is elusive, and the relationship of temperature to the fundamental dimensions of mass, length and time is much less obvious than for other common process measurements such as pressure, flow and level.

The concept of heat as a measurable quantity which flows from a hotter to a cooler body provided one of the first stepping stones to defining temperature scales, aided by Galileo's invention of the thermoscope in 1593 and the introduction of the first liquid-in-glass thermometers early in the seventeenth century. The concept of heat as a form of motion and temperature as a measure of the intensity of that motion became acceptable in the mid-nineteenth century when Lord Kelvin defined the absolute (or thermodynamic) scale of temperature in relation to an ideal reversible heat engine working between temperatures T_1 and T_2. If the heat provided by the source is Q_1 and that delivered to the sink is Q_2, these quantities are related by:

$$\varepsilon = \frac{Q_1 - Q_2}{Q_1} \tag{1}$$

where ε is the efficiency of the engine. These quantities are clearly functions of temperature, and Kelvin defined his scale by setting:

$$\varepsilon = \frac{Q_1 - Q_2}{Q_1} = \frac{T_1 - T_2}{T_1} \tag{2}$$

(note that T_1 is higher than T_2).

Numerical values can be obtained by defining a single fixed point. This is the triple point of water, which is fixed at 273.16 K. This particular figure was chosen so that the magnitude of the degree Celsius (°C) and the newly defined Kelvin (K) would be the same. The triple point of water is defined as that temperature at which the solid, liquid and vapor phases of water are in equilibrium and is 0.01 K above the melting point of ice originally used to define the Celsius scale. Thus:

$$t(°C) = T(K) - 273.15 \tag{3}$$

In principle then, an unknown temperature could be determined by measuring the efficiency of a reversible engine working between the unknown temperature and the triple point of water. This hypothetical frictionless engine would use a perfect gas as its working substance. This is incapable of practical realization, but it opens the way to using a gas thermometer containing a near perfect gas as a means of establishing a temperature scale. Gases of low atomic weight, such as hydrogen and helium at low pressures, are used in gas thermometers, since their behavior approximates closely to that of a perfect gas and their slight deviations from perfect gas behavior are well documented. Such thermometers can be used to determine temperatures on the thermodynamic scale. Some other techniques, such as noise thermometry and total radiation pyrometry, can also be traced directly to the thermodynamic scale and provide a useful cross-check on gas thermometry measurements.

These methods of determining temperature are extremely complex and cumbersome and are employed only at national standards laboratories. They are quite unsuitable for actual temperature measurement or even for the routine calibration of other types of thermometer. This is not only because of the size and complexity of the apparatus but also because, with every possible precaution, they are not capable of the precision required for industrial and scientific temperature measurement.

1. International Temperature Scale

Since the thermodynamic scale can neither be used directly nor transferred to other thermometers with sufficient repeatability, a practical temperature scale has been defined, based on the temperatures of well-defined changes of state of pure materials. Standards laboratories in various countries periodically made determinations of the thermodynamic temperatures at which these changes of state occur.

The present scale was adopted by the Comité International des Poids et Mesures (CIPM) in January 1990 and replaced the earlier scale which had been introduced in 1968.

These fixed points are highly reproducible (to a few thousandths of a degree in most cases) but may not correspond so closely to the same temperatures on the thermodynamic scale. For example, the scatter of the results obtained by different workers in assigning a value to the gold point in the 1960s was a substantial fraction of a degree; on the basis of these results, the figure was increased in 1968 by about 1.4 °C from that used on the 1948 scale, and has just been changed again by about a quarter of a degree.

A number of interpolating instruments are specified in ITS-90 to determined temperatures between the fixed points. From 0.65 K to 5 K helium vapor pressure thermometers are used; from temperatures in the

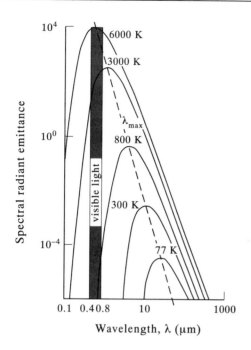

Figure 1
Blackbody radiator

range 3–5 K up to the triple point of neon at 24.6 K the constant volume gas thermometer is used; while from 14 K to the freezing point of silver (~962 °C) platinum resistance thermometers are used.

Above the silver point the scale is defined in terms of Planck's radiation equation:

$$N\lambda_b = C_1/\pi\,\Omega_0\lambda^5\,\{\exp(C_2\lambda^{-1}\,T^{-1}) - 1\} \qquad (4)$$

where $N\lambda_b$ is the spectral radiance of a blackbody, λ is the wavelength, C_1 and C_2 are constants, and Ω_0 is the unit solid angle. This relationship is shown in Fig. 1. The definition of the ITS includes a value for C_2, but C_1 is not specified, since the scale is defined by the ratio of the spectral radiances (both measured at the same wavelength) of a blackbody at temperature T and at the freezing point of silver, gold or copper.

ITS-90 defines, quite carefully, the characteristics required of the interpolating instruments and the interpolation equations and reference tables to be used with them. In the whole of the region from 14 K to the silver point, the interpolating resistance thermometer can also be used as a transfer standard to carry the scale from one laboratory to another. Above the silver point, however, the situation is different, since the types of radiation pyrometer used by the standards laboratories to maintain the scale are not readily transportable. Calibrations are transferred using tungsten strip lamps or blackbody lamps. The strip or tubular filaments of these lamps are heated by passing a stabilized direct current through them, and the process of calibration determines the relationship between the heating current and the apparent temperature of the strip determined by the pyrometer. Precious metal thermocouples can also be calibrated above the silver point by using secondary reference points, such as the freezing point of palladium or the freezing point of platinum.

In industrial practice any reference to a temperature unit (K, °C, °F, °R) is normally assumed to mean temperature as defined by the international temperature scale (ITS). In particular, expression of a temperature in Kelvin does not imply that the temperature is measured on the theoretical thermodynamic scale unless this is specifically indicated. Fortunately, for most practical process measurement and control applications the small discrepancies between the ITS and the thermodynamic scale, and between one version of ITS and the next, are insignificant.

2. Dissemination of the Temperature Scale

The ITS is maintained by certain national standards laboratories (US National Institute of Science and Technology, UK National Physical Laboratory, etc.). These laboratories calibrate suitable instruments supplied to them against the ITS and supply calibration certificates for them. These thermometers will generally be held by laboratories of the national calibration service in the country concerned, or by the laboratories of large user organizations, where they will be used to calibrate thermometric instruments for industry and commerce.

3. Types of Thermometer

3.1 Expansion Thermometer

The oldest form of thermometer, and that which is still the most common in everyday use, is the liquid-in-glass thermometer. While of little direct use in process control, these instruments cover a very wide range of temperatures and are readily available to measure temperatures as low as −80 °C or as high as 500 °C. Some versions are capable of very high accuracy and are therefore frequently used for the routine calibration of other thermometric instruments used in process measurement and control.

Various other types of expansion thermometer widely used in process applications include mercury-in-steel, vapor pressure and bimetallic devices. These are all capable of moving a diaphragm, a bellows or a shaft, which in turn may operate an electrical contact, a potentiometer, or a force balance system with either an electrical or a pneumatic output. Although these devices may be losing ground to electrical thermometers, they are still fitted in considerable numbers, and because they are simple and self-contained, often requiring no external power supply, they are frequently used in alarm and trip applications.

Vapor pressure thermometers are particularly useful for dealing with narrow ranges of temperature since the pressure of a vapor in contact with its liquid changes rapidly with temperature. In practice, a very rough approximation is that the vapor pressure will double for a 10 °C rise in temperature. It is important to note that the vapor pressure of the system will be that which corresponds to the temperature of the liquid–vapor interface. If the temperature of the measuring bulb should pass through the ambient temperature, the pressure-sensitive readout device will change from being full of gas to being full of liquid. The location of the interface may therefore change and the presence of a hydrostatic head of liquid may result in an error being introduced. Special designs are therefore required for instruments working around ambient temperature. The scale of a vapor pressure thermometer is highly nonlinear, and therefore a distinction is usually made between the total temperature range and the usable temperature range of a particular instrument.

3.2 Thermocouple Thermometers

Space precludes a detailed analysis of the theory of the thermoelectric effect, but here it may be sufficient to consider that if two metals are connected at one end, as shown in Fig. 2a, a potential difference may be measured at the open ends when a temperature difference exists between the points A and B. This potential difference results from the different work functions at the hot and cold junctions, and from the potential gradient which must accompany a temperature gradient in a conductor.

Almost any pair of dissimilar metals could be used to produce a thermocouple, but certain combinations have received a degree of international acceptance,

(a)

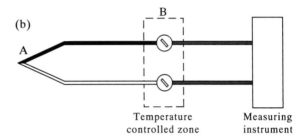

(b)

Figure 2
Thermocouple connections

and in normal industrial work it should very rarely be necessary to use thermocouples other than those listed in Table 1.

Most of the thermocouples listed in Table 1 have been in use for very many years, and emf temperature tables have been published by many national standards organizations for all of them. There were, however, slight discrepancies between the various tabulations, and in the early 1970s work was done at a number of national standards laboratories to resolve these differences. The resulting tabulations—the International Electrotechnical Commission Publication No. 584 (IEC 1977)—have been incorporated into

Table 1
Types of thermocouples most commonly used in industry

Type	Composition	Normal operating range (°C)	Maximum "spot" reading (°C)
E	Ni–Cr/Cu–Ni	−200–850	1100
J	Fe/Cu–Ni or Fe/Constantan	−200–850	1100
K	Ni–Cr/Ni–Al or Chromel/Alumel	−200–1100	1300
N	Ni–Cr–Si/Ni–Si or Nicrosil/Nisil	−200–1100	1300
T	Cu/Cu–Ni or Cu/Constantan	−250–400	400
B	Pt–30Rh/Pt–6Rh	0–1500	1700
R	Pt–13Rh/Pt	0–1400	1650
S	Pt–10Rh/Pt	0–1400	1650

national standards in many countries. They are in the process of being updated to take into account the publication of ITS-90.

The IEC tables include polynomial representations of the emf temperature relationships for use in computer data processing. These polynomials need to be used with caution; they contain up to 14 terms expressed to 11 significant figures, and cannot easily be truncated to produce an acceptable tabulation at a defined lower level of precision. Other, more convenient polynomials have been proposed which will provide, for particular temperature ranges, adequate representation of the tabulated figures for all industrial work (Coates 1978).

It is important, when using thermocouples, to remember that they always measure the difference between two temperatures since, in addition to the measuring junction, there is a second or reference junction (frequently called the "cold" junction), the temperature of which must be known or allowed for in determining the temperature of the measuring junction.

When the reference junction is maintained at 0 °C (by keeping it in an ice bath, for example), the measuring instrument should record the emf values tabulated in IEC 584 corresponding to the temperatures of the measuring junction. If, however, the reference junction is maintained at some other temperature, say 20 °C, the emf at the terminals of the measuring instrument will be lower than the tabulated value by an amount corresponding to the emf tabulated for 20 °C.

In industry, thermocouple wires are brought directly to the measuring instrument, and at its terminals—both of which are assumed to be at the same temperature —there are junctions between the thermocouple materials and the internal copper wiring of the instrument. The emf at the terminals will be that which corresponds to the temperature of the measuring junction less that which corresponds to the temperature of the terminals which form the reference junction.

In practice, deflection instruments may have bimetallic strips to move the zero of the pointer to compensate for ambient temperature changes, while electronic instruments may contain resistance thermometers or forward-biased semiconductor diodes in bridge circuits to provide an emf corresponding to the temperature of the terminals ("cold junction compensation").

In the arrangement shown in Fig. 2b, the change to copper wiring is made some distance away from the measuring instrument, and the reference junction is at this connection point. This arrangement is particularly useful in scanning and logging applications, since a large number of thermocouples may all have their reference junctions in the same temperature controlled zone, the wiring back to the scanning instrument being made entirely of copper. The output of the measuring

system has to take account of the known or measured temperature of the reference junctions.

In industrial practice one can seldom use a continuous length of wire made from one material all the way from the measuring point to the terminals of the indicator; extension or compensating leads are often required. Extension leads are of nominally the same material as the thermocouple leads, while compensating leads are made of alloys having a similar emf temperature relationship to that of the thermocouple, but only over a limited temperature range. Compensating leads are used mainly with thermocouples of precious metals: the high price of platinum alloys makes the use of these leads almost essential.

Each batch of wire is made to specified tolerances on its emf temperature relationship. The various joints inevitable in a long run of cable introduce errors into the total emf measured, and these must be evaluated.

The insulation of thermocouple wires and cables is usually color coded to permit the particular type of thermocouple in use to be identified. Unfortunately, the color codes have developed quite independently in various countries, and there are no points of similarity between those used, for example, in the UK, the USA and Germany.

The expected lifetime of a thermocouple in service depends on its type, the temperature at which it is used, the atmosphere around the thermocouple wires and the diameter of the wires. A long lifetime and high stability are features of precious-metal thermocouples provided that: (a) contamination from iron or its vapor is avoided at temperatures above 500 °C; and (b) conditions under which the oxide insulators might be reduced are not permitted. For base-metal thermocouples used within 100 °C or 200 °C of their upper temperature limit, a lifetime of a few thousand hours may be expected. With care, precious-metal thermocouples are capable of measurement accuracies of the order of ±1 °C at the gold point for very long periods; for base-metal thermocouples a typical tolerance is ±0.75% of the Celsius temperature or ±3 °C, whichever is greater. No significant improvement on these figures is likely to be possible under industrial conditions, because the thermal history of the thermocouple and temperature gradients in the cabling can influence the emf. If a calibration check is required it is very often best done *in situ*, using a different form of measuring instrument placed close to the thermocouple to be calibrated. Cycling a base-metal thermocouple above about 800 °C can cause calibration shifts of a few degrees Celsius.

The Nisil–Nicrosil thermocouple is similar to a nickel–Nichrome thermocouple, except that the two limbs contain a small proportion of silicon which oxidizes on the surface to give a self-passivating layer, largely eliminating drift through oxidation and inhomogeneity.

Thermocouples in swaged, mineral-insulated cables are generally less affected by inhomogeneity problems than are cables fabricated from wire and ceramic insulators.

3.3 Resistance Thermometers

Resistance thermometers are of two main types: thermistors, which are usually sintered mixtures of metal oxides with the characteristics of semiconductors, and units, which are based on the change in resistivity with temperature of pure metals or alloys. Thermistors are not yet in common use in industrial measurement and control, although they are widely used in laboratory work. The reasons for this are their highly nonlinear characteristics and the lack of standardization of their resistance–temperature relationships. However, they can have good stability; some suppliers can offer units which are interchangeable with others of their own manufacture to within a fraction of a degree. In some laboratory applications, the fact that thermistors can be made with extremely small dimensions is frequently useful, particularly in combination with their very high sensitivity over restricted ranges. (For more information on thermistors see Sachse (1975).)

While it might at first, appear that almost any metal could be used for resistance thermometry, only copper, nickel and platinum have found significant use.

The tendency of copper to oxidize limits its usefulness to below about 150 °C, while its very low resistivity means that very long lengths of fine wire are needed to produce a useful resistance; a copper thermometer is therefore normally rather bulky.

Nickel thermometers are frequently used in the USA. Nickel has the advantage of having a significantly higher resistivity than copper, and has a high temperature coefficient of resistance at room temperature. It can be used at higher temperatures than copper, although there are still problems with oxidation and corrosion, but its upper temperature limit is normally set by the peculiarity of its resistance–temperature curve in the neighborhood of the Curie point, at 385 °C, and by the instability of its resistivity when cycled through this temperature. Its wider acceptance is also hindered by the lack of international agreement on a resistance–temperature relationship for nickel thermometers.

Platinum is not, in most environments, subject to oxidation or corrosion; it is available as fine wire of extreme purity and its resistivity is higher than that of nickel. Platinum thermometers are capable of covering a wide range of temperatures: some industrial standards give resistance–temperature tabulations from −220 °C to +800 °C, while some devices have been used to even higher temperatures.

The industrial resistance thermometer element consists of a small glass or ceramic detector containing the platinum coil, which is then usually inserted into a metal sheath. More recently, detectors have been produced which have a film of platinum deposited onto a ceramic support taking the place of the wire.

There is now virtually worldwide agreement on the resistance–temperature relationship of industrial platinum resistance thermometers. This is defined in IEC 751, both by a tabulation and by a polynomial representation of that tabulation. This document specifies a resistance at the ice point of 100 Ω, and a resistance of 138.5 Ω at 100 °C. It also quotes two tolerance grades, classes A and B, roughly ±0.2% and ±0.5%, respectively, of the temperature in degrees Celsius. Special care is required in the use of resistance thermometers above about 500 °C because of the danger of contamination of the platinum by iron, or by metallic elements reduced from the glass or ceramic used in their construction. Provided suitable precautions are taken, the stability can be very good, amounting to changes of only a few hundredths of a degree Celsius after several years of use at 600 °C. The resistance of the thermometer element is usually measured with some form of bridge circuit, which must be designed to ignore the resistances of the connecting leads. Although the resistance–temperature relationship of platinum is slightly nonlinear, it can be represented adequately for industrial purposes by a simple quadratic equation over the greater part of its range, and by a simple quartic equation at low temperatures. A rough rule of thumb is that the terminal nonlinearity is about 0.4% per 100 °C span. Thus, a thermometer covering the range 0–100 °C would have a nonlinearity of 0.4% of 100 °C (or 0.4 °C); used over the span 0–300 °C it would have a nonlinearity of 1.2% of 300 °C (or 3.6 °C).

3.4 Radiation Thermometry (Pyrometry)

The temperature of a body can be determined by measuring the thermal radiation it emits. In the range of temperatures of general interest in industry, the wavelengths to be considered run from the far infrared down to the visible region, although work on plasma temperatures involves measurement in the ultraviolet region. All types of radiation pyrometer depend on Planck's radiation equation, but in different ways. A total-radiation pyrometer ideally measures radiation at all wavelengths emitted by the source. This corresponds to the area under the appropriate curve in Fig. 1, given by the well-known Stefan–Boltzmann law:

$$N_b = \frac{\sigma T^4}{\Omega_0 \pi} \qquad (5)$$

where N_b is the total radiance of a blackbody radiator at temperature $T(K)$, σ is the Stefan–Boltzmann constant and Ω_0 is the unit solid angle. However, since the detector of the pyrometer also radiates, the actual heat flux sensed by the instrument is proportional to $T_1^4 - T_2^4$ where T_1 is the source temperature and T_2 is the temperature of the detector.

In practice, such instruments tend to be somewhat slow in operation and to be limited in bandwidth by the material of the lens or window and by the

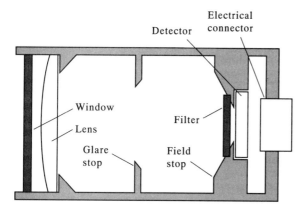

Figure 3
Typical industrial pyrometer

detector. The latter is usually a thin-film thermopile or bolometer and, because they use all of the available radiated power, these instruments can be used to measure comparatively low temperatures, down to −50 °C in some cases.

Narrow-band optical pyrometers use filters to limit the wavelengths passed to the detector to a narrow band, and hence respond according to Planck's radiation equation.

In two-color pyrometers the spectral radiances at two wavelengths are compared to identify the temperature. These devices are complex and not in common use in process control, and cannot be used if the emissivity of the target varies rapidly with wavelength.

The narrow-band pyrometer, particularly if operating in the infrared region, has a particular advantage when used to measure the temperature of optically transparent materials (such as plastics and glass), since it is possible to choose a wavelength in the infrared region at which the material has a high absorption, and is thus opaque.

Figure 3 shows a typical industrial pyrometer. The detector for a total-radiation pyrometer may be a thermopile or bolometer, or (if some method of chopping the radiation is available) a pyroelectric detector. In a narrow-band pyrometer the detector is an appropriate semiconductor photodetector, the type depending on the wavelengths of interest. The output of the detector is passed to a suitable electronic unit to amplify and linearize the signal.

A major problem in the use of pyrometers is the wide range of emissivities that may be exhibited by the bodies whose temperatures are to be measured. On a single material this may range from, say, 0.05 for polished aluminum to 0.4 for sand blasted aluminum, and from 0.07 for polished iron to 0.79 for rolled iron. It is sometimes possible to eliminate the problem by forming a blackbody in the surface whose temperature is to be measured, or by using a system of reflectors to form the equivalent of a blackbody around the part of the surface in question. If neither of these methods is

possible, the estimated emissivity may be used to make an appropriate adjustment to the gain of the instrument. A further problem lies in the possibility of some of the radiated energy being absorbed by, for example, fumes, water vapor or carbon dioxide in the atmosphere between the target and the pyrometer. Such absorption is minimized by choosing an appropriate wavelength, by moving the pyrometer closer to the target (in which case the pyrometer might have to be water cooled), or by using a fiber-optic light guide to conduct the radiation to the detector.

4. Use and Installation of Equipment

Modern temperature sensing instruments and their associated electronic readout devices are capable of very high accuracy and resolution. The ease with which such measurements can apparently now be made is frequently misleading since, at best, all they can indicate is the temperature of the sensor; this may differ significantly from the temperature of the body or fluid whose temperature it was intended to measure. This section concentrates on the most likely sources of error in temperature measurement, the methods of reducing them, and the considerations to be borne in mind when sensors are installed.

4.1 Conduction Errors (Cold End Effect)

If a thermometer is inserted into a body or fluid which is at a temperature higher than the ambient temperature, there will clearly be a heat flow along the thermometer stem from the sensing element to the connection head. Since some of this heat flux must pass through the sensing element, it follows that the element must be at a temperature slightly lower than that which it was intended to measure. Conversely, if the body or fluid is at a temperature much lower than the ambient temperature, the heat flow is in the other direction, and the detecting element will be at a temperature slightly higher than that which it should have measured. The magnitude of the effect depends on the heat transfer between the medium and the stem of the thermometer, and on the length and thermal conductivity of the thermometer stem. It is seldom necessary to carry out a very rigorous analysis, but a rough calculation of the heat transfer should always be made in order to establish whether the likely error is, say, only a fraction of a degree, or some tens of degrees. The worst cases are likely to be those in which the temperature of slowly moving gases is to be determined, since the heat transfer is then very poor. An example might be a thermometer screwed to the wall of a room which is intended to measure the air temperature; conduction to the wall along the stem of the thermometer and radiative transfer to other surfaces in the room may be expected to cause significant errors. Conduction errors can generally be minimized by employing sheaths which are as long as possible and constructed of thin material, and by lagging the connection head of the thermometer.

When, as is usual in process control, the thermometer is mounted in a pocket, or thermowell, the problem is often made much worse by the lack of good thermal contact between the thermometer and the pocket. Some improvement may be obtained by spring loading the thermometer into contact with at least the tip of the pocket, or by filling the gap between the thermometer and the pocket with a heat-conducting liquid or a grease.

4.2 Self-Heating

This particular source of error applies only to resistance thermometers (including thermistors), and is caused by the heating effect of the measuring current. A typical self-heating error for an industrial platinum resistance thermometer element is $0.03\,°\mathrm{C\,mW^{-1}}$ with the thermometer immersed in water or ice. In laboratory work with platinum resistance thermometers, 1 mA is a typical measuring current, while up to 5 mA would be normal for accurate industrial work.

In thermistors, measuring currents may have to be restricted to some tens of microamperes because of their higher resistance and generally smaller dimensions.

Some caution should be exercised when using published values for self-heating, since they have generally been obtained with the thermometer element immersed in water or in ice; much larger values are obtained from measurements in slowly moving gases.

4.3 Time Response

Clearly, any practical temperature measuring instrument will respond to a sudden change in temperature in a complex way, because of the different thermal resistances and thermal masses of which it is composed. However, it is normal in specifying the behavior of a thermometer to assume that it behaves as a single-lag system, and to take the time needed to achieve 63% of a step change as the time constant. The step change is usually produced by plunging the sensor into hot water moving at $1\,\mathrm{m\,s^{-1}}$. It should be noted that this value of the time constant applies only under these conditions, and that in air at the same velocity the time constant might be a hundred times larger. The use of a pocket or thermowell may also greatly increase the time constant, and it is not unusual to find that a thermometer with a time constant of a few seconds when plunged into water has this value increased to a few minutes when tested in a pocket.

4.4 Thermoelectric Potentials

The leads of a resistance thermometer or thermocouple are inevitably exposed to temperature gradients along their length, therefore any inhomogeneity in the connecting wires will produce stray thermoelectric potentials across the terminals.

For resistance thermometers, IEC 751 requires that this spurious emf determined in a particular test rig should be small compared with the interchangeability tolerance when the resistance is measured at about 1 mA. It is therefore not advisable to use currents much below 1 mA in dc resistance measurements on platinum resistance thermometers.

4.5 Total or Stagnation Temperatures

In a high-speed gas flow the temperature sensed, T_T, is higher than the normal static temperature, T_S. This is because, in addition to the random kinetic energy, which in effect defines the temperature of the gas, the gas molecules possess the additional kinetic energy of their common velocity. If the gas were brought adiabatically to rest on the surface of the sensor, the total temperature would be given by:

$$T_T = T_S\{1 + 1/2(\gamma - 1)\,(Ma)^2\} \qquad (6)$$

where γ is the ratio of specific heats for the gas and Ma is the Mach number.

For air at ambient temperature (25 °C) moving at Mach 0.1 (approximately $120\,\mathrm{km\,h^{-1}}$), this increase would amount to about 0.6 °C. An ordinary thermometer immersed in such a gas stream would give a reading somewhere between the static and the total temperature. If this error is likely to be serious, probes of special design must be used. This is naturally a particular consideration in aircraft instrumentation since, for example, at Concorde's maximum speed of Mach 2, the total temperature is 170 °C above the static temperature, which at the cruising altitude might be −60 °C.

4.6 Installation and Vibration

It is easy to forget that once the thermometer is installed in the body or is arranged to project into the fluid whose temperature is to be measured, it becomes part of the total structure and is subject to the same stresses as other parts of that structure. The most common reasons why temperature sensing elements fail are vibration and inadequate support of the element or its leads.

If the thermometer and its leads are rigidly attached to a solid body whose temperature is to be determined, they will be subject only to the vibration levels of the remainder of the structure. Most element manufacturers publish information on the vibration levels that their products will withstand, but it is often difficult to anticipate what levels of vibration and acceleration will be encountered in a particular installation.

A thermometer element projecting into a rapidly moving fluid is subject to different forces. Whether the thermometer is directly immersed or is inside a pocket or thermowell which is itself immersed in the fluid, similar considerations apply because the structure will typically be long and thin and cantilevered into the fluid from one end.

The first point (which is an obvious one, but which is occasionally overlooked) is that the sheath or pocket of the thermometer must be of sufficient strength to withstand both the pressure of the fluid and the sideways load of the hydraulic or aerodynamic drag forces exerted by the motion of the fluid.

If a thermometer is installed in pipework which is subject to vibration, the likely frequencies of vibration should be compared with the calculated resonant frequencies of the thermometer stem and/or that of the pocket, giving consideration to the effect of any mechanical resonance amplifying the vibration amplitude at the tip of the thermometer. Thermometer elements in pockets should always be spring loaded to prevent the tip from rattling in the pocket, which would result in very high impact accelerations and consequent early failure.

The commonest cause of vibration in flowing fluids, however, is vortex shedding from the sides of the probe. The vortex shedding frequency, given approximately by

$$f = \frac{V}{5d} \qquad (7)$$

where V is the fluid velocity and d is the diameter of the probe, should be calculated and compared with the calculated fundamental lateral resonant frequency of the probe. A potential danger can arise if the vortex frequency at the maximum flow rate is greater than the resonant frequency; this effect occurs most commonly with high-velocity steam lines.

The depth of immersion to be chosen for a thermometer projecting into a fluid, then, is a compromise involving an attempt to minimize the conduction error, while providing adequate strength and vibration resistance, without compromising response time or incurring unnecessary cost.

As well as these considerations, however, it is important to bear in mind the need to obtain a measurement which is representative of the average temperature of the fluid in the duct. The appropriate position for the probe can be determined only if there is some information available on the likely temperature distribution across the duct. If there is fully developed turbulent flow, and if the pipe is well lagged or if the fluid is at a temperature not too far from the ambient temperature, a single probe with its sensing portion projecting about one-sixth of the way into the pipe diameter may be adequate. In more difficult cases it may be necessary to make several measurements, or to survey the temperature distribution before deciding on the appropriate immersion depth.

In addition to choosing the materials and design of the sheath or thermowell in accordance with the physical constraints mentioned earlier, the chemical compatibility of these materials with the process fluid must also be considered. Suggested sheath materials for a wide variety of environments have been given by the American Society for Testing and Materials (1974).

4.7 Signal Conditioning

The various electrical thermometer elements are usually connected to an electronic circuit, in order to provide a standard output signal. In process control this signal is almost always a 4–20 mA standard transmission signal (although there is an increasing tendency to use digital transmission), and such instruments are usually known as temperature transmitters. These fall into two groups: field-mounted units, placed close to the point of measurement, and units in the control room.

A transmitter for use with a thermocouple will normally include provision for cold junction compensation and input–output isolation. The latter feature is necessary since the thermocouple may, in some applications, be connected to earth or left floating. Even if it floats it will have a low insulation resistance to earth at high temperatures. The low-drift amplification of the thermocouple voltage to the standard transmission signal may also incorporate linearization, although this is still the exception rather than the rule.

Transmitters for resistance thermometers are similar, except that isolation (though sometimes incorporated) is less necessary; the simple resistance–temperature relationship means that linearization is usually included. A similar situation applies to the transmitting electronics associated with radiation pyrometers, the manufacturers offering units with and without linearization.

Some, so-called "smart transmitters" containing microprocessors and significant memory have appeared recently. They are capable of accepting inputs from any of a wide range of resistance thermometers and thermocouples, and providing linearized analog of digital output signals.

4.8 System Considerations

In a large system the choice of sensing device will be influenced by considerations other than those of accuracy and robustness discussed earlier. In particular, the installation costs and the reliability required of the system will have to be taken into account. It must be remembered that the installation costs are usually much higher than the costs of the instruments, and may therefore be the deciding factor in the selection of sensing method.

Thus, while a thermocouple may be considerably cheaper than a resistance thermometer, the difference becomes far less significant when the cost of manufacture and installation of the thermowell or pocket, connection head and extension piece are taken into account. The thermocouple must be connected with the control room using expensive compensating or extension cable, while the resistance thermometer can be connected with copper wire. However, this difference can be virtually eliminated by using field-mounted temperature transmitters, since the connections back to the control room are copper in both cases. Unfortunately, field mounting the transmitters may result in the electronics being situated in an unfavorable environment, and in positions that may be inconvenient for servicing.

Further economies can be made by placing some sort of scanning device near a section of the plant so that signals from a number of thermometers may be selected in sequence and passed back along a single pair of conductors. This will generally be acceptable for data logging, but not for the primary control of safety signals from the plant, since a failure of the scanner would interrupt signals from all of the measuring points.

See also: Transducers, Thermal

Bibliography

American Society for Testing and Materials 1974 Manual On the Use of Thermocouples in Temperature Measurement. Special Technical Publication 470A. ASTM, Philadelphia, PA
Baker H D, Ryder E A, Baker N H 1961 *Temperature Measurement in Engineering*, Vols. 1 and 2. Wiley, New York
Barber C R 1964 *The Calibration of Thermometers*. Her Majesty's Stationery Office, London
British Standards Institution 1960 Specification for distant indicating thermometers for ships' refrigerated cargo spaces, BS 3273. BSI, London
British Standards Institution 1981 Specification for dimensions of temperature detecting elements and corresponding pockets, BS 2765. BSI, London
British Standards Institution 1981 Platinum–10% rhodium/platinum thermocouples: type S, BS 4937, Pt 1. BSI, London
British Standards Institution 1981 Platinum–13% rhodium/platinum thermocouples: type R, BS 4937, Pt 2. BSI, London
British Standards Institution 1981 Iron/copper–nickel thermocouples: type J, BS 4937, Pt 3. BSI, London
British Standards Institution 1981 Nickel–chromium/nickel – aluminium thermocouples: type K, BS 4937, Pt 4. BSI, London
British Standards Institution 1981 Copper/copper/nickel thermocouples: type T, BS 4937, Pt 5. BSI, London
British Standards Institution 1981 Nickel–chromium/copper/nickel thermocouples: type E, BS 4937, Pt 6. BSI, London
British Standards Institution 1981 Platinum–30% rhodium/platinum–6% rhodium thermocouples: type B, BS 4937, Pt 7. BSI, London
British Standards Institution 1986 Nickel–chromium-silicon/nickel-silicon thermocouples including composition: type N, BS 4937, Pt 8. BSI, London
British Standards Institution 1986 Code of Practice for instrumentation in process control systems: installation design and practice, BS 6739. BSI, London
British Standards Institution 1987 Colour code for twin

compensating cables for thermocouples, BS 1843. BSI, London
British Standards Institution 1988 Guide to selection and use of temperature/time indicators, BS 1041, Pt 7. BSI, London
British Standards Institution 1989 Industrial resistance thermometry, BS 1041, Pt 3. BSI, London
British Standards Institution 1989 Radiation pyrometers, BS 1041, Pt 5. BSI, London
British Standards Institution 1991 Specification for thermocouple tolerances, BS 4937, Pt 20. BSI, London
British Standards Institution 1984 Specification for industrial platinum resistance thermometer sensors, BS 1904. BSI, London
British Standards Institution 1992 Guide to selection and use of liquid-in-glass thermometers, Sect. 2.1, BS 1041. BSI, London
British Standards Institution 1992 Specification for dial-type expansion thermometers. BS 5235. BSI, London
British Standards Institution 1992 Thermocouples, BS 1041, Pt 4. BSI, London
Coates P B 1978 Functional approximation to the standard thermocouple reference tables. National Physical Laboratory Report QU 46. National Physical Laboratory, London
Heimann W, Mester U 1975 Non-contact determination of temperature by measuring the infrared radiation emitted from the surface of a target. *Inst. Phys. Conf. Ser.* **26**, 219
National Bureau of Standards 1986 Precision measurement and calibration—temperature. Special Publication 300, Vol. 2. US Department of Commerce, Washington, DC
International Electrotechnical Commission 1977 IEC Recommendation No. 584. IEC, Geneva
Liptak B G 1969 *Instrument Engineers Handbook*, Vol. 1. Chilton, Radnor, PA
Quinn T J 0000 *Temperature*. Academic Press, London
Sachse H B 1975 *Semiconducting Temperature Sensors and Their Applications*. Wiley, New York

J. S. Johnston
[Rosemount Ltd, Bognor Regis, UK]

Transducers: An Introduction

Humans are naturally inquisitive; we constantly seek to learn more about our existence by observing activities taking place around us which, when coupled with our own intellectual thought, result in creation of new knowledge. Using this knowledge we modify the nature and form of the world in which we live to suit our own intentions.

The key entities of our physical world that concern these pursuits relate to the two basic physical quantities—energy and mass. These entities are observed and controlled using another entity—information.

Information is conveyed from one place to another as variations, technically known as modulations, of either energy links or, far less common in automation, by the transportation of mass containing fixed sets of

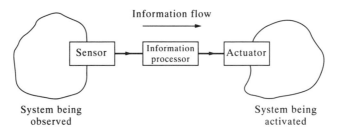

Figure 1
Role of sensor and actuator in a generalized instrument system

variations. Information does not, however, have the dimensions of mass or energy but nevertheless is always associated with them. The characteristics of information concern cognition, decisions and actions.

Information entities flowing in measurement and control systems are called signals (see *Signal Theory in Measurement and Instrumentation*).

Energy forms that can be employed in information systems include mechanical, electrical (including electronic), radiation (including thermal and optical), fluid flow and chemical types. Each has its special features that may make it suited to a given situation. A mixture of forms is often used to get from the original variable to the required output form.

1. The Transducer Device

A transducer, in the measurement and control usage of the term, is any device which converts an input signal of one energy form into an output signal of another form. The logically expected counterpart, the converter of mass, has little application because mass can rarely be transformed into other forms.

Two distinct types of transducer exist. An example of the first, the sensor, is the microphone that converts vibrations caused by an impinging sound wave into an electrical energy signal that represents the information contained in the vibrations. These create information

about existing energy activities.

The second kind is the actuator, of which an example is the loudspeaker device that converts given information into electrical energy variations that produce the sound waves we hear. These control energy or mass using given information.

A third, related, element is the transformer or modifier. This has its input and output in the same energy form. These could find application in either a sensing or actuation role. Examples are the electrical transformer or a mechanical lever.

The role played by a sensor and an actuator in a generalized situation is shown in Fig. 1 where a block diagram is given that represents an instrument system. The sensor is found as the interface between the system being observed and the information processor —called the front end. The actuator is found at the interface where the processed information is used to control the system being actuated—called the rear end.

Sensors gather suitable information about actions taking place. Actuators make use of information to bring about actions.

A simple example of the use of both is shown in Fig. 2 where a proprietary thick-film electrical resistance sensor, the flat square device, is mounted in the top part of the heater casing. Its role is to sense the temperature of the casing in order to control its

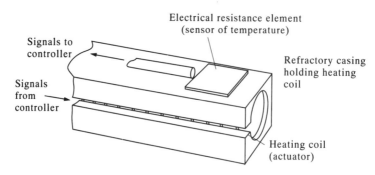

Figure 2
Thick-film type of electrical resistance temperature sensor placed on a refractory casing holding electrical resistance wire heating coils

temperature. Thermal energy is supplied in a controlled manner as information fed to the actuator heating the casing—in this case, the electrical resistance heating coils shown partially exposed at the side of the assembly.

This account now concentrates on sensors because they need more attention than actuators in systems and involve more features needing consideration.

2. Signals and Noise

It is the fluctuations of energy existing in a subject of interest that are detected by a sensor and thus used by the system signal processor to actuate the output.

Regardless of the energy form involved, unwanted energy fluctuations arise that do not represent meaningful information. These are known as noise. Because noise can have the same character as the desired signals they may be indistinguishable from those signals. The presence of noise means there always exists some level of error which leads to uncertainty about the truth of the information provided by the signal. A well-designed sensing system has errors small enough to be insignificant.

Noise produces less problems in the actuator stage, for by the time information has been processed to drive the actuator it has signal levels that are clean and far less influenced by noise.

3. Sensors

A sensor has the prime task of selecting specific information about an observable attribute of a subject. This is called the measurand. For example, the temperature sensor shown in Fig. 2 is placed on the casing in order to provide sufficiently error-free information about the state of the temperature of the casing.

In conceptual terms the sensor provides a transformation from a real physical world variable into an equivalent variable that is more convenient to use. The requirement is often best met by transducing the energy form originally present to another form with a known conversion factor. Current technology favors use of the electrical/electronic signal form wherever possible: however, optical techniques are now widely used.

As an example of a multistage transducer system consider a weighbridge used to obtain knowledge of the mass of a truck with its load. Figure 3a shows a diagrammatic organic layout of the physical parts of a typical weighbridge based on use of the electrical output load cell. This can be represented as a single-block diagram (Fig. 3b) or as a more detailed block diagram (Fig. 3c) that shows the various energy forms occurring in the successive stages that convert the input mass into the required electrical output.

In operation the mass of the truck induces a small

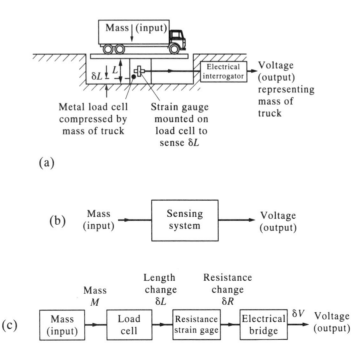

(a)

(b)

(c)

Figure 3
Weighing the mass of a truck using transducers: (a) physical layout of a typical weighbridge; (b) block diagram representation of the whole system; (c) breakdown of the various transducer stages

compressive length change, called strain, in the metal load cell on which the full mass of the truck rests. This strain is converted into an equivalent electrical signal using (several different methods could be used) special electrical resistance strain gauges that are interrogated by placing them in a suitable electrical circuit. It can be seen (Fig. 3c) that the energy form is sensed first as mechanical mass variation that is transduced into equivalent mechanical strain, then transduced into an electrical resistance change to be finally modified into an electrical voltage that can be recorded or displayed. In an automatic truck-filling situation the output signal would be used to control the flow load into the truck.

4. Characteristics of Sensors

Several important characteristics need to be under-stood about a sensor. First, it must provide correct transduction over long time periods. If not, then measurements made will be in error and incorrect records or controlling actions will take place. With a few exceptions all sensing systems need periodic calibration using special equipment to set the trans-duction factor to the correct, agreed value (see *Measurement: System of Scales and Units*).

The sensing process requires there to be a flow of energy from the observed subject to the sensor. When connected the sensor must, therefore, not significantly alter the value of the parameter it is measuring by significantly changing the energy flows taking place in the system under observation. This source of error is called loading. In sensing situations where loading influence is particularly exacting the energy flow at the time of each measurement is balanced to a zero state by introducing a counterflow, the balancing flow needed then representing the measured value.

In the early days of sensing technology development the physical transducer effects that could be detected above background noise were, by today's standards, relatively large energy level signals. As technology has improved so has the desire to sense increasingly smaller energy effects. An important characteristic of a sensor is its signal detection level. Much of modern sensing technology is concerned with system operation at very low energy levels. This has involved develop-ment of sophisticated signal processing methods that allow recovery of signals buried in a noise background that is often thousands of times larger than the signal.

5. Smart Sensors and Sensor Validation

Rising demands being placed on sensors have encour-aged development of additional features of a sensing system that ensure they carry out their task to expected standards over long periods of time. The performance of a simple sensor will usually be less than adequate for exacting applications. A class of

sensor that copes with many kinds of system defects has emerged. It has become known as the "smart" or intelligent sensor.

Operation of plant or activities that are life threa-tening or able to incur significant commercial losses requires the sensor data to be validated. Legal forces, on one hand, require the plant proprietor be able to prove due care has been taken in its operation. Commercial reasons require information about the degree of degradation of the signal so that trade-offs in settings can be made to maintain best overall plant operation. In both cases sensors are now recognized as key systems elements that need to be validated.

In supplying validation the operation of the sensor system requires further measurements and interpreta-tion be made of the ongoing performance of the sensor.

6. Application of Sensors

The earliest uses of transducers of the sensor kind were in scientific studies. Most sensors used in every-day application can be traced to a sequence of scientific research that took place for many decades beforehand. Commencement of the twentieth century saw the start of rapid application of sensors to the control of industrial processes and services that are too numerous to list. Among the first commercial uses was the application of the resistance thermometer (around 1990) to furnace temperature monitoring.

Today there are over 200 000 different proprietary sensors offered for sale that cater for just 100 different measurands found in process control. The variety arises for commercial reasons and because the various applications require widely different packaging. Those used in industrial and defense applications, for inst-ance, need to work reliably in harsh environments and thus require protective cases costing far more than the sensing parts they contain.

Today, sophisticated industrial processes rely on many sensors that monitor the process in order to provide output for operators and for sending signals to actuators that control its output. Testing of the first large commercial aircraft used around 100 measured points; current generations are developing systems with 15 000 measured points, many of which have response times of millisecond order. A modern alum-ina manufacturing plant will include 2000 or more controlled loops. Monitoring a large oil refinery for correct operation and maintenance reasons will in-volve measuring some 16 000 points each few minutes. Sensor uses in domestic and automotive applications are expanding.

7. Comparison of Actuators with Sensors

Actuators use the same physical principles as sensors but they have quite different requirement specifications. In contrast to sensors the operation of actuators, by virtue of their task of controlling available energy, involves substantial energy levels. A loudspeaker power output can range from milliwatts to hundreds of watts. A machine tool drive can involve thousands of watts. Power station equipment controls millions of watts.

The efficiency of energy use in actuators is important, not only because of the cost of wasted energy, but also because gross energy losses for inefficient applications can affect the overall performance of the system to a marked degree.

Often actuators are not required to control energy flows accurately because in their use in closed-loop operation the accuracy is obtained from sensors that assist correction.

These differences are seen by considering two different applications of the same concept—a lead screw component. When used in a precise inspection machine to measure the position of a measuring probe a lead screw must have a thread profile that has minimal backlash and it must be accurately cut. Energy efficiency in operation is not a key factor as the relatively low rate of use does not waste significant gross energy—a few watts. In contrast, the lead screw used to actuate (lift) a platform to a controlled height, as used to provision large aircraft, does not need precision but must be energy efficient to reduce overheating during its rapid lift. Losses of tens of kilowatts would overheat the lead screws jamming them.

In measurement and control systems the actuator is an important component but its design is less complex than that of sensors. The important parameters of actuator design are energy efficiency and ultimate breaking strength compared with the parameters of sensor design which need primarily to address fidelity of transduction and operation at low signal levels which means operation well below ultimate levels of strength.

See also: Instrument Elements: Models and Characteristics

Bibliography

Bentley J P 1983 *Principles of Measurement Systems.* Longman, London
Gopel W, Hesse J, Zemel J N 1991 *Sensors—A Comprehensive Survey.* VCH, Cambridge
Grattan K T V 1992 *Sensors: Technology: Systems and Applications.* Hilger, Bristol, UK
Jones B E 1987 *Current Advances in Sensors.* Hilger, Bristol, UK
Jones R V 1988 *Instrument and Experiences.* Wiley, Chichester, UK
Middelhoek S, Van der Spiegel J 1987 *Sensors and Actuators.* Elsevier, Lausanne, Switzerland
Neubert H K P 1975 *Instrument Transducers.* Oxford University Press, Oxford
Noltingk B E 1988 *Instrumentation Reference Book.* Butterworth-Heinemann, Oxford
Sydenham P H 1979 *Measuring Instruments: Tools of Knowledge and Control, History of Sensing.* Peregrinus, Stevenage, UK
Sydenham P H 1984 *Transducers for Measurement and Control.* Hilger, Bristol, UK
Sydenham P H, Hancock N H, Thorn R 1992 *Introduction to Measurement Science and Engineering.* Wiley, Chichester, UK
Sydenham P H, Thorn R 1982, 1983, 1992 *Handbook of Measurement Science*, Vols 1, 2, 3. Wiley, Chichester, UK

P. H. Sydenham
[University of South Australia,
Ingle Farm, Australia]

Transducers, Capacitive

A capacitor enables electrical energy to be stored when an electric field is applied to it. A simple capacitor is made from two conducting plates separated by an air gap or by a dielectric material. The amount of energy which can be stored is determined by the area of the plates, the gap between them and the dielectric constant of the material between the plates. Displacement of one of the plates or displacement of the material between the plates will cause a change in capacitance, which can be used to measure the displacement. Thus, capacitive devices are used both as displacement transducers and also as transductive elements in transducers used for measuring other quantities such as acceleration and pressure.

Proximity capacitive transducers operate by applying a voltage to a conducting plate which can then form a capacitor with any object which is connected mechanically to the ground; this capacitance varies with the distance between the plate and the object.

Because there is no contact between the moving and stationary parts in capacitive transducers, there is no friction or wear, therefore they have very long lifetimes—this has been quoted as 200 years!

The high-output impedances of capacitors which are appropriate for use in capacitive transducers make them very susceptible to noise; this also means that very high impedance instruments have to be used to measure the capacitance.

1. Signal Processing

The capacitor may be placed in series with a power supply and a resistor; when the plates are stationary, there is no current flow in the circuit and the voltage across the resistor is constant. A change in capacitance caused either by displacing one of the plates, or by moving the dielectric material between the plates, will produce a voltage across the resistor which is related to the amount of displacement. The values of the capacitance and the resistance can be chosen to give a reasonably linear relationship between the displacement and the output signal. This form of signal processing does not give a steady-state response.

A three-terminal capacitor can be used with a bridge circuit to give better linearity, as well as a steady-state response; this is effectively a differential system which measures the difference between the two capacitors. Such a system avoids problems caused by variations in environmental parameters, since both capacitors are subject to the same environment.

2. Dynamic Response

The dynamic response of capacitive transducers depends on their mechanical construction. One plate has to move relative to the other, in response to a change in the measurand; the rate at which this can occur is clearly determined by the weight of the moving plate and also the way in which it is mounted.

3. Linear Displacement Transducers

Displacement transducers may operate either by displacement of one of the plates or by displacement of the dielectric material between the plates.

A cylindrical capacitor where one plate moves inside the other may be used. One commercially available device which uses this arrangement has an additional fixed plate so that there is both a fixed and a variable capacitor giving a differential capacitor. The range of this transducer may be from 0–10 mm to 0–50 mm, with accuracies from ±0.001–±0.03 mm, with resolutions of up to 1 ppm. Another simpler commercial device has ranges from 0–25 mm to 0–150 mm. The accuracy of this transducer is limited by its linearity to 0.5% full scale output (FSO). Capacitive displacement transducers are produced which can measure displacements as small as 10^{-11} m.

4. Angular Displacement Transducers

Angular displacement transducers consist of a static and a rotating plate; as one plate rotates, the overlapping area changes so that the capacitance changes. In some cases, two static plates may be employed, giving a differential capacitor.

The angular range may be small (e.g., 0–10°), or it may have values approaching 360°. The accuracy is primarily limited by the linearity; quoted linearities vary from ±1% FSO to ±0.01% FSO.

5. Proximity Sensors

When a voltage is applied to a conductor, charges are generated on the plate, and an electric field is set up between the plate and the ground. When an object which is connected to the ground approaches the plate, it becomes polarized due to the electric field, thus effectively changing the capacitance of the system. The distance at which a target can be clearly detected depends on the material from which the target is made; conductors and dielectrics can be detected, but grounded metals give the largest detecting distance.

Generally, capacitive proximity devices incorporate switches which come into operation when the presence of the target is detected. The threshold at which the switch comes into operation can be altered to the required value.

6. Other Sensors

Differential pressure transducers are formed by using two capacitors which are made from three plates. One of the plates is a flexible diaphragm which separates regions of varying pressure. A bridge circuit gives a signal which varies as the pressure difference changes. One manufacturer produces pressure transducers with ranges from 0–300 kN m^{-2} to 0–8000 kN m^{-2}, with a linearity of ±1% FSO and an operating frequency range of 0–2.5 kHz.

Capacitive level sensors may be produced by placing two conductors in the liquid; a change in level will change the capacitive coupling of the plates.

Accelerometers are produced using capacitive elements where one of the plates acts as the seismic mass. Inclination can be measured by using a pendulum as one of the plates in a differential capacitor.

7. Future Developments

Developments in microelectronics are likely to lead to improved accuracy by incorporating corrections for nonlinearity. Microfabrication techniques will enable small capacitive sensors to be used as the transduction elements, which should greatly increase the dynamic range of these sensors.

See also: Bridges; Instrument Elements: Models and Characteristics; Resistance, Capacitance and Inductance Measurement; Transducers: An Introduction

Bibliography

Doebelin E O 1983 *Measurement Systems*. McGraw-Hill, New York
Mansfield P H 1973 *Electrical Transducers for Industrial Measurement*. Butterworth, London

Morris A S 1988 *Principles of Measurement and Instrumentation*. Prentice-Hall, New York

C. Wykes
[University of Nottingham, Nottingham, UK]

Transducers, Electrodynamic

Electrodynamic, or moving coil, transducers change a mechanical stimulus to an electrical signal. The basis of their operation is Maxwell's equations of electromagnetism and, as the name indicates, they provide an output signal for dynamic measurements, but give no output for static measurands. The basic equations of their operation and response will be discussed with regard to two particular applications for a force sensor and a velocity–vibration sensor. The dynamics of the sensor will also be discussed in order to determine the limitations and range of application of the device.

1. Operating Principle

The principle of the electrodynamic sensor can be summarized by Faraday's law of induction, which states that the induced electromotive force e across a conductor is related to the time-varying magnetic flux ϕ_M by

$$e = -\frac{d\phi_M}{dt} \qquad (1)$$

where the magnetic flux is given by the product of the magnetic flux density B and the area enclosed by the conductor. For a wire moving with a velocity v in a constant magnetic field this gives

$$e = -Blv \qquad (2)$$

where l is the length of conductor in the magnetic field. The other pertinent electrodynamic process is the force f, generated on a conductor of length l, in a constant magnetic field B, carrying a constant current i, which is given by

$$f = Bli \qquad (3)$$

These equations constitute the transfer characteristics of the ideal electrodynamic transducer.

2. Performance

The electrodynamic transducer, as the name suggests, gives an electrical output when subjected to a dynamic mechanical input. Static or quasistatic inputs produce no output. The dynamic response of the moving coil electrodynamic transducer will be discussed as it is one

of the most common. It is shown schematically in its ideal form in Fig. 1.

In studying the performance of the device it is important to remember that the transduction mechanism can operate in reverse. That is, an electrical input will give rise to a mechanical output. The complete transfer characteristics of the transducer then gives

$$f = Bli + Z_m v \qquad (4)$$

and

$$e = -Blv + Z_e i \qquad (5)$$

where Z_m and Z_e are the mechanical and electrical transducer impedances, respectively.

Equations (4) and (5) characterize the transducer completely, and may be written according to the particular transduction process. It is convenient to write them, separating the mechanical and electrical terms, as

$$F = \left(Bl + \frac{Z_m Z_e}{Bl}\right) i - \left(\frac{Z_m}{Bl}\right) e \qquad (6)$$

$$v = \left(\frac{Z_e}{Bl}\right) i - \left(\frac{1}{Bl}\right) e \qquad (7)$$

It is the detailed study of these equations that determines the response of the transducer.

2.1 Input Impedance

The mechanical input impedance $Z_m(\text{IN})$ is found by solving the transfer characteristics with e_o, the electrical excitation voltage equal to zero, that is, with an output short circuit. This gives

$$Z_m(\text{IN}) = \frac{f}{v} = Z_m + Z_{mo} + \frac{(Bl)^2}{(Z_e + Z_{eo})} \qquad (8)$$

where Z_{mo} and Z_{eo} represent the terminal mechanical and electrical impedances of the transducer,

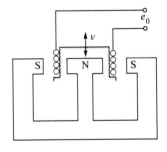

Figure 1
An ideal electrodynamic transducer

respectively. The terminal mechanical impedance can normally be neglected, and the electrical terminal resistor will normally be the indicator resistive load R_0. The electrical impedance of the transducer can be written

$$Z_e = R + j\omega L \qquad (9)$$

where ω is the circular frequency of the excitation, R is the resistance and L is the inductance of the moving coil. The mechanical impedance may be written in general terms as

$$Z_m = c + j\left(\omega m - \frac{k}{\omega}\right) \qquad (10)$$

where c is the mechanical damping factor, m is the mass of the dynamic element, k is the mechanical stiffness, and ω is the circular excitation frequency.

Substitution of these expressions into Eqn. (8), and assuming $R + R_0 \gg \omega L$ (i.e., the coil inductance is small) gives

$$Z_m(\text{IN}) = c + \frac{(Bl)^2}{(R + R_0)} + j$$

$$\left\{ \frac{k}{\omega}\left[\left(\frac{\omega}{\omega_0}\right)^2 - 1\right] - \omega L \left(\frac{Bl}{(R + R_0)}\right)^2 \right\} \qquad (11)$$

where $\omega_0^2 = k/m$ is the natural circular mechanical frequency. The electrical damping (the second term) often dominates.

2.2 Response of an Electrodynamic Force Sensor

If the input to the electrodynamic transducer, of the form shown in Fig. 2, is a force f_0 (or pressure $p_0 = f_0/A$, where A is the active area of the transducer), the transfer function in terms of the output current i is found by solving the transfer characteristics with e_o, the electrical excitation voltage, equal to zero. This transfer function then constitutes the frequency response of the sensor and is given by

$$\frac{i}{f_0} = \frac{Bl}{(Bl)^2 + (Z_m + Z_{mo})(Z_e + Z_{eo})} \qquad (12)$$

Substitution of the expressions for the impedances (assuming $\omega L \ll R + R_0$) then gives

$$\frac{i}{f_o} = \frac{Bl}{\dfrac{(Bl)^2}{R + R_o} + \left[c + j\left(\omega m - \dfrac{k}{\omega}\right)\right]} \qquad (13)$$

Analysis of Eqn. (13) can then be used to examine the

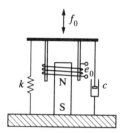

Figure 2
An electrodynamic force transducer: f_0 is the applied force, e_o is the output signal, k is the mechanical coupling constant and c is the mechanical damping constant

frequency response of the transducer, and specific regimes may be noted.

(a) At low frequencies the term k/ω in the denominator dominates. The output signal rises proportionally with frequency.

(b) Close to the resonant condition ($\omega m = k/\omega$) the response is determined by the damping term $[c + (Bl)^2/(R + R_0)]$. This is the most useful range of the electrodynamic force transducer and the actual shape of the response in this region can be made quite flat through suitable choice of damping via R_0.

(c) At high frequencies the inertia term (ωm) dominates and response falls off with increasing frequency.

(d) At higher frequencies still, where the inductive impedance becomes comparable with the resistive impedance, there is a further drop off in response.

From this it is clear that care must be taken in the design of a force transducer in order to achieve the required response in the required frequency range.

2.3 Response of an Electrodynamic Velocity Sensor

The electrodynamic vibration transducer is a velocity sensor. That is, an input velocity v_0 is applied to the transducer housing, within the operating frequency range, and an output signal is expected in proportion to the velocity. The system is as shown schematically in Fig. 3. From Fig. 3 it is seen that the velocity seen by the transduction process is the difference between the input velocity v_0 and the velocity of the seismic mass, v_m. Thus, the mechanical input conditions are very different from those of the electrodynamic force sensor and the system response is expected to differ.

The transfer Eqns. (6) and (7) are still applicable and the transfer function is (neglecting mechanical impedance) given by

$$\frac{i}{v_t} = \frac{Bl}{(Z_e + Z_{eo})} \qquad (14)$$

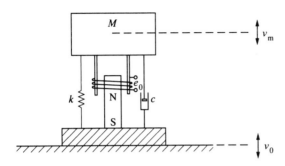

Figure 3
An electrodynamic velocity/vibration transducer: v_0 is the input velocity, e_o is the output signal, M is the seismic mass, k is the coupling constant to M and c is the mechanical damping constant

where

$$v_t = v_0 - v_m \tag{15}$$

Consideration of the forces invoked also gives

$$v_t \left(c - j \frac{k}{\omega} \right) + f_t = j\omega m v_m \tag{16}$$

where f_t is the reaction force generated by the electrical side of the transducer. Equation (16) can be written

$$v_0 = \left[1 - \left(\frac{\omega_0}{\omega} \right)^2 + \frac{c + f_t/v_t}{j\omega m} \right] v_m \tag{17}$$

From initial considerations, the ratio f_t/v_t is given by

$$\frac{f_t}{v_t} = \frac{(Bl)^2}{(Z_e + Z_{eo})} \tag{18}$$

Combining all of these gives a transfer function, which incorporates the mechanical impedance, of

$$\frac{i}{v_0} = \frac{Bl}{\dfrac{(Bl)^2}{j\omega m} + \left[1 - \left(\dfrac{\omega_0}{\omega} \right)^2 + \dfrac{c}{j\omega m} \right] (Z_e + Z_{eo})} \tag{19}$$

which, assuming R_0, the indicator resistance is much greater than the coil resistance and the inductive impedance (that is, $(Z_e + Z_{eo}) = R_0 + j\omega L$ and $R_0 \gg \omega L$) simplifies to

$$\frac{i}{v_0} = \frac{Bl}{\dfrac{(Bl)^2 + cR_0}{j\omega m} + \left[1 - \left(\dfrac{\omega_0}{\omega} \right)^2 \right] R_0} \tag{20}$$

Examination of the frequency response of the transfer function of Eqn. (20) shows a very different pattern from that of the electrodynamic force sensor.

(a) At low frequencies, the term $(\omega_0/\omega)^2$ dominates and the response increases as the square of frequency.

(b) Above resonance the damping term comes into effect and the response increases in proportion to frequency.

(c) Above the frequency $[cR_0 + (Bl)^2]/mR_0$ the response is constant and the transducer operates as an ideal velocity sensor.

(d) At higher frequencies where $\omega = R_0/L$, the response falls off.

In practical systems the presence of mechanical resonances may mask the transitions between the regimes.

3. Limitations

Other sources of damping and resonances in the system can cause serious detriment to the performance of an electrodynamic velocity sensor, and the potential sources of these must be considered and accommodated into the design process. These are briefly discussed.

3.1 Eddy Current Damping
Eddy current damping can be used to increase the damping in the transducer and is introduced by winding the transducer coil on a conductive metal former. Inductive coupling between the coil and the former can be modelled as an inductor, where the value of inductance is a function of the signal frequency. This is observed in transducers as a roll off of response at higher frequencies.

3.2 Secondary Resonances
Secondary resonances arise due to other resonance conditions in the mechanical system, and these tend to limit the upper frequency of vibration transducers. Possible origins of these resonances are as follows.

(a) The mass of the spring coupling the transducer and seismic mass—the coupling must then be considered as via a system which has compliance and mass distributed over its length.

(b) Secondary masses—a second mass attached to the main seismic mass by a second elastic medium which requires the coupling between the combined system to be considered.

(c) Contact resonance—the coupling of the vibration to the transducer housing may be via an elastic medium.

The details of these resonances are mathematically involved, and further information is available in more specialized texts (e.g., Neubert 1975).

See also: Analog Signal Conditioning and Processing; Instrument Elements: Models and Characteristics; Transducers: An Introduction

Bibliography

Neubert H K P 1975 *Instrument Transducers, An Introduction to Their Performance and Design.* Clarendon, Oxford

K. Weir
[City University, London, UK]

Transducers, Fiber-Optic

There is a continuing need for the development of new and improved instrumentation to enable accurate and reliable measurements to be carried out in an increasing number of sectors of industry. Fiber optics has offered such a promise for transducers for a number of years, although the appearance of actual transducers in the marketplace is more limited than would have been expected from the initial optimism generated by workers in the field in the early 1980s.

This field has been driven by the convergence of the technology developed for optical fiber communications in the 1970s and the increasing market for optically based consumer products in the 1980s, for example, the optical memory for computer systems, the laser printer and personal computer. In fiber-optic sensors the freedom from interference by radio-frequency fields and the absence of electrical currents flowing at the sensor head can provide benefits for their use in medical and related technologies. In addition, the inert nature of the transducer material means that there is intrinsic safety in biologically and chemically hazardous areas. The spillage of light due to fiber-optic cable fracture will set an upper limit on the optical power to be delivered in any application; such details are now well documented. The material itself is relatively inexpensive and the development costs of much of the technology have been met by the other industries using optically-related products. Thus, inexpensive fiber (developed for the communications industry) can be coupled to cheap laser diodes and detectors (developed for the compact disk player) in a fiber-optic sensor, such as a temperature transducer (Zhang *et al.* 1992). In addition, essentially unique sensing opportunities are afforded by fiber optics, for example, distributed sensing of temperature in a working environment in a single loop has been demonstrated, and is available as a commercial system (York Technology 1986). Transducers have been reported to sense, for example, temperature, pressure, fluid level, flow, position, vibration, chemical composition, current, voltage and rotation.

1. History and Background

Early work on the development of optical fiber sensors concentrated on the conversion to fiber use of a number of techniques which were familiar from conventional optical sensing. As an example, the infrared pyrometer, a device for the measurement of temperature in the higher region (500 °C and above) was extended using fiber optics to transmit the emitted radiation more conveniently (Dakin and Kahn 1977). At this early stage, work in the field generally fell into two separate and distinct categories. Simple devices were developed into commercial products, often by newly created small firms led by staff from major laboratories with relevant expertise, to tackle specialized markets, for example, the medical instrumentation field, or the monitoring of temperature using luminescence (Wickersheim and Alves 1982). The more complex areas of technology were pursued by firms with aerospace or defense interests, for example, the work of the Naval Research Laboratories (NRL) on hydrophones (Gaillorenzi *et al.* 1982) or fiber-optic gyroscopes (Udd 1986). The initial penetration of these devices on the market has been slower than had been expected, often due to their high component costs. However, the situation has been rectified by developments in the technology as shown in Table 1 (Udd 1991). The overall effect, combined with increasing reliability, is the beginning of market penetration. Perhaps the most dramatic of these is the cost of a typical laser diode, now a few dollars per item compared with one thousand times that figure in the early 1980s. In addition, complex components which were previously only laboratory devices, have now become commercially available, enhancing their applicability in fiber-optic sensors. As Table 1 illustrates, there is still some way to go before the devices are really competitive with other technology, for example, the sub-US$1000 target for the fiber-optic gyro will open up its use in direction location, as has been seen in the private marine field, and allow it to hold a mass market position.

2. Classification of Optical Fibers and Sensors

There are a number of potential classification schemes for optical fibers and the sensors based on them. One of the simplest is the classification of sensors in terms of those devices where the fiber itself is the sensing medium, the so-called intrinsic sensor, or those where the fiber merely acts as a medium to carry light to the transducer element, the extrinsic sensor. In the former case, the transmission characteristics of the fiber are modified by the action of an external effect, while in the latter the interaction ideally occurs only in the transducer element. Extrinsic sensors tend to be simpler and require less complex signal processing. Alternative classification schemes may be viewed in terms of modulation technique (Medlock 1986), ap-

Table 1
Critical components for fiber-optic sensors over the period 1980–2000

Year	1980	1990	2000
Laser diodes	US$3000 each (prototypes)	US$3 each (compact disk players)	
Single-mode fiber	US$5–10 m^{-1} (limited availability)	US$0.10 m^{-1} (standard telecom)	
Integrated optic modulators	laboratory devices	US$7000 each (prototypes)	US$50 each (fiber-optic gyros)
Fiber-optic gyros	laboratory devices	US$20 000 each (prototypes)	US$500–1000 each (low-cost navigation)

plication (Rogers 1986) or of a more fundamental nature, as introduced by Ning *et al.* (1991). In this article the simpler classification of intrinsic and extrinsic is used.

Figure 1 is a simple illustration of extrinsic fiber-optic sensors, together with details of the physical effect which can be employed, and the parameters which may be sensed using such an approach. Figure 2 shows intrinsic fiber-optic sensor devices, including the important subclassification of interferometric fiber-optic sensors. These are devices based on the well-known effect of optical interferometry; it can be seen that there is a wide variety of sensor devices which can be configured by using one basic class of technique. From Figs. 1 and 2, it can be seen that almost any effect can be transduced using either or both (and

sometimes many) of the effects discussed. A specific example is the sensing of temperature where many varied effects could be utilized. The skill in design is the utilization of the most appropriate effect, with efficiency and cost saving in mind, and the inclusion of compensation for such a temperature effect in other fiber-optic sensors (e.g., pressure). This is one of the particular skills of the optical fiber sensor designer.

Optical fibers conduct light by the process of total internal reflection in the silica medium. Light is internally reflected from the lower refractive index material, the cladding, to the core, with a forward propagation. The production of low-loss fibers has opened up the way for progress in communications and, thus, for high-performance optical fiber sensors. The most widely used type of fiber in optical fiber

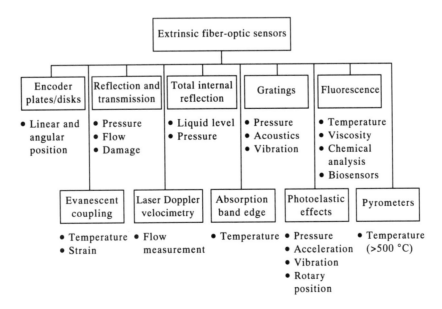

Figure 1
Extrinsic or hybrid fiber-optic sensors: light transits into and out of the fiber to reach the sensing region (after Udd 1991)

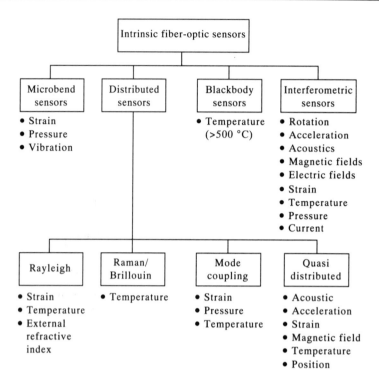

Figure 2
Intrinsic or all-fiber-optic sensors: the environmental effect is converted to a light signal within the fiber (after Udd 1991)

sensors is of the step-index configuration, where there is a sharp change between the refractive indices of core and cladding. Graded index fiber has a variable index across the core, with a consequent effect on the mode of propagation of the fiber. Fibers may be designed to be single mode (particularly for communications purposes), polarization preserving (to propagate a single linear polarization state), or doped with specific impurities, as in the fiber-optic laser. Such fibers can have implications for sensing, for example, temperature sensing (Farries *et al*. 1986). In all cases, attenuation by the fiber is so low as to be negligible, even for kilometer lengths of fiber, which is more than adequate for most fiber-optic sensor applications. However, care must be taken with the attenuation loss which can be induced due to bending and connecting, and for this reason intensity-based optical fiber sensors must be configured with a suitable reference to the intensity of light launched into the fiber.

An important factor in the development of optical fiber sensor technology has been the major improvements with optical fibers, and thus sensors, over recent years in optical sources, which are compatible with advancing technology. Light sources that are used to support such sensors produce radiation that is dominated by either spontaneous or stimulated emission. Thus, the familiar incandescent lamp is not well suited

to many such sensor applications as it is difficult to efficiently launch light from it into the fiber. This may be less of a problem with light-emitting diodes (LEDs) or laser diodes, which may often be "pig-tailed," with an attached fiber. High-power, monochromatic radiation is available from laser diodes, particularly in the red and near infrared part of the spectrum—regions where the optical transmission of the fiber material is optimized. Solid-state detectors, based on the same semiconductor technology, with high sensitivity in the same spectral regions, are now widely available at low cost, equally compatible with fibers. The field has recently been extended with the development of fiber-optic lasers, which offer a wider variety of wavelengths, potentially at high powers, with good coverage over the visible region, and with electrooptic conversion.

3. Illustrations of Fiber-Optic Sensors

In this article it is possible to illustrate only a few of the very many devices which have been discussed in the literature or commercially developed. A number of detailed reviews have been written, and several more detailed texts discuss the subject in detail (Gaillorenzi *et al*. 1982, Pitt *et al*. 1985, Udd 1991,

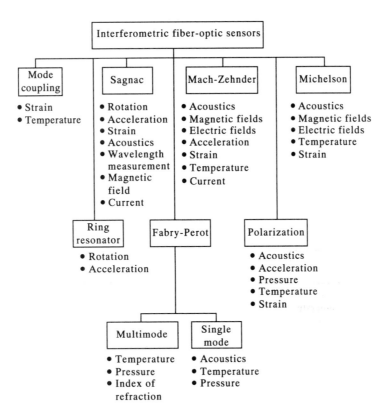

Figure 3
Interferometric fiber sensors (after Udd 1991)

Grattan and Meggitt 1994). Two simple illustrations are discussed in the following sections: the interferometric class of fiber-optic sensor, which embraces both intrinsic and extrinsic devices, and the distributed class of sensor.

3.1 Interferometric Fiber-Optic Sensors

The wide variety of interferometric fiber-optic sensors is shown in Fig. 3. At the head of each family is the type of interferometer involved (Sagnac, Michelson, Mach–Zehnder and mode-coupling) and below each are listed some of the measurands which can be determined using the technique. The principle of operation of the interferometer has been known for a long time and the history of Fabry–Perot sensors began at the turn of the twentieth century with derivates of the parallel-plate interferometer; even at that time sensors for current and voltage measurement using conventional optics were discussed by Fabry and Perot (Perot and Fabry 1898). The wide variety of measurands (including current and voltage sensors in fiber-optic form) encompasses acoustics, magnetic fields, electric fields, acceleration, strain, temperature,

rotation, and so on. At the moment, the Sagnac configuration of interferometer would appear to have the most significant commercial potential in the optical fiber gyroscope, which operates on the basis of a change of optical path length due to rotation. It is this change in path length which is measured in the interferometer. Issues of size, weight and reliability require further study, together with significant engineering issues. The versatility of the system is shown in the application of the Sagnac interferometer to acoustic sensing, strain sensing, wavelength measurements, and so on, and it promises to be the first high value commercial fiber-optic sensor product.

The multiplexing of fiber-optic sensors on a single optical cable represents an important advance in the practical implementation of such sensors, especially in view of the competition from microelectronic sensors. The formats of time-, frequency-, wavelength-, coherence-, and polarization-division multiplexing are all conceptually feasible for use with interferometric sensors, where signals from each specific sensor element in an array can be separated at the output (Dandridge *et al.* 1987). Various network topologies have been proposed to deal with such multiplexing schemes.

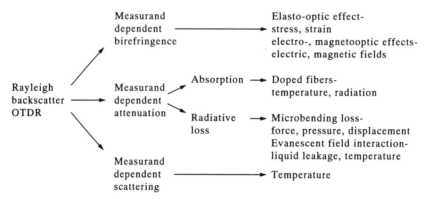

Figure 4
Rayleigh backscatter OTDR-based distributed sensing techniques and uses

3.2 Distributed Sensor Schemes

Multiplexing represents one approach to the problem of developing multisensor systems, using a series of discrete sensors on a single optical cable. However, intrinsically distributed sensors are particularly attractive for use in applications where monitoring of a single measurand is required at a large number of points, or continuously over the path of a fiber. Such application areas include structural monitoring (in civil engineering or increasingly in aerospace applications), temperature profiling (e.g., in transformers) or leakage detection (e.g., of cryogenic fluids for both safety and financial reasons). Quasidistributed sensors are distinguished from truly distributed sensors in that in the former the measurand is not monitored continuously but fiber optics offer the possibility of distributed sensing over the complete length of an optical fiber. The approach of optical time-domain reflectometry (OTDR) is frequently used, in which light scattered by Rayleigh scattering is backscattered at 180° to the propagation direction, and this light is monitored. It forms the basis of fault location in cables (where significant changes in the returned profile show the position of the fault). Short pulse sources are required and, for a pulse width of 10 ns, a 1 m resolution is obtained. Alternative ranging techniques include coherence, frequency techniques or pseudorandom encoding. Figure 4 shows the basic mechanisms that can be used in such intrinsic distributed sensors using Rayleigh techniques. Other approaches which may be employed include Raman scattering (for temperature sensing, in particular), mode coupling using frequency information to determine the location of the interaction which causes the scattering; simpler quasidistributed techniques may often be sufficient.

4. Summary

Fiber-optic sensors are gradually encroaching on conventional sensor markets, especially for test and evaluation purposes. The future of the technology looks bright, especially in conjunction with fiber-optic communications systems. Costs associated with the technology seem set to continue to drop, and they have shown their utility in high-temperature and other niche markets. The sensor field is so large that an inroad into a suitable niche, for example, the medical, environmental or distributed sensor areas will provide a secure future for the technology in the twenty-first century.

See also: Analog Signal Conditioning and Processing; Instrument Elements: Models and Characteristics; Optical Instruments; Optical Measurements; Transducers: An Introduction

Bibliography

Dakin J P, Kahn D A 1977 A novel fiber optic temperature probe. *Opt. Quantum Electron.* **9**, 540
Dandridge A, Tveten A B, Kersey A D, Yurek A M 1987 Multiplexing of interferometric sensors using phase generated carrier. *IEEE J. Lightwave Tech.* **5**, 947–51
Farries M C, Ferrman M E, Lamming R I, Poole S B, Payne D N, Leach A P 1986 Distributed temperature sensor using neodymium-doped optical fiber. *Electron. Lett.* **22**, 418
Gaillorenzi T G, Bucaro J A, Dandridge A, Sigel G, Cole J H, Rashley L, Priest R G 1982 Optical fiber sensor technology. *IEEE J. Quantum Electron.* **18**
Grattan K T V, Meggitt B T 1994 *Optical Fiber Sensor Technology.* Chapman and Hall, London
Medlock R S 1986 Review of modulation techniques for fibre optic sensors. *Int. J. Opt. Sensors* **1**, 47–68
Ning Y N, Grattan K T V, Wang W M, Palmer A W 1991 A systematic classification and identification of optical fibre sensors. *Sens. Actuat.* **29**, 21
Perot A, Fabry C H 1898 Sur un voltmetre electrostatique interferentiel pour etalonnage. *J. Phys.* **7**, 650–9
Pitt G D, Extance P, Neat R C, Batchelder D W, Jones

R E, Barnett J A, Pratt R H 1985 Optical fibre sensors. *Proc. IEE* **132**, 214–47

Rogers A J 1986 Distributed optical fibre sensors. *J. Phys. D* **19**, 2237

Udd E (ed.) 1986 Fiber optic gyros. 10th Anniversary Conference *Proc. Soc. Photo-Opt. Instrum. Eng.* **719**

Udd E 1991 *Fiber Optic Sensors.* Wiley, New York, p. 6

Wickersheim K A, Alves R V 1982 Fluoroptic thermometry: a new RF immune technology. *Biomedical Thermology.* Liss, New York, pp. 547–54

York Technology Ltd 1986 DTS System II. Manufacturer's Data. York Technology Ltd, Southampton, UK

Zhang Z, Grattan K T V, Palmer A W 1992 A fiber optic temperature sensor based on the fluorescence lifetime of alexandrite. *Rev. Sci. Instrum.* **63**, 3869

K. T. V. Grattan
[City University, London, UK]

Transducers, Inductive

Inductive transducers rely on the variation of the inductance of a coil due to an external measurand. This change is then observed in an electrical circuit. Inductive transduction processes are possible where, for example, the voltage is induced across a coil as it moves in a magnetic field (the electrodynamic transducer), or the magnetic properties of the inductor core material (using magnetostrictive material) change and are observed as a change in inductance. The most common type of variable inductive transducer uses changes in the reluctance of the magnetic path. The variation in inductive coupling between coils is also exploited in transformer-type transducers. These transducers are the most common and will be discussed in detail.

1. Operating Principle

The inductance L of a coil is given by

$$L = \frac{\mu_0 n^2}{\mathbb{R}} \tag{1}$$

where $\mathbb{R}$ is the reluctance of the magnetic path and is given by

$$\mathbb{R} = \frac{l}{\mu A} \tag{2}$$

where μ_0 is the permeability of free space, μ is the relative permeability of the core material, n is the number of turns on the uniformly wound coil, A is the cross-sectional area, and l is the length of the core. The reluctance is analogous to electrical impedance but for magnetic fields, and can be combined following the same procedures.

Clearly, from these equations, it is obvious which parameters affect the inductance of a coil. Changes in inductance will be observed if: the geometry of the coil is changed (n, A or l); the reluctance of the magnetic path is changed (e.g., by using different core materials and varying the proportions of these materials); or by changing the permeability of the core material. Changing the geometry is not normally used as a transduction method, but the other techniques are widely used in inductive transducers.

The measurement of the change in inductance for a particular transducer requires a detailed understanding of the response of an inductor to the interrogating excitation signal. It is driven by an ac signal (in the frequency range 400–40 000 Hz) and the dissipation factor in the inductor circuit is a function of this frequency. Contributions arise from:

(a) coil resistance (inversely proportional to frequency);

(b) eddy current resistance (reduced by laminating former, this is proportional to frequency); and

(c) hysteresis of magnetic flux and magnetic field (independent of frequency).

Thus, a frequency exists for minimum energy dissipation and can be calculated for any particular construction (Neubert 1975). Operation at this frequency maximizes sensitivity and reduces drift. The interrogation frequency does, however, limit the dynamic response (typically to a factor of ten less than the interrogation frequency). The variation in dissipation should also be considered with respect to the transduction process, as a variation in the measurand may change the dissipation factor. This is an important consideration in transducer design.

The self-capacitance of an inductor (arising from the coil and the capacitance of cables and connections) must also be considered as it changes the sensitivity of the transducer (the transducer must be calibrated with all the cabling connected). It may also introduce resonances at higher frequencies.

The design of the electrical circuit to transform the change in inductance to a change in current or voltage to drive the indicator device, is another aspect which must be considered in the use of inductive transducers. This transformation must be linear and not introduce any phase shift. Two common approaches are (Neubert 1975):

(a) an ac potentiometer, and

(b) an ac bridge (easy to incorporate push–pull transducer configuration).

The choice will depend on the particular response that is required and must be considered in conjunction with the overall transducer response.

Another inductive transduction mechanism is to vary the coupling between two (or more) inductive elements. The most important class of this transducer is the variation in coupling in the transformer type.

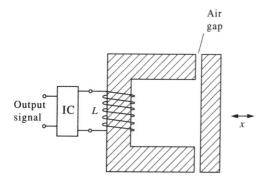

Figure 1
A variable air gap inductive transducer; IC is the interrogation circuit, *L* is the measured inductance

This is the basis of the linear variable differential transformer (LVDT) which will also be discussed.

1.1 Variable Reluctance Transducer

A variable reluctance transducer might consist of a core magnetic material incorporating an air gap, the length of which is arranged to change with the measurand. Such a scheme is shown schematically in Fig. 1. The total reluctance $\mathbb{R}_t$ of the magnetic path is then given by the sum of the reluctances of the core material (of relative permeability μ_c) and the air gap (relative permeability of unity). Neglecting fringing of the field and assuming the cross-sectional area of the air gap and the core material are equal, this gives

$$\mathbb{R}_t = \mathbb{R}_{core} + \mathbb{R}_{gap} = \frac{1}{A}\left(\frac{l-x}{\mu_c} + x\right) \quad (3)$$

where *l* is the length of the core and *x* is the length of the air gap. Thus, the measured inductance of the coil is given by (assuming $\mu_c \gg 1$)

$$L = \mu_0 n^2 A \frac{1}{x + l/\mu_c} \quad (4)$$

Therefore, for a finite decrease (increase) in air gap δx, there is a finite increase (decrease) in inductance δL. Substitution in Eqn. (4) then gives a fractional change in inductance of

$$\frac{\delta L}{L} = \frac{\delta x}{x} \frac{1}{1+l/(\mu_c x)} \frac{1}{1 \pm (\delta x/x)\,[1/(1+l/(\mu_c x))]} \quad (5)$$

where the negative sign corresponds to a decrease in

air gap, and the positive an increase. This can be developed into a series for

$$\left|\frac{dx}{x}\;\frac{1}{1+l/(\mu_c x)}\right| \ll 1 \quad (6)$$

to give

$$\frac{\delta L}{L} = \frac{dx}{x}\;\frac{1}{1+l/(\mu_c x)}$$

$$\left[1 \pm \frac{dx}{x}\frac{1}{1+l/(\mu_c x)} + \left(\frac{dx}{x}\frac{1}{1+l/(\mu_c x)}\right)^2 \pm \dots\right] \quad (7)$$

From Eqn. (7) it can be seen that there is a linear term and higher-order term which introduce nonlinearities. This situation can be improved. If the transducer is configured with two inductance coils in opposition (a push–pull configuration), then as the air gap changes the inductance in one coil will increase and in the other decrease, and if the output is taken as the sum of the inductances, the even order terms disappear and a more linear response is achieved.

In general, the sensitivity of such an inductance transducer can be increased by reducing the term $l/(\mu_c x)$. This is achieved with a short core length *l*, and a high relative permeability μ_c. This is limited when the term becomes small compared to unity, when the sensitivity can no longer be improved. In addition, to maintain good linearity the fractional change in air gap must be small to reduce the influence of the nonlinear terms. Typically, a fractional 0.1–0.2% change in inductance would be expected with a nonlinearity of 2–3% of full-scale displacement, depending on the specific design.

1.2 Moving-Core Inductive Transducers

This type of transducer can also be thought of as operating by varying reluctance. In this situation, however, this variation is achieved by moving a ferromagnetic material within the core of the coil, as shown in Fig. 2. A simplified analysis of the mechanism of transduction process can be obtained if a uniform magnetic field is assumed within the coil. The inductance of such a coil (with no core element) is given by

$$L = \mu_0 n^2 \frac{\pi r^2}{l} \quad (8)$$

When a ferromagnetic core of length l_c, radius r_c and relative permeability μ_c is introduced into the coil the inductance is then given by

$$L = \mu_0 n^2 \pi \frac{lr^2 + (\mu_c - 1)l_c r_c^2}{l} \quad (9)$$

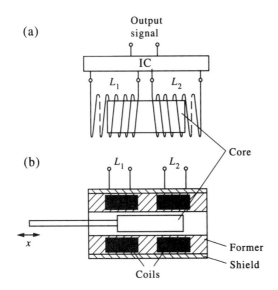

Figure 2
A moving core inductive transducer: (a) in schematic
form; (b) in cross section. IC is the interrogation circuit,
L are the measured inductances

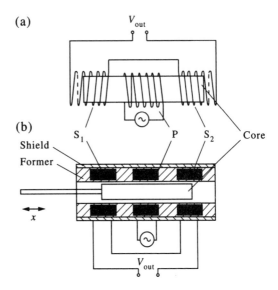

Figure 3
A linear variable differential transformer: (a) in
schematic form; (b) in cross section. P is the primary coil,
S are the secondary coils

The change in inductance δL can then be calculated
for a change in length of core material, for example, if
the core is pushed further into the coil by an amount
δl_c. The fractional change is the more important
parameter, as it represents the sensitivity of the
transducer, and is given by

$$\frac{\delta L}{L} = \frac{\delta l_c}{l_c} \frac{1}{1 + \left(\dfrac{l}{l_c}\right)\left(\dfrac{r}{r_c}\right)^2 \dfrac{1}{(\mu_c - 1)}} \tag{10}$$

If the transducer is arranged in a push–pull configuration, as indicated in Fig. 2, the second coil will suffer a
change of the same magnitude but opposing sign.

Equation (10) shows that the fractional change in
inductance is proportional to the fractional change in
core length multiplied by a factor smaller than unity.
This factor is maximized for large values of core
permeability μ_c, core length l_c and core radius r_c.

The reluctance of the coil is quite high (due to the
air path), therefore, for a given inductance a large
number of turns is required. This in turn increases
the coil self-capacitance, which at high frequencies
may result in resonance. The large air path in the coil
also gives rise to stray field, which makes the coil open
to pickup from external fields. Another problem is the
coupling between the two coils when a push–pull
configuration is used. Therefore, in the design, decoupling is provided by using a specially shaped
former to reduce mutual magnetic flux.

2. Transformer Inductive Transducers

Transformer inductive transducers rely on the variable
coupling between two (or more) coils of a transformer. A schematic representation of such a transducer
is shown in Fig. 3. The coil on the primary side is
driven by a signal of suitable frequency (in the range
400–40 000 Hz), and the secondary output voltage is
dependent on the coupling between the two coils. The
most common form of this type of transducer is the
linear variable differential transformer (Herceg 1972,
Neubert 1975). In this form there are two identical
secondary coils positioned symmetrically as shown in
Fig. 3. Movement of the core from the central position
will then increase the coupling to one of the coils,
while reducing coupling to the other. The difference
signal between the outputs from the secondary coils
has an amplitude in proportion to the displacement of
the core.

The output undergoes a phase change of 180° in
passing through the null position, and is normally out
of phase with the excitation signal. This is, however, a
function of the drive frequency and is chosen to
reduce this effect. Another problem is that at the null
position the output voltage is not exactly zero (due to
harmonics and the fact that the two secondary coils
will not be precisely identical), but careful manufacture has reduced this to less than 1% of full scale.

The LVDT has some unique advantages:

(a) there is no friction between the core and the coil
therefore the input impedance is very low;

(b) the absence of friction gives a (theoretical) infinite mechanical lifetime; and

(c) the LVDT offers infinite resolution, due to the inductive principle of operation and the absence of friction.

LVDTs are available with different measurement ranges, with a full-scale displacement of 1 mm to several meters. They are used extensively in weighing machines, pressure transducers and load cells. They can be used to measure dynamic processes, but this requires the excitation signal to be much larger than the frequency of the signal of interest (typically by a factor of ten).

See also: Analog Signal Conditioning and Processing; Bridges; Instrument Elements: Models and Characteristics; Resistance, Capacitance and Inductance Measurement; Transducers: An Introduction

Bibliography

Neubert H K P 1975 *Instrument Transducers, An Introduction to Their Performance and Design.* Clarendon, Oxford

Herceg E E 1972 *Schaevitz Handbook of Measurement and Control.* Schaevitz Engineering, Pennsauken, NJ

K. Weir
[City University, London, UK]

Transducers, Ionizing Radiation

The term "transducer" is not commonly used in connection with ionizing radiation measurements, but its application to sensors in this field is clearly suitable in view of the wide variety of energy transformations which can take place in the process of detection and measurement. Any form of ionizing radiation incident upon matter will transfer energy to the atoms of the medium, causing excitation and producing free electrons with a widely varying spread of kinetic energy, many of which can produce further ionization and excitation.

1. Basic Interactions

Heavy charged particles such as protons or alpha particles are rarely deflected by atomic nuclei and, consequently, each will move in a straight line. The path length is short, and the maximum energy imparted by an alpha particle in any single interchange is about 1/500th of its energy, so there is a small, highly ionized region formed around the track of the particle. The liberated electrons can recombine, cause further ionization or excitation or can be collected by applying a strong electric field.

Beta particles also ionize and excite atoms along their path, but interactions are less frequent than for heavy particles of the same energy due to their higher velocity. They are also easily scattered due to their small mass, so that their paths are erratic. Energy transfers range from very small to total energy transfer, so that the energy distribution within the absorbing medium is wider and more uniform. Consequently, collection of the free electrons without appreciable recombination is easier to achieve.

X and gamma rays are absorbed less readily than ionizing particles, and interact in a completely different way. At low energies each photon is completely absorbed in a single photoelectric interaction, one of the atomic electrons being ejected with the surplus photon energy. At higher energies the Compton process predominates, in which the photon is scattered at reduced energy, and an electron recoils with the remaining energy. At energies above 1.02 MeV photons can be involved in pair production, that is, the simultaneous creation of an electron–positron pair, which carry away any surplus energy as kinetic energy. Many scattered photons will be "lost" to the detector, although there will be some build-up due to scatter into the detector from the surrounding region. At all but the very lowest photon energies, the ionization density resulting from photon interactions will be much lower than from the same number of charged particles.

Detection systems can normally be designed either to register the number of primary events occurring within a sensor or to measure the energy deposited within it. The intrinsic efficiency ε of a detector is the number of events recorded by it, expressed as a fraction of the number of particles or photons incident upon it. The resolution of an energy-sensitive detector is the full-width half-maximum (FWHM) of the number–energy response curve for monoenergetic radiation, expressed as a percentage of the mean energy value registered. Both parameters will depend upon the type and quality of the incident radiation.

Another important characteristic is the dead time of a detector, that is, the time for which it is inoperative while recovering from the effect of an ionizing event. This is obviously linked with the basic detection process and the length and shape of pulses generated within the sensor. It can also be affected by subsequent electronic circuitry.

Comprehensive accounts of radiation detectors and their use in measuring ionizing radiation are given by Knoll (1979) and Tsoulfanidis (1983). Various types of radiation detectors are given in Table 1.

2. Gas Filled Detectors

The average energy w required to produce an electron–ion pair in a gas is 25–40 eV, depending on the

Table 1
Detectors and their uses

Detector	Main use
Ionization chamber	dosimetry, standardization
Proportional counter	low-energy alpha, beta and soft x-ray spectroscopy
Geiger counter	monitoring, general counting
Inorganic scintillator	gamma spectroscopy
Organic scintillator	electron spectroscopy, fast counting
Surface semiconductor	alpha, proton, heavy-ion spectroscopy
Ge(Li)	gamma spectroscopy
Si(Li)	electron and x-ray spectroscopy
HPGe	gamma spectroscopy

nature of the gas. The statistically generated limit to the resolution of a detector is given by $K/\sqrt{n}$, where K is a constant and n is the average number of charge carriers generated by an event. The limit to resolution for radiation of energy E is therefore $K\sqrt{(w/E)}$, which has a relatively high value for a gas counter. Another disadvantage of gaseous detectors is that the electron collection time is quite long, thus restricting their use to relatively low count rates.

2.1 Ionization Chambers

The simplest type of detector is the ionization chamber which basically consists of two electrodes within a gas chamber. A potential difference is maintained between the electrodes which is sufficient to collect all the free electrons generated in the gas without allowing recombination, but which is not high enough to permit multiplication by collision. The required field strength is usually between $10^4 \, V \, m^{-1}$ and $10^5 \, V \, m^{-1}$, at atmospheric pressure. The ionization current, integrated and smoothed in an RC circuit, is usually measured by means of a sensitive electrometer. Such currents are very small, often of the order of $10^{-12} \, A$ or less, therefore careful design of the chamber is essential. Guard rings are required to delineate the sensitive volume, and the pressure, type and purity of the gas used is very important.

Dosimeters, which are essentially integrating ionization chambers, are in frequent use. These are small air-filled chambers with an initially charged electrode inside. The subsequent partial discharge of this electrode is used as a measure of the number of free electrons produced by incident radiation. Some of these chambers have "air equivalent" walls, that is, they are designed so that radiation scattered into the chamber from the walls is equal in ionizing effect to that which would be scattered into it, if it were surrounded by an infinite volume of air also undergoing irradiation.

2.2 Proportional Counters

Free electrons generated in an ionization chamber can, in principle, be collected and used to register each event as a separate pulse, but a measure of amplification is required since the number of electrons involved is so small. Proportional counters provide this amplification by initiating carefully controlled electron multiplication within the counter. In simplest form they use cylindrical geometry, having a uniform axial wire as anode and the outer containing tube as cathode. This creates a field strength within the counter which increases rapidly near the anode to values in excess of $10^6 \, V \, m^{-1}$. When ionization takes place within the counter, electrons are accelerated towards the anode, causing further ionization. This process is repeated many times in the vicinity of the wire, multiplication of the original number of free electrons by a factor up to 10^5 being possible. Under strictly controlled conditions the multiplication factor M is constant so that the number of electrons collected will be proportional to the energy of the original particle. The value of M is very sensitive to the applied voltage, therefore a very smooth and stable voltage supply is essential. The geometrical design of the counter is critical, since variations in field strength near the central wire can result in local variations in M, thus leading to a deterioration in the resolution of the counter. M also depends upon the nature, purity and pressure of the contained gas. Values of M between 10^2 and 10^4 are most commonly used, since lower values give better resolution with fewer spurious pulses.

Proportional counters are favored for registering low-energy particles or soft x rays, especially when the source can be mounted within the counter, and the resolution for low-energy x rays can be quite good, for example, 5% at 20 keV.

Modern developments include the use of specialized multiwire proportional-type counters which can localize events to a high degree of spatial resolution. These have been used in combination with digital imaging techniques to give resolutions better than 100 μm.

2.3 Geiger–Müller Counters

Geiger counters, like proportional counters, use gas multiplication of each primary ionizing event within the counter, but higher field strengths are created which initiate avalanches further from the central electrode, and a chain reaction is set up which results in 10^9–10^{10} ion pairs being formed, regardless of the type or energy of the initial radiation which triggered the event. Once the required field strength is reached, the response of the Geiger tube is relatively insensitive to the applied voltage until breakdown conditions occur, which is usually 30–50% above the threshold value for counting conditions.

The anode wire is supported at one end only, and is located axially within the cylindrical metal cathode. Tubes commonly contain helium or argon at pressures of 10^4–2×10^4 Pa, together with a small percentage of quenching gas, for example, ethyl alcohol or a halogen, to prevent the discharge being maintained by energy liberated when positive ions strike the cathode. When required for alpha or beta detection, Geiger tubes have a thin mica or aluminum entrance window at one end. Gamma counting takes place mainly through interactions with the wall material, so the design for gamma detectors is windowless and more robust.

These counters are cheap, easy to use and do not require expensive or complex auxiliary equipment, but their response is not energy related and they do not differentiate between events initiated by different types of radiation.

3. Scintillation Counters

As a result of the liberation of electrons by ionizing radiation, and their subsequent recombination with holes at activator sites, certain insulators emit light in the ultraviolet and visible regions of the spectrum. The conversion efficiency f can be defined as the mean fraction of the absorbed energy which is converted into light, and this depends on the nature of the scintillator and also on the type and energy of the radiation incident upon it. On average about 20 eV of absorbed radiation energy is required to produce an electron–hole pair in these insulators, and the most efficient scintillators will emit about 4×10^4 photons as a result of the absorption of 1 MeV. This light is received on the photocathode of a photomultiplier and converted into a small pulse of electrons, the quantum efficiency being about 20%, that is, one electron liberated, on average, by five incident photons. These electrons are accelerated by the electric field and the number is multiplied as a result of successive collisions with a series of dynodes, each being maintained at a higher potential than the previous one, so that gains of up to 10^7 can be achieved.

For optimum use the scintillator should be chosen to suit the particular type of radiation to be measured. The photomultiplier should have maximum quantum efficiency in the wavelength region emitted by the scintillator and be operated at a voltage consistent with the best signal-to-noise ratio.

Since the signal relies upon light transmission rather than electron collection the resulting pulses are of relatively short rise time, but the decay time varies from a few nanoseconds to several microseconds in different scintillators.

Birks (1964) has written a comprehensive account of scintillators and their use.

3.1 Inorganic Scintillators

The light emission from inorganic scintillators depends on the energy states within the crystal lattice, and these are usually modified by the presence of small quantities of activators which alter the energy states at localized sites. Sodium iodide activated with thallium, NaI(Tl), is probably the most widely used scintillator, since it is one of the most efficient in light output ($f = 0.13$) and, having atoms of fairly high atomic number, it is a good absorber of gamma photons. However, it is rather fragile, has a long decay time and, being hygroscopic, must be encapsulated. CsI(Tl) and CsI(Na) have certain advantages for gamma counting and LiI(Eu), when enriched with ^{6}Li, can be used for neutron detection. Glass scintillators, containing ^{6}Li and activated with cerium can, similarly, be used for neutron measurements. They have also become popular for beta and gamma counting because of the variety of shapes, including fibers, in which they can be fabricated, and because they can withstand hostile environments. The decay time, usually 50–80 ns, is shorter than for most inorganic scintillators, but the conversion efficiency is low and this adversely affects the resolution.

3.2 Organic Scintillators

Organic scintillators cover a wide range of materials which rely upon the deexcitation of individual molecules for light emission. Pure organic materials such as anthracene or toluene can be used, but these are now more usually employed in solid or liquid solution, often in combination with other components which can act as wavelength shifters. Plastic scintillators using polystyrene and polyvinyl toluene as solvents are used a lot, since they can be machined into a variety of shapes, or formed into thin sheets or fibers. Liquid scintillators can be used to count low-activity or low-energy sources by immersion within the liquid.

The conversion efficiency of plastic scintillators is only a few percent, but for gamma photons the absorption coefficient is very low, which gives an advantage when detecting charged particles in the presence of gammas, since discrimination is good. Pulses are short, having rise and decay times of the order of nanoseconds, therefore they are very useful for fast counting.

In general, it is desirable that the deexcitation process in scintillators should be rapid, enabling high count rates to be accommodated. However, certain organic scintillators, for example, LiF or $CaSO_4(Mn)$, incorporate trapping centres from which electrons or holes can be released, with subsequent light emission, only at high temperatures. Consequently, an integrated dose of radiation can be assessed by measuring the total light output from such a thermoluminescent dosimeter (TLD), when it is heated to an appropriate temperature. The response can be linear, from 10^{-4}–10 Gy. These detectors can also be used for neutron dosimetry by enrichment with ^{6}Li.

4. Semiconductor Detectors

The gap between the valency and conduction bands in semiconductors is only about 1 eV, therefore ionizing radiation will readily produce electron–hole pairs within a semiconductor. These carriers can be collected under the influence of an applied field, as in a gas filled detector, but there are important differences. In order to avoid recombination, very high collecting fields must be applied. This can be achieved over very narrow regions of high resistivity with relatively low applied potentials by several different means. The sensitive volume of a semiconductor detector is therefore much smaller than that of a gaseous detector. Also, since the average energy required to produce an electron–hole pair is very low, about 3 eV, compared with the 30 eV required to produce an electron–ion pair in a gas, the statistical uncertainty of the pulse size is smaller and the energy resolution of spectroscopy much improved.

The capacitance of a semiconductor detector is small and does not remain constant, depending greatly on the effective voltage under operating conditions. It is therefore necessary to feed the collected charge into a charge-sensitive, low-noise preamplifier, before pulse shaping and further amplification.

4.1 Surface Detectors

In *p–n* junction detectors a junction is created near the surface of a silicon wafer by diffusing either a *p*-type impurity into *n*-type silicon or an *n*-type impurity into *p*-type silicon. When a reverse bias is applied to the wafer, a region depleted of carriers is formed at the junction and the applied voltage then operates across this narrow region, which may typically be less than 100 μm wide, creating a field strength of 10^5–10^6 V m^{-1}. Electron–hole pairs created within this region are swept apart and collected within a few nanoseconds.

Similarly, surface barrier detectors consist of a wafer of high-purity *n*-type silicon, with an oxidized layer on one surface. This has a thin gold layer evaporated onto it, which serves as a contact and is the window through which charged particles enter the detector. The operation is similar to that of a diffused junction detector, a reverse bias ranging from 10–2000 V, producing a sensitive region of depth from 10 μm to about 5 mm, depending upon the bulk resistivity.

These detectors can be used for detection and energy analysis of heavy charged particles or low-energy electrons, provided the depletion layer is deep enough to completely stop the particles within it.

4.2 Lithium Drifted Detectors

The sensitive region of the detectors previously described is limited because of the difficulty of producing pure, high-resistivity silicon, and the requirement not to exceed the breakdown voltage. Detectors with much deeper, sensitive regions can be made by diffusing lithium into *p*-type silicon or germanium to compensate for the acceptor impurities. When the process is complete, opposite ends of the wafer will have *n* and *p* surfaces, respectively, and a reverse bias of several thousand volts can be applied across compensated regions, typically of 5–10 mm depth. These are widely used in gamma- and x-ray spectroscopy. Detectors with a coaxial structure can also be made. Ge(Li) detectors have a high stopping power and the resolution is very good, being about 0.15% for ^{60}Co gammas, but they must be maintained permanently at 77 K in order to stabilize the lithium. Si(Li) detectors are less efficient and have rather poorer resolution, but are sometimes preferred for x-ray and beta spectroscopy, since they are less easily damaged by temperature fluctuations. For efficient photon counting large detectors are desirable, but unfortunately the resolution and collection time also increase with size.

Larger, high-purity, germanium detectors (HPGe) are now available, which need to be cooled only when in use to reduce thermal noise. CdTe, GaAs and HgI$_2$ detectors, which can be operated at room temperature, are also used. These have a higher efficiency but poorer resolution than germanium detectors and are smaller in size,

Aussel (1986) gives a comparative survey of the properties of semiconductor detectors.

5. Future Developments

Semiconductor detectors probably offer the greatest scope for future developments, both in the use of improved materials and in the fabrication of silicon-based devices. Position-sensitive silicon detectors are already in use, in which one contact is resistive so that pulse size is dependent upon the point of entry of the particle. Detectors are now being made with arrays of elements (pixels) which are miniature reverse bias diodes on high-resistivity silicon; the incorporation of multiple connections and circuitry within a single wafer is also being developed.

See also: Analog Signal Conditioning and Processing; Instrument Elements: Models and Characteristics; Ionizing Radiation Measurements; Transducers: An Introduction

Bibliography

Aussel J P 1986 Measurement of ionizing radiation semiconductor detectors, a review. Report No. CEA-R-5361. CEA, France

Birks J B 1964 *The Theory and Practice of Scintillation Counting.* Pergamon, Oxford

Knoll G F 1979 *Radiation Detection and Measurement.* Wiley, New York

Tsoulfanidis N 1983 *Measurement and Detection of Radiation*. McGraw-Hill, New York

R. B. J. Palmer
[City University, London, UK]

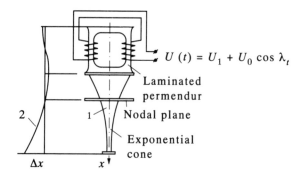

$$U(t) = U_1 + U_0 \cos \lambda_t$$

Laminated permendur

Nodal plane

Exponential cone

Figure 1
Ultrasonic transducers: (1) with velocity transformer, (2) amplitude distribution along the length of transducer

Transducers, Magnetostrictive

Magnetostrictive transducers are used to convert electrical energy to mechanical energy or vice versa and are based on the effect discovered in 1847 by Joule (and the converse effect discovered in 1868 by Villari), that when some materials are magnetized, a change in dimensions occurs; conversely, a mechanical stress causes a change in the magnetization. The best known materials are pure nickel, iron–cobalt alloys (e.g., Permendur), iron–rare-earth element alloys (terbium, dysprosium), and so on. This change can be either positive or negative in a direction parallel to the magnetic field, and is temperature dependent ceasing at the Curie temperature.

1. Properties

In linear magnetization theory the change in dimension is proportional to the applied field, and the change in magnetization under stress is proportional to the stress.

The energy coupling factor R_c is an important characteristic of a transducer material. Alloys of rare earths give high values, but cannot be laminated, so that they suffer eddy current losses at high frequencies. Hysteresis can complicate the field strain relationship of a transducer.

Magnetostrictive transducers usually perform better at cryogenic temperatures (2–77 K) than piezoelectric transducers.

2. Applications

With the increase in competition from piezoelectric transducers, the field of applications of magnetostrictive transducers has been reduced to underwater acoustics (high-power transmitters), industrial ultrasonic applications and a range of devices exploiting both direct and inverse magnetostrictive effects, for example, delay lines, filters, loudspeakers, microphones, oscillators, resonators, and so on. However, the development, at the end of the 1980s, of materials with extremely high magnetostrictive properties (e.g., Terfenol) has dramatically extended the area of applications.

2.1 High-Power Ultrasonics

The widest applications of magnetostrictive transducers are in industrial ultrasonics (atomization, cleaning, homogenization, soldering, welding, drilling, fatigue testing, some underwater sound applications, etc.), in which they are used to generate considerable acoustic or vibrational power. The most frequently used shape of a transducer is shown in Fig. 1, in which a large amount of vibrational energy is required to be radiated in a small area.

2.2 Magnetostrictive Transducers–Actuators with Unlimited Displacements

Magnetostrictive transducers–actuators are based on the transformation of mechanical oscillations into continuous motion with high resolution. In a magnetostrictive transducer–actuator (see Fig. 2a) the Wiedemann effect—the twisting of a ferromagnetic under the simultaneous action of circular, Φ_c, and longitudinal, Φ_l, magnetic fields—is used to generate axial and torsional resonant oscillations in a hollow cylinder, made from magnetostrictive material. Two component oscillations in the contact area are transformed into continuous motion of the rotor.

New iron–rare-earth element alloys (e.g., Terfenol) have made possible new actuators using travelling-wave-type oscillations. In them (see Fig. 2b, c) magnetic fields are generated by individual toroidal coils, energized sequentially, in order to pass a zone of lengthening and thinning along the rod, thus causing it to make a small step in the opposite direction.

2.3 Active and Adaptive Structures, Vibrations and Noise Suppression

Piezoactive (piezoceramic and magnetostrictive) transducers are incorporated into structures for vibration and noise control, for example, active mirror mounts, active decoupler pylons, to increase flutter margins, adaptive wings with adaptively modifiable wing cross section, and the use of embedded acoustic detecting

(a)

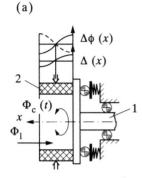

(b)

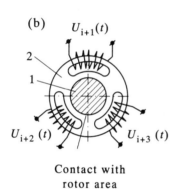

Contact with
rotor area

(c)

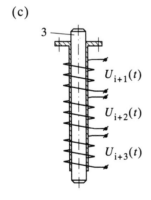

Figure 2
Magnetostrictive transducers–actuators with (a, b) rotary
and (c) translational motion. 1 is the rotor, 2 is the active
element, 3 is the slider made from Terfenol. $\Delta\phi(x)$ and
Δx are the distribution of the amplitudes of torsional and
axial displacements along axis x. $U_i(t) = U_0 + U_0'\cos(\lambda + 2\lambda i/k)$, $i = 0, 1, \ldots, k - 1$, $k \geq 3$, U_0 and U_0' are constants

fibers (nickel alloys) for monitoring acoustic signa-
tures.

See also: Analog Signal Conditioning and Processing;
Instrument Elements: Models and Characteristics; Transducers: An Introduction; Transducers, Piezoelectric

Bibliography

Brown B, Goodman J E 1965 *High Intensity Ultrasonics,
Industrial Applications*. Iliffe, London
Culshaw B, Gardiner P, McDonach A (eds.) 1992 *Proc.
1st Int. Conf. Smart Structures and Materials*. Strath-
clyde University, Glasgow, UK
Knowles G J (ed.) 1991 Active materials and adaptive
structures. *Proc. ADPA/AIAA/ASME/SPIE Conf.
Active Materials and Adaptive Structures*. Institute of
Physics, London
Lee C-K, Moon F C 1990 Modal sensors/actuators. *J.
Appl. Mech.* **57**, 434–41
1988 Magnetostriction: the competitive edge. *Sens. Rev.*
8(1), 41–3
Ragulskis K, Bansevicius R, Barauskas R, Kulvietis G
1988 *Vibromotors for Precision Microrobots*. Hemisphere, Washington, DC

J. H. Milner
[City University, London, UK]

R. Bansevicius
[Kaunas University of Technology,
Kaunas, Lithuania]

Transducers, Photoelectric

The rapid increase in the use of optical techniques,
both in open air-path and fiber-optic form, for
measurement and instrumentation, has resulted in a
concomitant need for suitable electronic detectors.
The field encompasses those detectors used tradi-
tionaliy in bulk optical instruments, such as the
spectrophotometer and fluorometer, together with
those developed in recent decades, primarily for use in
optical communications technology. The instrumental
engineers of traditional optical devices have not been
slow to exploit the high sensitivity and potential that
exists as a result of these developments for a wide
range of applications (Briggs *et al.* 1990).

A photodetector is a transducer (energy conversion)
device, which transforms electromagnetic radiation
into electrons, and subsequently into a current flow in
the output circuit of the device. The photoelectric
effect is widely used as an operating principle
(Agrawal 1992). The device must be sensitive to
radiation in the region of operation of the instrument
in which it is installed; devices can be configured to
operate in the ultraviolet (200–400 nm), visible
(400–700 nm) or near infrared (700–2500 nm or beyond),

by the use of suitable materials and arrangements. Many devices have sensitivities which extend over more than one spectral region, whereas others may be tailored to be insensitive over a specific region, for example, the visible spectrum, and thus be solar blind. This has advantages where low signal levels at regions outside the visible are to be detected, often in the presence of a preponderance of solar or lamp radiation in that region. In many transducer devices for photodetection, the photocurrent generated by the transducer will require amplification, either within or outside the device. This is particularly necessary when measuring the low levels of radiant energy often experienced in many spectroscopic experiments. Transducers may be categorized as single-element detectors, such as photovoltaic cells, solid-state photodetectors, photoemissive tubes and photomultipliers, and solid-state array devices which are multielement. As such, these transducers vary considerably, in their characteristics and, thus, in their selection for a particular application. Characteristics such as sensitivity, spectral coverage, size, speed of response, fragility, power source, need for subsequent amplification, noise effects and so on, are important, in addition to the usual economic factors.

1. Glass Envelope Transducers

1.1 Photovoltaic Cells

Photovoltaic, or self-generative cells are simple and rugged devices, which require no external power supply and may be connected directly to a microammeter or galvanometer to determine their output (Willard 1981). A typical construction is shown in Fig. 1. They consist of a metal plate used as one electrode, on which a thin layer of semiconductor, such as selenium, is deposited. A second electrode is formed from a thin silver or gold layer sputtered on the selenium. Operation of the device is by generation of electron–hole pairs at the silver–selenium interface, and the electrons pass to the electrode formed from the thin sputtered metal layer. The response time of the device is slow (≤ 100 Hz), with a high temperature coefficient. Thus, it is mainly used for simple, non-sophisticated applications, such as inexpensive cameras, or in situations where external power sources are not available.

1.2 Vacuum Photodiode and Photomultiplier

The vacuum photodiode is a photocathode–anode combination contained in an evacuated envelope, to protect the integrity of the cathode material. The single-stage device is termed a photodiode, the multistage configuration includes an integral amplification stage, and is a powerful and important tool in optical detection.

A typical phototube contains a light sensitive cathode in the form of a half-cylinder of metal, coated

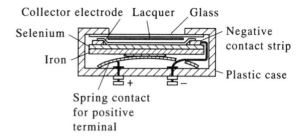

Figure 1
Construction of a photovoltaic cell

on its receiving surface with the layer on which the illumination falls. The anode wire is located along the axis of a cylinder or is a rectangular wire that frames the cathode. The transducer operates with a voltage (typically 100 V) applied between the anode and cathode; the electrons ejected as a result of radiation striking it are drawn to the anode, causing a photoelectric current i through a load resistor R_L of signal voltage iR_L. This current is directly proportional to the rate of photoelectron production, which is proportional, in turn, to the incident light flux. While they are sensitive devices, they are limited by the spurious emission of electrons caused by thermal energy (dark current) and by the low current level produced by low illumination levels. Photocurrents as low as 10 pA may be amplified in an external circuit, but there is difficulty in amplifying lower currents which may be smaller than the ohmic leakage across the tube envelope which shunts the load resistance: this affects the otherwise fast (submicrosecond) response time of the device in usual applications, with fast responses available for transient analysis.

The photomultiplier tube, or electron multiplier tube, is a combination of a photoemissive cathode with an internal electron multiplying chain of dynodes. The device operates in a similar way to the photodiode, that is, incident radiation causes the emission of photoelectrons from the cathode: these emitted electrons are focused by an electrostatic field and accelerated towards a curved surface, the first dynode, which is coated with a compound which ejects several electrons when impacted by a high-energy electron. This multiplying process is repeated several times within the phototube, over successive dynodes, each maintained at a successively higher voltage (with respect to the previous one) to produce an amplified current at the final electrode, the anode. The geometry of the dynodes and the internal construction is such as to converge the electron beam. To prevent deterioration of the dynode surfaces due to local heating, the anode current must be kept below 1 mA. Thus, the voltage between the final dynode and the anode should be restricted to ≤ 50 V, and a resistor chain is used to create a potential difference of typically 75–100 V between successive dynodes. Hence, a typ-

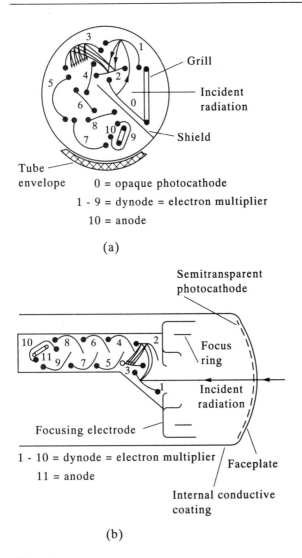

0 = opaque photocathode

1 - 9 = dynode = electron multiplier

10 = anode

(a)

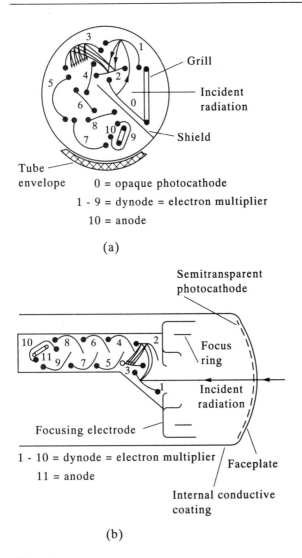

1 - 10 = dynode = electron multiplier

11 = anode

(b)

Figure 2
Photomultiplier design: (a) the circular-cage multiplier structure in a side-on tube; (b) the linear-multiplier structure in a head-on tube (courtesy of Radio Corporation of America)

ical device, with 12 stages, is operated at stabilized (for reliable performance) voltages between 1.0 kV and 1.5 kV. Two typical photomultiplier designs are shown in Fig. 2—the circular cage structure and the head-on tube—where the choice of configuration is determined by the nature of the application, the space available, power supply considerations, and so on.

The total gain of the device G, having n stages and a secondary electron emission factor ϕ per stage, is $G = (\phi)^n$. The value of ϕ is dependent on a number of factors, material, electrical and geometric, but it usually lies in the region between three and ten. The

device should be designed to have the highest possible absorption coefficient for the incident radiation on the photocathode, with the lowest possible energy absorption coefficient for photoelectrons. The surface material must also have a low work function in order to allow it to have the widest possible spectral coverage (at the long wavelength end). A wide range of photocathode types have been discussed and developed over the years. The most sensitive composition are bialkalis (K–Sb–Cs), while multialkalis (Na–K–Cs–Sb) are used for red and infrared response, with Ag–O–Cs being used to give a response up to about 1.1 μm. To operate in the ultraviolet region of the spectrum, where photon energies are higher, other materials are chosen, for example, Cs–Te cathodes, which are solar blind, are used in the region from the vacuum ultraviolet at 120 nm to 350 nm. Flat responses are available from Ga–As to cover the wide region from 200 nm to 940 nm. The wide range of such spectral responses and sensitivities of photocathodes are shown in Fig. 3, which have been taken from the catalog of a major manufacturer. The use of the device in the ultraviolet region requires the use of an appropriate window material for the envelope of the photomultiplier. Quartz is traditionally employed for ultraviolet transmission, with more exotic materials being required for transmission at wavelengths less than 200 nm.

The limitation on the use of the photomultiplier is the value of the dark current, the residual current which still flows due to thermionic emission, field emission, ohmic leakage and emission caused by natural radioactivity of the device itself. The dark current is also amplified by the dynode chain, and is

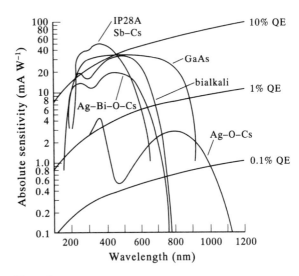

Figure 3
Spectral response curves of selected photoemissive surfaces: QE, quantum efficiency (courtesy of Radio Corporation of America)

usually reduced by physical cooling of the phototube with dry ice (solid carbon dioxide). It may be possible to automatically offset the dark current by an electronic zeroing arrangement.

2. Solid-State Transducers

2.1 p–n Photodiodes

Photodiodes operate on a completely different principle from the detectors previously discussed. The fundamental process behind the photodetection process is optical absorption in the semiconductor slab. If the energy $h\nu$ of the incident photon (where h is Planck's constant and ν is the frequency of the light) exceeds the bandgap of the material, an electron–hole pair is generated each time a photon is absorbed. Under the influence of an electric field set up by an applied voltage, an electric current is made to flow. The photocurrent i is directly proportional to the power of the incident radiation P, with the constant of proportionality ρ given in units of $A\,W^{-1}$ and representing the responsivity of the device. This is related to the quantum efficiency η, defined by:

$$\eta = \frac{h\nu\rho}{q}$$

where q is defined by the electron generation rate being given by i/q. The value of ρ may thus be given by $\eta\lambda/1.24$, in summary, where the wavelength λ is given in micrometers. This shows that the responsivity of the photodetector increases with wavelength simply because the same current can be generated with photons of reduced energy. The wavelength dependence of familar semiconductor materials, GaAs, germanium and silicon, and complex compounds, are shown in Fig. 4. The cutoff, where the quantum efficiency drops to zero, occurs where the photon energy becomes too small to generate electrons. This cutoff wavelength determines the useful range of the device.

Semiconductor photodetectors may also be classified as photovoltaic and photoconductive. A homogeneous semiconductor with ohmic contacts acts as a simple kind of photoconductive detector, where little current flows in the absence of light but, when illuminated, the conductivity increases due to electron–hole generation, in proportion to optical power. Photovoltaic detectors operate by a built-in electric field that opposes current flow in the absence of light. Electron–hole pairs generated by light absorption are swept across the device by the built-in electric field, resulting in current flow. Reverse-biased p–n junctions fall into this category, as photodiodes.

The construction of a simple planar diffused p–n junction diode is shown in Fig. 5. The response time is governed by the transit time of electron–hole pairs in the material. Typically, this is in the subnanosecond

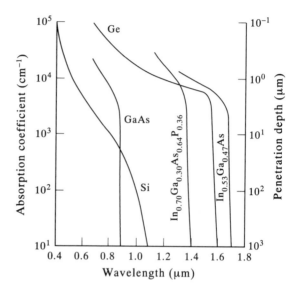

Figure 4
Wavelength dependence of the absorption coefficient for several semiconductor materials

region, and may be tailored for specific applications by device construction. The limiting factor for p–n photodiodes is the presence of a diffusive component in the photogenerated current, elevated by the absorption of light outside the depletion region. This may be optimized by decreasing the widths of the p- and n-regions and increasing the depletion region width so that most of the power is absorbed there. Such an approach is used in the p–i–n photodiode discussed in Sect. 2.2.

2.2 p–i–n Photodiode

In the p–i–n photodiode, a simple undoped (or lightly doped) region (i) is inserted between the p–n junction. This consists of nearly intrinsic material, offering a high resistance; most of the voltage drop occurs across it. The depletion region extends throughout the i-region and the dimension of the depletion width W can be controlled. Thus, the drift component of the detector current dominates (by comparison with the simple p–n junction), as most of the light is absorbed in this region. The choice of the value of W is determined by factors such as response time and quantum efficiency, and the bandwidth of the detector is limited by the slow response associated with the relatively slow transit time across an i-layer, typically 20–50 μm wide.

The performance of such devices can be improved considerably with a more sophisticated construction, which uses a double-heterostructure design. In these, the middle i-layer is sandwiched between the p- and n-type layers of a different semiconductor, whose bandgap is chosen such that the light is absorbed only in the middle i-layer. In the application of this

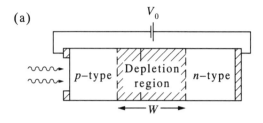

(a)

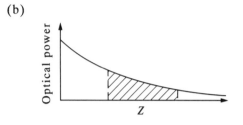

(b)

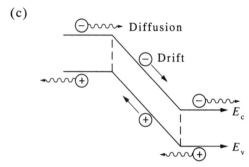

(c)

Figure 5
(a) *p–n* photodiode and the associated depletion region under reverse bias; (b) variation of optical power inside the photodiode; (c) energy-band diagram showing carrier movement by drift inside the depletion region and diffusion outside it

principle to an InGaAs device, the bandgap of InP is 1.35 eV, and it is transparent to light with $\lambda \geq 920$ nm. The bandgap of the InGaAs material is 0.75 eV, which corresponds to a cutoff wavelength of 1650 nm. Thus, the middle InGaAs layer absorbs strongly in the wavelength region 1.3–1.6 μm. The diffusive component of the detector current is absent because photons are only absorbed inside the depletion layer. Further, the quantum efficiency can be made almost 100%. Such devices are particularly useful in the optical communications field, but also for spectroscopic purposes where vacuum devices tend to be particularly inefficient.

2.3 Avalanche Photodiodes

There is a requirement that detectors have a minimum current in order to operate reliably; this translates into a minimum power $P' = i/\rho$. Thus, detectors with a large value of ρ are preferred with a lower power requirement P'. The responsivity of photodiodes is given by $q/h\nu$, for unit quantum efficiency. Avalanche

photodiodes (APDs) have much larger values of ρ, as they are designed to provide an internal current gain in a similar way to the photomultiplier, operating under the phenomenon of impact ionization. An accelerating electron may acquire enough energy to create a new electron–hole pair in this process, and a similar process may occur for holes. The magnitude of the process is determined by the impact-ionization coefficient α, of the order of 1×10^{-4} cm^{-1} for electric fields in the range 20–40 μV cm^{-1}. The application of a high voltage (≥ 100 V) can create such fields. The APD differs from the *p–i–n* device in that an additional layer is included to allow for the generation of secondary electron–hole pairs by impact ionization. Under reverse bias, a high electric field exists in the *p*-type layer sandwiched between the *i*-type and *n*-type layers. This is referred to as the gain or multiplication region; the *i*-type region still acts as the depletion or radiation absorption region. Electrons generated in the *i*-type region crossing the gain region generate secondary electron–hole pairs and give a current gain.

The device has the drawback of a longer response time, due to the multiplication process and the gain itself decreases at higher frequencies because of this. Silicon devices can be designed to have a high performance, for operation in the 800 nm wavelength region. A typical value of the gain factor is 100, resulting in a diode with low noise and relatively large bandwidth.

2.4 Advanced APD Structures

It is difficult to make high-quality APD devices which operate in the 1.3–1.6 μm region; this is the region where vacuum photomultipliers tend to have a very poor performance. An advanced structure of APD for use in this important region is one based on InGaAs, which uses an InP layer in a heterostructure device for the gain region. This overcomes the difficulty of tunnelling breakdown and the low bandwidth performance of InGaAs alone. Since the absorption and multiplication layers in such a structure are separate, this structure is known as a separate absorption and multiplication (SAM) APD. The complexity of the device is increased by the inclusion of an additional layer between InP and InGaAs, of intermediate value bandgap, to overcome the slow response and small bandwidth of such a device. The use of multiple quantum well (MQW) or superlattice structures is a different approach to the development of high-response, high-bandwidth photodiodes. This requires more advanced production techniques with alternate layers (≤ 10 nm thickness) of different bandgap semiconductor materials. Such advances have only come about in recent years with the increased sophistication of the production of semiconductor structures.

2.5 Charge-Coupled Devices

The metal–oxide–semiconductor (MOS) transistor is an important device for implementing integrated circuitry, especially in the computer industry. The metal–insulator–semiconductor (MIS) capacitor is a related device; the charge-coupled device (CCD) is an important application of this technology. The detailed operation of the device is discussed by Ferendici (1991). In summary, the CCD concept is based on the storing of charges (minority carriers) in a potential well, generated in a MIS capacitor. By manipulating these potentials, this charge is transferred from one well to an adjacent one. The concept may be easily understood, by analogy, in terms of the transfer of fluid from one bucket to an adjacent one, in a bucket brigade. Charge injection can take place by optical illumination, on a semiconductor surface next to an insulator, in the surface transfer CCD. The actual CCD comprises a number of MOS capacitors placed next to each other in a linear or planar array. In operation, the charge is finally collected in an external circuit, by a diode producing a pulsed current, depending on the charge collected and thus the degree of illumination.

The diode array spectrometer uses such CCD devices; this has enabled the simplification of the conventional double-beam device, with the added benefit of the ability to capture a complete spectrum at any one time.

See also: Analog Signal Conditioning and Processing; Instrument Elements: Models and Characteristics; Optical Instruments; Transducers: An Introduction

Bibliography

Agrawal G P 1992 *Fiber Optic Communication Systems*
Briggs R, Grattan K T V, Mouaziz Z, Elvidge A 1990 On-line monitoring of residual chlorine. *Proc. 15th IAWPRC Biennial Conf. Water Research & Control.* Pergamon, Oxford, pp. 27–38
Ferendeci A M 1991 *Solid State and Electron Devices.* McGraw-Hill, New York
Kasper B L 1988 *Optical Fiber Transmission II.* In: Miller S E, Kaminow I P (eds.) Academic Press, London
Willard H H, Merritt L L, Dean J A, Settle F A 1981 *Instrumental Methods of Analysis.* Wadsworth, CA

K. T. V. Grattan
[City University, London, UK]

Transducers, Piezoelectric

Piezoelectric transducers operate using the piezoelectric effects—direct conversion of mechanical energy to electrical energy via a material which changes its state of polarization when stressed (direct piezoeffect), or which changes its shape when its ambient electric field changes (reverse piezoeffect). The direct effect was discovered by the Curie brothers in 1880, and the reverse effect was predicted by Lippmann in 1881. The range of application of these transducers is immense and is still increasing, especially with developments in adaptive, smart or intelligent structures.

1. Working Principle

Most applications exploit direct or reverse piezoelectric effects, some of them use both, simultaneously. The most important characteristic of a transducer material is the coupling factor, which is the square root of the ratio of electrical (mechanical) work that can be done under ideal conditions to the total energy stored from the mechanical (electrical) source. Piezoelectric materials are characterized by their permittivity, and elastic and piezoelectric constants which are related by two equivalent tensor equations:

$$S_i = s_{ij}^E T_j + d_{mi} E_m, \qquad T_i = c_{ij}^E S_j - e_{mi} E_m$$

where S is the strain, T is the stress, E_m is the electric field, s_{ij} and c_{ij} are the stiffness and compliance constants respectively, and e_{mi} and d_{mi} are the piezoelectric constants (e.g., $e_{33} = (-\partial T_3/\partial E_3)_S = (\partial D_3/\partial S_3)_E$. In high-frequency modes usually the second equation is used.

The important coefficients are: (a) the charge output coefficient d—the amount of charge density $(C\,m^{-1})$ caused by a given stress $(N\,m^{-1})$, typical values 60–500 pC N^{-1}; and (b) the field output coefficient g—the field $(V\,m^{-1})$ produced by a given stress $(N\,m^{-1})$, typical values, 10–40 mV N^{-1}.

A wide range of piezoceramics based on lead titanate zirconate (PZT) are being used in various applications. Transducers exploiting the direct piezo-effect (generator) use so-called soft ceramics, with enhanced permittivity and lower aging with poorer linearity. Hard ceramics with lower permittivity but enhanced linearity are used in transducers exploiting the reverse piezoeffect (motor). Dielectric and mechanical loss factors are amplitude dependent; peak dynamic stress defines the type of material used.

Piezoelectric transducers offer the possibility of controlling the structure of a system in two unique ways: (a) by changing the connections of the electrodes to generate different displacement patterns; and (b) by varying the direction of the polarization vector to generate multicomponent motion, three-dimensional trajectories, nonsinusoidal oscillations, and so on.

An illustration of a transducer using a variable polarization vector is shown in Fig. 1. A cylindrical rod made from PZT ceramic is initially radially polarized using a central electrode (in a form of a wire) and an electrode on the outer cylindrical surface of the rod. After the poling is complete, the outer electrode is divided into three sections. The application of electric potential to the right and left groups of

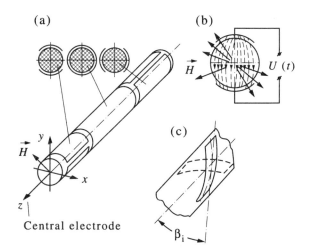

(a)

(b)

$\vec{H}$ $U(t)$

$\vec{H}$

y

x

z

Central electrode

(c)

β_i

Figure 1
Schematic representation of the piezoelectric transducer with variable vector of polarization H: (a) the form of electrodes; (b) distribution of electric potential in a cross section of the transducer; (c) the form of electrode in case of torsional oscillations ($\beta_t = \pi/4$)

electrodes results in a field which is only fully aligned with the direction of polarization near the electroded surface. This gives the possibility of generating the following types of motion in the rod: (a) longitudinal displacements or oscillations along the z axis; (b) flexural displacements or oscillations in the xz or yz planes; and (c) torsional displacements or oscillations (changing the central outer electrode in a way shown in Fig. 1c).

2. Applications

Piezoelectric transducers for converting mechanical energy to electrical energy and vice versa have, over the years, been one of the most important applications in precision engineering. Recently, new areas of application have emerged, for example, in precision structures, damage assessment and health monitoring, adaptive optics, advanced avionics and adaptive space structures, ensuring both enhanced and more reliable systems.

2.1 Transducers with Direct Piezoeffect
Conversion of mechanical to electrical power requires high electromechanical coupling, good time and temperature stability, high permittivity (especially for low-frequency applications), and high mechanical strength. Most of these applications are in the nonresonant operating mode, for example, accelerometers, photoflash actuators, ignition systems, hydrophones, microphones, and so on. Resonant operating mode devices are mainly for direct conversion of mechanical

energy to electrical energy, for example, low-cost piezoactive materials in combination with high-impedance light sources, panels from piezoactive materials in which flutter-type self-induced oscillations are generated by wind, and so on.

2.2 Transducers with Reverse Piezoeffect
Transducers with a reverse piezoeffect represent a very large group and require high mechanical strength, good linearity, high mechanical coupling, high permittivity for low-frequency transducers, and good time and temperature stability.

Both resonant and nonresonant operating modes are used, which include a great number of well-known transducers for measuring and nondestructive testing (devices with various types of waves (longitudinal, transverse, shear, lamb, Rayleigh, etc.)), metallurgical, chemical and biological applications (atomization, degassing, precipitation, crystallization, etc.), buzzers, loudspeakers, machine-tool drivers, and also for adaptive mechanics and smart structures, for example, error compensation in optical devices and active truss elements for control of precision structures.

2.3 Motion Transformers
Motion transformers are a new and important class of piezoelectric transducers, comprising piezoelectric drives, piezoelectric actuators, piezoelectric (ultrasonic) motors, and vibromotors. According to the dynamic processes taking place in the transducer, they can be divided into two groups.

(*a*) *Actuators with small displacements*. These are based on the deformation of piezoelectric transducers. The number of degrees of mobility of such actuators is unlimited and depends only on a number of control channels.

(*b*) *Actuators with unlimited displacements*. These are based on the transformation of mechanical oscillations into continuous motion. The maximum number of degrees of mobility is five, but in the case of transducers made from flexible piezoactive materials (polyvinylidine fluoride) film (PVDF), piezorubber (PZR), it is unlimited. Oscillating transducers can interact with liquids; in this case the maximum number of degrees of mobility is six (as in a "piezoelectric fish").

In Fig. 2 an example of an actuator with three degrees of mobility is shown. The direction and velocity of rotation of mirror 1 depends on the distribution in the contact area (between the transducer and mirror) of the three vector components of the high-frequency oscillations, which are controlled by activating certain zones of the transducer.

The main feature of piezoelectric transducer–actuators, which are of importance to precision engineering, is multifunctional application. This is the ability to integrate a larger part of a system on a single transducer through the combination of electronic,

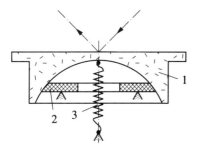

Figure 2
Piezoelectric scanner: 1 is the mirror, 2 is the piezoceramic disc with segmented electrodes, 3 is the spring

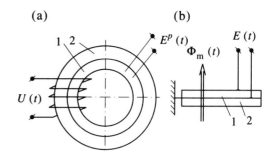

Figure 4
Transducers–sensors, using both (1) piezoelectric and (2) magnetostrictive materials: (a) coupled transducer with low-input and high-output impedance; (b) magnetic field sensor

optoelectronic and mechanical functions, in order to match the application requirements in terms of functionality, performance, reliability and flexibility. This feature is a characteristic of all piezoelectric transducers and is related to structure control. The simplest example is given in Fig. 3, where an "intelligent" gripper is actuated by transducer 1 with several zones of electrodes. Transformation of high-frequency, resonant, nonharmonical oscillations results in a gripping motion, whereas signals from the other electrodes are correlated with the force, acting on the object, touch and slip sense. Applying two excitation signals to the electrodes gives

$$U_1(t) = U_{10}\cos\gamma t$$
$$U_2(t) = U_{20}\cos 2(\gamma + \Delta\gamma)t$$

which results in low-frequency beat oscillations of the gripper, the frequency being equal to $\Delta\gamma/2\pi$ Hz, the mode of operation used in the assembly.

Other features of piezoelectric transducer–actuators are as follows.

(a) Ruggedness, simplicity, small size per unit force, extreme precision, good dynamic response, positioning accuracy, no backlash, wide range of

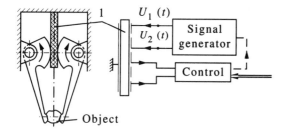

Figure 3
Piezoelectric gripper—the illustration of multifunctionality in piezoelectric transducers. The gripper is actuated when $U_1(t) = U_{10}\cos(\gamma t \pm \pi)$, $U_2(t) = U_{20}\cos 2\gamma t$

speed control, wide temperature range (up to the Curie point of the piezoactive material).

(b) Possibility of miniaturization—piezoelectric actuators will fill the gap between the usual drives and the subminiature devices manufactured by silicon technology.

2.4 Coupled Piezoelectric and Magnetostrictive Transducers

Coupled piezoelectric and magnetostrictive transducers are systems which can transform magnetic energy to electrical energy and vice versa or electrical energy to electrical energy with a large impedance ratio via mechanical coupling between the two transducers (see Fig. 4).

See also: Analog Signal Conditioning and Processing; Instrument Elements: Models and Characteristics; Transducers; An Introduction; Transducers, Magnetostrictive

Bibliography

Burfoot J C, Taylor G W 1979 *Polar Dielectrics and Their Applications*. University of California Press, Berkeley, CA
Brown B, Goodman J E 1965 *High Intensity Ultrasonics, Industrial Applications*. Iliffe, London
Cady W G 1964 *Piezoelectricity*. Dover, New York
Culshaw B, Gardiner P, McDonach A (eds.) 1992 *Proc. 1st Int. Conf. Smart Structures and Materials*. Strathclyde University, UK
Fleischer M, Meixner H 1991 Ultrasonic motors. *Mechatronics* 1(4), 403–15
Jaffe B, Cook W, Jaffe H 1971 *Piezoelectric Ceramics*. Academic Press, London
Jebb A, Bansevicius R 1991 Actuator/sensor mechanisms for microrobots. *Proc. 8th World Congr. Theory of Machines and Mechanisms*, Vol. 3. Society of Czechoslovakia Mathematicians and Physicists, Prague, pp. 875–8
Knowles G J (ed.) 1991 Active materials and adaptive structures, *Proc. Meeting in Alexandria, VA*. Grumman Corporation, Bethpage, NY

Moulson A J, Herbert J M 1990 *Electroceramics*. Chapman and Hall, London

Lee C-K, Moon F C 1990 Modal sensors/actuators. *J. Appl. Mech.* **57**, 434–41

Preumont A, Dufour J-P, Malkian Chr 1992 Active damping by a local force feedback with piezoelectric actuators. *J. Guid. Control Dyn.* **15**, 390–5

Ragulskis K, Bansevicius R, Barauskas R, Kulvietis G 1988 *Vibromotors for Precision Microrobots*. Hemisphere, Washington, DC

van Randeraat J, Setterington R E (eds.) 1974 *Piezoelectric Ceramics*. Mullard, London

R. Bansevicius
[Kaunas University of Technology,
Kaunas, Lithuania]

J. H. Milner
[City University, London, UK]

Transducers, Pneumatic

Although pneumatic instrumentation is now being superseded by the equivalent electronic equipment, it has played a major role in the evolution of the oil, gas and chemical industries over the past six decades, as the only technology that could function safely and reliably in hazardous environments. Also, diaphragm motors and air cylinders provide abundant safe power for actuating the final control elements.

It is important to appreciate that, in the process industries, measurement of temperature, pressure, flow, level and density fulfills at least 95% of the operational requirements. Pneumatic transducers, which have been subjected to several decades of development and refinement, are available for all these process parameters. The modern electronic counterparts have benefited from the evolution of pneumatic transducers but, from a measurement viewpoint, only recently have significant improvements been achieved.

1. Basic Pneumatic Sensor

All pneumatic transducers depend for their operation on a flapper/nozzle system to sense a very small deflection or movement. The essential components, shown in Fig. 1, are a nozzle formed from a truncated 90° cone with an axial hole or throat, typically between 0.2 mm and 1.0 mm in diameter, which is supplied via a restriction (which is equivalent to a pneumatic resistance) with compressed air, typically at a pressure of 120 kPa. The flapper is a smooth flat surface which is positioned close to the nozzle and arranged so that it can move with respect to the throat of the nozzle. In an ideal system, the pressure in the section between the restrictor and the nozzle rises to

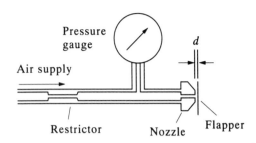

Figure 1
Basic components of a flapper/nozzle system

the full value of the supply when the flapper comes into contact with the nozzle, as shown in Fig. 2. If the flapper is moved away from the nozzle, this pressure falls, initially very rapidly and then more gradually, towards an asymptotic value determined by the supply pressure, the characteristics of the restrictor and the dimensions of the throat in the nozzle.

To make use of this sensor for measurement purposes, the pressure change in the section between the nozzle and the restrictor must be amplified; this is

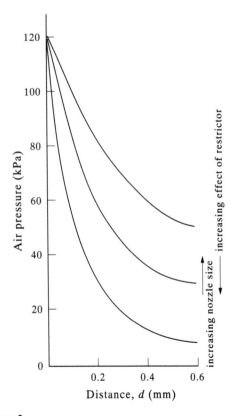

Figure 2
Basic characteristic of a flapper/nozzle system

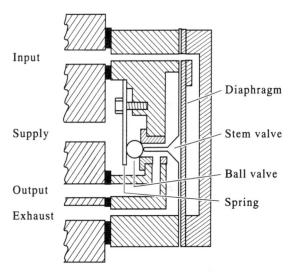

Figure 3
Arrangement of a pneumatic relay (amplifier) (courtesy of Foxboro Company)

effected by a relay (which is the pneumatic equivalent of an amplifier), a typical example of which is shown in Fig. 3. Its input is taken from the section between the restriction and the nozzle so that the pressure is applied to a diaphragm on which a conical stem valve is mounted.

The majority of the early pneumatic measuring instruments involved a motion balance system such as that shown in Fig. 4. The motion, due to the sensing element responding to a change in the measured variable, is applied to pivot A, causing it to move, for example, to the left. Initially, this reduces the gap

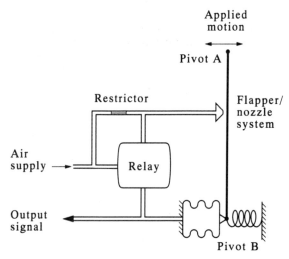

Figure 4
Arrangement of a pneumatic motion balance system

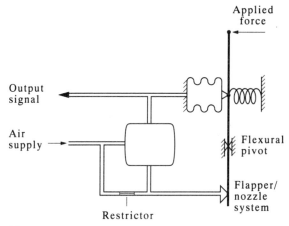

Figure 5
Arrangement of a pneumatic force balance system

between the flapper and the nozzle so that the back pressure increases. This change is amplified by the relay and applied to a bellows, causing it to expand until the link, pivoted at B, moves the flapper to a new position with respect to the nozzle, which reestablishes equilibrium.

The corresponding force balance system is shown in Fig. 5. The rigid force bar is pivoted near its centre, and the force to be measured is applied at one end. The flapper/nozzle system is mounted at the opposite end, adjacent to a bellows and an opposing spring which, as before, provides an initial bias or zero adjustment. The signal from the flapper/nozzle is amplified by the relay and then applied to a bellows. In operation, when a force is applied as shown in the diagram, the gap between the flapper and nozzle is reduced slightly, causing the back pressure to increase. This change is amplified by the relay and applied to the bellows so that a new equilibrium position is reached in which the input force is balanced.

As mentioned previously, motion balance transducers are more sensitive to vibration than force balance transducers. Consequently, relatively few of the pneumatic transducers now manufactured are based on this principle but it is quite widely used in pneumatic controllers. Because of the wide application of the basic force balance mechanism, it is described in detail in Sect. 2.

2. Principal Measurements

There are seven physical parameters for which pneumatic transducers are commercially available, namely: temperature, pressure, flow, level, density, speed of rotation and motion or position and, in addition, equipment for integrating, multiplying and dividing pneumatic signals has been developed as well

Table 1
Characteristics of pneumatic temperature transducers

	SAMA class		
	IA, IB	IIA, B, C, D	IIIB
Principle of operation	volume change	pressure change	pressure change
Temperature range	fairly low to moderate	low to moderate	very low to high
Spans	moderate	moderate	moderate
Response	slow	moderate	moderate
Scale	linear	nonlinear	linear
Overrange Capabability	medium	least	greatest
Fill medium	liquid	vapor	gas
Sensor size	small	medium	large
Ambient temperature compensation	A: full B: case	not required	B: case

as ancillary equipment such as actuators, pneumatic-to-current and current-to-pneumatic converters and positioners.

2.1 Temperature Transducers

All pneumatic temperature transducers are based on filled thermal systems, the principal features of which are set out in Table 1.

In the Scientific Apparatus Manufacturers Association (SAMA) classification set out in Table 1, class I covers liquid expansion systems which are characterized by good accuracy, narrow spans, the smallest bulbs, uniform scales and the capability of being applied in differential measurements. Class IA devices are fully compensated for the effects of changes in ambient temperature on the accuracy of measurement, and have a greater overrange protection than other devices. In class IB devices, the compensation for variations in ambient temperature is applied to the motion sensing device but not to the connecting capillary. Class II systems are simple and reasonably acccurate devices which do not require any compensation for the effect of changes in ambient temperature. Class IIC systems can cross the ambient temperature and IID systems can operate above, below or at ambient temperature. Class III covers the gas-filled systems which, in general, involve larger sensor bulbs than those required for class I and class II systems, but the bulb can be shaped to suit specific requirements. Class III sensors are usually operated in conjunction with a force balance mechanism and this enables spans as narrow as 25 °C to be obtained.

Compared with the liquid- and vapor-filled systems, the gas-filled system is characterized by its basic linear response and by its ability to be set to operate with both narrower or wider spans. The span setting is determined by the pressure of the gas filling which cannot be adjusted subsequently without complete recalibration.

2.2 Pressure Transducers

All the pneumatic pressure transmitters depend on transducing the pressure to be measured into a force or a motion for their operation; there are two principal methods for achieving this, one based on diaphragms and the other on the Bourdon tubes. Absolute pressure transmitters based on motion balance use a double spiral Bourdon tube. One of the Bourdon tubes is evacuated and sealed while the pressure to be measured is applied to the other. Changes in the ambient pressure are sensed by the sealed tube and transferred by a system of links and levers to correct the signal from the primary sensing tube. This type of sensor is used for transmitters having upper range values of 100 kPa to 700 kPa.

The gauge pressure transmitters based on the force balance system use either bellows or the basic Bourdon tube, as opposed to the helical or spiral versions. A typical transmitter based on a bellows is shown in Fig. 6.

From the viewpoint of process industries, differential pressure transmitters are particularly important because they provide the means for transducing the measurements of flow, level and density into a pneumatic signal. Their widest use is in measuring the differential pressure created by the flow of fluid through a head producing device such as an orifice plate, a Venturi or a nozzle, but they can also be used to determine liquid level by measuring the static pressure at a point in or near the base of a storage

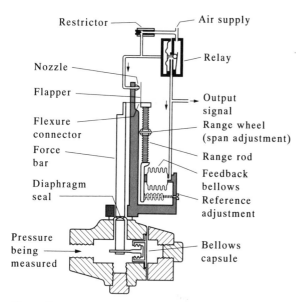

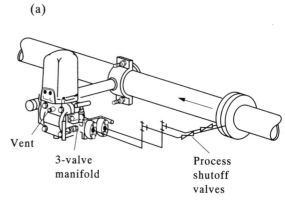

(a)

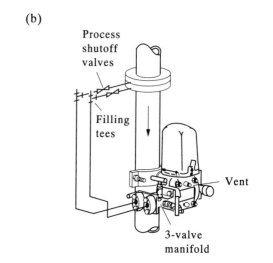

(b)

Figure 6
Force balance mechanism operated in conjunction with a medium-pressure range bellows assembly (courtesy of Foxboro Company)

vessel, as well as to determine liquid density by measuring the difference in static pressure at two points a known distance apart in a liquid.

Figure 7
Typical installations of: (a) a force balance transmitter; (b) an orifice plate system for flow measurement (courtesy of Foxboro Company)

2.3 Flow Transducers

The majority of flow measurements in process industries are made with an orifice plate and a differential pressure transmitter, of the type described in Sect. 2.2. A typical arrangement is shown in Fig. 7, but more comprehensive information regarding the method is to be found in national and international standards, such as BS 1042 and ISO 5167.

An alternative to the orifice plate method of applying Bernoulli's equation to the measurement of flow is to be found in the target or drag plate flowmeter. In it, a circular target is located centrally in the pipe so that the flowing fluid is forced through the annular gap, whereas with an orifice plate the fluid is forced through the central hole. The resultant force exerted on target is proportional to the square of the flow rate.

The principal advantages of this method are that it can handle viscous and dirty liquids, because the problems of blocked connections to the sensor, which occur with orifice plate–differential pressure systems, are avoided.

2.4 Level Transducers

The majority of level measurements which involve pneumatic instruments are based on static pressure measurements, and use pressure transducers of the

type previously described, but modified to enable them to be bolted onto the side of a vessel. A basic system is shown in Fig. 8.

2.5 Density Transducers

There are two principal methods of measuring the density of a liquid, one being based on the measurement of the static pressure at two levels a known distance apart and the other on the measurement of the buoyancy force on a submerged displacer. The basic system for the former is shown in Fig. 9.

2.6 Speed Transducers

The rotational speed of a shaft can be measured pneumatically by first transducing the rotary motion into a force, then measuring the force with another variation of the force balance system.

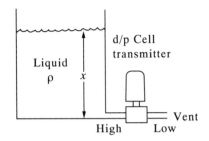

Figure 8
Arrangement of a differential pressure transmitter for measuring liquid level, basic scheme: d/p, differential pressure; ρ, density

2.7 Motion/Position Transducers

The principal application of these transducers is for positioning the stems of control valves in accordance with pneumatic input signals, irrespective of stem friction and the operating pressure of the line in which the associated valve is installed. An example of such a transducer is shown in Fig. 10. The unit is usually mounted on the valve yoke with a peg on the valve stem located in an elongated slot in the feedback arm to convert the stem position into an angular rotation of a shaft.

2.8 Pneumatic-to-Current Converters

These are a particular type of electronic pressure transmitter which is used to provide the interface between pneumatic and electronic control systems. They are characterized by the fact that the input signal is "clean" air as opposed to a hostile process fluid, almost invariably in the range 20 kPa to 100 kPa, or the equivalent in alternative units, and they are located in a sheltered environment. The output is usually 4 mA to 20 mA. The majority are based on

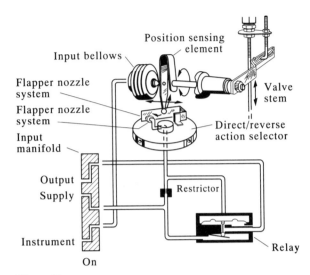

Figure 10
Pneumatic (valve stem) positioner (courtesy of Foxboro Company)

silicon or metal strain gauge technology and are assembled in compact multiple units.

2.9 Current-to-Pneumatic Converters and Positioners

Although the majority of modern control loops are based on electronic systems, the final control element is most often actuated by a diaphragm motor or an air cylinder. Current-to-pneumatic converters and positioners are required for this purpose.

The essential components of a current-to-pneumatic converter are shown in Fig. 11. The input current is applied to a rectangular coil mounted on cross flexures which allow it relative freedom to rotate about its axis, but prevent any axial movement. The coil is sus-

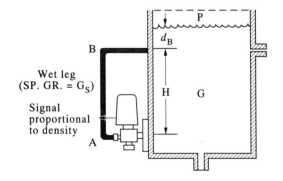

Figure 9
Arrangement of differential pressure transmitter for measuring liquid density: SP.GR, specific gravity; d_B, minimum level of liquid above level B required when measuring density

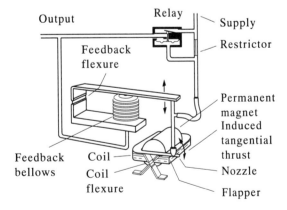

Figure 11
Current-to-pneumatic converter (courtesy of Foxboro Company)

pended around a cylindrical permanent magnet which is magnetized diametrically and, in conjunction with the surrounding hollow cylinder, provides a powerful magnetic field. At one end, the coil carries a flapper which is positioned close to a nozzle which, in turn, is mounted on an arm suspended from the main body of the unit via a flexural pivot. When the direct current flowing through the coil changes, it rotates slightly, and in doing so the relative position of the flapper and nozzle also changes. This, in turn, causes a change in the back pressure at the nozzle which is then amplified by the relay and applied to the feedback bellows so that the relative position of the flapper and nozzle is restored virtually to its previous value. In operation, the flapper/nozzle system is adjusted so that when a current of 4 mA is applied to the coil, the output signal is 20 kPa and when the current is 20 mA the output is 100 kPa.

See also: Instrument Elements: Models and Characteristics; Transducers: An Introduction

E. H. Higham
[Sussex University, Falmer, UK]

Transducers, Potentiometric

The potentiometric transducer converts displacement into a voltage change by the change in position of a wiper on a resistive element with a fixed voltage held across it, as shown schematically in Fig. 1. Simple designs may use a linear wire or multiturn helical wire as the resistive element and thus linear or angular displacement can be measured. Other measurands require a second transduction process to transform the measurement to a displacement or rotation. This article will discuss the operation of the potentiometric transducer and will concentrate on the metal wire-wound potentiometer as it offers high-precision performance.

1. Operating Principle

1.1 Sensitivity

From simple analysis, the circuit of Fig. 1 operates as a voltage divider, with the open circuit output voltage V_0 given by

$$V_0 = V_s \frac{R_x}{R_T} \qquad (1)$$

where R_x is the resistance (and the output impedance) corresponding to a relative wiper displacement x, R_T is the total resistance of, and V_s is the voltage across, the potentiometer windings. For a uniformly wound

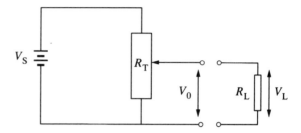

Figure 1
Schematic representation of the electrical circuit of a potentiometric transducer

potentiometer the relative wiper displacement and ratio of resistances $R_x : R_T$ are equal.

However, the voltage measurement device will have an associated load resistance R_L, and circuit analysis (see Fig. 1) then gives an output voltage V_L of

$$V_L = \frac{x}{1 + \dfrac{R_T}{R_L} x(1-x)} V_s \qquad (2)$$

For open circuit ($R_L = \infty$) the response is linear, but for finite values of load resistance the output is nonlinear. Examination of the response function shows that good linearity (1–2% of full scale) can be achieved if, for a given load resistance, a low-resistance potentiometer is chosen. This is in conflict with the desire for high sensitivity. The sensitivity of an unloaded potentiometer is given in volts for a full-scale travel of the wiper, which ideally is the voltage across the potentiometer windings. Sensitivity is thus in direct proportion to the voltage across the potentiometer, and suggests that increasing this voltage increases sensitivity. However, this voltage is limited by the power rating of the potentiometer and its heat dissipation. For dissipation of the resistive heating the maximum voltage across the potentiometer is given by

$$V_s = \sqrt{P R_T} \qquad (3)$$

where P is the power dissipation in watts. Thus, low potentiometer resistance limits the applied voltage and, therefore, the sensitivity. In order to achieve a sufficiently high resistance for a given wiper stroke, the wire-wound construction is used, and it is this that provides the high sensitivity offered by this type of potentiometer over resistive-film devices.

The nonlinear response can be overcome in other ways: by designing the potentiometer to be nonlinear, perhaps by using a former of varying cross section, or by using a second nonlinear resistor in series with the load resistor, which may itself be a nonlinear potentio-

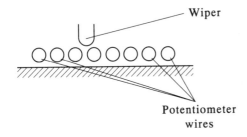

Figure 2
Potentiometer wiper shorts adjacent turns

meter with its wiper attached to that of the transduction potentiometer.

1.2 Resolution and Noise

For a voltage V_s across a precision, wire-wound potentiometer of n windings, the voltage resolution ΔV is expected to be

$$\Delta V = \frac{V_s}{n} \qquad (4)$$

That is, there are discrete steps in resolution. The situation is further complicated by the fact that in practical systems the wiper blade may short adjacent turns of the potentiometer, as shown in Fig. 2. Magnitude and duration (as the wiper moves) of these steps in resolution depends on the design and geometry of the device. For example, in a new potentiometer, the radius of the wire and the radius tip must be carefully considered; a wiper which is too sharp may rip the track and one which is too blunt may short more than two wires. Further, as the device is used, its characteristics change as flats develop, but the potentiometer must operate within specification over at least 10^6 complete cycles of movement.

One obvious source of noise in a potentiometer is thermal (or Johnson) noise associated with the potentiometer resistance, which gives rise to a root mean square variation in voltage across the potentiometer, ΔV, of

$$\Delta V = \sqrt{4 k T \Delta f R_T} \qquad (5)$$

where Δf is the system bandwidth, k is Boltzmann's constant and T is the absolute temperature. Clearly, this is negligible unless operated under high-temperature, high-bandwidth or high-resistance regimes.

Other possible sources of noise are variations in contact resistance between the windings and wiper, due to variation in contact area and pressure fluctuations, and foreign particles on the track. This is an important factor in potentiometer noise as it invariably increases as the device gets older due to wear of the track and oxidization of the metal components. The

choice of metals used must also be made carefully, as rubbing of different metals may generate small voltages, as may thermoelectric effects. Combinations of metals which reduce these effects have been identified and are commonly used (e.g., gold wiper and constantan winding).

A final serious noise problem arises due to vibration or high-velocity movement of the potentiometer wiper when contact with the coil may be temporarily lost. This is potentially very serious as its magnitude is comparable with the full-scale voltage variation. The design of potentiometers usually addresses this problem by carefully examining the contact pressure.

2. Summary

In principle, the potentiometer is suitable for both static and dynamic measurements. As a second transduction process is often required to transform the measurand to displacement, the response is determined by the combined transducer response. The potentiometer does, however, introduce certain limitations in dynamic measurements. The input impedance of the potentiometer depends on the mass of the wiper, the friction between the wiper and windings, and the friction in the bearing. This can be reduced by careful engineering, but only at increased expense. The input impedance generally limits high-frequency response.

The potentiometric transducer is primarily used in applications where a linear response and large dynamic range are required from a transducer that can provide a high-output power.

See also: Analog Signal Conditioning and Processing; Instrument Elements: Models and Characteristics; Transducers: An Introduction

K. Weir
[City University, London, UK]

Transducers, Solid-State

Solid-state transducers may be developed in all five signal domains (radiant, thermal, magnetic, mechanical and chemical). The term solid-state transducers usually implies silicon sensors, as only silicon technology is mature enough to consider the integration of sensors with readout circuits. Advantages are: (a) minimum dimensions of the complete measurement system (all the functions are integrated in the smallest possible unit); (b) optimum signal-to-noise ratio (the leads from sensing element to amplifier or impedance converter are as short as possible); (c) optimum effectiveness of compensation techniques (accurately matched compensating elements that are at the same temperature can be realized to compensate for undesirable characteristics such as offset); and (d) on-chip multiplexing. Moreover, an analog-to-digital control electronics can be incorporated into the chip to allow a connection to a digital sensor bus. Sensors complying with such features are usually referred to as "smart" sensors.

1. Radiant Sensors

Silicon has an indirect bandgap of 1.12 eV, which prohibits its use as a photondetector in the infrared range. Photondetectors in the 1–10 μm range are usually based on lead salts. Silicon infrared detectors are thermal radiation detectors, using a thermopile with the hot junction covered with a black absorber to allow the conversion of incident radiation into heat. Silicon is a suitable material for photon detection in the visible part of the spectrum. Both the photoconduction and the photojunction effect can be applied. In photoconductors the excess generated charge carriers cause a change in the resistivity of a lightly doped slab of silicon. The highest sensitivity can be obtained using two interdigitated meander shaped contacts. The disadvantage of the photoconductor is its relatively long response time. In photojunction detectors the photogenerated electron–hole pairs are separated in *p–n* junctions with the holes drifting towards the *p*-type layer. In a solar cell the detector operates in the photovoltaic mode (external short-circuiting of the junction) and the detector supplies a photocurrent proportional to the intensity of the incident light and a wavelength dependent efficiency factor. An improvement of the long-wavelength response is possible using a PIN structure as shown in Fig. 1, with an antireflective coating to minimize the reflectance that originates from the refractive index of silicon (3.5 at 700 nm). A very high sensitivity can be obtained by operating the diode at a reverse bias close to the avalanche multiplication voltage; the avalanche gain can be increased to over 100 times the photovoltaic response.

Another important solid-state radiant detector is the position sensitive detector (PSD). The operation is based on the lateral photoelectric effect.

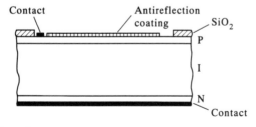

Figure 1
PIN photodiode with a high responsivity

A doped silicon layer with two opposite contacts, spaced a certain distance apart, can be used to sense the position of a light spot in between the contacts. The position of the light spot is proportional to the normalized difference in the photocurrents. Semiconductor photodetectors can be easily combined with selection and readout circuits to form high-density two-dimensional solid-state cameras.

2. Thermal Sensors

Temperature is a measure of the amount of heat stored in a system or medium. Temperature sensors are either of the contact type, in which the sensor comes into thermal equilibrium with the substance whose temperature is being measured, or noncontact type, in which the temperature is measured using the radiation laws. The latter category are considered to be radiant sensors and include the infrared sensors. A key issue for contact type temperature sensors is the measurement error introduced by the heat capacity of the sensing device and the heat leakage. In the case of temperature sensing in a system with a small heat capacity the equilibrium temperature will be determined by the original object and sensor temperatures, and by the ratio between their heat capacities. Therefore, for temperature sensing in small systems a small temperature sensor and effective isolation from the surroundings is mandatory, leading to silicon integrated sensors. Temperature sensors can be classified as sensors that utilize a different intermediate signal domain: tandem transducers and thermal sensors that directly provide an electrical output signal. Usually the mechanical domain is used in thermal tandem transducers.

Another class of direct temperature sensors utilizes the Seebeck effect. Such a thermocouple is comprised of a closed electrical circuit of two different metals, in which a potential is generated that is proportional to the temperature difference between the two junctions. An output voltage can be measured proportional to the temperature difference. A disadvantage of temperature difference measurements is that one of the junctions should be connected to a reference temperature for absolute temperature measurement. The sensitivity can be increased by alternately placing several thermocouple leads in series, creating a thermopile.

A large number of Si–Al junctions can be produced in series using lithographic techniques, resulting in a device as shown in Fig. 2, with sensitivities exceeding 100 mV K^{-1}.

3. Magnetic Sensors

The principle of operation of silicon magnetic sensors is based on the Lorentz force: the deflection of charge carriers in a direction perpendicular to both

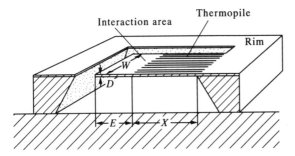

Figure 2
Cantilever beam thermopile temperature difference
sensor

the current and the magnetic field is proportional to the
product of the velocity of the charge carriers and
the magnetic field strength. Classification of magnetic
sensors can be made according to the underlying
mechanism determining the sensitivity: the Hall effect
or current deflection. A voltage is measured in Hall
effect devices. The direction of the current is fixed and
the applied magnetic field will rotate the electric field
vector. In current-deflection measuring devices, the
electric field direction is fixed and the current direction
rotates, depending on the applied magnetic field.
Silicon is a particularly suitable material for the
fabrication of magnetic sensors, despite its relatively
low mobility in comparison with gallium arsenide
(GaAs). GaAs can be operated up to 250 °C, silicon
up to 150 °C, but silicon is a much better choice in
terms of dissipated power. A Hall plate is a rectangular sheet of resistive material with four symmetrically
positioned contacts at the plate boundaries (see Fig.
3). A magnetotransistor (magnistor) is a bipolar
junction transistor (BJT) or a field-effect transistor
(FET), modified such that the deflection of the current
in either the base, collector or channel can be measured.

Finally, there are other magnetic field sensors such
as the carrier-domain magnetometer and the magneto-
diode. A magnetic field sensitive current domain in
carrier-domain magnetometers moves continuously
through the device, resulting in a frequency-dependent
output signal. Charge carriers in magnetodiodes are
deflected between a low- and high-recombination area,
resulting in magnetic field dependent diode characteristics. Since all magnetic sensors measure some kind of
differential signal, careful design is needed to minimize the imbalance. Contactless switching as used in
keyboards and brushless dc motors is one of the major
mass production applications of Hall plates.

4. Mechanical Sensors

Silicon is used as a mechanical material with a Young's
modulus of $2 \times 10^7\,\mathrm{N\,cm^{-2}}$. It exhibits a high

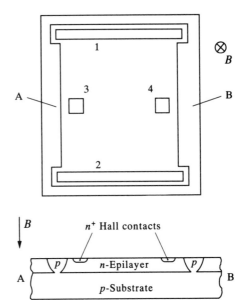

Figure 3
Integrated Hall plate

piezoresistive effect with both positive and negative
gauge factors up to 150, depending on the crystal
orientation and the type of doping of the layer. Silicon
is not piezoelectric; however, piezoelectric layers, such
as ZnO, can be deposited on top of a silicon layer.
This technique is frequently used in surface acoustic
wave (SAW) devices. Integrated resistors are used to
measure the internal strain in a micromechanical
structure, which is the result of an external mechanical
quantity. Anisotropic etching along a preferential
crystal axis is used to realize selective thinning of only
one of the dimensions of a silicon wafer. In this way
membranes and cantilevers are constructed. Deposition of thin layers and the selective removal of some is
also used to realize mechanical sensors (surface micromachining). Maximum sensitivity can be tailored by
placing the piezoresistor in a position where the strain
developed by the mechanical quantity (pressure, acceleration or tactile force) is at a maximum. In a pressure
sensor the piezoresistors are placed in a bridge configuration in a membrane, to compensate for the
temperature dependence of the piezoresistive effect.
Channels are etched around a seismic mass, apart
from a few small suspension beams, to realize integrated acceleration sensors. The piezoresistors are
placed in the beams to sense the acceleration induced
strain in the beams. Pressure and acceleration sensors
are also realized by measuring the deflection of a
membrane capacitively, as shown in Fig. 4. Air and
fluid flow is measured indirectly using thermal and
micromechanical methods.

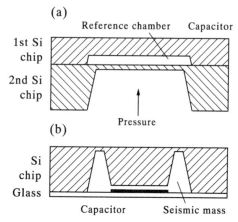

Figure 4
(a) Capacitive pressure sensor composed of two silicon chips to form both the capacitor and the reference chamber; (b) Capacitive acceleration sensor bonded to a glass substrate

5. Chemical Sensors

Chemical sensors respond to the molecular or ionic concentration of a specified component in a gas or liquid. The main types of chemical sensors are: (a) potentiometric sensors, that generate an electrochemical potential related to the concentration of the material in solution; (b) amperometric sensors, that produce an electric current due to an electrochemical reaction (proportional to the concentration of the material in solution); (c) electrical admittance sensors, that change their conductivity or permittivity upon exposure to a gas; (d) catalytic sensors, with which the heat liberated in a controlled chemical reaction is measured; and (e) mass sensors, that measure the mass of a gas or liquid absorbed by a specific absorbant.

Silicon is not a suitable sensor material for chemical signals. The major role of silicon in chemical sensors is to act as the substrate for a chemical interface between the substance and an electronic component. Silicon sensors with a completely integrated transduction are still rare.

The most extensively investigated silicon-based chemical sensor is the ion-sensitive field-effect transistor (ISFET). This potentiometric device is basically a metal–oxide–semiconductor field-effect transistor (MOSFET), without a metal gate. The gate insulator makes direct contact with the electrolyte; the surface potential at the oxide–electrolyte interface depends on the pH and determines the drain current of the device. The measurement of the potential difference between the sensor and the liquid requires a separate, stable reference electrode. The response of an ISFET is not fully in accordance with the Nernst equation ($U = c + 59$ mV/pH unit, where U is the electrode potential and c is the standard potential), due to the particular interaction between the inorganic insulator and the electrolyte.

Sensors of the conductivity type often use a properly doped metal oxide, whose resistivity changes with the uptake of a gas. A commercial type is the tin oxide gas sensor, which is sensitive to CO. The operating temperature is high, over 300 °C. Many organic materials change their conductivity upon exposure to certain gases; the mechanisms underlying these effects are not yet very well understood.

The concentration of flammable gases (such as hydrocarbons, and also CO) can be measured with catalytic devices. Such transducers consist of a catalyst (to sustain the reaction at reasonable temperatures), a temperature sensor (to measure the temperature rise due to the heat of reaction) and a heater (to maintain the catalyst at the operating temperature). Common catalysts are platinum and palladium. Operating temperatures are high (typically 500 °C). Catalytic sensors are not gas specific.

MOSFETs with a catalytic metal gate are sensitive to hydrogen-containing molecules such as H_2, NH_3, H_2S, and so on. The gas diffuses through the metal and is adsorbed at the metal–oxide interface. This results in a change of the threshold voltage of the MOSFET, that depends on the gas concentration at the surface. The oldest type is the palladium MOSFET for H_2. Concentrations down to 0.1 ppm can be detected. The operating temperature is between 60 °C and 150 °C. The response time is of the order of seconds. Practical Pd-gate MOSFETs have an integrated heating resistor and a temperature sensor.

Mass sensors for gas detection are based either on a vibrating crystal or on surface acoustic waves, both coated with a gas-specific absorber.

An important class of humidity sensors is that based on the absorption of water from the substance under test. Such absorption sensors use the relationship between a characteristic property of hygroscopic materials and the amount of absorbed water at the absorption equilibrium.

The most popular material for use as an absorption sensor is Al_2O_3. When aluminum is electrochemically oxidized (anodization), a porous layer of Al_2O_3 is created on the aluminum surface. When exposed to a humid atmosphere, this layer absorbs water molecules, partially filling the pores with liquid water by capillary condensation. Both the resistivity and the dielectric constant of the layer change according to the amount of absorbed water, which is, in turn, related to the relative humidity. The construction of a sensor is completed by the deposition of a metal layer on top of the Al_2O_3 (mostly gold for its chemical resistivity), thin enough to allow water molecules to penetrate into the pores. The structure acts as either a capacitor or a resistor, both varying with relative humidity.

Some polymers have a relative permittivity that changes with water absorption and can, therefore, be used as a dielectric material for capacitive humidity sensors. The most thoroughly investigated polymers for this purpose are cellulose acetate buthyrate (CAB) and polyimide. The sensors are produced as flat capacitors, mounted on a glass or ceramic substrate. The top electrode consists of a very thin metal layer, or it has a digitated structure, to allow the uptake of water by the polymer film.

Current research on absorption sensors is directed to the use of other porous ceramics and compounds (for instance, $MgCr_2O_4–TiO_2$) and other polymers.

Accurate humidity measurements are obtained with the dew-point method. This method is based on maintaining equilibrium between evaporation and condensation of the water on a cooled surface. This equilibrium occurs, by definition, at the dew-point temperature, which is uniquely related to the water vapor content of the test gas. Optical dew detectors use a polished metal mirror. Dew on the cooled mirror is detected by an electrooptical system, responding to scattering of a light beam by the dew drops. Capacitive dew detectors consist of a flat body with an electrically isolating top layer (for instance, oxidized silicon) on which a pair of interdigitated electrodes is deposited (for instance, aluminum or tantalum).

See also: Analog Signal Processing and Conditioning; Instrument Elements: Models and Characteristics; Transducers: An Introduction

R. F. Wolffenbuttel, P. J. A. Munter and
P. P. L. Regtien
[Delft University of Technology,
Delft, The Netherlands]

Transducers: Strain Gauges

Strain gauges are resistive elements made from conductor material in convenient forms which, when strained, react by varying their nominal resistance values. They can measure the strain of a structure, to which they are bonded by suitable techniques, by measuring the variations in their resistance. Besides monitoring the strain state of the structure, the gauges are used in instrumentation for the construction of transducers of physical quantities, such as force, pressure, acceleration, and so on, which are converted to strain by suitable mechanical devices.

1. Principles of Operation

Traditionally, a strain gauge, in its simplest form, is made from very thin metal wire (typically some tenths of a micrometer thick) bonded on an isolating support of the same thickness, so that it appears as shown in the grid in Fig. 1. It is a pattern that takes advantage

of the maximum sensitivity distribution along direction x. The wire is generally welded onto two thicker metallic pads. It is possible to solder the cables onto them, which are used for external connection with the measuring instrument. To a first approximation, since the wires are mainly aligned in the x direction, it can be assumed that the wire undergoes the same strain as the surface to which it is attached.

2. Sensitivity

A metallic wire under stress, within its elastic field, is stretched and at the same time there is a decrease in its cross-sectional area; in particular, for a circular section, considering $\varepsilon = \Delta l/l$, where ε is the strain, its diameter D undergoes a variation equal to $(-\nu\varepsilon)$, where ν is Poisson's ratio. The relationship between the relative resistance variation and the strain can be easily calculated by remembering that the resistance of a conductor is:

$$R = \rho \, \frac{l}{S} \tag{1}$$

where ρ is the resistivity of the material, l is the length of the wire and S is the area of its section ($\pi D^2/4$ for a circular section). By taking the logarithms, differentiating, and adopting differentials to represent the small variations of physical quantities, from Eqn. (1) we have:

$$\frac{\Delta R}{R} = \frac{\Delta \rho}{\rho} + \varepsilon - 2 \, \frac{\Delta D}{D} \tag{2}$$

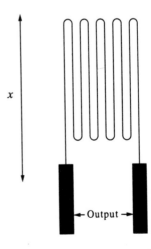

Figure 1
Example of the geometric disposition of a metallic wire strain gauge

and introducing Poisson's ratio:

$$\frac{\Delta R}{R} = \frac{\Delta \rho}{\rho} + (1 + 2\nu)\varepsilon \qquad (3)$$

Evidently the variation of resistance depends on two terms; one takes the geometrical variations of the conductor into consideration, while the other takes the resistivity variations of the conductor material into consideration. This latter phenomenon is called the piezoresistive effect.

Sensitivity to strain is defined by the relationship: $K = \Delta R/R\varepsilon$, where K is the gauge factor. Manufacturing firms usually supply the value of K and R, so that from the measurement of ΔR, as previously stated, the measurement of ε can be deduced.

3. Causes for Error and Influencing Factors

The proportionality relationship between the relative resistance variation and the strain is based on a few simple hypotheses and on the assumption that the strain of a structure is transmitted unchanged to the sensitive element of the strain gauge. The actual devices deviate considerably from the ideal behavior, for different reasons; some of these are:

(a) the hysteresis characteristic of the sensitive element material;

(b) the dependence of K on ε; and

(c) the dependence of K on the transversal sectional strain (lateral effect).

Besides these causes for error, there are those that are derived from the technical production of strain gauges, from the aging, and from the way they are bonded to the structure whose strain is being measured. In addition to the effects deriving from nonlinear behavior, other effects must be evaluated: those derived from parasitic influencing factors during resistance measuring. The most important are as follows.

(a) *Temperature.* This is responsible for expansion or contraction of a structure, resistance variations of the resistive element, variations in the gauge factor and thermoelectric effects on the junction between the strain gauge and the cables.

(b) *Magnetic fields.* These may cause induced voltages, when adequately linked to the resistive element, and determine resistive variations, which depend on the resistive material used, both for magnetoresistive and magnetostriction effects.

Both these factors influence the measurement accuracy of strain gauges. However, other possibilities exist, which enable improvement of accuracy; these include

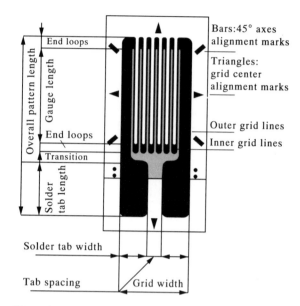

Figure 2
Actual metallic strain gauge foil and related terminology

the correct choice of materials and technologies used in the production of strain gauges. Accurate and suitable bonding techniques for strain gauges, the use of convenient instruments for resistance, and resistance variation measurements all compensate for the effects described.

4. Strain Gauge Technologies

4.1 Metal Strain Gauges

The gauges normally used for strain measurement usually comprise a thin metal grid on a support, commonly a film of epoxide material (reinforced by glass wires for high-temperature applications), as shown in Fig. 2. The technique is similar to that for printed circuits (which uses film deposition and chemical etching to accurately define geometry). The most commonly used metals for the grid are constantan, certain modified alloys of nickel–chromium, an alloy of iron–chromium–molybdenum and, depending on temperature, platinum–tungsten. The choice of these materials depends on minimizing the errors indicated earlier. The value of the unstrained resistance is, for the most part, equal to 120 Ω. The gauge factor depends almost exclusively on the geometrical relative change, since values vary from 2 to 3.5 depending on the metal used. The gauge factor is given by:

$$K = (1 + 2\nu) + \frac{\Delta \rho}{\rho \varepsilon}$$

and because $\nu = 0.3$

$$K = 1.6 + \frac{\Delta \rho}{\rho \varepsilon}$$

The fundamental structure described is then used to make strain gauges that are even more complex. Using the same support, more grids are set up at different angles (rosette at 90°, 120°, etc.), suitable for different types of measurement, or to accomplish the suitable compensation. In the construction of force and pressure transducers, when pressure deflects thin steel plates, the traditional gauge has been substituted by resistors deposited directly onto the elastic elements. This thin-film technology thus eliminates problems of bonding which are typical of film strain gauges.

4.2 Semiconductor Strain Gauge

The semiconductor materials silicon and germanium are known to have a high piezoresistivity effect and therefore have a very high sensitivity, which is 65 to 130 times higher than that of metals. However, the increased sensitivity has some drawbacks, such as: (a) dependence of the sensitivity on the highly nonlinear crystallographical direction; (b) strong dependency of R and K on the temperature; (c) high fragility; (d) limited range of measurable strain; and (e) high cost. Essentially, the semiconductor strain gauge is realized with the same techniques used for the metal strain gauge, at least considering its use as a single element for bonding onto structures. Semiconductor strain gauges are particularly interesting for the construction of solid-state transducers of physical quantities such as pressure, acceleration, and so on.

The advantages of this technology are the possibility of miniaturizing the transducers and the possibility of integrating the measuring circuit, the compensation network and the sensitive element to the same semiconductor chip. The sensitive element is an integrating and inseparable part of the strain element (there are no problems associated with bonding to the elastic structure).

4.3 Thick-Film Strain Gauges

Other materials which are suitable for the production of strain gauges, due to their piezoresistive properties, are resistive silk screen pastes for hybrid circuits. Their K value can be between 10 and 30; that is, between the K values of metals and semiconductors. They maintain their properties even with regard to temperature variations, as do metal strain gauges. The use of these strain gauges is limited to the construction of transducers of physical quantities such as pressure, force, acceleration, and so on, in which the elements sensitive to induced strain are deposited for silk-screening onto a strained substrate, usually alumina. With this technology it is possible to make transducers of reduced dimensions, incorporating electronic circuits, at costs slightly lower than those of semiconductors.

See also: Analog Signal Conditioning and Processing; Instrument Elements: Models and Characteristics; Mass, Force and Weight Measurement; Transducers: An Introduction

Bibliography

Allocca J A, Stuart A 1984 *Transducers, Theory and Applications*. Prentice-Hall, Englewood Cliffs, NJ, pp. 29–68
Anderson J C 1986 Thin film transducers and sensors. *J. Vac. Sci. Technol. A* **4**, 610–16
Avril J 1974 *Encyclopedie Vishay d'Analyse des Contraintes*. Vishay-Micromesures, Paris, pp. 169–216
Beckwith T G, Buck N L 1973 *Mechanical Measurements*, 2nd edn. Addison-Wesley, Reading, MA, pp. 287–313
Canali C, Malavasi D, Morten B, Prudenziati M, Taroni A 1980 Piezoresistive effects in thick film resistor. *J. Appl. Phys.* **51**, 3282
Canali C, Malavasi D, Morten B, Prudenziati M, Taroni A 1980 Strain sensitivity in thick film resistors. *IEEE Trans. Comp. Hybrid Manuf. Tech.* **3**, 241
Germer W, Todt W 1983 Low-cost pressure/force transducer with silicon thin film strain gauges. *Sens. Actuat.* **4**, 183–9
Giles A C M, Somers G H J 1973 Miniature pressure transducer with a silicon diaphragm. *Philips Tech. Rev.* **33**, 14–20
Prudenziati M 1984 Strain gauge a film sottile e loro applicazioni. *Fisica e Tecnologia* **7**, 191–206

D. Marioli
[Università degli Studi di Brescia, Brescia, Italy]

Transducers, Thermal

Thermal transducers are devices used for measuring radiant heat flux. The incident radiant power heats the transducer to a temperature T_D(°C) which is above the surrounding temperature T_S(°C). The transducer is usually a resistive, thermoelectric or pyroelectric detector which gives a resistance, voltage or current output: this depends on the temperature difference $T_D - T_S$ and, thus, incident power. Thermal transducers respond equally to all wavelengths in the incident radiation and are therefore broadband devices; this is in contrast to photoelectric transducers which only respond to a narrow band of wavelengths and are narrowband devices.

In order to examine the factors affecting both the sensitivity and the time constant, the heat balance equation for a thermal transducer is required; this is as follows:

$$W - UA(T_D - T_S) = MC\frac{dT}{dt} \qquad (1)$$

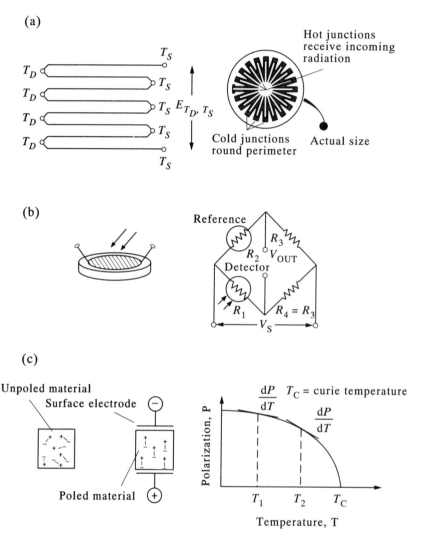

Figure 1
Thermal transducers: (a) thermopile, (b) bolometer, (c) pyroelectric

where W is the incident radiant power (W), M is the transducer mass (kg), C is the transducer specific heat $(J\,kg^{-1}\,°C^{-1})$, A is the transducer surface area (m^2), and U is the heat transfer coefficient $(W\,m^{-2}\,°C^{-1})$.

Rearranging Eqn. (1) gives:

$$T_D + \tau \frac{dT_D}{dt} = \frac{1}{UA} W + T_S \qquad (2)$$

where the time constant $\tau = MC/UA$. In the steady state $dT_D/dt = 0$, and Eqn. (2) reduces to:

$$T_D - T_S = \frac{1}{UA} W \qquad (3)$$

showing that temperature difference is proportional to incident power.

Commonly used thermal transducers are thermopiles (see Fig. 1a), which consist of a large number of thermocouples in series, and bolometers, which are metal or semiconductor (thermistor) resistance material in the form of thin films or flakes (see Fig. 1b). In both cases the surface of the detector is blackened to maximize the absorption of incoming radiation. For a thermopile consisting of n thermocouples in series, with a hot junction at T_D and a reference junction at T_S, the emf is given approximately by:

$$E_{T_D,T_S} \simeq na_1(T_D - T_S) \qquad (4)$$

Substituting for $(T_D - T_S)$ from Eqn. (3) gives:

$$E_{T_D, T_S} \simeq \left(\frac{na_1}{UA}\right) W \qquad (5)$$

Therefore, the transducer sensitivity or responsivity K is given by:

$$K = \frac{na_1}{UA} \cdot \frac{mV}{W}$$

The bridge circuit shown in Fig. 1b gives an output voltage approximately proportional to power W. Here R_1 is a radiation detecting bolometer at temperature T_D, R_2 is a reference bolometer at temperature T_S and R_3 and R_4 are fixed equal resistors. If R_1 and R_2 are metal resistive elements with resistance R_0 at $0\,°C$ and temperature coefficient $\alpha\,°C^{-1}$ then:

$$R_1 = R_0(1 + \alpha T_D), \qquad R_2 = R_0(1 + \alpha T_S) \qquad (6)$$

Provided $R_3 \gg R_0$, the bridge output voltage is given approximately by:

$$V_{OUT} \simeq V_S \frac{R_0}{R_3} \alpha (T_D - T_S) \qquad (7)$$

Using Eqn. (3) gives:

$$V_{OUT} \simeq \frac{V_S \alpha R_0}{UA R_3} \cdot W \qquad (8)$$

Therefore, system output voltage is approximately proportional to W and responsivity K, is given by:

$$K = \frac{V_S \alpha R_0}{UA R_3} \cdot \frac{V}{W}$$

The time constant τ of a thermal transducer is minimized by using thin flakes or films which have a large area-to-volume ratio, that is, a small value of M/A. However, τ cannot be reduced much below a few milliseconds because of the low heat transfer coefficient U between the transducer and the surrounding air.

Pyroelectric thermal transducers are coming into wider use. These are man-made ferroelectric ceramics which also show piezoelectric properties. The principle of ferroelectricity is shown in Fig. 1c. The ceramic is composed of a mass of minute crystallites; provided the ceramic is below the Curie temperature, each crystallite behaves as a small electric dipole. Normally this material is unpoled, that is, the electric dipoles are randomly oriented with respect to each other. The material can be poled, that is, the dipoles are lined up, by applying an electric field when the ceramic is just below the Curie temperature. After the material has cooled and the applied field has been removed, the dipoles remain lined up, leaving the ceramic with a residual polarization P. The pyroelectric effect arises because the incident radiant power causes the ceramic temperature T to increase; P decreases with T according to the nonlinear relationship shown in Fig. 1c. This reduction in P causes a reduction in capture surface charge in the material and an excess of induced change on the electrodes. If Δq is the excess charge caused by a temperature rise ΔT then

$$\Delta q = \left(\frac{dP}{dT}\right) A \Delta T \qquad (9)$$

where A is the area of the electrodes and dP/dT is the slope of the P–T characteristics. The electrodes and the rectangular block of dielectric ceramic form a parallel plate capacitor C_N. The ceramic can either be regarded as a charge generator, Δq in parallel with C_N, or a Norton current source, i_N in parallel with C_N, where:

$$i_N = \frac{dq}{dt} = A \frac{dP}{dT} \cdot \frac{dT}{dt} \qquad (10)$$

A typical pyroelectric transducer has a diameter of 2 mm, a responsivity or sensitivity of $250\ V\ W^{-1}$ and a noise equivalent power of $2 \times 10^{-9}\ W\ Hz^{-1}$ at $25\,°C$. When used with a silicon window the wavelength response is reduced to between 1 and 15.

See also: Analog Signal Conditioning and Processing; Instrument Elements: Models and Characteristics; Temperature Measurement; Transducers: An Introduction

J. P. Bentley
[University of Teesside,
Middlesbrough, UK]

V

Velocity and Acceleration Measurement

The four quantities, linear speed, linear acceleration, angular speed and angular acceleration are measured in many areas of technology.

1. Linear Acceleration

Acceleration is the rate of change of speed, which is the rate of change of position. Accelerometers—devices which measure acceleration are very widely used. Typical applications are:

(a) vibration measurement to assess wear in machines,

(b) the investigation of structures under impact and shock conditions;

(c) in guidance systems for missiles and torpedoes;

(d) on stable platforms for spacecraft and ships; and

(e) the measurement of acceleration and deceleration of mass transit systems.

There are several commonly used types of accelerometer, described and compared in the following sections.

1.1 Piezoelectric Accelerometers

When piezoelectric material, in the form shown in Fig. 1, is subjected to acceleration, a force is exerted on it which produces a charge or voltage. There is generally a linear relation between the charge produced and the change in applied force:

$$Q = D\Delta F$$

where Q is the induced surface charge, D is the coefficient of proportionality and ΔF is the change in applied force. The values of D and some other important parameters for piezoelectric materials are shown in Table 1.

The device acts like a parallel-plate capacitor of capacitance

$$C = \frac{\varepsilon_0 \varepsilon_r A}{d}$$

where A is the area, d is the plate separation, ε_0 is the permittivity of free space and ε_r is the relative permittivity. The change in potential can be calculated from

$$\Delta V = \frac{D\Delta Fd}{\varepsilon_0 \varepsilon_r A}$$

For example, a piece of barium titanate, $1\,cm^2$ in area and $1\,mm$ thick, subjected to a force from a $1\,g$ weight, would develop a potential difference of $1.4\,mV$ across its thickness.

Table 1 shows the resistivity of some piezoelectric materials. The device shown in Fig. 1 has a mechanical resonance, which means that for a reasonably linear response the frequency should not be more than about one third of the resonant frequency. Piezoelectric accelerometers are usable over a frequency range between their low- and high-frequency limits.

The transverse sensitivity is usually about 2% of the main axis sensitivity. Piezoelectric accelerometers are available with a wide range of sensitivities and frequency responses suitable for many applications.

1.2 Strain Gauge Accelerometers

These devices usually consist of a full or half bridge of elements, the resistances of which depend strongly on strain, mounted on a cantilever beam and carrying a seismic mass as shown in Fig. 2. The beam is often located in a suitable damping fluid.

The low-output impedance of the transducer makes signal conditioning easy. The transverse sensitivity is about 2%, the same as for the piezoelectric accelerometer.

1.3 Micromachined Accelerometers

These accelerometers use silicon technology and consist of a micromachined silicon mass suspended by a number of beams from a silicon frame. Piezoresistors

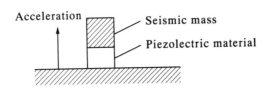

Figure 1
Piezoelectric accelerometer

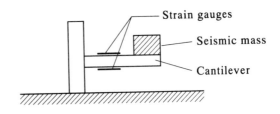

Figure 2
Strain gauge accelerometer

Table 1
Properties of some piezoelectric materials

	Coefficient D (10^{-12} C N^{-1})	Relative dielectric constant	Young's modulus (GN m^{-2})	Maximum safe stress (MN m^{-2})	Electrical resistivity (GΩ m)
Quartz	2.3	4.5	80	98	>1000
Barium titanate ceramic	140	1200	110	80	>100
Lead zirconate titanate	105	1600	55		>30
Lead niobate	200	1500	88	20	>100

in the beams detect the deflection of the beams with acceleration. They are low profile and low mass and are made at a low cost. They are available in surface mount package form.

1.4 Servo Accelerometers

Servo accelerometers are closed-loop force balance transducers. A typical construction is shown in Fig. 3. Any change in position of the mass is detected by the position sensor, the amplified output of which is applied to the torque motor, which opposes the torque produced in the mass by acceleration. The current applied to the torque motor is nearly proportional to the acceleration, and the output in the form of a voltage is usually produced across a resistor in series with the motor.

Servo accelerometers are typically used for relatively low acceleration measurements, below g (the acceleration due to gravity), and at low frequencies, typically below 50 Hz. The resolution is high and the transverse sensitivity is low, typically 0.2%. Since the mass of the transducer is fairly high, it is unsuitable for mounting on small objects, for which the small size of the piezoelectric or piezoresistive transducer is attractive. It is therefore used in guidance systems, for stabilizing platforms and in mass transit systems.

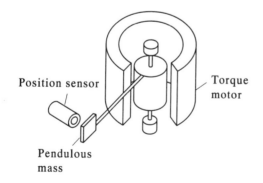

Figure 3
Servo accelerometer

2. Angular Speed

So many industrial processes depend on having a controlled or known speed of rotation, that it is not surprising that many methods of measuring angular speed have been developed for various applications. Some of the more important methods are described below.

2.1 Stroboscopes

This device has a negligible effect on the rotating object and can be used to measure the angular speed of otherwise inaccessible objects. It flashes light on the object, producing an illusion that the object is stationary when the rate of flashing is simply related to the speed of rotation. The stationary illusion occurs when the object rotates at n times the flashing rate, where n is an integer. The reading required from the stroboscope scale is at $n = 1$; the existence of the other stationary displays can cause some confusion, and the highest flashing rate to produce a stationary pattern should be chosen.

2.2 Tachometers

The word tachometer was coined in the early nineteenth century from the Greek and literally means speed meter, but in current usage the meaning is limited to a device to measure the speed of rotation of a machine.

(*a*) *DC brush tachometer*. This device is a simple dc machine as shown in Fig. 4. The voltage generated is converted by the commutator to a unidirectional but pulsating voltage. The ripple can be made small by using a number of spatially distributed coils. The output is typically high, 200 rpm (revolutions per minute) giving 1 V. The device is usable up to several thousand rpm and has good linearity, the deviation from linearity being about 0.1%.

(*b*) *AC tachometer (rotating magnet)*. In this device, unlike the dc tachometer, the coil is stationary and the magnet rotates, as shown in Fig. 5. This arrangement is useful in an environment likely to contaminate a brush tachometer, or in applications where brush replacement is difficult.

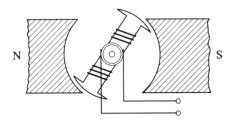

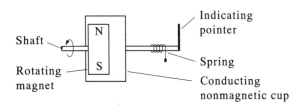

Figure 6
Drag cup tachometer

Figure 4
DC tachogenerator

(*c*) *AC tachometer (three-phase)*. This device is a three-phase electric generator together with a three-phase rectifier. Again the absence of brushes is an advantage in some applications. At low speeds, the low voltage drop across the rectifier produces nonlinearity. The speed range is therefore more restricted than that of the dc brush device.

(*d*) *Eddy current drag cup tachometer*. This device is based on one of the earliest electromagnetic discoveries, Arago's disk, the principles of which were first made clear by Michael Faraday. A typical construction is shown in Fig. 6. The movement of the permanent magnet induces eddy currents in the conducting cup, producing a tendency for it to be pulled round with the bar magnet. This motion is opposed by a spring and the resultant deflection depends on the speed of rotation of the shaft. This type of transducer can be used up to 50 000 rpm.

(*e*) *Reluctance tachometer*. The speed of a rotating shaft can be measured by a noncontact method using the change in reluctance of a path. The change in reluctance can be produced either by nonuniformities of shape inherent in the rotating object, or by the addition of a gear wheel with teeth specifically formed to produce the desired pulse shape. The device is illustrated in Fig. 7.

(*f*) *Hall effect transducer*. A Hall effect device can be used as a tachometer by inserting it between a permanent magnet and irregularities on a rotating object. With constant current through the device, the

voltage produced is in the form of pulses, the amplitude of which does not depend on speed. These transducers are therefore suitable for tachometers where a very wide speed range is required.

(*g*) *Capacitative speed transducer*. An earthed slotted disk attached to a rotating shaft can be used to alter the capacitance between the two electrodes of the three-terminal capacitor shown in Fig. 8. Impedance measurement gives precise speed measurement.

(*h*) *Optoelectronic methods*. Optoelectronic light sources and detectors are widely used to measure speed of rotation. The light emitter and detector are commonly mounted on opposite sides of a gap which is interrupted by a suitable disk attached to the rotating shaft. Most sensors use infrared light-emitting diodes (LEDs), since the most readily available photodiodes and phototransistors are most sensitive to wavelengths in that region.

Between 500 and 5000 slits are made in the extremely low-inertia rotating disk. The light detector intercepts the light passing through a total of about 20 adjacent slits for easy alignment and to reduce the effect of disk imperfections. The arrangement often used is shown in Fig. 9. Incremental light interrupters usually consist of a stationary mask and moving disk, both having a grid of opaque and transparent areas. As the disk moves past the mask, a triangular waveform is produced by the detector. This output is converted into pulses whose frequency gives a measure of the angular speed.

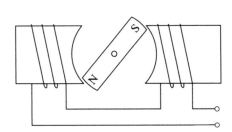

Figure 5
AC tachometer

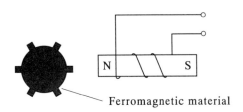

Figure 7
Reluctance tachometer

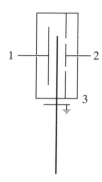

Figure 8
Capacitative tachometer (terminals denoted 1, 2, 3)

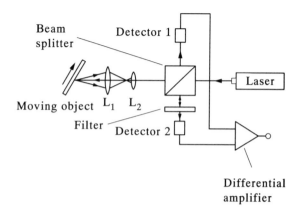

Figure 10
Laser Doppler velocimeter

For safety, to avoid electrical interference or to save space, fiber-optic cables can be used to lead light to and away from this type of device.

3. Linear Speed

Where linear motion is produced by a rotating device, as is very often the case, probably the easiest method of measuring the linear speed is to convert from a measurement of angular speed by any of the methods described in Sect. 2. Another method is to integrate the output from an accelerometer. This is preferable to differentiating the output of a displacement transducer, because of the noise problems associated with differentiation.

Linear speed can always be measured by timing the movement of an object between two points a known distance apart. LED and photodetector arrangements form a convenient means of detecting position, although capacitative and inductive methods are also applicable. Fiber-optic cable is attractive in some environments.

Linear velocity can also be measured by converting the linear motion of a measuring cable connected to the moving object into a rotation by means of a precision cable drum. A constant force spring motor provides the torque for cable retraction. This method has the advantage that cable misalignment has little effect on accuracy, therefore low-cost installation is possible. These devices are suitable for measuring velocity, where large linear displacements occur and precise alignment is difficult.

Laser Doppler velocimeters, used in flow measurement, can also be employed for measuring the linear speed of certain objects, such as wire (Saarimaa 1979). A typical scheme is shown in Fig. 10.

4. Angular Acceleration

Low angular acceleration can be accurately measured using the type of servo accelerometer described in Sect. 1. Piezoresistive angular accelerometers have also been made; they are a modified version of the piezoresistive accelerometer described in Sect. 1. Another method, which can be used for high angular accelerations, is to process the information given by optoelectronic methods using a shaft encoder as described in Sect. 2.

See also: Angular Rate Sensing; Position Inertial Systems

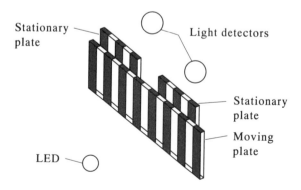

Figure 9
Optoelectronic tachometer

Bibliography

Figliola R S, Beasley D F 1991 *Theory and design for mechanical measurements.* Wiley, New York
Middelhoek S, Audet S A 1989 *Silicon Sensors.* Academic Press, London
Morris A S 1988 *Principles of Measurement and Instrumentation.* Prentice-Hall, Hemel Hempstead, UK
Muller R S, Howe R T, Senturia S D, Smith R L, White

R M 1991 Acceleration sensors. *Microsensors*. Institute of Electrical and Electronics Engineers, New York, Sect. 5.2

Pallas-Areny R, Webster J G 1991 *Sensors and Signal Conditioning*. Wiley, New York

Rietmuller W, Benecke W, Schnakenberg U, Wagner B 1991 A smart accelerator with on-chip electronics fabricated by a commercial CMOS process. *Sens. Actuat. A* **31**, 121–4

Saarimaa R 1979 A laser doppler velocimeter for surface velocity measurement. *J. Phys. E* **12**, 600–3

M. J. Cunningham
[University of Manchester, Manchester, UK]

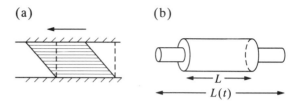

Figure 1
(a) Shear and (b) extensional flows

Viscosity Measurement

Many of the fluids used in industry have peculiar flow properties. When pumped or even stretched they do not continue to show the thickness, or viscosity, they have at rest, but change in a dramatic way that can result in them becoming much thinner, changing with time or even, in extreme cases, acting like solids. The science of rheology is devoted to the study of deformation of such materials. By and large, fluids made up of small molecules will show no change with rate of flow and are Newtonian. High molecular weight fluids, such as molten plastics, being formed from very large molecules, have viscosities that can change with flow rate and so are non-Newtonian. Those that are elastic at the same time are called viscoelastic.

A very important group of fluids industrially are those made up of suspensions or emulsions, for example, printing inks, paints and drilling muds. Often the suspended particles can form a temporary structure in the fluid that will break down with flow and reform at rest. These "thixotropic" characteristics have been put to good use in the manufacture of such commercially important materials as nondrip paints.

Shear flow, which takes place when a liquid travels along a pipe or channel, is the most commonly encountered form of deformation. The measurement of shear viscosity and elasticity is well understood and is the subject of a large body of literature.

1. Shear Flow

In the majority of instruments the fluid under examination is subjected to either a controlled shear rate ($\dot{\gamma}$) or shear stress (σ_{21}), (see Fig. 1). The response of the fluid in the form of a stress distribution can be written as

$$\sigma_{21} = \dot{\gamma}\eta(\dot{\gamma}) \tag{1}$$

$$\sigma_{11} - \sigma_{22} = N_1 \tag{2}$$

$$\sigma_{22} - \sigma_{33} = N_2 \tag{3}$$

where $\eta(\dot{\gamma})$ is the apparent shear viscosity at a given shear rate $\dot{\gamma}$, and N_1 and N_2 are the first and second normal stress differences, respectively, at shear rate $\dot{\gamma}$. Generally, N_2 is of little significance and can be measured only with considerable difficulty. The first normal stress difference N_1 is of major importance, being directly related to elasticity. When, for example, a viscoelastic fluid is extruded from a tube or die it flows to take up a diameter greater than that of the pipe. This die swell is related directly to N_1.

1.1 Viscosity Units

From Eqn. (1), apparent viscosity can be written as

$$\eta = \frac{\sigma}{\dot{\gamma}} \tag{4}$$

The SI unit of viscosity is the pascal second (Pa s). An alternative centimeter-gram-second (CGS) unit still widely used is the poise, where 10 poise = 1 Pa s.

In many industrial situations kinematic viscosity ν is used

$$\nu = \frac{\eta(\dot{\gamma})}{\rho} \tag{5}$$

where ρ is the fluid density.

1.2 Rheological Characteristics of Fluids

Figure 2 shows typical flow curves for a series of different fluids. Figure 2 also illustrates a most important property of fluids, which is that their rheological behavior can vary depending on their rate of deformation. All three fluids are Newtonian at low shear rate but fluid (b) then goes through a power law region, obeying the relationship

$$\eta = K\dot{\gamma}^{(n-1)} \tag{6}$$

where K is a constant and n has a value less than unity. It then reaches an upper Newtonian regime. On the other hand, fluid (c) becomes rheopectic or shear thickening at higher shear rates. Strictly speaking one should not refer to fluids as being Newtonian or non-Newtonian, but describe them as having these types of behavior over a given range of deformation rates. The concept is formalized by the Deborah number (De). This is defined as the ratio of the

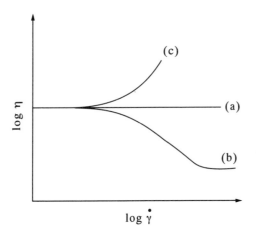

Figure 2
Flow curves of (a) Newtonian, (b) shear thinning, and (c) shear thickening fluids

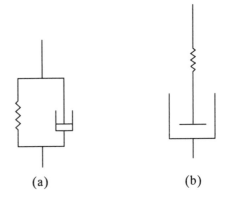

Figure 3
(a) The Kelvin model, (b) the Maxwell model

relaxation time of the fluid to a characteristic flow time. As De increases the fluid changes from having liquidlike to solidlike behavior.

Perhaps the most dramatic changes in shear flow are those shown by thixotropic and yield value fluids. In the former, viscosity falls with time of flow, and recovers at rest. In the latter, the fluid has solidlike behavior up to a yield stress σ_y, whereupon it changes abruptly to a liquid that can be Newtonian or non-Newtonian. Recent research has thrown some doubt on the concept of yield value. It would appear that, provided a sensitive enough instrument is used, flow can be detected at any stress. Nevertheless, the concept of yield value σ_y, still has considerable use in many practical applications. A number of equations have been developed which include σ_y. Examples are:

(a) the Bingham equation

$$\frac{(\sigma_{21} - \sigma_y)}{\dot\gamma} = \eta \qquad (7)$$

which describes a fluid that is Newtonian once yield value has been exceeded;

(b) the Herschel–Bulkley equation

$$\sigma_{21} - \sigma_y = k\,\dot\gamma^n \qquad (8)$$

where the fluid has power law characteristics above σ_y; and

(c) the Casson equation

$$\eta = \eta_0 + \left[\left(\frac{(\eta_\infty - \eta_0)}{(1 + \dot\gamma/\dot\gamma_c)^n} \right) \right] \qquad (9)$$

Note that Eqn. (9) incorporates the concept of an upper Newtonian viscosity η_∞. At high shear rate,

with no further structural changes taking place, the fluid reverts to a constant viscosity.

The fluids described in Eqns. (7)–(9) may be inelastic. Elastic fluids show partial recovery at the cessation of shear and the flow is therefore time dependent. A viscoelastic fluid will have a retarded recovery and display a finite relaxation time. One method of illustrating the effect of the viscous and elastic components of flow is by the use of mechanical models (see Fig. 3). These spring and dashpot components correspond to the elastic and viscous (frictional) parts of the flow, respectively, and are used to describe complex rheological behavior.

1.3 Measurement of Shear Viscosity

Ideally all the fluid under examination should be subjected to the same shear conditions. It is also necessary to ensure accurate temperature control since viscosity is highly temperature dependent, obeying the Arrhenius-type equation

$$\eta = \frac{A \exp(E_{vis})}{RT} \qquad (10)$$

where A is a preexponential function, E_{vis} is the activation energy for viscous flow, T is the temperature (K), and R is the gas constant.

1.4 Rotational Viscometers

A relatively small number of geometries exist, which allow a practical instrument to be designed in which the fluid is subjected to the same shear rate throughout. These (see Fig. 4) include cone and plate, and cone and cone. Concentric cylinders, providing the gap is very small compared with the cylinder radii, approximate to the ideal. Parallel plate viscometers do not produce a uniform shear rate, although the formulae for their use in oscillation have been precisely determined. Formulae for cone and plate and coaxial cylinder viscometers are given in Table 1.

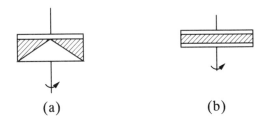

(a)   (b)

Figure 4
(a) Cone and plate and (b) parallel plate geometries

The main limitation of such instruments is that very high shear rates cannot be achieved due to breakdown of adhesion between the sample and the shearing surfaces. In addition, centrifugal and inertial forces can induce secondary flows and viscous heating. Nevertheless, they can be used to characterize a fluid precisely at low and moderate shear rates. A further important advantage of the cone and plate system is that it lends itself to the determination of normal stress, and hence elasticity. This is achieved by measuring the thrust developed normal to the direction of shear. Rotational viscometers are also frequently used to measure the change of viscosity with time (Whorlow 1980).

Undoubtedly the fastest growing area of rotational rheometry is in the use of controlled stress instruments. Instead of applying a steady shear rate, a controlled stress is applied to the fluid. The response of the sample is then monitored. The great advantage of the technique is that if a structure exists in the fluid, as is the case with many gels, suspensions and emulsions, as well as for polymeric fluids, the structure can be examined rheologically without destroying it.

Further, using oscillatory flow analysis, it is possible to calculate dynamic viscosity, storage and loss modulus, as well as yield values and relaxation times from creep experiments, virtually instantaneously.

1.5 Capillary Rheometers

Since the Poiseuille flow generated in a capillary creates a hyperbolic velocity profile, shear rate is greatest at the wall of the capillary and zero along the

Table 1
Formulas for shear stress and shear rate for cone and plate and coaxial cylinder geometries

Cone and plate (and cone and cone)	Coaxial cylinder (very narrow gap)
Shear stress $3G/2\pi R^3$	$G/2\pi R_a^2$
Shear rate Ω/ψ	$R_a\Omega/(R_2 - R_1)$

G, torque per unit length; R, cone radius; Ω, rate of rotation in (rad s^{-1}); ψ, cone angle (rad); $R_2 - R_1$, coaxial cylinder gap; R_a, average radius to center of fluid in coaxial cylinder

center line. The maximum shear rate at the wall $\dot{\gamma}$ is given by:

$$\dot{\gamma} = \frac{4Q}{\pi R^3} \qquad (11)$$

where R is the capillary radius and Q is the volume rate of flow. Shear stress σ is given by

$$\sigma = \frac{PR}{2L} \qquad (12)$$

where P is the pressure drop along the capillary.

A fluid having a yield value may give a further complication, in that a flat velocity profile can be created at the point in the capillary at which the stress falls below the yield value. This plug flow can also arise as a result of slip at the wall due to the presence of a low-viscosity layer of fluid.

It should be noted that large errors can result from entrance and exit pressure losses in the capillary instrument. These can be found by using capillaries of different length and plotting pressure against capillary length. Extrapolation to zero length gives the entrance and exit pressure losses. It has also been found that the entrance pressure loss is closely related to elongational flow. At least one commercial instrument uses the data to obtain a measure of elongational viscosity.

For process control applications capillary viscometers can be used for both low- and high-viscosity fluids. In the former the fluid is forced through the capillary by an extruder or gear pump at constant rate. The pressure drop along the capillary is then measured by transducers or similar devices. With low-viscosity liquids, providing the flow rate is a constant, the rate of movement of a float or even a rising bubble can be used to assess viscosity.

2. Oscillatory Shear Flow

It is often an advantage with fluids that show yield values or thixotropy to be able to measure the rheological behavior when the structure is undisturbed. Low-amplitude oscillatory methods (Walters 1975) offer such an opportunity and can also be used for non-structure-forming fluids.

The theory relies on measurements being made within the linear viscoelastic region when the stress generated will be proportional to the oscillation amplitude. The fluid, usually between parallel plates, is subjected to an applied oscillation of amplitude θ_2 via one plate, while the other, restrained by a torsion bar, responds with amplitude θ_1. The phase lag ϕ between the plates will depend on both dynamic viscosity η' and dynamic rigidity G' (elasticity):

$$\eta' = \frac{-S\,\theta_r \sin\phi}{(\theta_r^2 - 2\theta_r \cos\phi + 1)} \qquad (13)$$

where $\theta_r = \theta_1/\theta_2$, and

$$S = \frac{2h(K - I\omega^2)}{\pi R^4 \omega} \tag{14}$$

$$G = \frac{\omega S(\cos\phi - \theta_r)}{(\theta_r^2 - 2\theta_r\cos\phi + 1)} \tag{15}$$

where h is the sample thickness, ω is the frequency, K is the restoring constant of the torsion bar, R is the plate radius, and I is the inertia of the plate restrained by the torsion bar. For cone and plate flow a similar expression can be used, with S corrected for cone angle.

3. Extensional Flow

If a body (see Fig. 1) of length $L(t)$ (assuming a time-dependent flow) is elongated, the rate of strain $\dot{\varepsilon}$ can be defined as:

$$\dot{\varepsilon} = \left(\frac{1}{L}\right) dL/dt = -\left(\frac{1}{A}\right) dA/dt \tag{16}$$

where A is the cross-sectional area. The elongational stress σ_E is:

$$\sigma_E = \frac{\text{Load}}{A} \tag{17}$$

and extensional viscosity η_E is:

$$\eta_E = \frac{\sigma_E}{\dot{\varepsilon}_E} \tag{18}$$

The Trouton ratio, the ratio of extensional to shear viscosity, has a value of three for Newtonian flow. Some non-Newtonian fluids have been found to have Trouton ratios up to several thousand.

The difficulty in carrying out extensional flow measurements, particularly for low-viscosity fluids, has resulted in developments being considerably retarded. The techniques used fall into two main categories, equilibrium and nonequilibrium. The former, which approach closest to the ideal from a theoretical standpoint, can be applied to highly viscous fluids. Low-viscosity fluids have to be examined using the latter method, and it is only at the beginning of the 1990s that a theoretical interpretation of the data is beginning to emerge.

3.1 Equilibrium Techniques

(a) Constant strain rate measurements. A number of instruments have been designed to extend molten polymers at a constant rate and measure the stress produced. One of the best known of these is described by Meissner (1969). The polymer in the form of a thin strip or monofilament is gripped between two sets of rollers immersed in a thermostatically controlled bath. One set of rollers is held by a spring arrangement whose deflection, when the polymer strip is extended, can be related to the extensional stress.

(b) Constant stress measurements. These essentially carry out creep measurements by applying a load which decreases as the sample cross-sectional area decreases on extension, keeping the stress constant. The most notable instrument in this category is that manufactured by Rheometrics Inc., based on Munstedt's original design (Munstedt 1975).

3.2 Nonequilibrium Techniques

The instrument described by Ferguson and Hudson (1975) involves the extrusion of a low-viscosity (below 50 Pa s) fluid from a nozzle attached to a thin-walled stainless steel tube. The fluid is caught and extended by a rotating drum. The load generated is measured by the deflection of the thin-walled tube. The liquid filament is simultaneously scanned using a TV imaging system, and from the diameter of the filament and its rate of change the extensional stress and rate of extension are calculated.

Another easily set up experiment, although less easy to interpret, is the opposing jet method (see Fig. 5). Fluid is sucked into the two jets at the same rate by applying a vacuum. The force needed to produce a given flow rate is measured by deflection of one of the tubes attached to the nozzle. A stagnant region is created along the axis of the nozzles where a very strong extensional stress is generated, and this results in a flaring of birefringence at high extension rates.

4. Industrial Viscometers

A large number of process control viscometers are now commercially available. These operate not only

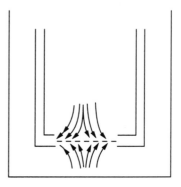

Figure 5
The opposing jet technique for extensional viscosity measurement

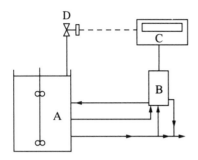

Figure 6
On-line control loop for a process control viscometer: A, mixing vessel; B, control viscometer; C, data logging microprocessor; D, control valve

on the capillary principle but also use rotational geometries such as rotating spindles, bobs, concentric cylinders, disks, paddles, cones and plates, and parallel plates. For on-line viscometry it is essential to keep the temperature constant or build in a compensation factor to allow for variations. Regardless of the exact method being used for on-line measurement using an industrial viscometer, it is necessary to have a complete laboratory assessment of the rheological characteristics of the fluid. The use of a sampling system is advisable, and a common fault with many commercial instruments is that there is no positive displacement of the fluid between the measurement surfaces to allow fresh sample to enter. A further problem that can lead to a progressive drift in the results is the build up of the fluid components on the shearing surfaces (McCullough and Andrew 1974).

A common feature of many industrial viscometers is that they operate at only one shear rate, or frequency. This is a serious fault since, as will now be clear, a non-Newtonian fluid cannot be characterized by a single point measurement. A number of instruments are, however, now available which employ capillary viscometry and oscillation methods, as well as concentric cylinder techniques to obtain a flow curve which can be compared with the ideal flow curve for on-line control. A diagrammatic view of a closed-loop control system using viscometry is shown in Fig. 6. In this case, a reactor or mixing vessel can be continuously sampled and the viscosity data from the control viscometer fed to a data logging microprocessor. This compares the input signal with the programmed value, and drives the control valve to alter the relative concentrations of the components or reagents in the mixing vessel.

See also: Density and Consistency Measurement

Bibliography

Ferguson J, Hudson N 1975 A new viscometer for the measurement of apparent elongational viscosity. *J.*
Phys. E. **8**, 526–30
McCullough R L, Andrew W G 1974 Analytical instruments. In: Andrew W G (ed.) *Applied Instrumentation in the Process Industries*, Vol. 1. Gulf Publishing, Houston, TX, pp. 200–38
Meissner J 1969 A rheometer for investigation of deformation–mechanical properties of plastic melts under defined extensional straining. *Rheol. Acta* **8**, 78–88
Munstedt H 1975 Viscoelasticity of polystyrene melts in tensile creep experiments. *Rheol. Acta* **14**, 1077–88
Walters K 1975 *Rheometry*. Chapman and Hall, London
Whorlow R W 1980 *Rheological Techniques*. Ellis Horwood, Chichester, UK

J. Ferguson
[University of Strathclyde, Glasgow, UK]

Voltage Measurement

Of all electrical quantities, voltage is the most commonly measured and, fortunately, one of the most accurately measurable. Many other parameters are determined by first converting them into an equivalent voltage. Measurement becomes more difficult as the frequency rises, so many ac signals are determined in terms of a dc equivalent. As there is no absolute value of an alternating signal many instruments respond to one quantity but are calibrated to indicate another, resulting in errors when the waveshape changes. The ideal voltmeter would present an infinite input impedance to the voltage being measured. A potentiometric system provide this feature, and therefore represents the ideal technique for measuring voltage.

1. Potentiometric Methods

1.1 Manual Potentiometers

Although the use of potentiometers has declined in recent years as digital voltmeters (DVMs) have become more accurate and more sensitive, potentiometers still have advantages as the voltage comparison circuit. This circuit is entirely passive, depending on the ratio between resistors, and is therefore dependent only on the ratio of the resistor values; the system can therefore be extremely stable and reliable.

The basic technique is to balance an unknown voltage against a known variable voltage generally derived from a tapped chain of resistors carrying a constant current (see Fig. 1). The linearity of the system is a function of the resistors, and the accuracy with which the voltage may be measured is a function of the current flowing. Before use a reference voltage is compared with the same nominal voltage derived from the potentiometer, and the potentiometer current adjusted until these two voltages are equal. With suitably stable supplies and circuit design, accuracies of better than 10 ppm can be achieved. At balance no

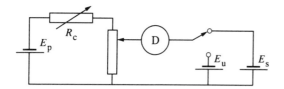

Figure 1
Principle of the manual potentiometer; E_p is the potentiometer supply potential, R_c the current control rheostat, E_s the standard source and E_u the unknown voltage

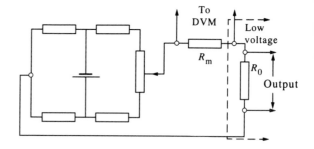

Figure 3
Modified constant-resistance–variable-current potentiometer

current is drawn from the unknown, effectively giving an infinite input impedance for the potentiometric measuring circuit.

1.2 Constant-Resistance Potentiometers

The potentiometer circuit described above is a constant-current–variable-resistance system. One of the problems associated with this design is the large number of sliding contacts between dissimilar materials in the measuring circuit. These can result in variable resistances and thermal emfs in the measuring circuit, and the latter are especially undesirable when measuring small voltages.

The circuit shown in Fig. 2 incorporates the simplest form of constant-resistance–variable-current potentiometer. The resistor R_v controls the current in—and hence the voltage across—the resistor R. As a milliammeter will probably not give a sufficiently accurate measurement of the current, it can be replaced by a standard resistor and a DVM. This simple circuit gives a nonlinear relationship between R_v and I (the current through the milliammeter), and cannot give both positive and negative values. Both these objections can be overcome by using the system shown in Fig. 3. Typically the resistors R_m and R_0 are in the ratio 1000:1 which, for an output of $\pm 100\,\mu V$ gives a DVM input of $\pm 100\,mV$. There should then be little difficulty in obtaining a variable $100\,\mu V$ output known to within $\pm 0.01\%$. It is important, though, to minimize thermal emfs in the low-voltage section of the circuit by using untinned copper leads, copper terminals and solder with a low thermal voltage with respect to copper.

These constant-resistance systems are available as

(a) potentiometers for measuring voltages in the millivolt region, for example, for measuring thermocouple voltages;

(b) part of other systems, for example, a reference circuit in ac–dc transfer standards; or

(c) separate units for generating small dc voltages.

1.3 Differential Voltmeters

These instruments first appeared in the mid-1950s and are essentially electronic potentiometers. As shown in Figs. 4 and 5, they consist of:

(a) a reference voltage from a Zener diode circuit, giving probably 1 V and 10 V as reference levels;

(b) a Kelvin–Varley divider network used to derive the required voltage, from 1 V or 10 V; and

(c) an input attenuator to give higher-range inputs, probably 100 and 1000 V.

The Kelvin–Varley divider is usually adjusted manually, and the built-in high-impedance null detector (which is sensitive to voltage) indicates the difference between the dialled voltage and the input voltage. Its scale shows the percentage of the range value, and the differential voltmeter is therefore an ideal instrument for measuring voltage changes. (The detector signal is available via terminals with sufficient power to drive a pen recorder.)

Infinite input resistance at balance can be achieved on the direct input ranges—usually 1 V and 10 V—but not on the attenuated ranges. Off-balance on the

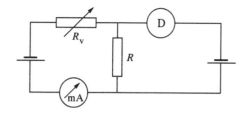

Figure 2
Constant-resistance–variable-current potentiometer; D indicates detector

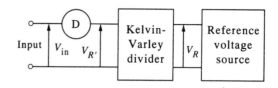

Figure 4
Differential voltmeter

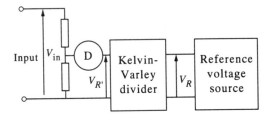

Figure 5
Variant differential voltmeter

direct input ranges, the input resistance is approximately that of the null detector—typically 1 MΩ.

2. Voltage Standards

Although periods of warranty are increasing (up to five years from some manufacturers), some thought must be given by the user to the problem of servicing and calibrating equipment. With short-term accuracies of 5 ppm or less being claimed for the latest instruments, it is becoming increasingly difficult to provide suitable calibration sources. A user may not be able to use his digital multimeter (DMM) to its lowest uncertainty unless he has ready access to a suitable voltage standard.

2.1 Standard Cell Enclosures

The most important property for any standard is stability, both short and long term. A standard cell is badly affected by changes in temperature, movement, vibration and any electrical loading placed upon it; once disturbed, manually or electrically, it may be many days before it resettles. A typical cell has an internal resistance of about 500 Ω, so drawing a current of only 1 μA (which is small enough not to affect the cell's stability) will create a voltage drop across the cell of 0.5 mV. With an emf of approximately 1 V, this represents a fall of 0.05%. For maximum accuracy, therefore, no current larger than about 1 nA can be taken, which corresponds to an effective load circuit resistance of 10^9 Ω (1000 MΩ). Any significant current drain may permanently change the cell's emf.

The best voltage standard consists of up to 12 cells contained in a common temperature-controlled enclosure. In the event of a mains power failure, an emergency battery supply automatically maintains the temperature of the enclosure. If possible, the enclosure should not be moved. If movement is essential, great care must be taken to avoid vibration and bumping. The individual cell stabilities are checked by routine intercomparison of the cells in the bank. To calibrate the bank an additional transportable unit is used to act as a transfer device between the laboratory and its reference authority (for many laboratories in the UK this will be the National Physical Laboratory

or the Services Electrical Standards Centre). With proper care, stabilities of 1 μV per year and 0.1 μV over a short timescale can be achieved. There seems to be agreement between knowledgeable users that the latest standard cell banks are less stable than their predecessors.

2.2 Commercial Electronic Voltage Standards

Commercial voltage sources based on Zener diodes, producing a variety of outputs near 1 V or 10 V, have been available for some years. Many have provided a stability of about 1 μV or 2 μV per month, that is, they have been ten times less stable than a good standard cell source. However, the best modern solid-state voltage references exhibit a random fluctuation of about 0.2 ppm and an annual drift rate of 2 ppm or better, which is usually steady and predictable.

Some solid-state voltage references contain a single reference device: others contain up to four devices. The higher grade units have the reference devices enclosed within temperature-controlled enclosures. Internal batteries are sometimes provided to maintain temperature when a mains supply is not available, the batteries being automatically recharged when a mains supply again becomes available. In some cases the operating current through the voltage reference device is maintained during the period when a mains supply is not available.

The voltage reference devices usually operate at about 7 V, and this is sometimes made available at front panel terminals. Divider networks provide outputs at a Weston cell voltage (1.018 V and 1.0 V). An output of 10 V is also usual, obtained via a gain stabilized amplifier. This 10 V output is very useful for setting up the basic range of high-grade digital voltmeters, often nominally 10 V. This 10 V output has the additional advantage that it reduces the effects of thermal emfs and noise voltages by a factor of about 10 compared with a Weston cell.

Solid-state voltage references are also incorporated into what amount to digitally set regulated power supplies called calibrators.

Calibrators, often bus controlled, can be used for setting up and calibrating all but the most accurate of digital voltmeters: for these, other techniques are required.

3. Alternating Waveforms

An alternating voltage, since it is always changing, cannot be described by a single value unless some convention is adopted. The usual convention is "that voltage which would give rise to the same amount of heating in a nonreactive resistor as would a direct voltage of the same value." Since heating in a given resistor is proportional to the square of the current (and therefore to the square of the voltage), the parameter of interest is the rms value of the voltage.

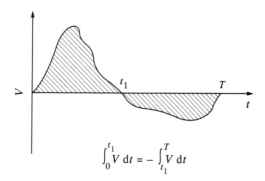

Figure 6
One cycle of an alternating signal

This is valid also for nonsinusoidal voltages and currents.

Some instruments (usually the less accurate ones) employ a rectifier and a dc responding indicator (such as a moving coil milliammeter) to indicate the mean value of the current so produced. This corresponds to the rectified mean value of the ac quantity. However, these instruments are scaled to indicate the rms value on the assumption that the signal is sinusoidal. Errors are thus produced if the input is not a sinusoid.

The more accurate instruments use some form of (often electronic) circuit element having an overall square-law response. They thus respond to the rms value of the signal but, because the electronics unavoidably has a limited dynamic range, they may not respond fully to high short-duration signal peaks. In addition, very high peak values may cause damage by a flashover. Moreover, very high rates of change of the signal may exceed the frequency response of the electronics.

The ratio of the maximum short-duration peak value to the rms value is called the crest factor. A typical value is three.

An alternating quantity is one which acts in alternate directions and whose magnitude undergoes a definite cycle of change in definite intervals of time. The positive and negative half-cycles are not necessarily similar, but the time integral over a complete cycle is zero; that is, the positive and negative are equal and there is no dc level (see Fig. 6).

One cycle is completed in the periodic time T, and the frequency f is the number of cycles per second. The average value of an alternating quantity is zero, but the mean value I_{mean} of a current after full-wave rectification is of great practical importance; the peak or maximum value I_{max} may also be important. The rms value I of an alternating current i is that direct current which would produce the same average heating effect in a pure resistance, and is given by

$$I = \left(\frac{1}{T} \int_0^T i^2 \, dt \right)^{1/2}$$

We can also define the form factor

$$F = \frac{I}{I_{mean}}$$

and the crest or peak factor

$$P = \frac{I_{max}}{I}$$

For a sinusoidal current it may be shown that

$$I_{mean} = \frac{2I_{max}}{\pi}, \quad I = \frac{I_{max}}{\sqrt{2}}$$

Hence, for a sinusoidal wave (see Fig. 7)

$$F = \pi/(2\sqrt{2}) = 1.111$$
$$P = \sqrt{2} = 1.414$$

An alternating waveform which is not sinusoidal is said to be complex. A complex waveform can be represented as a basic or fundamental wave plus one or more harmonic waves whose frequencies are integral multiples of the frequency of the fundamental. The addition of harmonics to a fundamental always results in an increase in the rms value of the waveform, whatever the phase relationship. The peak value of a waveform may be increased or decreased by the addition of a harmonic. The maximum possible change is equal to the peak value of the harmonic, and occurs when the peak of the original waveform and the peak of the added harmonic coincide.

Depending on the phase relationship, the addition of an odd harmonic may change the resulting mean value in either direction by up to $(x/n)\%$, where x is the magnitude of the nth harmonic expressed as a percentage of the fundamental. The addition of even harmonics may increase the mean value, or may leave it unaltered.

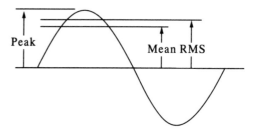

Figure 7
A sinusoidal signal

4. The Decibel Scale

In many measuring systems, relative signal levels may be of greater significance than absolute levels. The decibel system of specifying relative signal levels, originally used in line communication systems, is in widespread use throughout the electronics industry. The transfer of power is fundamental in all electrical systems, and consequently the basic statement of the decibel scale is in terms of power.

If the power delivered into the circuit element shown in Fig. 8 is P_i, and the power delivered into the terminating resistance R_o is P_o, then the gain in decibels is defined as

$$G_{dB} = 10 \log_{10}(P_o/P_i) \tag{1}$$

A positive value indicates a gain—amplification, while a negative value indicates a loss—attenuation. While this is the basic definition, any direct practical application requires the measurement of power, which is costly and certainly less convenient than the measurement of voltage.

Equation (1) can be rewritten as

$$G_{dB} = 10 \log_{10}\left(\frac{V_o^2}{R_o} \middle/ \frac{V_i^2}{R_i}\right)$$

$$G_{dB} = 10 \log_{10}\left(\frac{V_o^2}{V_i^2} \times \frac{R_i}{R_o}\right)$$

If $R_i = R_o$, then

$$G_{dB} = 20 \log_{10}(V_o/V_i) \tag{2}$$

(Note that the requirement that the input resistance of the circuit element equal the output resistance should not be confused with the requirement for maximum power transfer (matching), which is that the resistance of a signal source equal the resistance of the load connected to it.) Equation (2) allows the decibel scale to be applied to voltage ratios and also allows voltmeters to be produced with decibel scales. As the voltmeter indicates only voltage and not voltage ratio, a suitable reference voltage has to be chosen. To this

end, zero decibels (0 dB) is sometimes defined as 1 mW into 600 Ω, but other values are also used. 1 mW dissipated in 600 Ω corresponds to 0.775 V. When using decibel scales it is essential to know whether a power or a voltage ratio is being used.

There are two main advantages of using the decibel scale. First, it allows the overall gain in a system to be determined by algebraic addition of the gains (or losses) in the component parts. Second, being a logarithmic scale, large numbers are reduced to more easily handled sizes; that is, a rejection ratio of 120 dB implies a reduction in signal level of 1 000 000 : 1.

5. Voltage Dividers—Attenuators

The basic and therefore most accurate ranges on most voltmeters are those with nominal full-scale values of 1 V and 10 V, these being determined by the reference voltages of commercial Zener diodes. Lower voltages are amplified and higher voltages attenuated before measurement. The accuracy of higher voltage measurements depends in part on the attenuation uncertainties.

The traditional dc low-frequency attenuator consists of a chain of resistors, and it is assumed that the input–output voltage ratio is equal to the input–output resistance ratio. This is true only if the load presented to the output stage is high enough (ideally it should be infinite).

Choosing the resistor values for a particular voltage divider is always an uneasy compromise; for example, a divider is required to reduce a 100 V signal to 1 V (see Fig. 9). Some possible combinations are shown in Table 1.

The 10 MΩ input resistance gives an adequately low power dissipation, but if the accuracy required is 0.01% the input resistance of the next stage which is loading the output of the voltage divider will need to be at least 10^{10} Ω for there to be a negligible loading effect.

As the value of a resistor varies with age and temperature (both ambient temperature and the local temperature produced by Joule heating), the uncertainties introduced by a resistive attenuator are always a major contribution to the total inaccuracies in a measuring system.

On dc, an input resistance of 10 MΩ is usual, but

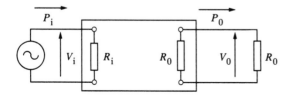

Figure 8
Power into and out of a circuit

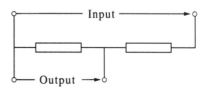

Figure 9
An attenuator

Table 1
Possible combinations of components to produce an attenuation of a 100 V signal to 1 V

Input resistance	Output resistance	Current	Power dissipated
10 MΩ	100 kΩ	10 μA	1 mW
100 kΩ	1 kΩ	1 μA	0.1 W
1 kΩ	10 Ω	100 mμA	10 W

because of the unavoidable stray capacitances which are difficult to control, audio-frequency ac dividers usually have an input impedance of 1 MΩ.

6. Passive Analog Meters

Although superseded in many respects by digital systems, moving coil multimeters are still used in large numbers. Not only cheap and truly independent of any power supply, they are also largely immune to errors caused by ac interference signals. They are current sensitive (i.e., respond to the current flowing not the voltage applied) and have, therefore, a relatively low input resistance. This is unlikely to be greater than 20 kΩ V^{-1} (for example, input resistance of 200 kΩ on the 10 V range), which compares unfavorably with the 10 MΩ or greater for a typical electronic voltmeter.

Alternating-signal ranges are provided by using a bridge rectifier circuit. The indication is therefore proportional to the mean value of the rectified signal, smoothing being achieved by the mechanical inertia of the system. As the user will require the instrument to be scaled in terms of the equivalent rms value a multiplying factor of 1.111 is included in the scale (1.111 being the ratio of rms to mean value for a sine wave). Should the waveform being applied to the instrument not be sinusoidal the instrument indication will be in error.

The input voltage–output current characteristic of a rectifier is not linear at low voltages: below about 0.3 V the current is less than a linear relationship would predict. A low ac voltage range (say 5 V or 10 V full scale) therefore has a scale that is cramped towards its lower end. If the same indicating instrument is also used on higher ac voltage ranges and on dc, the scale nonlinearities then become smaller. More than one scale arc may therefore have to be used. Another approach, used in some dc–ac multimeters, is to apply the input on the low ac voltage ranges to the current transformer (which is incorporated to measure alternating current), used as a step-up voltage transformer so that the voltage applied to the rectifier (via a multiplier resistor) is much higher, and so a linear scale can be used. The penalty is a reduced input impedance with quite a high-input current.

Voltage ranging is achieved by resistive attenuators but current ranging will require a current transformer (as a rectifier circuit is nonlinear). Any direct component in the alternating current will have serious effects on the accuracy of the indication.

Other analog systems can be used for both dc and ac voltage measurements (for example, electrodynamic, electrostatic or thermal methods). Reference to BS 89 (British Standards Institution 1990) will give details of the specification of all passive analog instruments.

7. Electronic Analog Meters

Electronic analog systems are designed to overcome the limitations of passive devices, namely low input impedance, low sensitivity and poor frequency response. The first two limitations are dealt with by the inclusion of amplifiers and the third by choice of components and suitable amplifier techniques.

7.1 DC Voltmeters

The measurement of direct voltage represents a simple application of electronics in which a dc amplifier precedes the passive analog meter movement. Two types of amplifier are used.

(a) Direct coupled amplifiers are cheap and are therefore used in inexpensive instruments where the amplifier provides greater sensitivity and increased input impedance. The amplifier can be designed to saturate when overloaded and so protect the meter movement. However, the drift may be unacceptably high.

(b) Where sensitivity down to microvolt levels and greater stability are required a chopper-stabilized amplifier may be used. The direct voltage to be measured is chopped to create a square wave which is then amplified using an ac amplifier, before being converted back into an equivalent direct voltage for display.

7.2 AC Voltmeters

These may be classified into the three following types.

(*a*) *Average responding instrument.* This is virtually a passive system preceded by an ac amplifier, the rectifier and meter circuit being included in a feedback loop to improve the system linearity. It is common to achieve sensitivities up to 1 mV at full-scale deflection, input impedances of 10 MΩ up to audio frequencies and, with care, 1% accuracy up to a few megahertz.

(*b*) *Peak responding system.* The input signal is fed via a diode to a capacitor which is charged to the peak value of the applied voltage. The capacitor then slowly discharges through the high-input impedance of the dc amplifier until recharged by the next peak value (see Fig. 10). It is common to place the input circuit diode

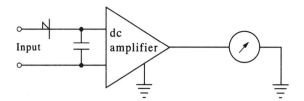

Figure 10
The peak responding system

and capacitor into a probe, so limiting problems due to lead inductance and capacitance. As the signal amplified is dc, the only frequency limitation problems are due to the input diode circuit, and accuracies of 2% can be achieved up to several hundred megahertz for inputs above about 0.5 V. Peak responding voltmeters are more sensitive to harmonic distortion than other types.

(c) *RMS responding meters.* These use a matched pair of thermocouples (see Fig. 11). The amplified ac signal to be measured is passed through one thermocouple, while a dc signal is passed through the other. The difference between the two thermocouple output voltages is used as the input to the dc amplifier and this maintains a dc current in the balancing thermocouple which is the rms equivalent of the ac current in the measuring thermocouple. The output of the dc amplifier is displayed. The system response is linear but, unfortunately, the accuracy is limited to about 3% of full scale. The scale is square law and therefore cramped at the lower end. The upper frequency is limited by the ac amplifier performance to about 1 MHz.

The typical mean or rms instrument will have an input impedance of 10 MΩ shunted by perhaps 15 pF. At 100 kHz the input capacitive reactance has already fallen to 100 kΩ. It will be primarily the capacitive reactance which determines the loading effect of the voltmeter.

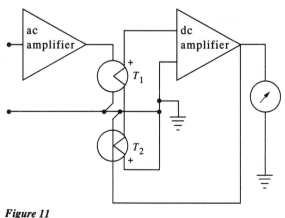

Figure 11
RMS responding meter

8. Digital Instruments

The properties of the ideal voltmeter include high resolution, accuracy and sensitivity, very high input impedance, rapid reading rate and an automatic measurement and control facility. As all these features can be available in a digital instrument, it is not surprising that they have almost completely eclipsed analog meters in the measurement of dc and low-frequency (lf) signals.

The heart of a digital instrument is a direct voltage measuring module, and many techniques have been developed each with their own advantages and disadvantages which must be carefully considered by a prospective user. The two systems described below are chosen to illustrate the widely differing characteristics available.

8.1 Potentiometric Systems

In the early development of digital voltmeters (DVMs) many potentiometric systems were marketed, but as other techniques have been developed, only the successive approximation logic has remained. The heart of the system is a Zener diode reference voltage supplying a binary coded decimal (BCD) sequence voltage divider (see Fig. 12). The balance is achieved by holding or rejecting voltage sections until the total equals the unknown being measured. For example, if 5.84 V is to be measured, the following sequence would be effected: 8 V tried and rejected, 4 V added and held, 2 V added and rejected, 1 V added and held, 0.8 V added and held, 0.4 V added and rejected, 0.2 V added and rejected, 0.1 V added and rejected, and so on.

To measure with a resolution of 1 in 10^n, $4n$ voltages are added sequentially whatever the value of the input voltage. The measuring time is therefore constant, and being dependent on the speed of the logic and solid-state switching, can be very short (many readings per millisecond). This theoretically makes the technique an ideal part of an automated system. However, if there is any change of one least significant digit, or greater, in the input voltage the logic system "sees an

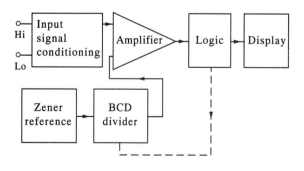

Figure 12
Potentiometric system

error" in the balance and rebalances starting at the most significant digit. The system is therefore badly affected by any superimposed (interference) ac signals. Placing a low-pass filter in front of the instrument may remove the problem but with a disastrous loss of reading rate. For example, suppose a DVM claiming an accuracy of $\pm 0.02\%$ is preceded by a filter of time constant $0.1\,\text{s}$. Assuming that the filter must not introduce any significant additional uncertainty the difference between the filter input and output must not exceed 0.002%. Therefore,

$$\exp\left(\frac{-t}{T}\right) = 0.000\,02$$

giving $t/T = 6.2$, that is, $t = 0.6$ (maximum reading rate is approximately $2\,\text{s}^{-1}$).

8.2 Dual-Slope Systems

The majority of DVMs are based on voltage to time interval or voltage to frequency conversion. Of these the most popular is the dual-slope integrating system (see Fig. 13). The voltage to be measured is used to charge a capacitor for a fixed sampling time. The capacitor is then discharged by connecting it to a fixed internal reference voltage of opposite polarity (see Fig. 14). The time taken to discharge (encoding time) is then proportional to the unknown voltage. The attraction of this system is not only that the long-term stabilities of the integrating components are of little importance, but also that large interference rejection ratios can be achieved.

If the sampling time is made equal to a whole number of periods of any interference signal (usually emanating from the power supply) the average value of the interference signal over this time is zero

$$\int_0^T \sin(\omega t + \theta)\,\mathrm{d}t = 0$$

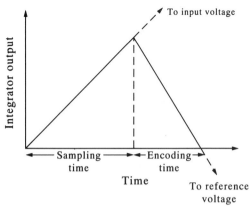

Figure 14
Fixed internal reference voltage of opposite polarity

and it therefore has no effect on the direct-signal measurement. As the power supply frequency will vary, the internal clock pulse generator can, with advantage, be locked to the power supply. As this controls the sampling time and provides the pulses which are counted internally in order to measure the encoding time, variations in the power supply will have no effect on the accuracy of the dc measurement provided the frequency is acceptably constant for the few milliseconds during which a measurement is made.

The system is rather slow reading, as for 50 Hz supply systems the sampling time is at least 20 ms, making the maximum reading rate about $20\,\text{s}^{-1}$.

The very large rejection ratios claimed for these devices must be interpreted with some care. For a $4\frac{1}{2}$-digit instrument on the 20 V range (20.000) a 140 dB rejection ratio implies an ac interference signal of 10 kV would result in an error of one digit (1 mV). This 10 kV ac signal would, if nothing else, saturate the input amplifiers and create nonlinearities.

8.3 Alternating Voltage Measurement

In all ac indicating DVMs the ac signal is converted into an equivalent dc value before measurement. Earlier systems employed precision rectification circuits with the inherent problems of indicating rms values using a system that responds inherently to mean values. The accuracy was limited by the nonlinearity of the rectifier characteristic, and the frequency range by the amplifier feeding the current rectification circuit.

Modern systems use rms computation circuits for ac signal measurements. The simplest method is shown in Fig. 15, but unfortunately the sensitivity is inversely proportional to the square of the input voltage, which results in a loss of sensitivity at low signal levels and a corresponding loss of accuracy. This problem of lack of sensitivity is overcome by using a logarithmic feedback technique (see Fig. 16). The computation block is arranged as shown in Fig. 17.

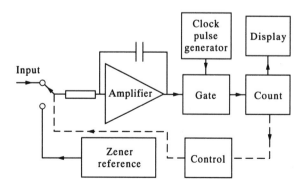

Figure 13
Dual-slope integrating system

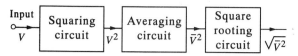

Figure 15
Simple alternating voltage measurement

Digital systems using this technique claim accuracies very little worse than that possible using the best standards laboratory passive equipment.

9. Interference Signals—Guarding

Since the 1980s the resolution, sensitivity and input impedance of electronic systems have steadily increased. At the same time there has been a reduction in operating time constant. Although these features are generally welcomed, it does mean that instruments have become more sensitive to unwanted signals. The sensitivity of an instrument to these signals is expressed in terms of series and common mode rejection ratios.

9.1 Series Mode Interference

An unwanted potential difference across the input terminals of the DVM is termed series mode interference. This may result directly from an unwanted emf being present in series with the potential being measured. For example, thermal emfs at switch or relay contacts and 50 Hz emfs induced via magnetic coupling with power frequency supplies and equipment (see Fig. 18).

Direct emfs are particularly serious as they are indistinguishable from the voltage under test in a dc measuring system and there is therefore no chance of removing them once they are established (methods such as using double contact switches can be used to equalize thermal voltages).

The only means available to prevent ac series mode signals reaching the input terminals is to provide a suitable filter.

The series mode rejection ratio is expressed (in dB) as

$$20 \log_{10} \times \frac{\text{series mode voltage present}}{\text{indication produced in instrument}}$$

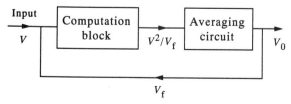

Figure 16
Logarithmic feedback technique

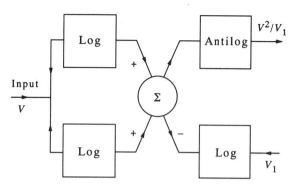

Figure 17
Computation block

9.2 Common Mode Interference

This is an unwanted signal which is common to both Hi and Lo input leads (see Fig. 19). The voltage V_C itself will cause no signal between the input terminals. However, due to the finite impedance of leakage paths, circulating currents will be produced and the consequent voltage drops result in an unwanted potential at the instrument terminals.

The effective dc circuit of an unguarded instrument is shown in Fig. 20. If the lead resistances R_{Hi} and R_{Lo} were equal, and if the leakage resistances R_{Hi-E} and R_{Lo-E} were also equal then no resulting potential would appear across R_{IN}.

In practice, $R_{Hi-E} > R_{Lo-E}$ and we must be prepared for imbalance in the effective lead resistance (this can be created not merely by the physical leads but by resistance associated with the source being measured). The effect of this imbalance will be greatest when all the resistance is concentrated in the low lead (assumed to be 1 kΩ when specifying the rejection ratio).

The common mode rejection ratio is expressed (in dB) as

$$20 \log_{10} \times \frac{\text{common mode voltage present}}{\substack{\text{resulting voltage produced} \\ \text{across input terminals}}}$$

9.3 Guard Techniques

In a high-input-impedance device both the input circuitry and the input leads may be guarded. The aim is to create an equipotential surface placed physically

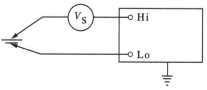

Figure 18
Series mode interference

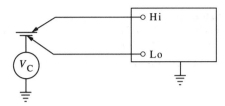

Figure 19
Common mode interference

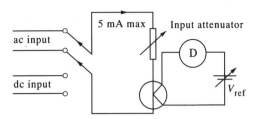

Figure 21
Voltage transfer measurement

between the circuits being guarded and earth so that leakage currents to earth are prevented. To achieve this effectively:

(a) the guard lead resistance would have to be low;

(b) the impedance between the guard box and earth must be high; and

(c) the guard must be connected to the source of common mode signal.

Even though these idealized conditions may be achieved, infinite common mode rejection (CMR) will not be realized due to leakage paths that still exist between Hi and Lo terminals and earth. The information gained from the input circuits has to be displayed or recorded and in spite of careful isolation techniques the impedance to ground is not infinite. The effect is not only to reduce the CMR but to create variations in the CMR depending on how the guard is connected. If the guard is connected to the instrument Lo input terminal the CMR will be equal to that of an unguarded instrument. One should never connect the guard to earth as it will produce a CMR less than that for an instrument with no guard.

10. AC/DC Transfer Systems

High-accuracy voltage and current measurements require a comparison between the unknown and a standard signal. The only standard available is direct voltage—either a standard cell or Zener diode system. No standard ac sources are available and it is therefore necessary to refer an ac measurement to a dc standard

via equipment known to respond equally (within the uncertainty of the system) to ac and dc signals. Such a transfer system needs to be rms responding, and this indicates the possibility of using electrodynamic, electrostatic or thermal systems. At low accuracies and frequencies an electrodynamic ammeter or voltmeter may be calibrated when carrying direct current and then used for measuring alternating quantities. For accuracies better than $\pm 0.2\%$ a thermal system is used.

The heart of the system is a thermoelement whose ac/dc response is known (calibrated by a higher laboratory to give traceability to ac measurements). It is assumed that the thermocouple output will be the same when

(a) carrying ac, or

(b) carrying dc which is the rms equivalent of (a).

Although this system is basically a current transfer system (heater impedance typically $100\,\Omega$) most commercial systems are designed for voltage transfer measurements (see Fig. 21).

The ac to be measured is connected to the thermocouple heater element and V_{ref} adjusted until the detector indicates a null. The ac is removed and variable dc connected to the same input and adjusted until the detector again indicates a null. Provided V_{ref} has remained unchanged the dc input is the rms equivalent of the ac being measured. The dc input is determined using any technique having acceptable accuracy, for example, a good quality DVM. The output of the thermocouple will be only a few millivolts and there would be great difficulty in measuring this with sufficient accuracy. Hence, the need for a balancing technique where only resolution and stability and not accuracy are required.

Most thermocouples produce a slightly different couple output when the direction of the current through the heater is reversed. When measuring an alternating signal, reversal occurs every half cycle. Means therefore has to be provided to reverse the direction of the direct current through the heater. This is achieved using a reversing switch. The value of the alternating voltage is taken to be the mean of the forward and reverse direct voltages when they are adjusted to obtain a null on the detector.

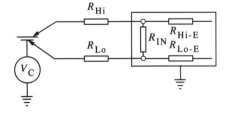

Figure 20
Effective dc circuit of an unguarded instrument

400

Accuracies up to 0.01% and frequency ranges up to a few megahertz can be achieved. Currents can be measured by replacing the input attenuator with suitable resistance shunts. Automatic systems are available. Even using this transfer technique it is difficult to achieve the performance necessary to calibrate reliably the latest generation ac responding digital multimeters.

11. Further Information for the User

Details on the measurement of current and voltage can be obtained from the technical publications of many instrument system manufacturers including Cropico, Datron, Fluke, Guildline, Hewlett Packard, Philips, Solartron and Tinsley. In addition, useful information can be found in the Measurement Vacation School Notes published by the Institution of Electrical Engineers, London.

See also: Current Measurement

Bibliography

British Standards Institution 1990 *Specifications for Direct Acting Indicating Electrical Measuring Instruments and their Accessories*, BS 89. BSI, London

P. M. Clifford
[City University, London, UK]

R. Walker
[University of Portsmouth, Portsmouth, UK]

LIST OF CONTRIBUTORS

Contributors are listed in alphabetical order together with their addresses. Titles of articles that they have authored follow in alphabetical order. Where articles are coauthored, this has been indicated by an asterisk preceding the title.

Bajek, W. A.
UOP
Building C
Des Plaines, IL 60017-5017
USA
Installation and Commissioning

Bansevicius, R.
Department of Mechanical Engineering
Kaunas University of Technology
Kestucio 27
Kaunas 3004
Lithuania
Transducers, Magnetostrictive
Transducers, Piezoelectric

Barker, H. A.
Department of Electrical and Electronic Engineering
University of Swansea
Singleton Park
Swansea
SA2 8PP
UK
Signal Transmission
Signal Processing

Batchelor, B.
Department of Computing Mathematics
University of Wales College of Cardiff
Cardiff
CF1 1XL
UK
Image Processing and its Industrial Applications

Beck, M. S.
Department of Instrumentation and Analytical Science
UMIST
PO Box 88
Manchester
M60 1QD
UK
Correlation in Measurement and Instrumentation

Bentley, J. P.
Division of Instrumentation and Control
School of Information Engineering
University of Teesside
Borough Road
Middlesbrough

TS1 3BA
UK
Transducers, Thermal

Bogue, R. W.
Kingston House
Tuckermarsh
Bere Alston
Devon
PL20 7HB
UK
Environmental Measurement and Instrumentation

Boyle, W.
School of Electrical, Electronic and Information
    Engineering
City University
Northampton Square
London
EC1V 0HB
UK
Optical Instruments

Bridge, B.
School of Electrical, Electronic and Information
    Engineering
South Bank University
103 Borough Road
London
SE1 0AA
UK
Nondestructive Testing, Electromagnetic
Nondestructive Testing: General Principles
Nondestructive Testing, Radiographic

Carr-Brion, K. G.
Department of Fluid Engineering and Instrumentation
School of Mechanical Engineering
Cranfield Institute of Technology
Cranfield
Bedfordshire
MK43 0AL
UK
Chemical Analysis, Instrumental

Charlwood, F. J.
Department of Systems Science
City University

Northampton Square
London
EC1V 0HB
UK
Reliability and Maintenance
Safety

Clarke, J. R. P.
13 Old Wool Lane
Cheadle Hulme
Cheshire
SK8 5JB
UK
Process Instrumentation Applications

Clifford, P. M.
School of Electrical, Electronic and Information
   Engineering
City University
Northampton Square
London
EC1V 0HB
UK
*Bridges
*Current Measurement
*Frequency and Time Measurement
*Voltage Measurement

Cripps, M. D.
School of Electrical, Electronic and Information
   Engineering
City University
Northampton Square
London
EC1V 0HB
UK
Digital Instruments

Cunningham, M. J.
Department of Electrical Engineering
University of Manchester
Manchester
M13 9PL
UK
Velocity and Acceleration Measurement

Erdem, U.
Negretti Automation Ltd
Stocklake
Aylesbury
Buckinghamshire
HP20 1DR
UK
Mass, Force and Weight Measurement

Eykhoff, P.
Department of Electrical Engineering
Eindhoven University of Technology

PO Box 513
NL-5600 MB Eindhoven
The Netherlands
Identification in Measurement and Instrumentation

Ferrero, C.
Istituto di Metrologia G. Colonnetti
Strada delle Cacce 73
I-10135 Torino
Italy
Force and Dimensions: Tactile Sensors
*Mechanical and Robotics Applications: Multicomponent
   Force Sensors*

Ferguson, J.
Department of Pure and Applied Chemistry
University of Strathclyde
Thomas Graham Building
295 Cathedral Street
Glasgow
G1 1XL
UK
Viscosity Measurement

Finkelstein, A. C. W.
Imperial College of Science, Technology and Medicine
University of London
180 Queen's Gate
London
SW7 2BZ
UK
*Design Principles for Instrument Systems
*Life Cycle

Finkelstein, L.
School of Electrical, Electronic and Information
   Engineering
City University
Northampton Square
London
EC1V 0HB
UK
*Artificial Intelligence in Measurement and
   Instrumentation*
*Design Principles for Instrument Systems
Errors and Uncertainty
Errors: Avoidance and Compensation
Instrument Elements: Models and Characteristics
Instrument Systems: General Requirements
Instruments: Models and Characteristics
Instruments: Performance Characteristics
*Life Cycle
Measurement: Fundamental Principles
Measurement Science

Fletcher, K. S.
The Foxboro Company EMO
PO Box 500

600 North Bedford Street
East Bridgewater, MA 02333
USA
Electrochemical Measurements

Goode, C. H.
KDG Mobery
Crompton Way
Crawley
West Sussex
RH10 2YZ
UK
*Level Measurement

Grattan, K. T. V.
School of Electrical Engineering and Applied Physics
City University
Northampton Square
London
EC1V 0HB
UK
*Analytical Physical Measurements: Principles and
    Practice*
*Density and Consistency Measurement
*Level Measurement
Optical Measurements
Transducers, Fiber-Optic
Transducers, Photoelectric

Higham, E. H.
University of Sussex
Falmer
Brighton
BN1 9QT
UK
Transducers, Pneumatic

Hofmann, D.
Department of Technology for Science Instrumentation
Friedrich Schiller University
E Thaelmann Ring 32
D-6900 Jena
Germany
Displacement and Dimension Measurement

Hopkins, S. W. J.
8 Blenheim Gardens
Sanderstead
South Croydon
CR2 9AA
UK
Operational Environments

Johnston, J. S.
Rosemount Ltd

Heath Place
Bognor Regis
Sussex
PO22 9SH
UK
Temperature Measurement

Jones, B. E.
The Brunel Centre for Manufacturing Metrology
Brunel University
Uxbridge
Middlesex
UB8 3PH
UK
Instrument Systems: Functional Architectures

Jones, D. W.
Department of Chemistry and Chemical Technology
University of Bradford
Bradford
BD7 1DP
UK
Nuclear Magnetic Resonance

Karcanias, N.
School of Electrical Engineering and Applied Physics
City University
Northampton Square
London
EC1V 0HB
UK
Instrumentation in Systems Design and Control

Kirkbright, G. F.
[deceased; late of UMIST, Manchester, UK]
*Spectroscopy: Fundamentals and Applications

Kortela, U.
Control Engineering Laboratory
Helsinki University of Technology
SF-02150 Espoo 15
Finland
*Density and Consistency Measurement

Lang, T.
École Supérieure d'Électricité
Plateau de Moulon
F-91190 Gif-sur-Yvette
France
Analog Signal Conditioning and Processing
Analog-to-Digital and Digital-to-Analog Conversion

Marioli, D.
Dipartimento di Elettronica per l'Automazione
Università degli Studi di Brescia

via Branze
I-25123 Brescia
Italy
Transducers: Strain Gauges

Medlock, R. S.
[deceased; late of ABB Kent, Luton, UK]
Flow Measurement: Applications and Instrumentation
   Selection
Flow Measurement: Economic Factors, Markets and
   Developments
Flow Measurement: Principles and Techniques

Milner, J. H.
School of Electrical, Electronic and Information
   Engineering
City University
Northampton Square
London
EC1V 0HB
UK
Transducers, Magnetostrictive
Transducers, Piezoelectric

Munter, P. J. A.
Department of Electrical Engineering
Delft University of Technology
Mekelweg 4
NL-2628 CD Delft
The Netherlands
Transducers, Solid-State

Palmer, R. B. J.
15 Sundridge Avenue
Welling
Kent
DA16 2SR
UK
Ionizing Radiation Measurements
Transducers, Ionizing Radiation

Payne, P. A.
Department of Instrumentation and Analytical Science
UMIST
PO Box 88
Manchester
M60 1QD
UK
Biological and Biomedical Measurement Systems

Paterson, I. W. F.
Whessoe Systems and Controls
Brinkborn Road
Darlington
DL3 6DS
UK
Level Measurement

Plackmann, D. G.
UOP
Building C
Des Plaines, IL 60017-5017
USA
Installation and Commissioning

Plaskowski, A.
Department of Instrumentation and Analytical Science
UMIST
PO Box 88
Manchester
M60 1QD
UK
Correlation in Measurement and Instrumentation

Regtien, P. P. L.
Department of Electrical Engineering
Delft University of Technology
Mekelweg 4
NL-2628 CD Delft
The Netherlands
Transducers, Solid-State

Rizk, M.
School of Electrical Engineering and Applied Physics
City University
Northampton Square
London
EC1V 0HB
UK
Microwave Measurements

Sanderson, M. L.
Department of Electrical Engineering
Salford University
Salford
M5 4WT
UK
Bridges
Amplifiers

Schmidt, G. T.
Charles Stark Draper Laboratory
555 Technology Square
Cambridge, MA 02139-3563
USA
Position Inertial Systems

Sloman, M. S.
Department of Computing
Imperial College of Science, Technology and Medicine
University of London
180 Queen's Gate
London
SW7 2BZ

UK
Networks of Instruments

Spillane, D. E. M.
School of Applied Chemistry
University of North London
166 Holloway Road
London
N7 8DB
UK
Spectroscopy: Fundamentals and Applications

Stuart, P. R.
National Physical Laboratory
Queens Road
Teddington
Middlesex
TW11 0LW
UK
Pressure Measurement
Pressure Measurement: Vacuum

Sutcliffe, A.
School of Electrical, Electronic and Information
    Engineering
City University
Northampton Square
London
EC1V 0HB
UK
Operator–Instrument Interface

Sutcliffe, D. S.
Division of Electrical Science
National Physical Laboratory
Queens Road
Teddington
Middlesex
TW11 0LW
UK
Frequency and Time Measurement

Sydenham, P. H.
School of Electronic Engineering
University of South Australia
PO Box 1
Ingle Farm
South Australia 5001
Australia
Construction and Manufacture of Instrument Systems
Measurement and Instrumentation: History
Transducers: An Introduction

Timonen, E.
Technical Research Center Finland
PO Box 181
SF-90101 Oulu

Finland
Density and Consistency Measurement

Turner, J. D.
Mechanical Engineering Department
University of Southampton
Highfield
Southampton
SO9 5NH
UK
Automotive Applications of Measurement and
    Instrumentation

Usher, M. J.
Department of Cybernetics
School of Information and Engineering Sciences
University of Reading
Whiteknights
PO Box 225
Reading
RG6 2AY
UK
Noise: Physical Sources and Characteristics

Walker, R.
Department of Electrical Engineering
University of Portsmouth
Anglesea Building
Portsmouth
PO1 3DJ
UK
Current Measurement
Voltage Measurement

Weight, J. P.
School of Electrical Engineering and Applied Physics
City University
Northampton Square
London
EC1V 0HB
UK
Nondestructive Testing, Ultrasonic

Weiler, J.
ETH Zentrum/ETL
CH-8092 Zürich
Switzerland
Resistance, Capacitance and Inductance Measurement

Weinberg, M. S.
Charles Stark Draper Laboratory
555 Technology Square
Cambridge, MA 02139-3563
USA
Angular Rate Sensing

Weir, K.
School of Electrical, Electronic and Information
   Engineering
City University
Northampton Square
London
EC1V 0HB
UK
Transducers, Electrodynamic
Transducers, Inductive
Transducers, Potentiometric

Wolffenbuttel, R. F.
Department of Electrical Engineering
Delft University of Technology
Mekelweg 4
NL-2628 CD Delft
The Netherlands
Transducers, Solid-State

Woschni, E.-G.
Technische Universität Chemnitz-Zwickau
Fachbereich Elektrotechnik Lehrstuhl Nachrichtentechnik
Postschliessfach 964
D-9009 Chemnitz-Zwickau
Germany
Information Theory in Measurement and Instrumentation
Signal Theory in Measurement and Instrumentation

Wykes, C.
Department of Production Engineering and Production
   Management
University of Nottingham
University Park
Nottingham
NG7 2RD
UK
Transducers, Capacitive

Zhang, Z. Y.
School of Electrical, Electronic and Information
   Engineering
City University
Northampton Square
London
EC1V 0HB
UK
*Amplifiers

SUBJECT INDEX

The Subject Index has been compiled to assist the reader in locating all references to a particular topic in the Encyclopedia. Entries may have up to three levels of heading. Where there is a substantive discussion of the topic, the page numbers appear in ***bold italic*** type. As a further aid to the reader, cross-references have also been given to terms of related interest. These can be found at the bottom of the entry for the first-level term to which they apply. Every effort has been made to make the index as comprehensive as possible and to standardize the terms used.